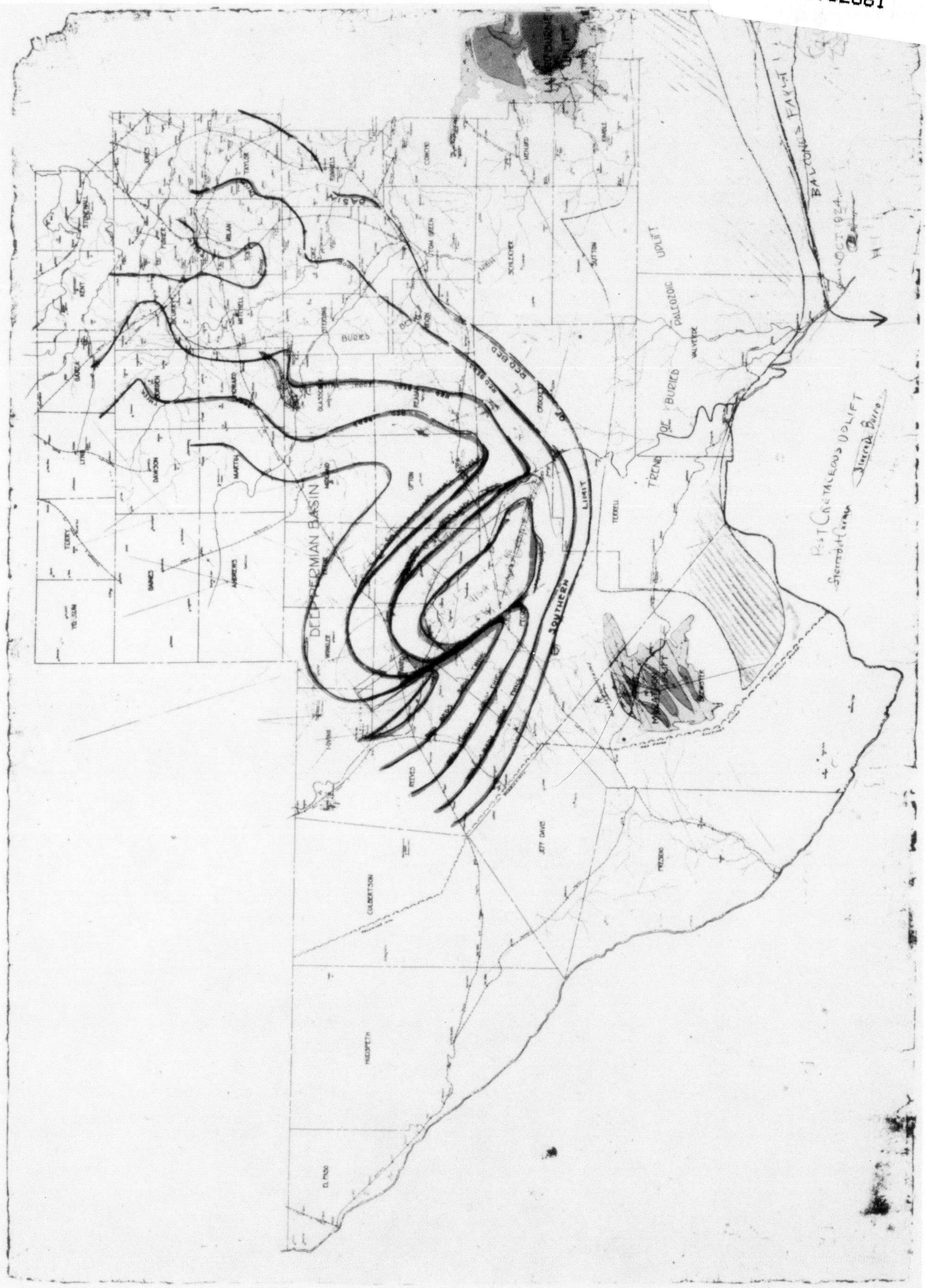
DEEP PERMIAN BASIN
SOUTHERN LIMIT
BURIED
TREND OF BURIED PALEOZOIC UPLIFT
BALCONES FAULT
OCT 1924
EL PASO
HUDSPETH
CULBERTSON
REEVES
JEFF DAVIS
PRESIDIO
TERRELL
CROCKETT
SUTTON
SCHLEICHER
TOM GREEN
VALVERDE
DAWSON
ANDREWS
UPTON

Giant Oil and Gas Fields

A Core Workshop

Volume 2

Organized and Compiled
By
Anthony J. Lomando
and
Paul M. Harris

SEPM Core Workshop No. 12
Houston, March 19-20, 1988

Copyright 1988 By Society of Economic Paleontologists and Mineralogists

ISBN #0-918985-72-2

Additional copies of this publication may be ordered from SEPM. Send your order to:

> SEPM
> Post Office Box 4756
> Tulsa, OK 74159-0756
> U.S.A.

© Copyright 1988 by
Society of Economic Paleontologists and Mineralogists
Printed in the United States of America

PREFACE

Giant fields tend to be grouped as a distinctive class of hydrocarbon target, but giant reservoirs are vastly different from one another with varying factors controlling rock type, facies, porosity evolution, and trap mechanism, even in adjoining reservoir zones. As such, the ideas generated from the geological analyses of these giants can be applied in many basins of the world to exploration and production targets of any size.

The papers presented in these volumes are examples of giant fields from North America, the North Sea, Middle East, and Indonesia (Figure 1). We have been fortunate enough to put together a group of papers which spans the major range of geologic time (Figure 2) and have organized them accordingly. This sense of diversity extends through the major characteristics represented by these examples of giant reservoirs. A good mix of siliciclastic and carbonate rock types deposited in fluvial and supratidal settings down to deep marine environments can be found among these papers. Trap types range from simple and complex structural to stratigraphic and combination traps. We hope that this group of papers will inform and stimulate the reader to develop new ideas and approaches to explore for and develop fields of all types and sizes.

The core workshop and publication were made possible with the help of many people. The SEPM staff and Continuing Education Committee supported the concept of the workshop and handled the logistical preparations. We thank Chevron Overseas Petroleum Inc. and Chevron Oil Field Research Company for supporting our efforts in organizing and presenting the workshop.

A. J. LOMANDO   P. M. HARRIS

Figure 1
Location Map for Fields (Numbers From Table of Contents).

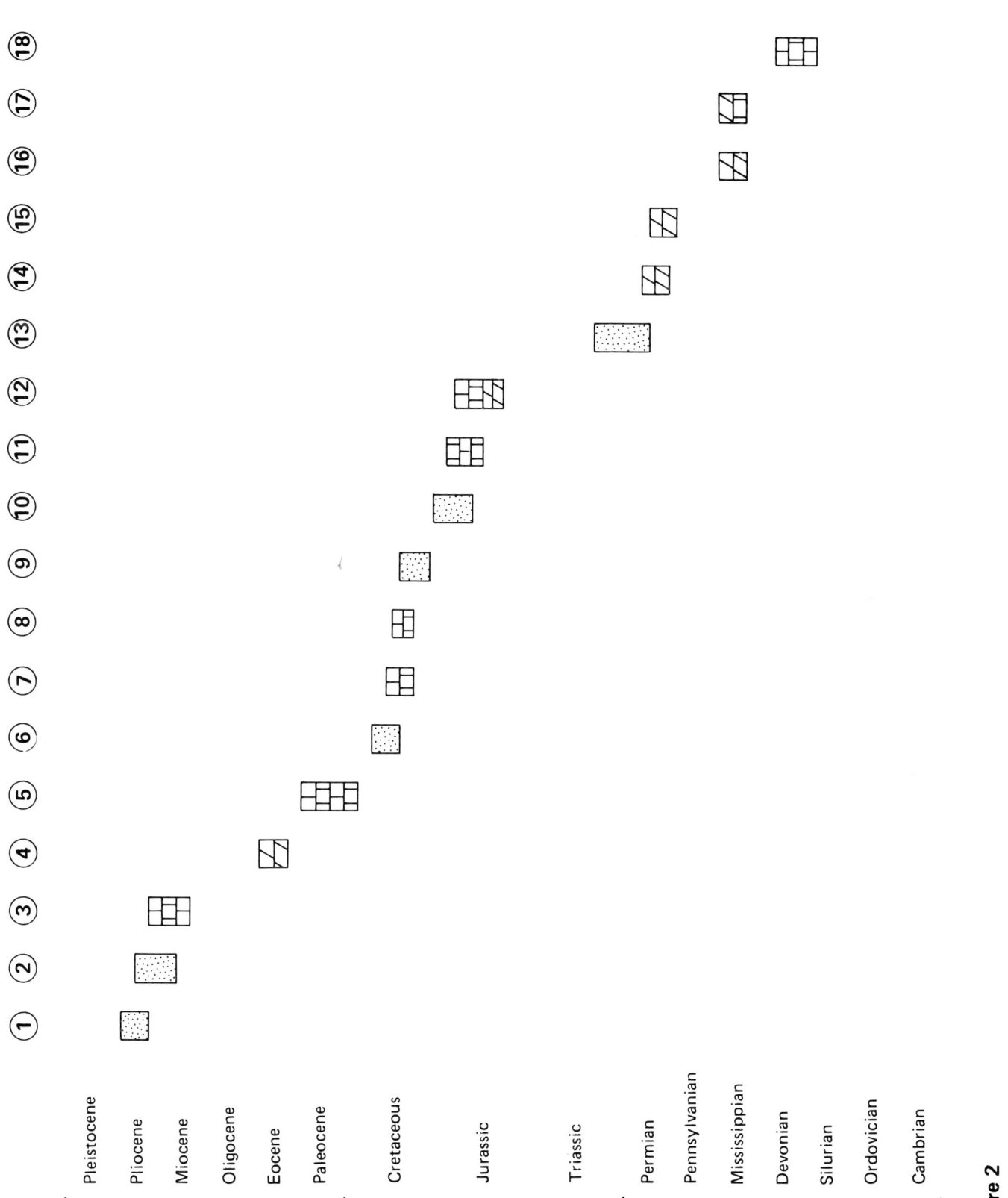

Figure 2
Age and Lithology for Fields (Numbers From Table of Contents).

TABLE OF CONTENTS

VOLUME 1

Page

CENOZOIC EXAMPLES

1. LITHOFACIES, DEPOSITIONAL ENVIRONMENTS, AND RESERVOIR QUALITY OF PLIOCENE DEEP-SEA FAN SEDIMENTS, INGLEWOOD FIELD, LOS ANGELES BASIN, CALIFORNIA
   by W. J. Schweller, J. Gidman, C. W. Grant, and A. A. Reed 1

2. DEPOSITIONAL FACIES, PALEOENVIRONMENTS, RESERVOIR QUALITY, AND WELL LOG CHARACTERISTICS OF MIO-PLIOCENE DEEP-WATER SANDS, LONG BEACH UNIT, WILMINGTON FIELD, CALIFORNIA
   by R. M. Slatt, J. M. Boak, G. T. Goodrich, M. B. Lagoe, C. L. Vavra, J. M. Bishop, and S. M. Zucker 31

3. LITHOFACIES ANALYSIS OF THE ARUN RESERVOIR, NORTH SUMATRA, INDONESIA
   by C. F. Jordan, Jr., and M. Abdullah 89

4. THE EOCENE RESERVOIRS OF WAFRA FIELD, KUWAIT/SAUDI ARABIA PARTITIONED NEUTRAL ZONE
   by H. M. C. Danielli 119

5. CHALK FROM THE EKOFISK AREA, NORTH SEA: NANNOFOSSILS + MICROPORES = GIANT FIELDS
   by C. T. Feazel and H. E. Farrell 155

MESOZOIC EXAMPLES

6. THE DISTRIBUTION OF RESERVOIR SANDSTONE IN THE LOWER CRETACEOUS MUDDY SANDSTONE, HILIGHT FIELD, POWDER RIVER BASIN, WYOMING
   by D. M. Wheeler, E. R. Gustason, and M. J. Furst 179

7. GIANT GAS ACCUMULATION IN "CHALKY"-TEXTURED MICRITIC LIMESTONE, LOWER CRETACEOUS SHUAIBA FORMATION, EASTERN UNITED ARAB EMIRATES
   by S. O. Moshier, C. R. Handford, R. W. Scott, and R. D. Boutell . 229

8. EVOLUTION OF THE LOWER CRETACEOUS RATAWI OOLITE RESERVOIR, WAFRA FIELD, KUWAIT-SAUDI ARABIA PARTITIONED NEUTRAL ZONE
   by S. A. Longacre and E. P. Ginger 273

9. SHELF DEPOSITIONAL ENVIRONMENTS AND RESERVOIR CHARACTERISTICS OF THE KUPARUK RIVER FORMATION (LOWER CRETACEOUS), KUPARUK FIELD, NORTH SLOPE, ALASKA
   by G. C. Gaynor and M. H. Scheihing 333

10. LATE JURASSIC SEDIMENTATION AND TECTONICS, MAIN AREA CLAYMORE RESERVOIR, NORTH SEA
    by S. D. Harker and C. E. Maher 391

VOLUME 2

Page

MESOZOIC EXAMPLES (contd.)

11. LITHOFACIES, DIAGENESIS, AND DEPOSITIONAL SEQUENCE; ARAB-D MEMBER, GHAWAR FIELD, SAUDI ARABIA
by J. C. Mitchell, P. J. Lehman, D. L. Cantrell, I. A. Al-Jallal, and M. A. R. Al-Thagafy 459

12. PALEOENVIRONMENTAL AND DIAGENETIC RESERVOIR CHARACTERIZATION OF THE SMACKOVER FORMATION, JAY FIELD, WEST FLORIDA
by D. M. Bliefnick and P. A. Mariotti 515

13. SEDIMENTOLOGY AND DEPOSITIONAL ENVIRONMENTS OF THE IVISHAK SANDSTONE, PRUDHOE BAY FIELD, NORTH SLOPE, ALASKA
by C. D. Atkinson, P. N. Trumbly, and M. C. Kremer 561

PALEOZOIC EXAMPLES

14. DEPOSITIONAL FACIES AND POROSITY DISTRIBUTION, PERMIAN (GUADALUPIAN) SAN ANDRES AND GRAYBURG FORMATIONS, P.J.W.D.M. FIELD COMPLEX, CENTRAL BASIN PLATFORM, WEST TEXAS
by R. P. Major, D. G. Bebout, and F. J. Lucia 615

15. STRATIGRAPHY AND LITHOFACIES OF THE SAN ANDRES FORMATION, C. S. DEAN "A", XIT, AND SW LEVELLAND UNITS OF LEVELLAND-SLAUGHTER FIELD, PERMIAN BASIN
by P. M. Harris and E. L. Stoudt 649

16. MISSION CANYON (MISSISSIPPIAN) RESERVOIR STUDY, WHITNEY CANYON-CARTER CREEK FIELD, SOUTHWESTERN WYOMING
by P. M. Harris, P. E. Flynn, and J. L. Sieverding 695

17. STRUCTURAL HISTORY AND RESERVOIR CHARACTERISTICS (MISSISSIPPIAN) OF NESSON ANTICLINE, NORTH DAKOTA
by R. F. Lindsay, S. B. Anderson, J. A. LeFever, L. C. Gerhard, and R. D. LeFever 741

18. THE DEVONIAN SWAN HILLS FORMATION AT SWAN HILLS FIELD AND ADJACENT AREAS, CENTRAL ALBERTA, CANADA
by C. A. Viau 803

LITHOFACIES, DIAGENESIS AND DEPOSITIONAL SEQUENCE; ARAB-D MEMBER, GHAWAR FIELD, SAUDI ARABIA

J. C. MITCHELL
P. J. LEHMANN
D. L. CANTRELL
EXXON PRODUCTION RESEARCH COMPANY
HOUSTON, TEXAS   77001

I. A. AL-JALLAL
M. A. R. AL-THAGAFY
ARABIAN AMERICAN OIL COMPANY
DHAHRAN, SAUDI ARABIA

ABSTRACT

Ghawar, the world's largest oil field, is located in the Eastern Province of the Kingdom of Saudi Arabia. This giant field is formed by an elongate northeast to southwest trending anticline. Production comes from Arab-D Member carbonates of the Upper Jurassic Arab Formation, which consists of four geographically-widespread carbonate/evaporite members. The Arab-D comprises two major shoaling upward cycles deposited during a relative highstand in sea level. These cycles are composed of smaller scale upward shoaling cycles and are comprised of a variety of skeletal grainstones and packstones with ooid grainstones locally common in the uppermost Arab-D. The Arab-D is further subdivided into time-stratigraphic reservoir zones and subzones that are based largely on porosity log pattern correlation. The first cycle comprises Zone 3 of the Arab-D. The abundance of grain-supported textures (packstone, mud-lean packstone, and grainstone) stays fairly constant in Zone 3B, but increases upwards through Zone 3A. A second shoaling-upward cycle beginning near the base of Zone 2B resulted in the deposition of virtually mud-free shoal-water deposited sediments seen in Zone 2A and culminated in the deposition of thin subtidal to intertidal/supratidal cycles and sabkha evaporites of Zone 1 and the lower portion of the Arab-D Anhydrite. The overlying upper Arab-D Anhydrite comprises sabkha evaporites and subaqueous evaporites with thin carbonate interbeds that can be traced for hundreds of kilometers. The overall pattern of sedimentation seen at Ghawar is that of a thinning carbonate section and a thickening evaporite section going from north to south, even though the overall thickness of the Arab-D Member remains fairly constant.

The major diagenetic processes active in the Arab-D include dolomitization, leaching and recrystallization, cementation, compaction and fracturing. In general, interparticle porosity is abundant and moldic porosity is common in the reservoir, whereas intrapraticle, fracture, burrow and shelter porosity are much less common or rare. Intercrystal pores are common in dolomites. Microporosity is present throughout the reservoir in limestones and dolomites. It occurs as microporous skeletal and non-skeletal grains, microporous matrix and micropores between cement crystals.

LOCATION AND GEOLOGIC SETTING

Ghawar field is located in the Eastern Province of the Kingdom of Saudi Arabia. This giant oil field is formed by an elongate northeast to southwest trending anticline. Production comes from the D Member of the Upper Jurassic (Kimmeridgian to Tithonian) Arab Formation, which consists of four geographically-widespread carbonate/evaporite members (D, C, B, and A, from oldest to youngest). The Arab-D Member is the focus of this paper.

Deposition of the Arab Formation was initiated in the shallow Tethys Sea, on a broad, tectonically stable shelf or platform. Ghawar is located on the eastern side of the Arabian Shield, west of the Qatar-Surmeh High (Wilson, 1975). The Basrah Basin lies to the north of Ghawar and the Rub'al Khali Basin lies to the south (Figure 1). Regional studies (Powers, 1962; Leeder and Zeidan, 1972; and Wilson, 1975) indicate that the Arab anhydrite intervals thin from west to east and from south to north. The paleoclimate was hot and arid, probably much like today's climate in the Arabian Gulf area.

UPPER JURASSIC STRATIGRAPHY

The following discussion is based on detailed core descriptions and wireline log interpretation of wells in Ghawar. Sediments comprising the Hanifa, Jubaila, and Arab formations were deposited as a upward-shoaling package of sediments during Oxfordian, Kimmeridgian and Tithonian time. This package is subdivided into six major upward shoaling cycles. The Hanifa cycle was initiated in the deepest waters and shoaled to near sea level. The cycle comprising the Jubaila through the lower Arab-D was initiated in deep subtidal water that shoaled to sea level. The cycles constituting the upper Arab-D and the three overlying Arab members were subsequently deposited as upward shoaling carbonates that are overlain by evaporites.

Each of the successive sea level rises that gave rise to these deposits was of less magnitude than the preceding rise. Although small-scale variations

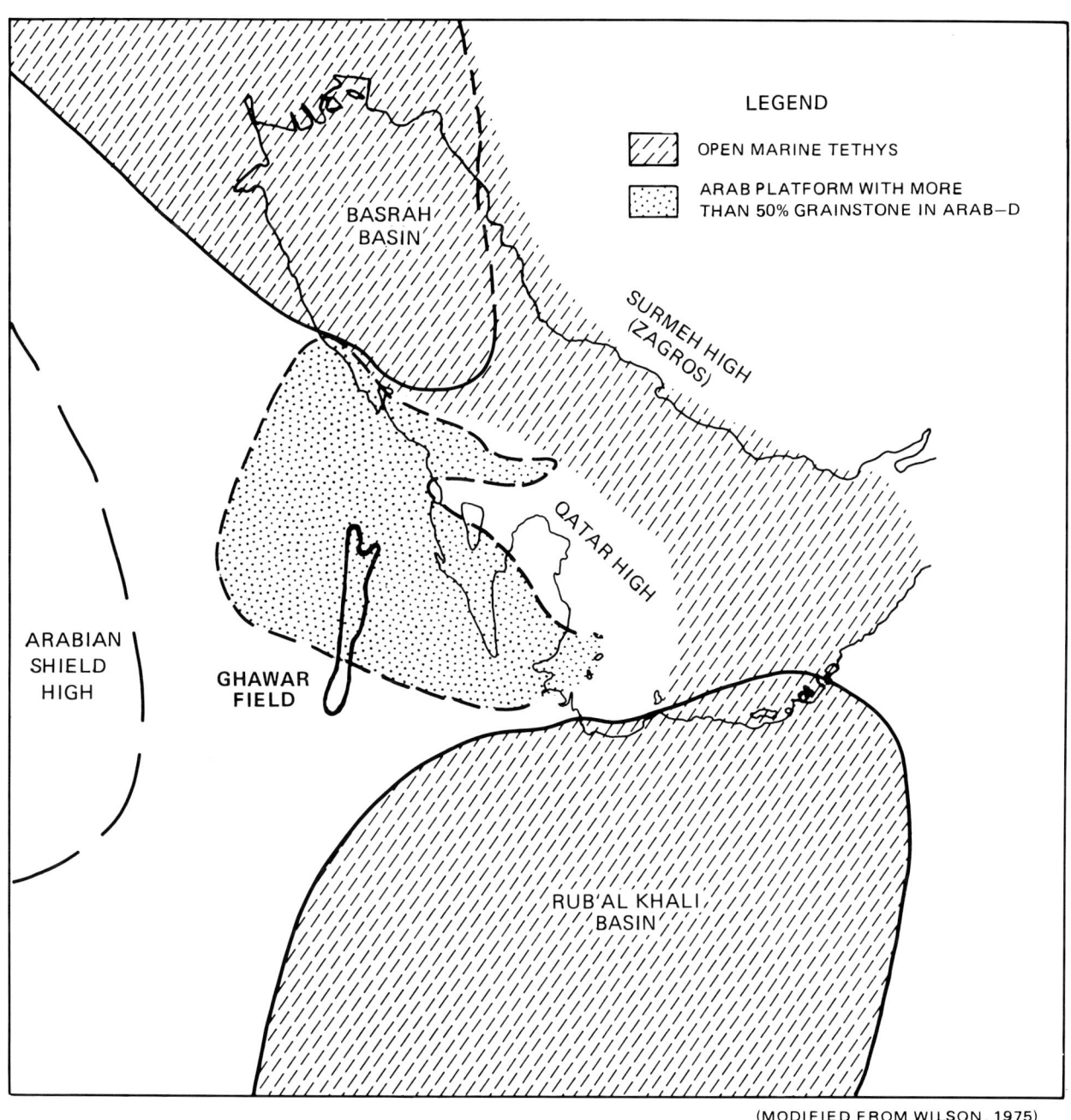

(MODIFIED FROM WILSON, 1975)

Figure 1. Location map, Ghawar field and Upper Jurassic paleogeography, eastern Saudi Arabia.

occur both from well to well and vertically within single wells, the overall patterns of sedimentation that resulted in areally extensive correlatable carbonate and evaporite units probably reflect eustatic sea level controls on this shelfal carbonate complex.

The Arab-D comprises two major shoaling upward cycles that are composed of many smaller scale upward shoaling cycles. These upward shoaling cycles are comprised of a variety of skeletal grainstones and packstones with ooid grainstones locally common in the uppermost Arab-D. In Ghawar, Arab-D Member carbonates are further subdivided into reservoir zones and subzones that are based largely on porosity log pattern correlation and detailed lithofacies studies.

The top of the Arab-D carbonate is characterized by thin subtidal to intertidal/ supratidal carbonates with sabkha evaporites. The upper portion of the Arab-D Member Anhydrite comprises sabkha evaporites and subaqueous evaporites with thin, carbonate interbeds that can be traced for hundreds of kilometers. Both palmate and bedded fabrics are preserved in these evaporites. The top of the Arab-D Anhydrite is marked by a sharp flooding surface.

OVERVIEW OF ARAB-D RESERVOIR AT GHAWAR FIELD

The Arab-D reservoir at Ghawar field is divisible into five major zones that are based largely on porosity log pattern correlation (Figure 2). Major zone boundaries separate field-wide lithologic and porosity breaks and are time-stratigraphic except for the tops of Zones 1 and 2A. These two time-transgressive zones follow the lateral facies change from the non-porous lower Arab-D Anhydrite to porous carbonates at the top of the Arab-D carbonate.

In Ghawar, the Arab-D is a grain-dominated reservoir consisting mainly of grainstones and packstones, although wackestones and mudstones do occur in the lower portion of the reservoir. Mud-rich sediments of Zone 3 of the Arab-D grade upward into the high porosity, low mud packstones and grainstones of Zone 2. In general, interparticle porosity is abundant and moldic porosity is common in the reservoir, whereas intraparticle, fracture, burrow and shelter porosity are much less common or rare. Intercrystal pores are common in the dolomites. Microporosity is present throughout the Arab-D in limestones and dolomites. It occurs as microporous skeletal and non-skeletal grains, microporous matrix, and micropores between cement crystals.

Both skeletal and non-skeletal grains are common in the Arab-D. Skeletal grains, especially benthic foraminifers, and dasycladacean green algae, are the dominant grain types in the Arab-D, although non-skeletal grains may be locally common. Other common skeletal grains include red algae, corals, stromatoporoids (including Cladocoropsis), brachiopods, bivalves, gastropods, echinoderms and ostracodes. Non-skeletal grains include ooids, coated grains with irregular concentric cryptalgal laminations, composite grains, intraclasts, peloids, and pellets.

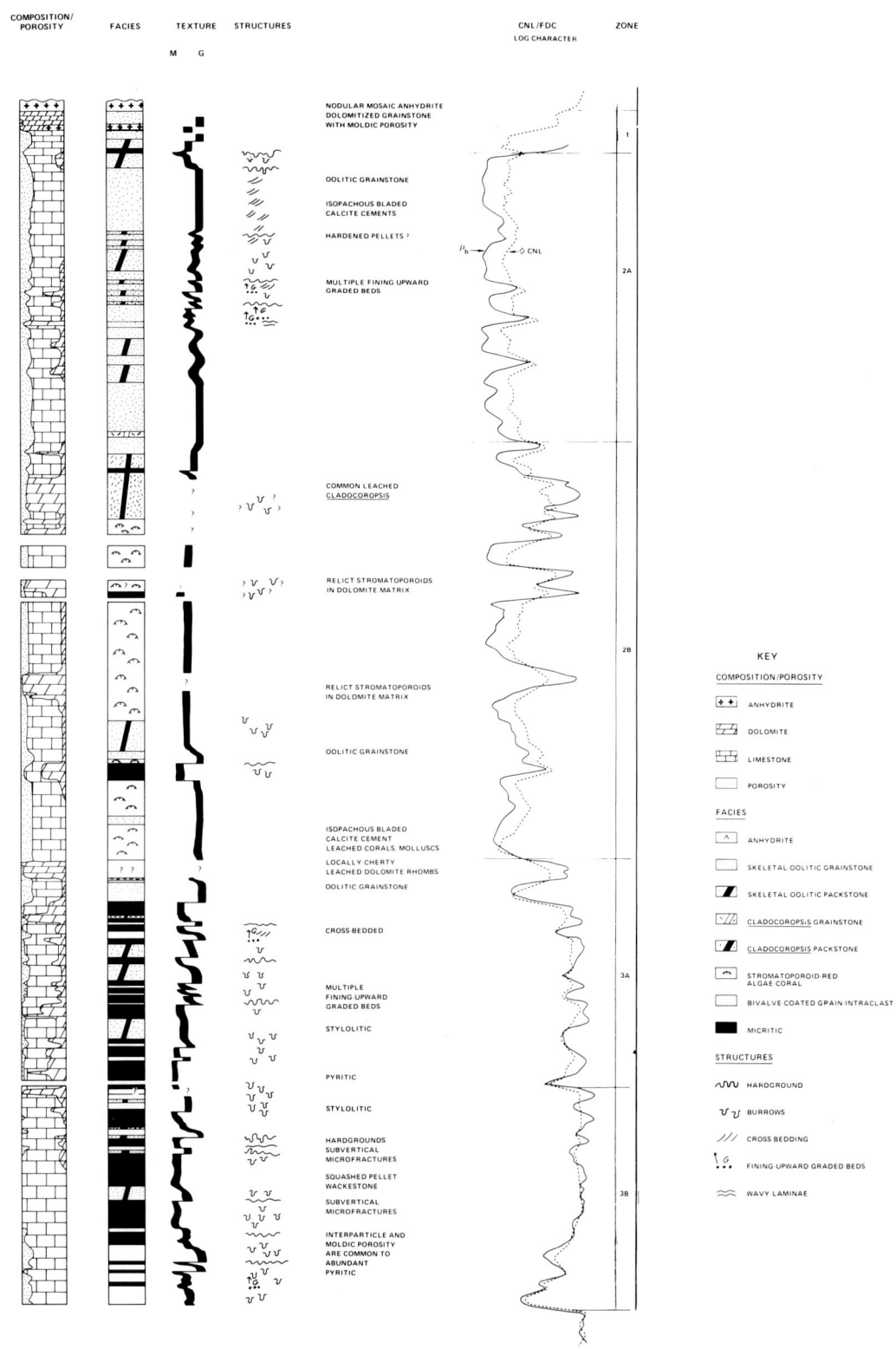

Figure 2. Typical Arab-D reservoir section, Well "X", 'Uthmamiyah area, Ghawar field.

GHAWAR ARAB-D LITHOFACIES

A classification scheme developed in order to organize Ghawar Arab-D rocks into genetically meaningful packages divides these rocks into six depositional lithofacies and one diagenetic lithofacies. These depositional lithofacies include one anhydrite and five carbonate lithofacies. The carbonate lithofacies are distinguished on the basis of their typical depositional components, and include: (1) skeletal-oolitic limestones, (2) Cladocoropsis limestones, (3) stromatoporoid-red algae-coral limestones, (4) bivalve-coated grain-intraclast limestones, and (5) micritic limestones. Lithofacies 1, 2, and 5 are commonly dolomitized to various degrees. Since dolomitization frequently destroys all evidence of original depositional lithofacies, another carbonate lithofacies is recognized: the diagenetic lithofacies dolomite. Characteristics of Ghawar Arab-D carbonate lithofacies are summarized in Figure 3. Vertical succession of lithofacies in a typical Ghawar well is shown in Figure 2.

Anhydrite

A variety of fabrics and structures occur in anhydrites interbedded with the upper part of the Arab-D reservoir, and in the overlying Arab-D Anhydrite (Figure 4). Nodular, bedded nodular, nodular mosaic, vertically-aligned mosaic, bedded mosaic, and massive fabrics are all seen.

Most anhydrite fabrics (nodular, nodular-mosaic, bedded nodular mosaic) seen in the portion of the Arab-D Anhydrite that immediately overlies the Arab-D reservoir (Figure 4 A, B) suggest deposition in a sabkha setting (Maiklem, et al., 1969; Loucks and Longman, 1982). Nodular anhydrites typically occur interbedded with fine grained, stromatolitic carbonates. Locally, in-situ growth of nodular anhydrite has distorted bedding. Fabrics seen above the Arab-D cycle in the Arab-D Anhydrite indicate a subaqueous origin (Figure 4 C). They include vertically-aligned mosaic, bedded mosaic, and massive anhydrite. In thin section, anhydrite generally exhibits a "felted" texture of very small, oriented, needle-like

LITHOFACIES	SKELETAL-OOLITIC	CLADOCOROPSIS	STROMATOPOROID-RED ALGAE-CORAL	BIVALVE-COATED GRAIN-INTRACLAST	MICRITIC	DOLOMITE
DEPOSITIONAL TYPES	GRAINSTONE, MUD-LEAN PACKSTONE, PACKSTONE	MUD-LEAN PACKSTONE, GRAINSTONE, PACKSTONE	MUD LEAN PACKSTONE, GRAINSTONE, PACKSTONE, BOUNDSTONE	PACKSTONE, MUD LEAN PACKSTONE, GRAINSTONE, WACKESTONE	WACKESTONE, MUDSTONE, PACKSTONE	INDETERMINATE
MAJOR GRAIN TYPES	MICRITIZED GRAINS, FORAMINIFERS INCLUDING MILIOLIDS), DASYCLADACEAN ALGAE, OOIDS, BIVALVES	MICRITIZED GRAINS, CLADOCOROPSIS, DASYCLADACEAN ALGAE, FORAMINIFERS (INCLUDING MILIOLIDS)	MICRITIZED GRAINS, STROMATOPOROIDS, CORALS, FORAMINIFERS (INCLUDING MILIOLIDS)	MICRITIZED GRAINS, BIVALVES, COATED GRAINS, INTRACLASTS, FORAMINIFERS	MICRITIZED GRAINS, BIVALVES, FORAMINIFERS (INCLUDING KURNUBIA), INTRACLASTS	ANHEDRAL TO EUHEDRAL DOLOMITE RHOMBS
MINOR GRAIN TYPES	ECHINODERMS, STROMATOPOROIDS, CORALS, CLADOCOROPSIS, GASTROPODS, COMPOSITE GRAINS, INTRACLASTS, OSTRACODES, RED ALGAE, BRACHIOPODS	STROMATOPOROIDS, ECHINODERMS, BIVALVES, CORALS, OOIDS, GASTROPODS, BRACHIOPODS, COMPOSITE GRAINS, RED ALGAE	CLADOCOROPSIS, BIVALVES, ECHINODERMS, DASYCLADACEAN ALGAE, INTRACLASTS, COATED GRAINS, COMPOSITE GRAINS, GASTROPODS	MILIOLID FORAMINIFERS, CORALS, STROMATOPOROIDS, DASYCLADACEAN ALGAE, ECHINODERMS, GASTROPODS, CLADOCOROPSIS	COATED GRAINS, MILIOLID FORAMINIFERS, DASYCLADACEAN ALGAE, INTRACLASTS, OSTRACODES, ECHINODERMS, GASTROPODS, STROMATOPOROIDS, CORALS	RELICT AND LEACHED BIVALVES, CLADOCOROPSIS, STROMATOPOROIDS, INTRACLASTS, ECHNINODERMS, IND. GRAINS
SEDIMENTARY STRUCTURES	CROSS BEDDING, BURROWS, HARDGROUNDS, FINING UPWARD GRADED BEDS, BORINGS, HOR. LAMINAE	BURROWS, HARDGROUNDS, CROSS BEDDING, HOR. LAMINAE	BURROWS, BORINGS, HARDGROUNDS, FINING UPWARD GRADED BEDS, CROSS BEDDING	BURROWS, HARDGROUNDS, BORING, FINING UPWARD GRADED BEDS, CROSS BEDDING	BURROWS, HARDGROUNDS, WAVY LAMINAE, HOR. LAMINAE, BORINGS, FINING UPWARD GRADED BEDS	RELICT BURROWS?, HARD GROUNDS? CARBONANCEOUS? LAMINAE
PORE TYPES	INTERPARTICLE, MOLDIC, INTRAPARTICLE, INTERCRYSTALLINE, FRACTURE	INTERPARTICLE, INTRAPARTICLE, MOLDIC, INTERCRYSTALLINE	INTERPARTICLE, MOLDIC, INTRAPARTICLE, INTERCRYSTALLINE, FRACTURE	INTERPARTICLE, MOLDIC, INTRAPARTICLE, INTERCRYSTALLINE, FRACTURE	INTERPARTICLE (WITHIN BURROW FILLS), INTRAPARTICLE, MOLDIC, INTERCRYSTALLINE, FRACTURE, VUG?	INTERCRYSTALLINE, MOLDIC, FRACTURE
DIAGENETIC MODIFICATION	LEACHING AND RECRYSTALIZATION, ISOPACHOUS BLADED CALCITE CEMENT, DOLOMITIZATION, PHYSICAL COMPACTION, STYLOLITIZATION, EQUANT CALCITE CEMENT, KAOLINITE EMPLACEMENT, ANHYDRITE EMPLACEMENT/REPLACEMENT, SILICIFICATION	LEACHING AND RECRYSTALIZATION, DOLOMITIZATION, EQUANT CALCITE CEMENT, STYLOLITIZATION, ANHYDRITE EMPLACEMENT, SILICIFICATION	LEACHING AND RECRYSTALIZATION, DOLOMITIZATION, ANHYDRITE EMPLACEMENT, STYLOLITIZATION, ISOPACHOUS BLADED CALCITE CEMENT, EQUANT CALCITE CEMENT, PYRITE, DEDOLOMITIZATION	LEACHING AND RECRYSTALIZATION, DOLOMITIZATION, STYLOLITIZATION, EQUANT CALCITE CEMENT, ANHYDRITE EMPLACEMENT, ISOPACHOUS BLADED CALCITE CEMENT, PYRITE	LEACHING AND RECRYSTALIZATION, DOLOMITIZATION, STYLOLITIZATION, ANHYDRITE EMPLACEMENT, PYRITE, SILICIFICATION, EQUANT CALCITE CEMENT, KAOLINITE	DOLOMITIZATION, LEACHING, ANHYDRITE EMPLACEMENT, EQUANT CALCITE CEMENT, STYLOLITIZATION, KAOLINITE, DEDOLOMITIZATION
DISTRIBUTION	1, 2A, 2B, 3A, 3B	2A, 2B	2A, 2B, 3A, 3B, JUBAILA	2B, 3A, 3B, JUBAILA	1, 2A, 2B, 3A, 3B, JUBAILA	1, 2A, 2B, 3A

Figure 3. Summary of characteristics and distribution of carbonate lithofacies, Arab-D reservoir, Ghawar field.

Figure 4. Slabbed core photographs and thin section photomicrograph of anhydrite, Arab-D reservoir and Arab-D Anhydrite.

    A. Interbedded anhydrite and carbonate are typical of the carbonate-anhydrite transition in Zone 1 of the Arab-D. In this example, anhydrite nodules (A) appear horizontally aligned to produce a bedded nodular fabric. Dolomitization has preserved fine details of the original fabric of this relict skeletal-oolitic grainstone. Slabbed core photograph, 'Uthmaniyah area.

    B. Nodular mosaic anhydrite typical of the lower Arab-D Anhydrite often contains carbonaceous inclusions (arrows) between anhydrite nodules. This carbonaceous material generally decreases upward into the anhydrite. Overall porosity is very low, with this anhydrite serving as the seal for the Arab-D reservoir. Slabbed core photograph, 'Uthmaniyah area.

    C. Vertically aligned (elongated) mosaic anhydrite was probably originally deposited as subaqueous palmate gypsum structures, and were subsequently transformed to anhydrite during burial. Slabbed core photograph, Hawiyah area, Arab-D Anhydrite.

    D. In thin section, nodular anhydrite exhibits a "felted" texture of oriented, neddle-like anhydrite fibers. Cross-polarized light, 'Uthmaniyah area, Arab-D Anhydrite.

Figure 5. Slabbed core photographs and thin section photomicrographs of skeletal-oolitic lithofacies.

    A. Hardgrounds (arrows) are locally common in skeletal-oolitic limestones, especially near the top of the Arab-D reservoir. Planar laminations are also present near the base of this sample. Slabbed core photograph, 'Uthmaniyah area, Zone 2A.

    B. This detail view of a hardground similar to the example in A above shows the abrupt change in rock types across this surface (outlined). Here, a well-cemented grainstone occurs below the hardground while a muddy, dolomitic packstone occurs above. Plane-polarized light, 'Uthmaniyah area, Zone 2A.

    C. Cross-bedded skeletal-oolitic grainstone, Zone 2A, Shedgum area.

    D. Skeletal-oolitic grainstone, Zone 2A. Grains, including ooids (OO), milliolid foraminifers (MF), micritized grains (MG) and the dasycladacean alga Salpingoporella (Dsp) and Clypeina (*) are surrounded by a thin fringe (arrow) of bladed calcite cement (probable syndepositional submarine cement). Minor monocrystalline overgrowths syntaxial with echinoderm debris are present. Interparticle porosity is abundant, moldic porosity (MO) is common. Thin section photomicrograph. Plane-polarized light. Bar for scale = 0.5 mm. Shedgum area.

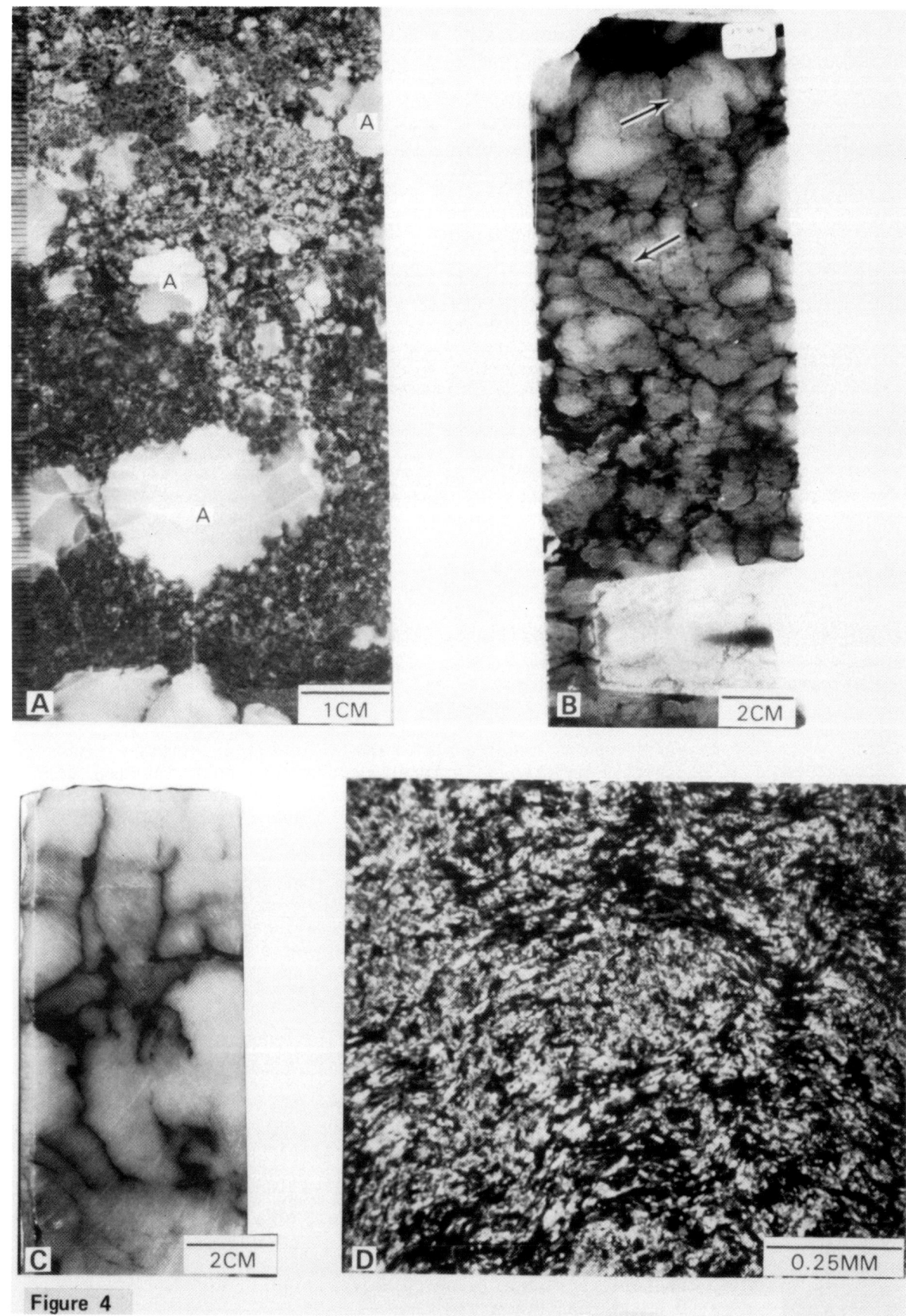

Figure 4

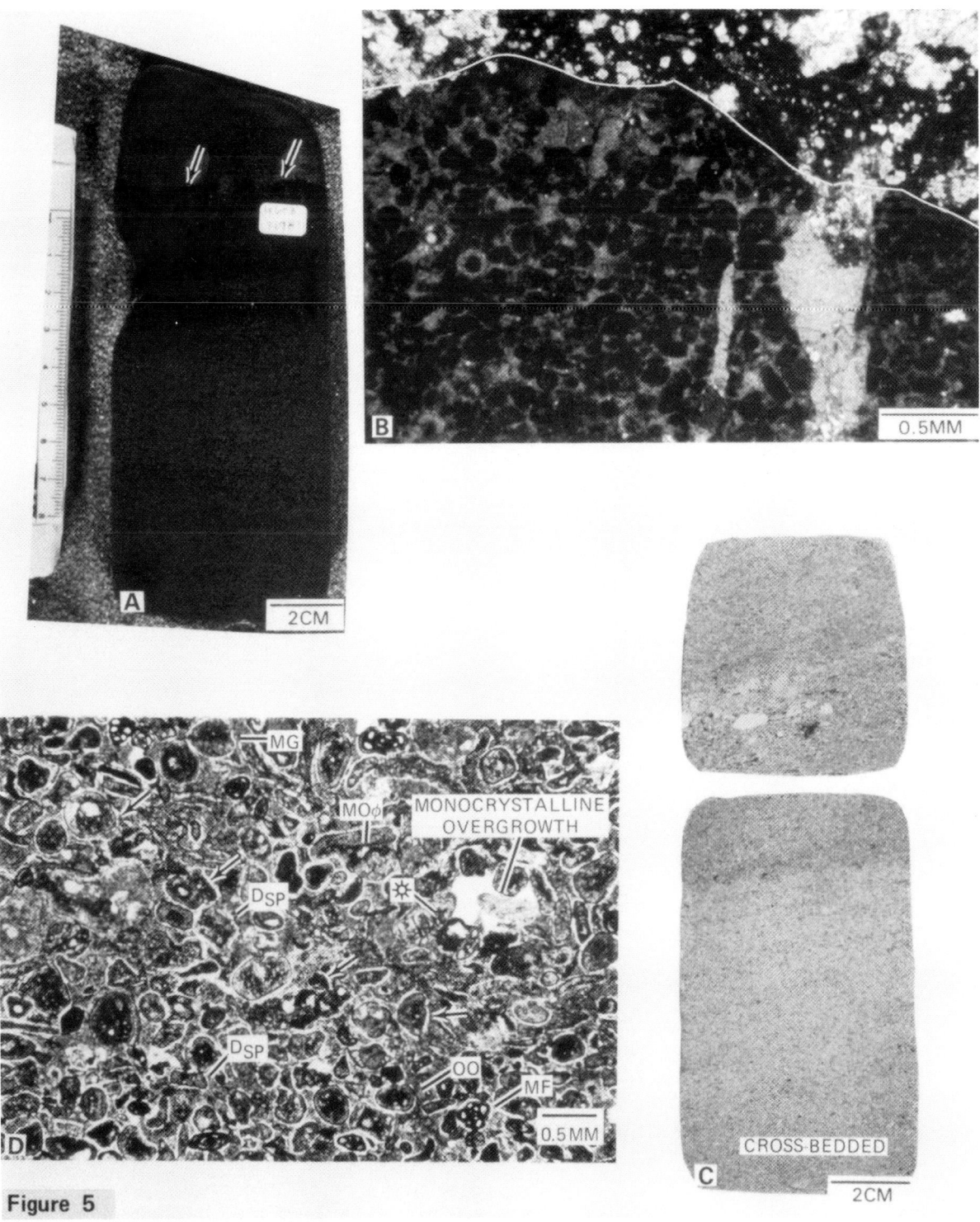

Figure 5

anhydrite crystals (Figure 4 D), although rare clear, coarser, rectangular laths of diagenetic anhydrite do occur (Figure 20 C). While anhydrite itself is non-porous, porous stringers of carbonate (especially dolomite) occur interbedded with this lithofacies (Figure 4 B).

Skeletal-Oolitic

This lithofacies comprises grainstones, mud-lean packstones, and packstones (Figures 5, 6). It constitutes most of the best Arab-D reservoir rock typically with the highest porosity and permeability. Although it occurs throughout the Arab-D, it is most common in core from the upper Arab-D (Zone 2A). Immediately beneath the Arab-D Anhydrite, this lithofacies is often completely replaced by fine-grained (silt-sized) dolomite. The fabric of the precursor limestone is preserved, in spite of complete dolomitization (Figure 14 A, B). Elsewhere in the reservoir, the skeletal-oolitic lithofacies locally is dolomitized. Packstones are typically more dolomitic than mud-lean packstones and grainstones (Figure 6 A, B, C). The degree of dolomitization ranges from a few rhombs scattered throughout interparticle and moldic pores to extensive replacement of original constituents.

Grainstones and mud-lean packstones are locally cross-bedded (Figure 5 C). Discrete burrows are common in packstones. Fining-upward graded beds occur throughout the upper portion of the reservoir, grainstones and mud-lean packstones grade up into packstones. Hardgrounds (Figure 5 A, B) commonly separate each graded bed.

Major grain types recognized are: micritized grains, benthic foraminifers, echinoderm debris, dasycladacean green algae, and thin-shelled bivalves. Other, less common grains include: calcareous red algae, intraclasts, Cladocoropsis, coral fragments and composite grains. Ooids are locally common to abundant in Zone 2A of the Arab-D.

In general, porosity type and abundance varies with texture and composition. Interparticle porosity is common to abundant in skeletal-oolitic grainstones (Figures 6 A, 17 A), rare to common in dolomitic skeletal-oolitic packstones. Intraparticle porosity is rare where present in skeletal-oolitic grainstones and dolomitic packstones alike. Intercrystalline porosity is rare to locally common in dolomitic packstones. Moldic porosity is rare to common in grainstones, mud-lean packstone and the dolomitic packstones. Moldic grains include: bivalves, gastropods, corals, and dasycladacean algae. Microporosity is common in ooids, foraminifers, red algae, micritized grains, and lime mud (Figures 6 E, F; 17 D, E). Intercrystalline, moldic and relict intraparticle and interparticle porosity are all present in skeletal-oolitic dolomites (Figure 14 A, B). The skeletal-oolitic lithofacies represents deposition in shoal water, high energy environments. Abundant oolitic intervals, cross-bedding and numerous hardgrounds all reflect the high water energy of this environment. This lithofacies was probably deposited as a series of sand shoals and tidal deltas in very shallow subtidal to intertidal water.

Cladocoropsis

This lithofacies (Figures 7, 13 B) is largely restricted to Zones 2B and 2A of the Arab-D reservoir. It comprises packstones, mud-lean packstones, and grainstones that have more than 10 percent Cladocoropsis. This stick-like stromatoporoid (Champetier and Fourcade, 1966) typically occurs in a matrix of sand-sized micritized grains, benthic foraminifers and calcareous algae. Locally, the matrix is nearly entirely miliolid foraminifers, a foraminifer indicative of a somewhat restrictive, lagoonal environment (Heckel, 1972). Echinoderm debris, bivalves, corals, fecal pellets, and other stromatoporoids are minor constituents. Cements include overgrowths syntaxial with echinoderm pieces and isopachous bladed calcite cement. Interparticle and intraparticle porosity are the major pore types present in Cladocoropsis limestone (Figure 7 B). Lime mud, foraminifers, micritized grains, pellets, and rarely Cladocoropsis are microporous. Where this lithofacies is completely dolomitized, the relatively fine-grained matrix is replaced by dolomite and the coarse-grained Cladocoropsis is leached (Figure 13 B).

Figure 6. Thin section photomicrographs and scanning electron microscope images of skeletal-oolitic lithofacies.

    A. Skeletal-oolitic carbonate (grainstone). Grains, including foraminifers (F), dasycladacean algae (Salpingoporella, SP) and micritized grains (MG) are surrounded by a thin fringe of bladed calcite cement (syndepositional submarine cement-arrows). Interparticle (BP) porosity is abundant, moldic (MO) porosity is common. Minor dolomite (D) is present in moldic pores. Hawiyah area, Zone 2A. Plane-polarized light. Bar for scale = 0.25 mm.

    B. Skeletal-oolitic carbonate (packstone). Microporosity is present in grains and matrix and is recognized by an indistinct blue haze. Grains are compacted because isopachous bladed calcite is not present, hence porosity and permeability are low, relative to samples with isopachous cements. Hawiyah area, Zone 3A. Plane-polarized light. Bar for scale = 1.0 mm.

    C. Partly-dolomitized skeletal-oolitic carbonate (packstone). Dolomite rhombs (arrows) embay and replace grains. Dolomite crystals have "dirty" centers (inclusions? relict grains?) that are surrounded by "clean" rims. Hawiyah area, Zone 2B. Plane-polarized light. Bar for scale = 0.5 mm.

    D. Thin section photomicrograph of skeletal-oolitic carbonate. "Microleached" and micritized skeletal grains (arrows) are abundant, giving the rock a "chalky" appearance. 'Ain Dar area, Zone 2A. Bar for scale = 0.2 mm. Plane-polarized light.

    E. Scanning electron microscope (SEM) image of pore cast (epoxy-impregnated and acid-etched rock chip). Epoxy-filled interparticle (BP) and micromoldic (MO) porosity stand out in relief. Original carbonate minerals are completely etched. 'Ain Dar area, Zone 2A. Bar for scale = 40 microns.

    F. Scanning electron microscope (SEM) image of epoxy pore cast showing detail of area outlined in Figure E. Epoxy-filled micromoldic porosity within skeletal grain stands in relief. 'Ain Dar area, Zone 2A. Bar for scale = 4 microns.

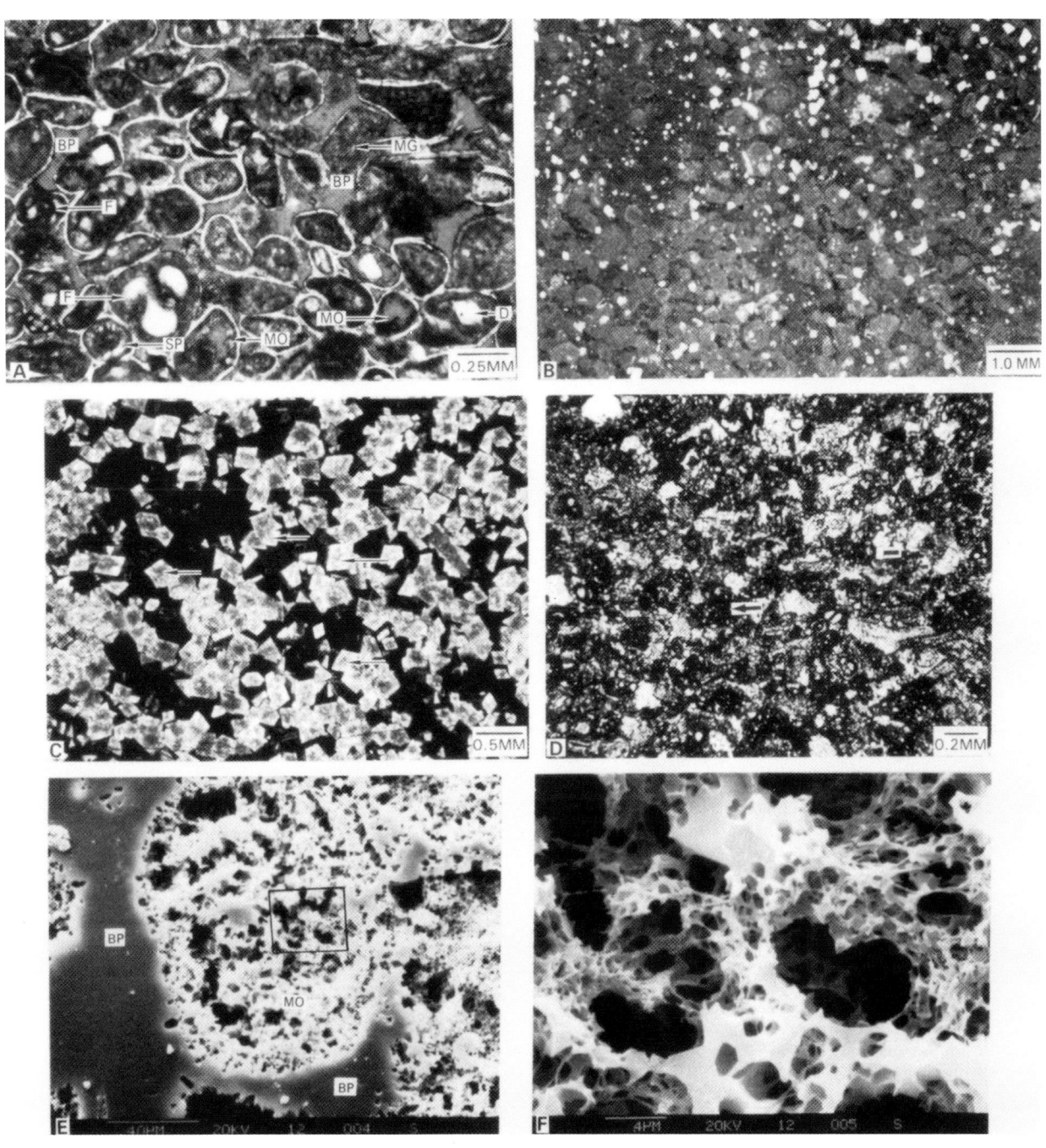

Figure 7. Slabbed core photograph and thin section photomicrographs of Cladocoropsis lithofacies.

    A. Cladocoropsis carbonate. Photograph of slabbed and polished core. Cladocoropsis (arrows) constitutes approximately twenty percent of this mud-lean packstone to grainstone, occuring in a matrix that is composed mainly of sand-sized skeletal grains. Interparticle and intraparticle porosity are common, but difficult to discern at scale of photograph. 'Ain Dar area, Zone 2B. Bar for scale = 2 cm.

    B. Thin section photomicrograph of Cladocoropsis carbonate (mud-lean packstone to grainstone). Cladocoropsis (arrows) occurs in a matrix that consists mainly of foraminifers, micritized skeletal grains, and Clypeina (dasycladacean algae) fragments. Dolomite is common, partly filling interparticle porosity and replacing skeletal grains. Interparticle (BP) and intraparticle (WP) pores are common, intercrystalline (BC) porosity is rare. 'Ain Dar area, Zone 2B. Bar for scale = 2 mm. Plane-polarized light.

    C. Thin section photomicrograph of preferentially dolomitized burrow-fill in Cladocoropsis carbonate. Burrow-fill is coarser-grained and less muddy relative to surrounding matrix. Monocrystalline calcite overgrowths (arrow) that are syntaxial with echinoderms are conspicuous but volumetrically insignificant. 'Ain Dar area, Zone 2B. Bar for scale = 2 mm. Plane-polarized light.

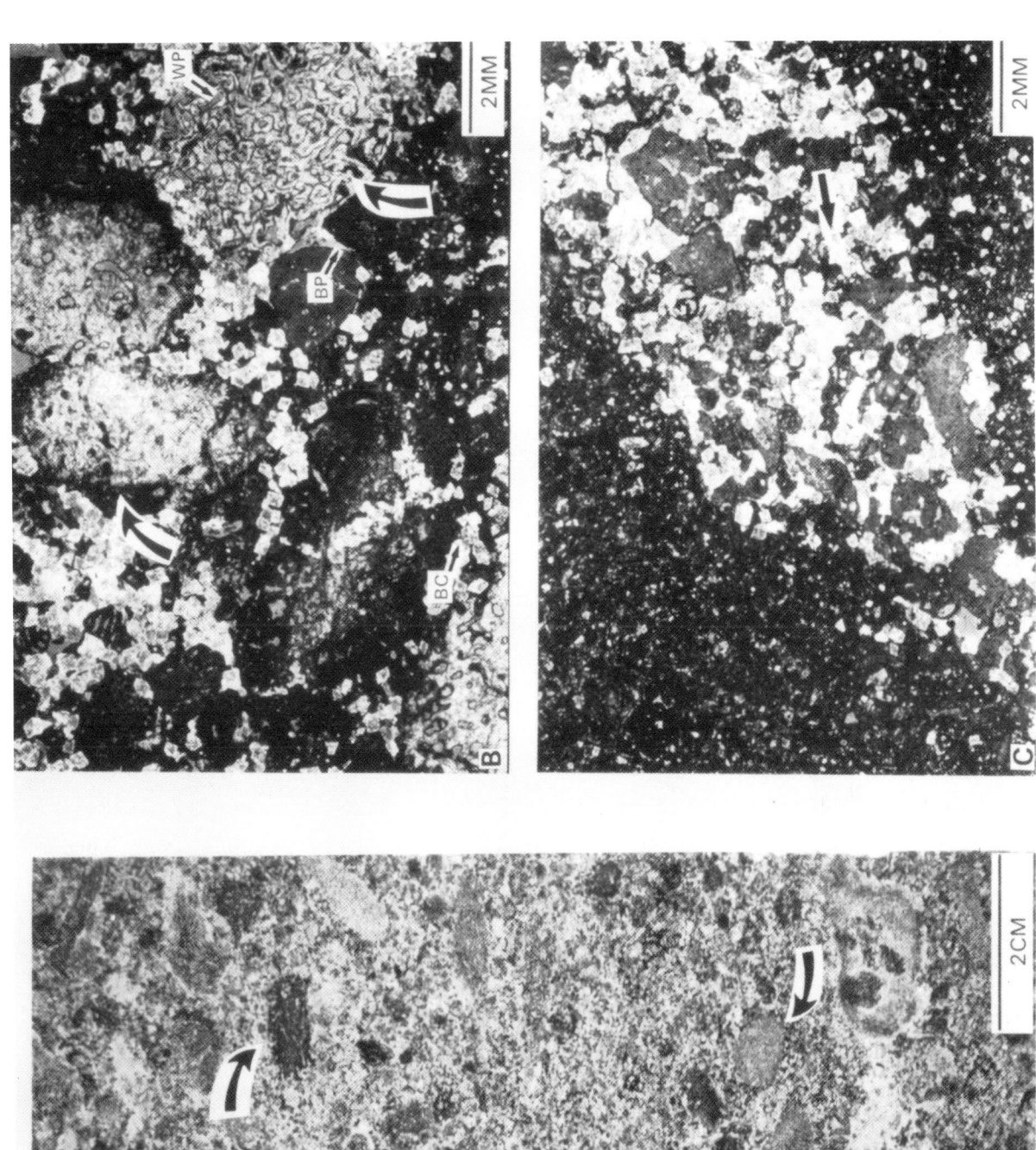

Intercrystalline and moldic pores are the dominant pore types in Cladocoropsis dolomite. Centimeter-sized Cladocoropsis molds are often well-connected yielding a rock with multidarcy horizontal permeability.

The Cladocoropsis lithofacies was deposited under marine subtidal conditions of moderate wave and current activity. Previous workers (Turnsek et al., 1981) have documented Cladocoropsis zones within dark-gray, massive micritic limestones in the Upper Jurassic of northwestern Yugoslavia, which they interpret to have formed in a quiet, somewhat restricted back-reef area. While Cladocoropsis zones in the Arab-D can be muddy (especially in the Cladocoropsis dolomites), most Cladocoropsis limestones are grain-supported, low mud rocks deposited under conditions of moderate wave and current activity. These zones represent lower energy conditions than other grainy lithofacies of the upper Arab-D, even though they are more grain-rich than the Cladocoropsis zones of Turnsek et al. (1981).

Stromatoporoid-Red Algae-Coral

This lithofacies (Figures 8, 9) consists of packstones, mud-lean packstones, and grainstones. Boundstones are recognized locally. This lithofacies is most abundant in Zone 2B and is locally present near the base of the Arab-D reservoir. Locally, it is gradational into the bivalve-coated grain-intraclast facies. Major grain types include stromatoporoids, pieces of colonial corals and micritized grains. Cladocoropsis, foraminifers, red algae, intraclasts, molluscs, coated grains and echinoderm debris are less common. Cements include isopachous bladed calcite, equant calcite and calcite overgrowths that are syntaxial with echinoderm debris (Figure 8 C).

The degree of dolomitization ranges from minor replacement of grains and plugging of interparticle pore space to the complete replacement of original rock constituents. "Ghosts" of relict stromatoporoids are often recognizable in spite of complete dolomitization (Figure 9 B).

Figure 8. Slabbed core photograph and thin section photomicrographs of stromatoporoid-red algae-coral lithofacies.

  A. Stromatoporoid - red algae - coral carbonate with grainstone texture. Photograph of slabbed and polished core. Stromatoporoids (S), leached molluscs (M) and partly-leached corals (C) occur throughout a matrix that is composed mostly of micritized skeletal grains. Moldic pores up to 15 mm in diameter are most conspicuous (arrow); interparticle and intraparticle porosity are also common, but not so easily discerned at scale of photograph. 'Ain Dar area, Zone 2B. Bar for scale = 2 cm.

  B. Thin section photomicrograph of stromatoporoid - red algae - coral carbonate (grainstone). Conspicuous leached coral debris (C) and Cladocoropsis (arrow) occur in a matrix composed mostly of sand-sized skeletal grains. Moldic porosity is abundant (MO), interparticle and intraparticle porosity are less common. Minor replacement of skeletal grains and plugging of pore space by dolomite occurs (D). 'Ain Dar area, Zone 2B. Bar for scale = 2 mm. Plane-polarized light.

  C. Thin section photomicrograph of stromatoporoid - red algae - coral carbonate (grainstone). Large partly-leached and partly-dolomitized coral occurs in matrix that is composed mainly of micritized skeletal grains. Interparticle (BP) and moldic porosity (MO) are common. Isopachous bladed calcite cement is present in matrix (arrow). 'Ain Dar area, Zone 3A. Bar for scale = 2 mm. Plane-polarized light.

Figure 9. Slabbed core photographs and thin section photomicrograph of stromatoporoid - red algae-coral lithofacies.

  A. Stromatoporoid-red algae-coral limestones are generally poorly sorted, which also increases heterogeneity. In this example, a very large (at least 8 cm in diameter), head-shaped stromatoporoid (S) is present in a finer-grained matrix. Locally, stromatoporoid-coral organic buildups are present in this lithofacies. The scar in photograph center is due to core plugging. Slabbed core photograph, 'Uthmaniyah area, Zone 2B.

  B. Fractured, dolomitized stromatoporoid - red algae-coral lithofacies. Stromatoporoids are recognizable in spite of complete replacement by dolomite. Hawiyah area, Zone 2B.

  C. Moldic porosity is common to abundant in stromatoporoid-red algae-coral limestones and contributes to the high degree of heterogeneity of this lithofacies. Moldic pores may result from the leaching of fragments of scleractinian corals (C). Slabbed core photograph, 'Ain Dar area, Zone 2B.

  D. Stromatoporoid boundstones are locally common in this lithofacies. Plane-polarized light, Hawiyah area, Zone 2B.

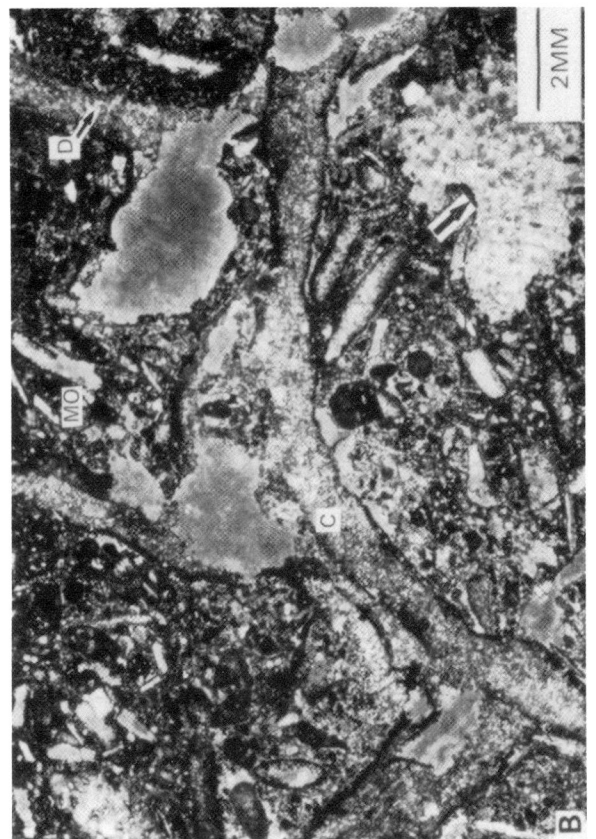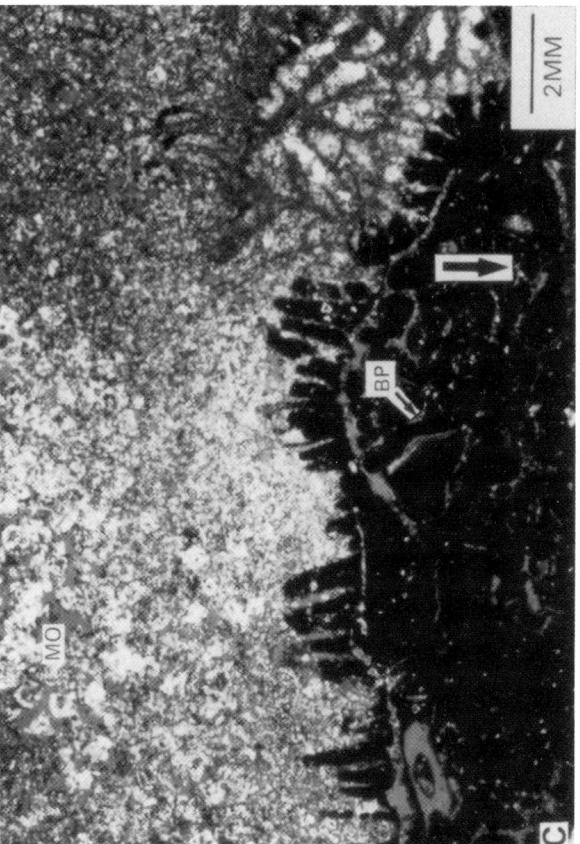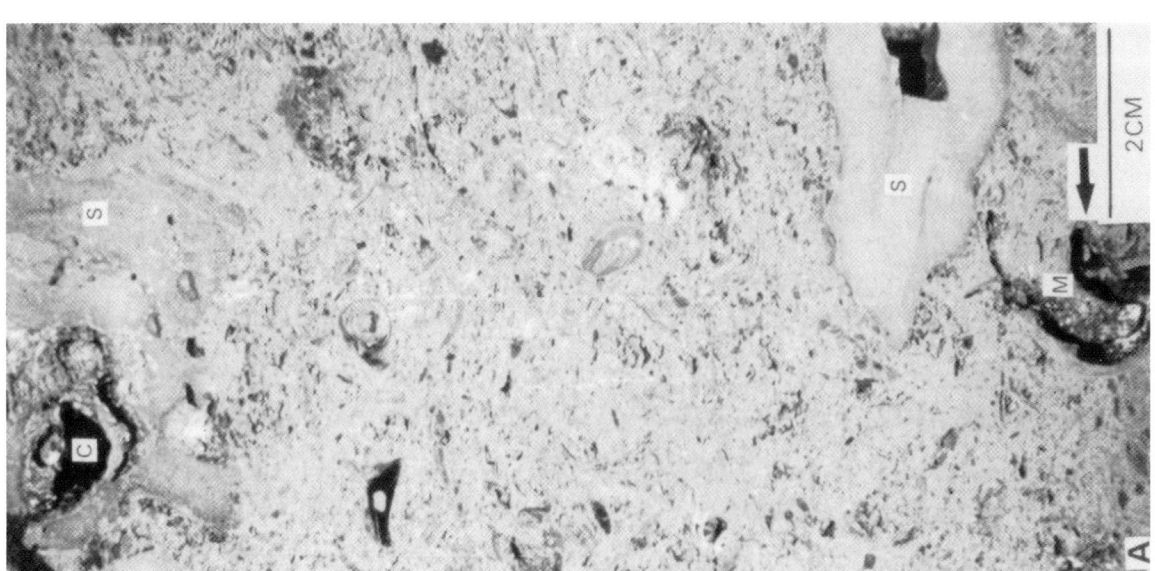

Figure 8

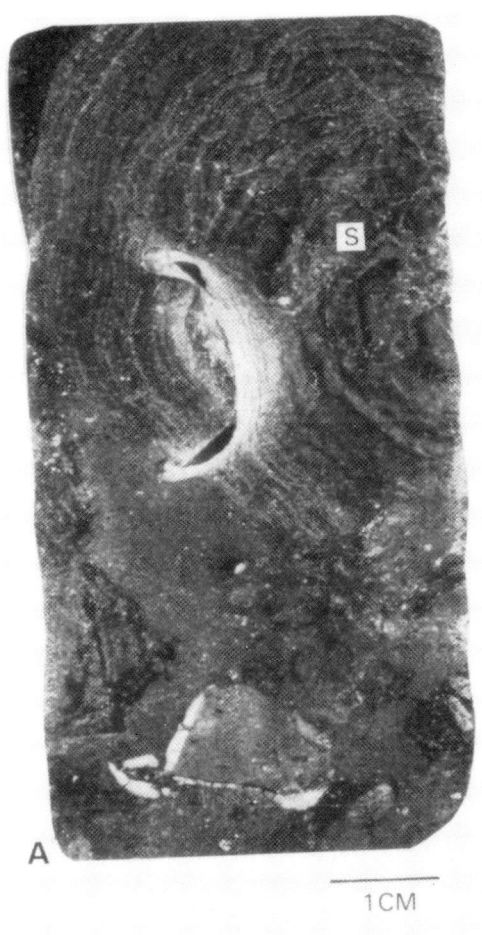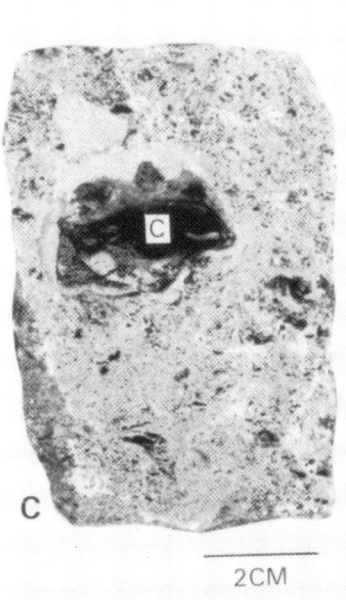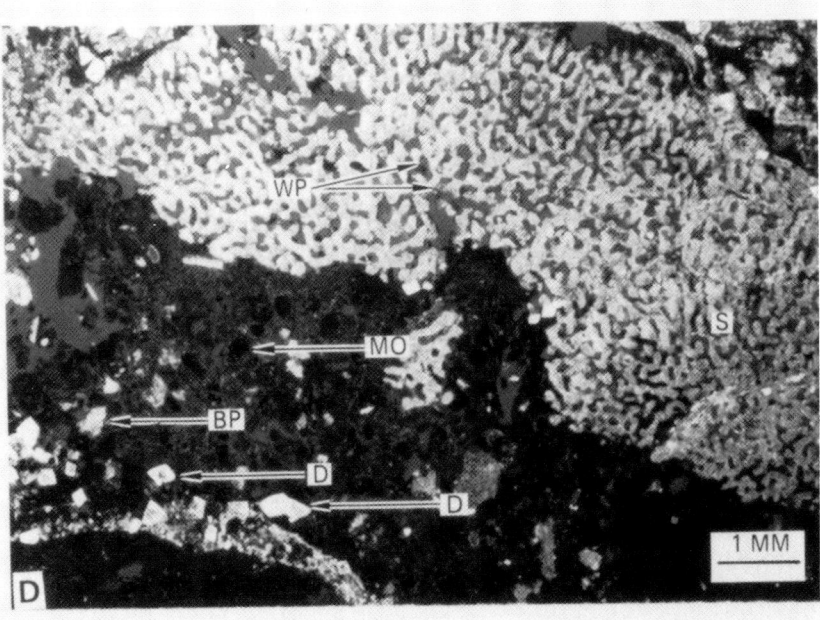

Figure 9

Interparticle porosity and intraparticle porosity are rare to common, intercrystalline porosity is rare. Moldic porosity is rare to locally abundant. Leached corals and molluscs yield pores that have maximum dimensions of several centimeters (Figures 8 A, C; 9 C). Mud matrix, micritized grains, red algae, intraclasts, and coated grains are microporous.

Stromatoporoid-red algae-coral limestones record subtidal deposition under normal marine conditions of generally moderate wave and current activity. Algae (and hermatypic corals) in this lithofacies imply deposition within the photic zone, while the presence of such stenotopic organisms as corals indicates normal marine conditions. Locally, this lithofacies contains organic buildups or mounds, which may have experienced high wave and current energy. While a relatively small number of in-place corals or stromatoporoids are noted in this facies, the overall abundance of these reefal organisms suggests that wave-resistant boundstones may have been significant.

Bivalve-Coated Grain-Intraclast

This highly variable lithofacies comprises packstones, mud-lean packstones, and grainstones (Figures 10, 11). It is largely restricted to Zone 3 of the Arab-D reservoir where it often occurs interbedded with the fine-grained micritic lithofacies. It locally is gradational with the stromatoporoid-red algae-coral lithofacies.

Basal contacts of most beds of this lithofacies are sharp and are marked by the presence of hardgrounds (submarine erosional surfaces) and/or stylolites (Figure 10 A, C, D). Grain-size decreases upwards, typically, the upper portion of each bed is gradational into muddier carbonates.

Major grain types are micritized grains, bivalves, coated grains and intraclasts (Figure 11 A-E). Coated grains typically comprise a partly-leached bivalve nuclei that is surrounded by a vaguely laminated algal coating. Intraclasts typically are fragments of micritic limestone.

Some intraclasts contain quartz sand. Echinoderm debris, foraminifers, corals, gastropods, and stromatoporoids are less common. Intraclasts, foraminifers, micritized grains, coated grains, and lime mud are microporous.

Overall, dolomite is a minor constituent; it partly fills interparticle and moldic pores as well as replacing grains and mud (Figure 10 B, C, 11 D). Cements include coarse equant calcite, calcite overgrowths syntaxial with echinoderm pieces, pyrite, and isopachous bladed calcite.

Moldic porosity is common. Grains with moldic porosity include bivalves, coated grains, and corals (Figures 10 B; 11 C, D, E). Interparticle porosity is rare to common and is locally abundant. Intraparticle porosity and intercrystalline porosity are rare.

The texture, the dominant pore type, and the dominant grain type and size varies widely. Where leached bivalves and corals are present, pores are often 15-20 mm in maximum dimension; where absent, pore size is typically less than 1 mm in maximum dimension.

Bivalve-coated grain-intraclast limestones were deposited in clear water of generally moderate turbulence. The presence of algae coating grains indicates deposition within the photic zone, in water of sufficient turbulence to move grains around and achieve relatively even algal coatings on grains. Also, the presence of muddy intraclasts ripped up from some nearby area of mud deposition probably indicates that brief, high energy episodic (storm ?) events were common.

Figure 10. Slabbed core photograph and thin section photomicrographs of bivalve-coated grain-intraclast lithofacies.

    A. Bivalve-coated grain-intraclast limestone and micritic limestone commonly occur immediately adjacent to each other. Here, low porosity micritic mudstone is capped by a hardground (outlined) and overlain by high porosity bivalve-coated grain-intraclast mud-lean packstone. Note that burrows (arrows) in micritic mudstone are filled by bivalve-coated grain-intraclast sediments that have filtered down from above. Plug scar is present near the center of this sample. Slabbed core photograph, 'Uthmaniyah area, Zone 3B.

    B. Micritic limestone (wackestone). Partly dolomitized (20-30 percent dolomite) micritic limestone with coarser-grained probable burrow-fill. Little porosity is apparent except for rare interparticle and moldic pores in burrow-fill. Thin section photomicrograph. Plane-polarized light. Bar for scale = 1.0 mm. Shedgum area, Zone 2B.

    C. Stylolitized contact between micritic limestone and bivalve-coated grain-intraclast lithofacies, base of Zone 2B. Dolomite completely fills interparticle pore space in bivalve-coated grain-intraclast limestone. Note the presence of a calcite-filled natural fracture within the micritic limestone (possibly related to stylolitization). Unfilled fracture in center of photograph is either drilling-induced or an artifact of thin section preparation. Thin section photomicrograph. Plane-polarized light. Bar for scale = 1.0 mm. Shedgum area.

    D. Discontinuity surface at contact between two fining-upward subcycles, Zone 2B. Micritic limestone lithofacies with conspicuous moldic porosity (MO) is overlain by bivalve-coated grain-intraclast limestone. Vague burrow-fill is present in micritic limestone at contact. Thin section photomicrograph. Plane-polarized light. Bar for scale = 1.0 mm. Shedgum area.

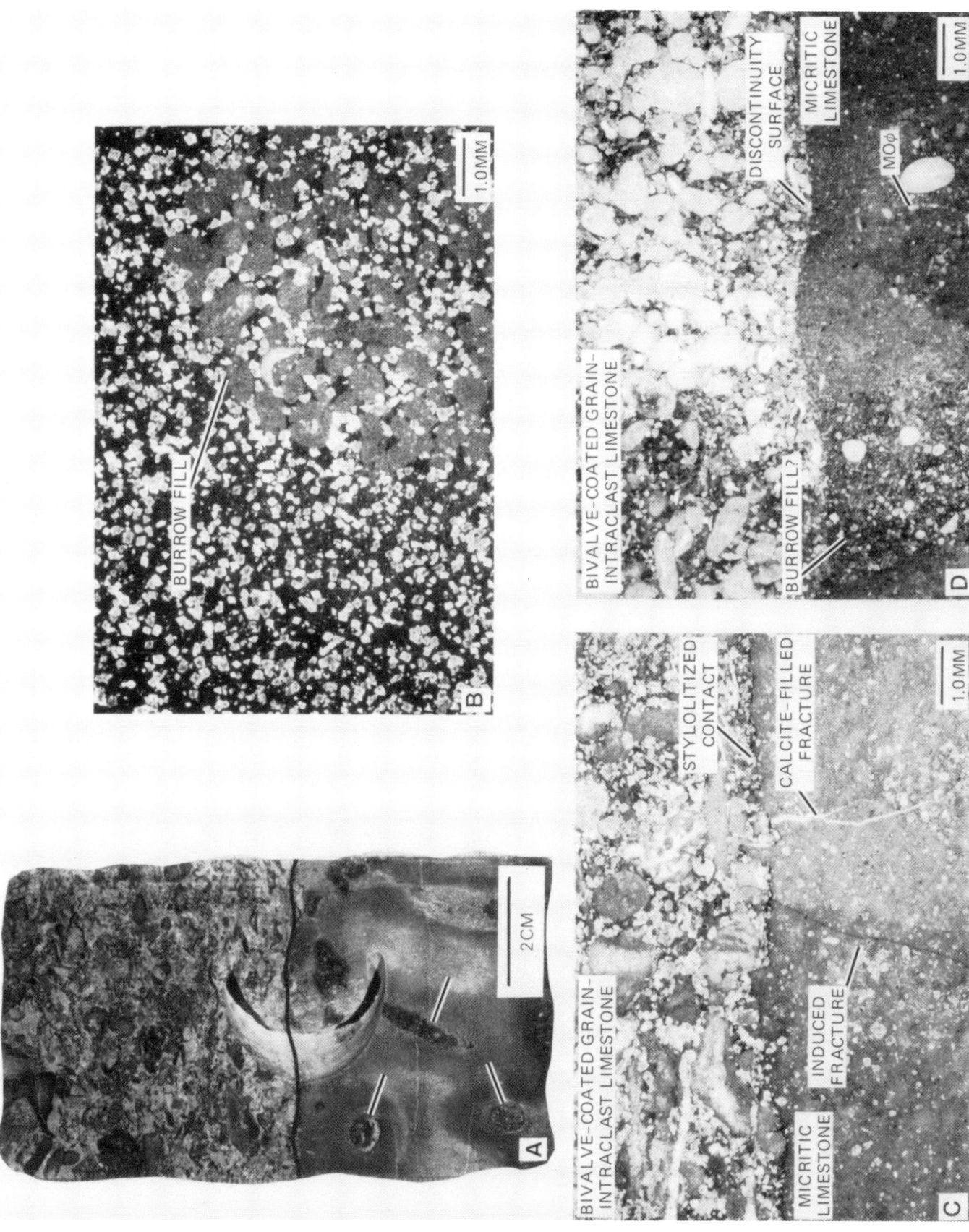

Figure 11. Slabbed core photographs and thin section photomicrographs of bivalve-coated grain-intraclast lithofacies.

    A. Slabbed core photograph of bivalve-coated grain-intraclast carbonate (packstone). Large intraclasts are conspicuous. Hawiyah area, Zone 3A.

    B. Bivalve-coated grain-intraclast carbonate (packstone). Constituents include intraclasts (I), micritized grains (MG) and coated grains (arrows). Interparticle and moldic porosity are present, but not abundant; microporosity is recognized by a vague blue haze. Hawiyah area, Zone 3A. Plane-polarized light. Bar for scale = 1.0 mm.

    C. Thin section photomicrograph of bivalve-coated grain-intraclast carbonate (packstone). Major grains are coated grains (GN) and bivalves (B). Dolomite makes up approximately 40% of this sample by bulk volume, plugging interparticle porosity and completely replacing coated grains. 'Ain Dar area, Zone 3A. Bar for scale = 2 mm. plane-polarized light.

    D. Bivalve-coated grain-intraclast carbonate (packstone). Photograph of slabbed and polished core. Coated grains (arrows) are the most abundant grain type. Dolomite (dark brown in photograph) constitutes approximately 30 percent of rock, partly replacing grains and plugging porosity. Large, irregular gaps in rock are scars due to cutting of core plugs. 'Ain Dar area, Zone 3A. Bar for scale = 2 cm.

    E. Bivalve-coated grain-intraclast limestone (mud-lean packstone to grainstone), Zone 3A. Abundant, large coated grains (CG) in a peloidal matrix (arrows). Most coated grains comprise a vaguely laminated coating (algal?) that surrounds a bivalve fragment (now leached). Interparticle porosity (BP) and moldic porosity (MO) are abundant. Thin section photomicrograph. Plane-polarized light. Bar for scale = 1.0 mm. Shedgum area.

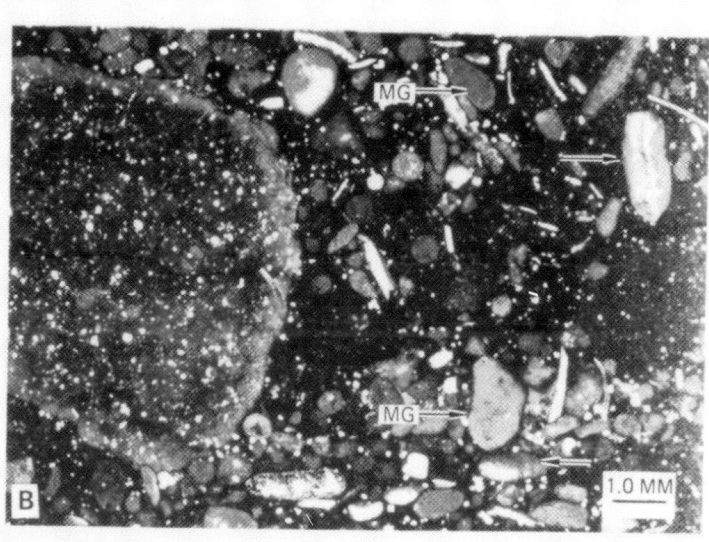
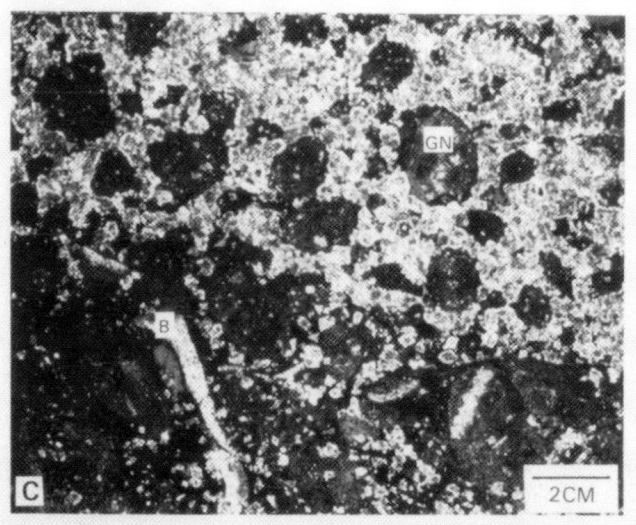
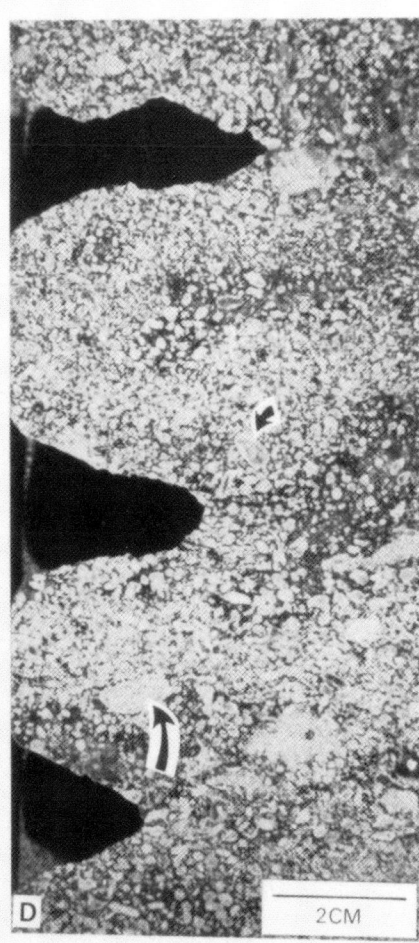
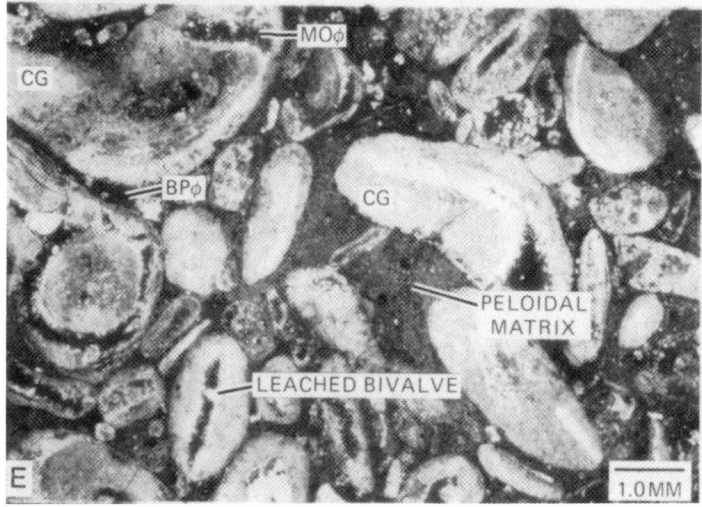

485

Micritic

The micritic lithofacies (Figures 10, 12) comprises wackestones, mudstones, and rare packstones. It occurs throughout the reservoir, but is more abundant in the lower half of the Arab-D in Zone 3, where it occurs interbedded with coarser, grain-supported carbonates. Dolomitization is common (Figure 12 D), the amount of dolomite present in a sample can vary widely over a short vertical distance. In partly dolomitized micritic limestone, dolomitization commonly is more extensive in burrow-fills than in the surrounding matrix. Thin section examination reveal that the micritic lithofacies commonly is vaguely peloidal (Figure 12 D). Major grain types present in addition to peloids include micritized skeletal grains, bivalves, and foraminifers.

Most components (grains and matrix) are microporous. Moldic pores and intraparticle porosity are rare to common (Figure 12 C). Interparticle porosity is rare and is typically restricted to burrow fills. There is no visible porosity in many samples (Figure 12 A, B, D), microporosity accounts for whatever porosity is present.

Micritic limestones and bivalve-coated grain-intraclast limestones typically occur together in fining-upward cycles up to one to two feet thick. These cycles begin with coarse-grained bivalve-coated grain-intraclast grainstones or mud-lean packstones deposited atop an underlying hardground surface; this hardground is typically extensively burrowed and bored, with bivalve-coated grain-intraclast sediments from above filling these borings. Bivalve-coated grain-intraclast grainstones fine upward into bivalve-coated grain-intraclast (and locally skeletal-oolitic) packstones, and finally into micritic wackestones and mudstones. This overall fining-upward cycle is capped by another hardground surface (and perhaps by another fining-upward cycle) at the top. This generalized cycle may not go to completion, however, and the next cycle may begin before the previous fining-upward cycle has graded up into micritic limestone. Such partial cycles result in amalgamated beds of bivalve-coated grain-intraclast grainstones or mud-lean

Figure 12. Slabbed core photograph and thin section photomicrographs of micritic lithofacies.

    A. This micritic mudstone is typical of much of the lower Arab-D reservoir. Porosity in this lithofacies is generally low. Burrows (arrows) are locally common. A plug scar is visible in the upper portion of this sample. Slabbed core photograph, Haradh area, Zone 3A.

    B. Slightly dolomitic micritic limestone with wackestone texture. Dolomite is rare, recognizable grains include foraminifers and echinoderm debris. Plane-polarized light, bar for scale = 0.4 mm. Haradh area, Zone 3A.

    C. Porous micritic limestone with wackestone texture. Moldic pores (MO) are common, intraparticle pores (WP) are rare (chambers within foraminifers). Plane-polarized light, bar for scale = 0.1 mm. Haradh area, Zone 3A.

    D. Micritic carbonate (packstone). Micritized grains (arrows) partly surrounded by "necklaces" of dolomite. Grains are sutured, no interparticle porosity is preserved. Hawiyah area, Zone 3A. Plane-polarized light. Bar for scale = 0.5 mm.

Figure 13. Thin section photomicrographs, slabbed core photographs, and scanning electron microscope image of dolomite lithofacies.

    A. Sucrosic dolomite (dolomite with intercrystalline porosity, arrows) commonly occurs immediately adjacent to areas of mosaic dolomite (dolomite without intercrystalline porosity). The large, irregularly shaped pores in this example are probably relict moldic pores. Slabbed core photograph, 'Uthmaniyah area, Zone 2B.

    B. _Cladocoropsis_ dolomite typically has large moldic pores (arrows) that result from the selective leaching of _Cladocoropsis_ grains during dolomitization. If these pores were originally abundant enough to have been touching, the resulting pores will be interconnected giving a rock with multidarcy horizontal permeability. Slabbed core photograph, Shedgum area, Zone 2B.

    C. Rare intercrystalline porosity in mosaic dolomite is partly filled with kaolinite. Most dolomite crystals have "dirty" (inclusion-rich?) centers and clear rims; a few dolomite crystals that fill pore space (arrows) lack "dirty" centers and may have precipitated from more dilute pore fluids. Plane-polarized light, bar for scale = 0.1 mm. Haradh area, Zone 2B.

    D. Dolomite, Zone 2B. Original depositional texture has been obliterated by dolomitization and is not recognizable. Probable burrow (right hand side of photograph) with well developed network of intercrystalline porosity relative to surrounding matrix. Thin section photomicrograph. Plane-polarized light. Bar for scale = 1.0 mm. Shedgum area.

    E. Dolomite centers are leached by later diagenetic fluids to produce "gutted" dolomite crystals. Scanning electron microscope image, Hawiyah area, Zone 3A.

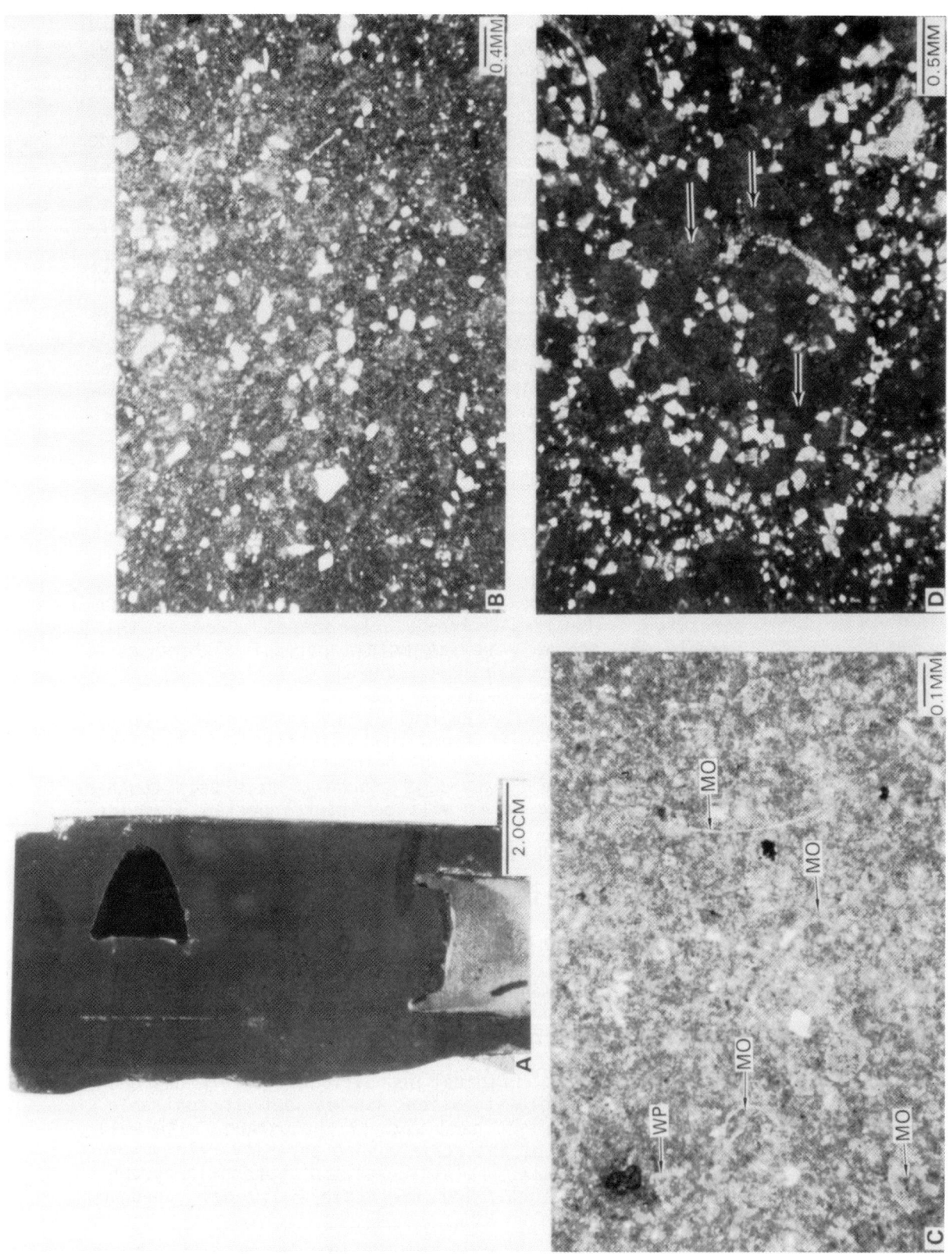

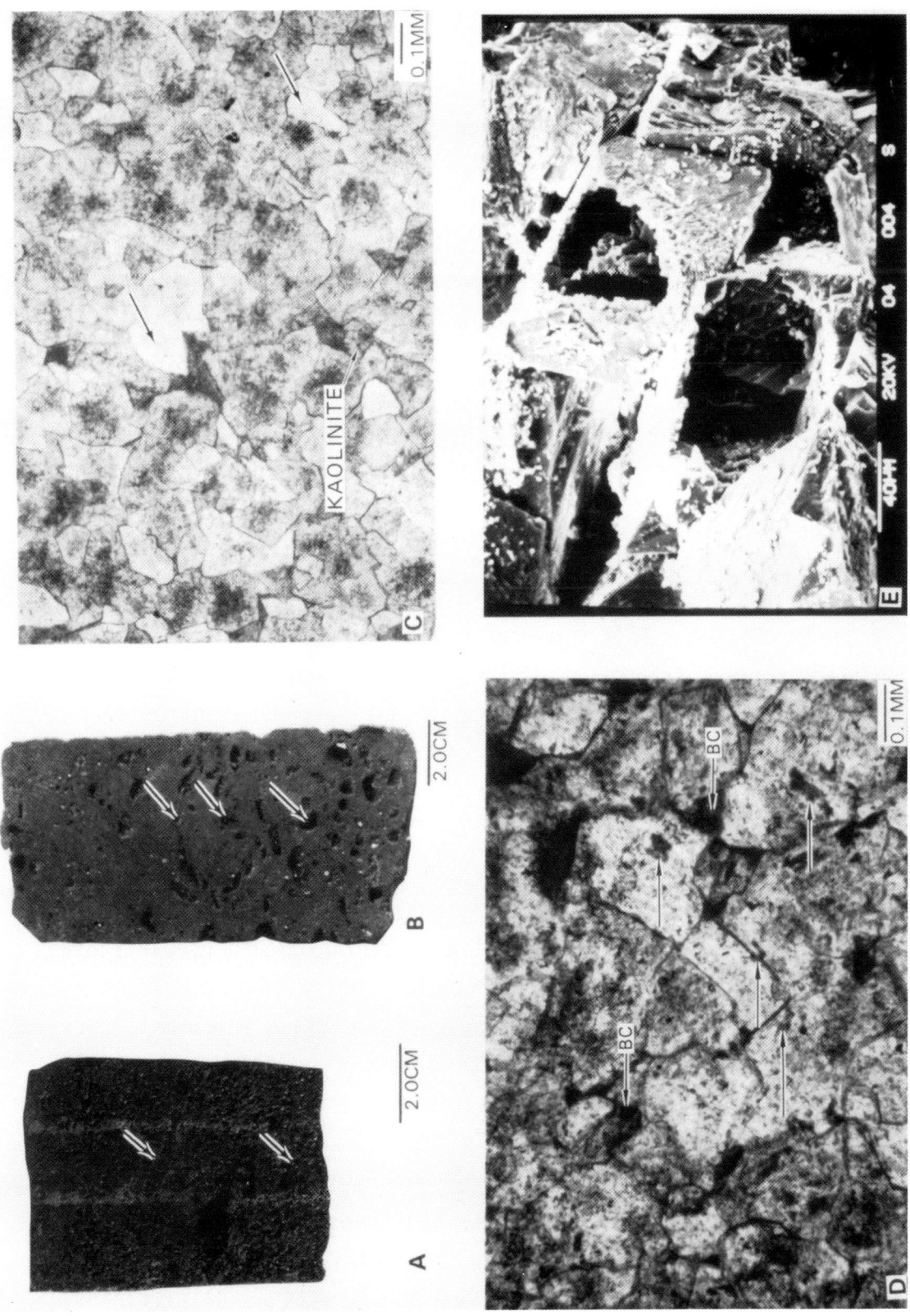

packstones that are thick (>4 feet thick) and show only a slight fining-upward character.

The micritic lithofacies, in contrast to the other lithofacies, was deposited under relatively low energy conditions. Sediments of this lithofacies are dominantly mud-supported and contain a sparse, low diversity biota. Burrowing organisms (infauna) are common, and either create grain-rich burrows in overall grain-poor muds or completely homogenize the sediment. Preserved sedimentary structures except for burrows are rare. All these factors argue for sediment accumulation in quiet, low energy conditions generally protected from strong current action or wave energy.

Dolomite

Coarse, (30-50 microns) non-fabric preserving dolomite is the most common dolomite seen in the Arab-D, occurring throughout the reservoir (Figures 13 A, C, D, E; 14 D). This type of dolomite typically obliterates precursor rock fabric, although relict grains and sedimentary structures (burrows and hardgrounds) are recognizable in a few cases (Figures 13 B; 14 A, B, C). Intercrystalline and moldic pores are most common; fracture porosity is present, but rare. "Mosaic" (non-porous) and "rhombic" (porous) end-member fabrics are recognized. Transitions from low porosity-low permeability mosaic fabrics to high porosity-high permeability rhombic fabrics are recognized. These transitions occur over a short distance and are commonly seen within a single core plug and probably are controlled by changes in original rock fabric. Dolomite crystals often appear zoned; some have a cloudy center surrounded by a clear rim (Figure 13 C), others are comprised of multiple alternating cloudy and clear zones. In many cases, centers and certain zones of the crystals are leached (Figure 13 D, E).

Fine-grained dolomite that replaces, yet preserves, the fabric of the precursor limestone is commonly seen immediately below the Arab-D Anhydrite at the top of the Arab-D carbonate (Figure 14 A, B). Original

Figure 14. Slabbed core photograph and thin section photomicrographs of dolomite lithofacies.

    A. Unlike typical dolomites in Zones 2 and 3, Zone 1 dolomites immediately underlying the Arab-D Anhydrite can preserve fine fabric details. Here, details of relict grains, pore types and even cements are clearly visible. These fine-grained dolomites have porosity/permeability characteristics that are more typical of the limestones they replace than of dolomites. Plane-polarized light, Shedgum area.

    B. Detail of the previous photograph. Relict grains, such as the gastropod (G) fragments, relict interparticle (BP), intraparticle (WP) and moldic (MO) porosity and even relict isopachous bladed cement (arrows) can be observed. Plane-polarized light, Shedgum area.

    C. Burrowing of a precursor limestone has been preserved in spite of dolomitization. In this photograph, areas of sucrosic dolomite probably reflect original grain-rich burrow-fills (arrows) while surrounding mosaic dolomite probably result from by dolomitization of unburrowed, mud-rich limestone. Note again the abrupt transition from dense mosaic dolomite to porous sucrosic dolomite. Slabbed-core photograph, Hawiyah area, Zone 2B.

    D. Within a single dolomite sample, a dolomite fabric can vary from sucrosic (on the right) with abundant intercrystalline (BC) porosity to mosaic (on the left) with little visible porosity. Abrupt variations in fabric probably reflect some original fabric inhomogeneity such as a burrow or bed boundary in the precursor limestone. Plane-polarized light, Shedgum area, Zone 2B.

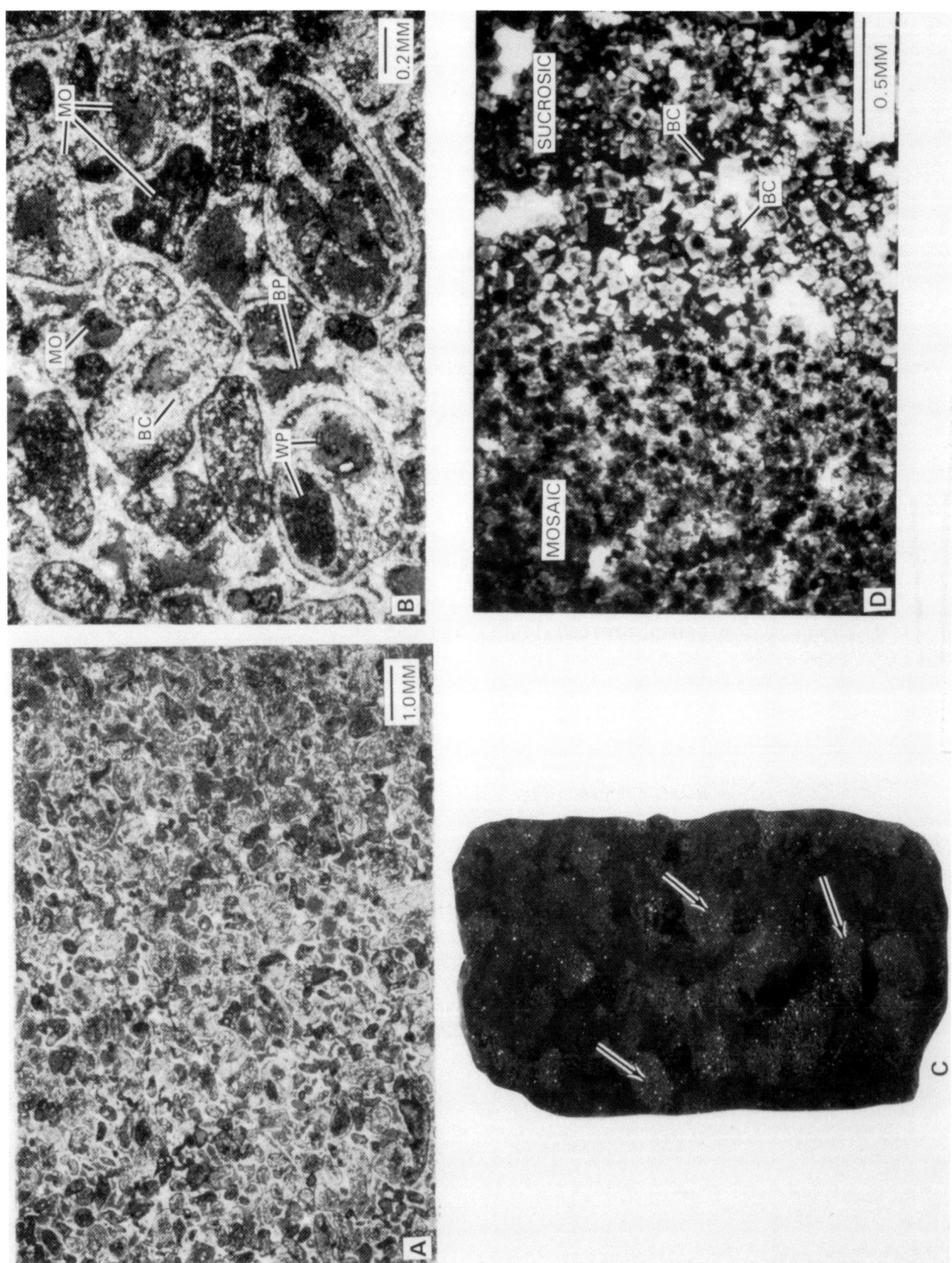

depositional texture is preserved in spite of complete dolomitization, hence this dolomite variety is assigned to one of the five other carbonate lithofacies, most typically the skeletal-oolitic lithofacies. Interparticle, moldic, intraparticle, and intercrystalline porosity are all present in the fine-grained, fabric preserving dolomite. Locally, anhydrite or calcite cement partly fills pores.

GHAWAR ARAB-D DEPOSITIONAL SEQUENCE

Two "shoaling-upward" cycles deposited during a relative highstand in sea level constitute the bulk of the Arab-D. The first cycle comprises Zone 3 of the Arab-D. The abundance of grain-supported textures (packstone, mud-lean packstone, grainstone) stays fairly constant in Zone 3B, but increases upwards through Zone 3A. A second shoaling-upward cycle beginning near the base of Zone 2B resulted in the deposition of the virtually mud-free shoal-water deposited sediments seen in Zone 2A and culminated in the deposition of thin subtidal to intertidal/supratidal cycles and sabkha evaporites of Zone 1 and the lower portion of the overlying Arab-D Anhydrite.

Within the first cycle, the lower half (Zone 3B) consists largely of the low porosity, mud-supported micritic lithofacies. This lithofacies is interbedded with the porous grain-supported bivalve-coated grain-intraclast lithofacies. The skeletal-oolitic and stromatoporoid-red algae-coral lithofacies are present, yet rare. They occur in only a few wells. No _Cladocoropsis_ limestone is seen. Dolomites are extremely rare.

Although the micritic lithofacies is still predominant in the upper half of this cycle (Zone 3A), porous skeletal-oolitic grainstones and packstones become increasingly common. The porous skeletal-oolitic and bivalve-coated grain-intraclast lithofacies are interbedded with the relatively non-porous micritic lithofacies. The stromatoporoid-red algae-coral lithofacies is a minor element, the _Cladocoropsis_ lithofacies is absent. Dolomites are common to abundant. Bivalve-coated grain-intraclast limestones are intimately related to micritic limestones throughout this portion of the lower Arab-D. These two lithofacies characteristically occur interbedded in packages or cycles that are apparent over two scales of observation. Throughout most of Zone 3 relatively thick (5 to 10 feet or more) accumulations of bivalve-coated grain-intraclast limestone alternate with thick micritic limestones. These large-scale cycles probably form in response to repeated cyclic variations in sea level with the thick intervals of bivalve-coated grain-intraclast limestone deposited during periods of subtidal open-marine circulation while the micritic limestones were deposited during more restricted, very shallow, quiet water conditions. Within these large-scale cycles, smaller-scale fining-upward episodic cycles also occur that consist of very thin (usually several inches thick) basal coarse-grained bivalve-coated grain-intraclast limestone that fines upward into micritic limestone. These cycles have been interpreted as storm deposits on the basis on the thinness of basal coarse-grained skeletal carbonates and the abrupt upward transition from coarse to finer-grained sediment. Other workers (Aigner, 1982; Kreisa and Bamback, 1982) interpret similar cycles in other areas as having a storm origin.

A distinct change in lithofacies occurs at the base of the Zone 2B. In contrast to the underlying fairly low porosity and muddy rocks, Zone 2B comprises a wide variety of lithofacies, all of which are predominantly grain-supported. The abundance of the predominantly mud-supported micritic lithofacies decreases upwards throughout Zone 2B. The upper third of Zone 2B is extensively dolomitized.

Zone 2A is almost exclusively skeletal-oolitic grainstone, mud-lean packstone, and packstone. The _Cladocoropsis_, stromatoporoid-red algae-coral, and micritic lithofacies are present, but are not quantitatively important. Dolomite is essentially absent. Anhydrite is locally present.

Zone 1, the interval immediately underlying the Arab-D Anhydrite, comprises skeletal-oolitic, micritic and anhydrite lithofacies. Much of this zone is dolomite that has completely replaced, yet preserved, the depositional fabric of the original sediment.

The upward increase in abundance of grainy carbonates seen in the lower Arab-D (Zone 3) records an upward increase in the energy of the depositional environment. Sediments laid down in widespread, predominantly low-energy, deeper subtidal settings pass upward into sediments laid down in a higher-energy, more shallow subtidal setting. Variations in the energy regime, caused by minor fluctuations in relative sea level and by storm events within the overall shoaling-upward cycle, resulted in the deposition of alternating intervals of low-porosity muddy carbonates and higher-porosity grainy carbonates that characterize this portion of the reservoir.

Zones 2B, 2A and 1 together are interpreted as a second upward-shoaling package of sediments deposited following a relative rise in sea level. Zone 2B sediments were deposited in an open-marine, subtidal shelf setting. At times variations in wave and current energy across the shelf resulted in a complex mosaic of grainy shoals, muddier lagoons, and localized organic buildups or patch reefs. As a result, considerable lateral changes in lithofacies, depositional texture and thickness characterize much of this portion of the reservoir. At other times during deposition of Zone 2B sediments, the environment across the shelf was more uniform, resulting in the widespread deposition of sediments that do not show much variation in lithofacies, texture or thickness throughout the study area.

Zone 2A was laid down in a widespread, high-energy shallow shelf environment. Tide and wave-generated currents swept throughout the area. Most carbonate mud was winnowed, mud-lean grain-supported sediments are predominant. Stabilized grain flats comprised of micritized skeletal and non-skeletal grains as well as active oolitic grain flats constitute most of this zone. The lower portion of this interval is fairly uniform in thickness. The upper portion of this interval shows considerable thickness variation resulting from accretion of carbonate beds at the top of the reservoir, local thickening of grain flats, and erosion of the top of Zone 2A.

The position of Zone 1 immediately underlying the Arab-D Anhydrite within the stratigraphic section, its restricted biota (dasycladacean algae, gastropods, bivalves, and ostracodes) and presence of algal stromatolites and interbedded nodular anhydrite all suggest that it was deposited in a complex of shallow subtidal, intertidal and locally supratidal environments. Zone 1 records the transition from the high-energy shallow water grainstones of Zone 2A to the sabkha evaporites of the lower portion of the Arab-D Anhydrite, ending the second cycle.

LATERAL FACIES VARIATIONS

While the overall pattern of sedimentation seen in the Arab-D is fairly uniform and has resulted in areally extensive carbonate and evaporite units, significant lateral variability in the distribution of lithofacies does occur (and is summarized in Figure 15). Generally, the basic pattern across Ghawar field is one of a thinning carbonate section and a thickening evaporite section going from north to south even though the overall thickness of the Arab-D Member remains relatively constant. This thickness change of the carbonate section to the south is thought to have occurred by a gradual facies change from carbonate to evaporite at the top of the Arab-D reservoir; with deposition of primarily shoal-water skeletal-oolitic limestones in the north being contemporaneous with anhydrite deposition (and sabkha conditions) to the south.

Within this framework of an overall thinning carbonate section to the south, then, lateral variations are also observed to occur within the depositional carbonate lithofacies of the Arab-D reservoir. Generally, lateral variability is least pronounced in the lower Arab-D (Zone 3). As previously noted, matrix-supported micritic limestone is the dominant lithofacies near the base of the Arab-D reservoir. Within this overall quiet-water, mud-rich regime, individual beds of coarser-grained

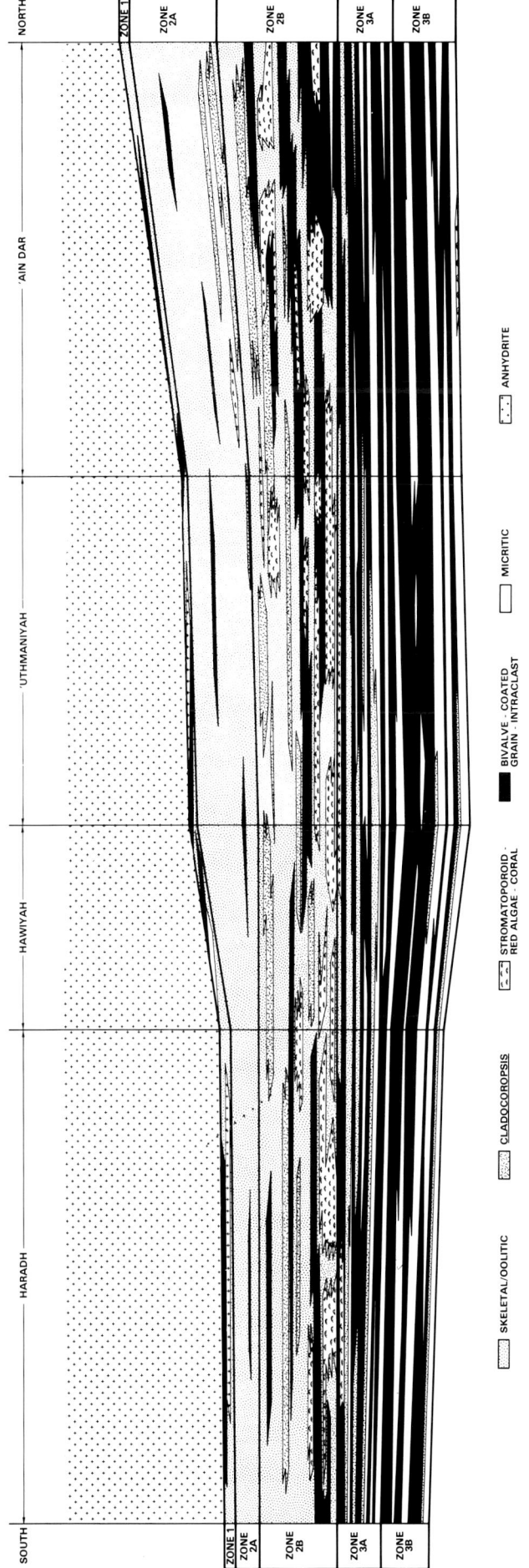

Figure 15. Arab-D depositional lithofacies, south-north cross section, Ghawar field.

bivalve-coated grain-intraclast or skeletal-oolitic limestone occur and typically are traceable for tens of kilometers. As earlier stated, these thin (usually one foot or less thick) interbeds of coarser-grained material may represent either cyclic variations in sea level and/or episodic storm deposition. In this lower poriton of the Arab-D, porosity is typically limited to these laterally continuous beds of coarser-grained limestone causing reservoir stratification. Middle and upper portions of the reservoir (Zones 2B, 2A), in contrast, are dominated by porous skeletal-oolitic limestone, but also contain subordinate amounts of stromatoporoid-red algae-coral, Cladocoropsis and micritic limestone. Typically, lateral continuity of these subordinate lithofacies is low, with facies changes often occurring within a kilometer or less. Locally, thick accumulations of stromatoporoid-red algae-coral limestone are present that contain in-place corals and stromatoporoid boundstones. These areas appear as local "thicks" on isopach maps of this portion of the reservoir, and have been interpreted to be carbonate build-ups or mounds. In general, all these subordinate lithofacies in the middle and upper Arab-D (especially Cladocoropsis and micritic lithofacies) migrate up-section going to the north. These facies changes probably reflect more open, normal marine (and perhaps deeper water) conditions that were present in the north relative to contemporaneous conditions present in the south.

DIAGENESIS

Diagenesis modifies primary sediment textures and fabrics of the original lime sediment and has great impact on the final reservoir quality of the rock. The major diagenetic processes active in the Arab-D include: dolomitization, leaching and recrystallization, cementation, compaction and fracturing. Despite the variety of diagenetic processes that have been active, much of the original texture of the Arab-D still is preserved. Figure 16 summarizes diagenesis and porosity evolution of the Arab-D at Ghawar.

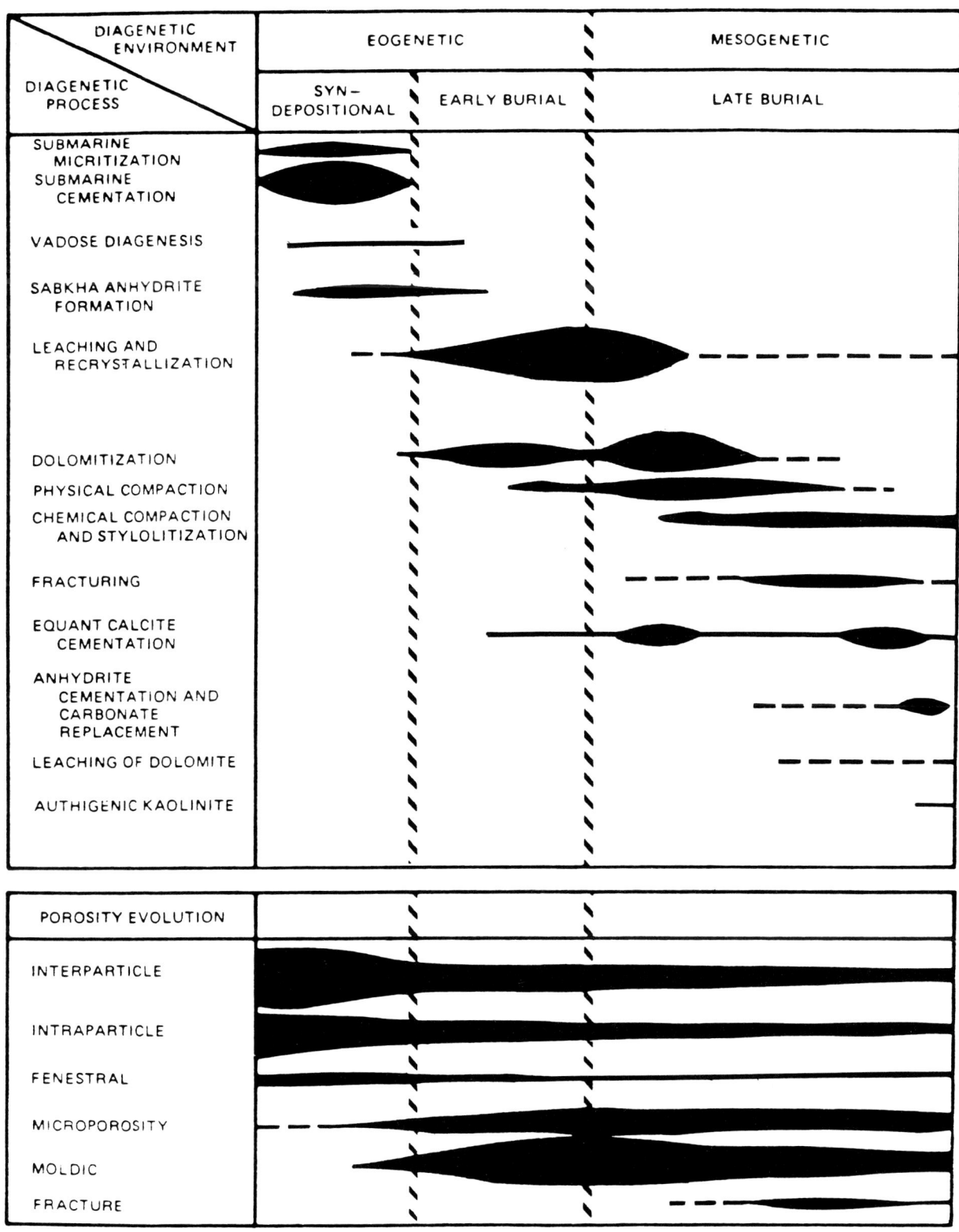

Figure 16. Paragenesis and porosity evolution, Arab-D reservoir, Ghawar field.

Dolomitization

Dolomite typically occurs throughout the Arab-D as thin, stratigraphically-concordant beds one to 15 feet thick. Stratigraphically, dolomite units are most common within Zone 2B and the top of Zone 2A, the rest of the reservoir contain only minor amounts of dolomite. Thick sections of "pervasive" or non-stratigraphic dolomite also occur in the Arab-D at Ghawar.

Petrographically, dolomite crystals exhibit several different textures. These crystals range in size from about 30 to 150 microns (Figure 13 C, D, E; 14 D). As previously noted, dolomite crystals may contain cloudy, (calcite) inclusion-rich centers and clear, limpid rims. Cloudy centers are interpreted to result from dolomitization by fluids that were near saturation with repsect to calcite, forming many small calcite crystals which were included in the centers of dolomite rhombs. Later dolomitizing fluids became undersaturated with respect to calcite, and further dolomite crystal growth was inclusion-free (Sibley, 1980; Land et al., 1975). These later fluids may also be responsible for dissolving previously-formed calcite inclusions (Figure 13 D, E), to form partially leached dolomite crystal centers (Sibley, 1982). In addition to the cloudy-center clear-rim dolomite crystals, clear inclusion-free crystals are also common (Figure 13 C). These crystals represent dolomite cement that has been passively precipitated into pre-existing pores. Finally, rare baroque (late diagenetic, high temperature) dolomite occurs primarily as fracture-filling cement.

Leaching and Recrystallization

Leaching and recrystallization are pervasive throughout the Arab-D (Figure 17). Recrystallization generally has little impact on the reservoir. Leaching, in contrast, has significantly affected the reservoir in all the carbonate lithofacies.

In dolomite, grains that were significantly larger than the surrounding matrix are typically leached (Figure 13 A, B). During dolomitization, the larger grains resist dolomitization because of their low surface area-to-volume ratios. These grains are later leached during more advanced dolomitization as a result of their mineralogic instability with respect to dolomite. In contrast, leaching in limestones is dependent upon the original grain mineralogy and possibly grain microstructure. Original aragonitic grains (especially dasycladacean algae, molluscs, corals and probably ooids) in Arab-D limestone dissolve to form moldic pores that are easily compactible (Figure 17 A). Grains originally composed of high magnesium calcite (such as most foraminifers, red algae, composite grains, coated grains, intraclasts, and micritized grains) are either moldic or microporous.

Microporosity is common to all the limestone lithofacies of the Arab-D, and even occurs rarely in dolomite. Most microporosity occurs as micro-moldic or micro-intercrystalline pores in either grains, matrix or cement (Figures 6 E, F; 17 B, D, E). Microporous grains and matrix are comprised of a dense maze of highly interconnected pore throats generally several microns or less in size that intersect at more equant (often tetrahedral-shaped) pores. These pores result from the leaching and incomplete reprecipitation of metastable carbonate during diagenesis. Microporous cements, in contrast, exhibit thin tubular to sheet-like pores between cement crystals and do not involve leaching.

Cementation

Cements within Arab-D limestones are predominantly marine phreatic and meteoric phreatic. Other types of carbonate cement (such as vadose or deep burial cements) are not observed in the Arab-D in Ghawar. Anhydrite and dolomite cements are discussed elsewhere in this section.

Marine phreatic cements occur primarily as isopachous fibrous to bladed cement. Isopachous fibrous to bladed cement (Figure 18 A) is common throughout the grainy facies of the Arab-D and appears as a thin cement rind 10 to 50 microns thick that surrounds and separates grains (Figure

Figure 17. Thin section photomicrographs and scanning electron microscope images of pore types in Arab-D reservoir.

    A. Thin section photomicrograph of skeletal-oolitic carbonate (grainstone). Grains include ooids (OO), bivalves (B), foraminifers (F), micritized skeletal grains (MG) and echinoderm debris (O). Interparticle (BP) and moldic (MO) porosity are abundant. A thin fringe of isopachous bladed-calcite cement (arrows) surrounds most grains. Monocrystalline calcite overgrowths surrounding echinoderm debris are conspicuous but volumetrically insignificant. 'Ain Dar area, Zone 2A. Bar for scale = 0.5 mm. Plane-polarized light.

    B. Microporous Thaumatoporella (TP), a red alga, foraminifers (F) and indeterminate micritized grains from Cladocoropsis limestone. Plane-polarized light, bar for scale = 0.1 mm. Haradh area, Zone 2B.

    C. "Intracrystalline" porosity (arrows) within dolomite rhombs that partly fill porosity and replace grains in Cladocoropsis limestone. Cladocoropsis (lower right corner of photograph) is microporous (MP). Plane-polarized light, bar for scale = 0.2 mm. Haradh area, Zone 2B.

    D. Several types of pores are visible in this pore cast. Macroporosity is present in the form of interparticle (BP) and moldic (MO) porosity. Also, microporosity is present and occurs both as microporosity in grains (MC) and as microporosity between cement crystals (arrows) that isopachously coated grains in this sample. Scanning electron microscope image.

    E. Detail of previous photograph shows the fine, "hair-like" pore network within a microporous grain, as well as the thin tubular micropores (arrows) between cement crystals etched during sample preparation. Scanning electron microsocpe image.

    F. Large, irregular-shaped pore (vug?) that is partly filled with kaolinite. Kaolinite and euhedral authigenic quartz crystals occur throughout samples of Arab-D rock from the Haradh area. Source of silica may be from dissolution of locally abundant siliceous sponge spicules. Plane-polarized light, bar for scale = 0.5 mm. Zone 3A.

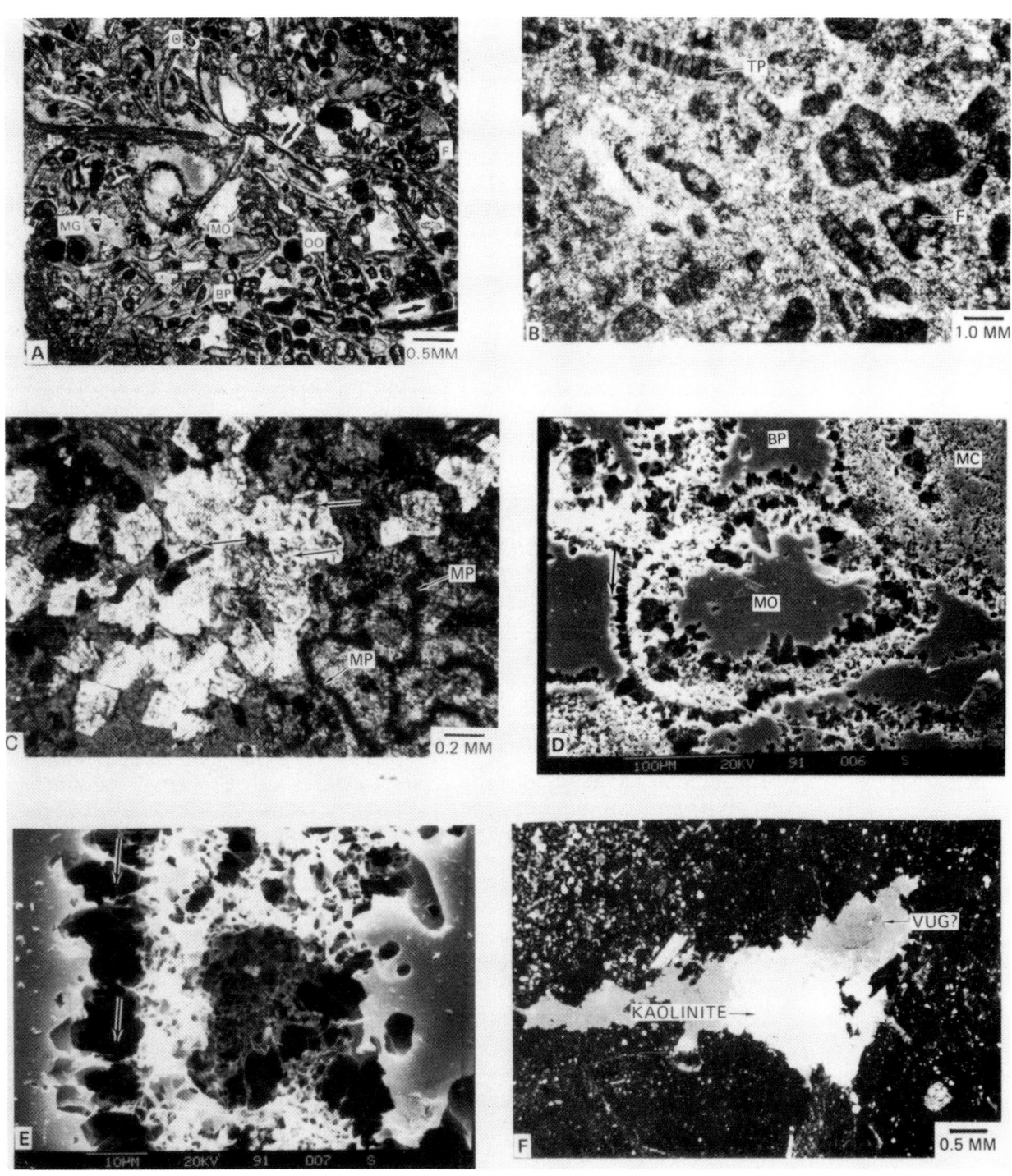

Figure 18. Thin section photomicrographs showing cement fabrics and mechanical compaction.

 A. Isopachous fibrous to bladed cement (arrows) is abundant in this Arab-D grainstone. This cement surrounds and separates grains and was probably precipitated relatively early in the diagenetic history of the rock. While the cement may occlude some interparticle porosity, it may strengthen the rock and impede later compaction. In this example, rare sparry equant calcite cement (EC) partially fills some moldic and interparticle porosity. Plane-polarized light, 'Uthmaniyah area, Zone 2A.

 B. Isopachous fibrous to bladed calcite cement (FC) is present and partially occludes interparticle porosity, while equant calcite cement (EC) occurs in a large moldic pore (outlined). Plane-polarized light, 'Uthmaniyah area, Zone 2A.

 C. Isopachous fibrous to bladed cement (FC) is present as a first cement generation, while equant cement (EC) fills secondary moldic pores and a monocrystalline calcite overgrowth (MO) syntaxial with echinoderm debris acts as a second, pore-central cement generation. Plane-polarized light, 'Uthmaniyah area, Zone 2A.

 D. Equant calcite cement (EC) has extensively occluded interparticle porosity. In this sample, no isopachous fibrous to bladed cement is present as a first cement generation, so equant cement both coats grains and fills porosity. Plane-polarized light, 'Uthmaniyah area, Zone 3B.

 E. Mechanical compaction in the form of physical grain breakage (arrows) is common for elongate grains in lightly cemented mud-poor rocks. Plane-polarized light, Hawiyah area, Zone 2A.

 F. While mechanical compaction is generally rare in mud-supported rocks, some grain squashing may occur. Here, a foraminifer (F) has been crushed. Plane-polarized light, 'Uthmaniyah area, Zone 3B.

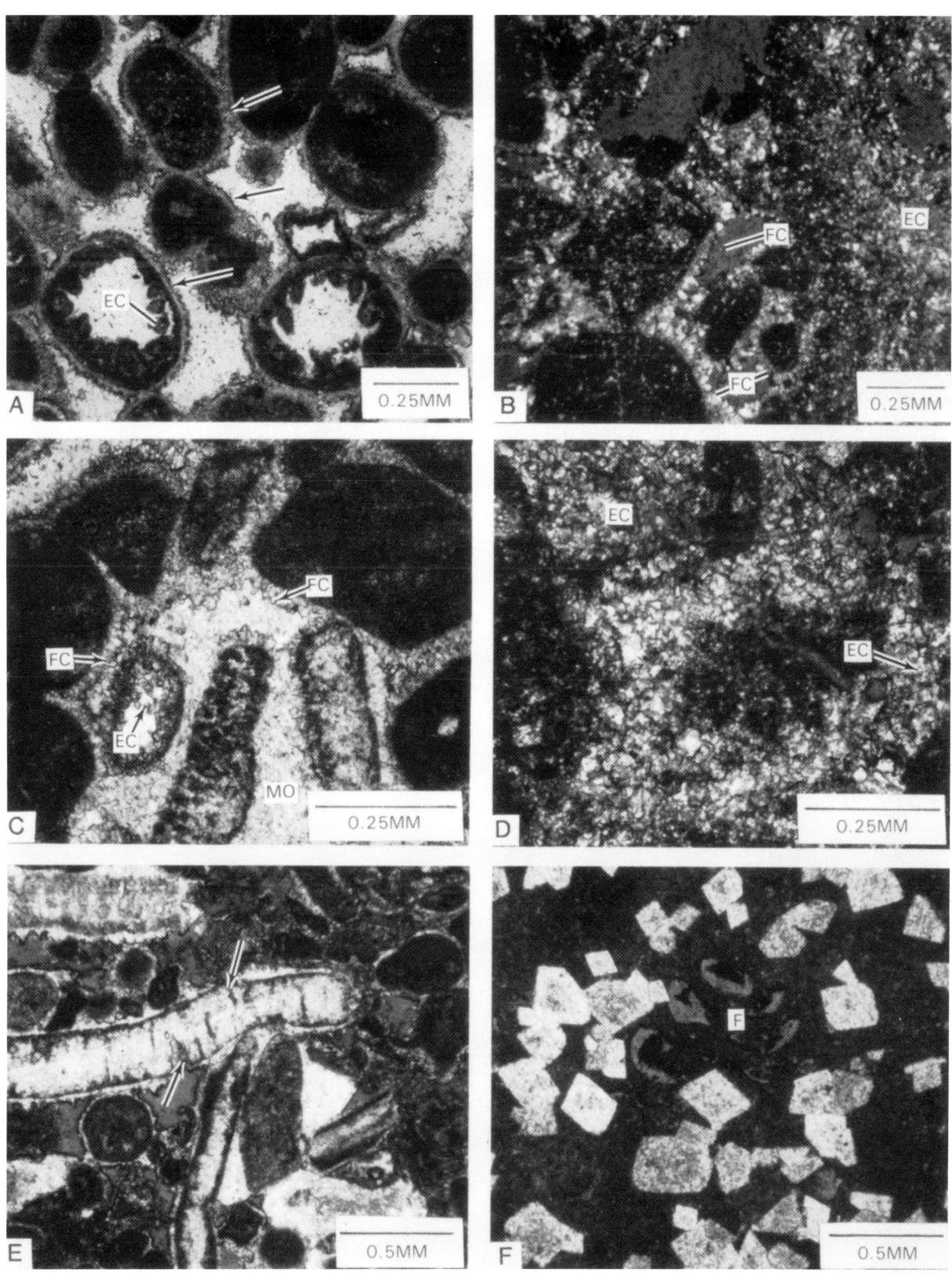

18 A). In general, this probable syndepositional cement rarely occludes an appreciable portion of interparticle pore space.

Even though isopachous cement may only slightly occlude interparticle porosity, isopachous fibrous to bladed cement strengthens the rock fabric shortly after deposition, and thereby aids in preserving the original open packing of the sediment. Although interparticle porosity is slightly reduced by cement, rocks with abundant isopachous cement tend to be less affected by later compaction than those without (compare, for example, Figure 18 A, and F). Also, isopachous fibrous to bladed cement are abundant enough to form local zones of submarine lithification, or hardgrounds (Figure 5 A, B). Isopachous cement is typically most abundant near the surface of a hardground and diminished downward. While such lithified zones are unlikely to persist laterally (Dravis, 1979), hardgrounds may form thin low porosity/low permeability zones that act as local impediments or baffles to vertical flow. These zones are most effective as impediments to flow where several hardgrounds occur stacked within a short vertical distance in the section.

Meteoric phreatic cements occur as sparry equant calcite cement (Figure 18 B) and as monocrystalline overgrowths (Figure 18 C). Overall, meteoric cements are volumetrically minor in the Arab-D in Ghawar and are only locally effective in occluding porosity. Where syndepositional marine phreatic cements are present, meteoric cements represent a second, pore-central cement generation (Figure 18 C). Where present, equant calcite cement fills both primary porosity (Figure 18 A, B, D) and secondary leached moldic porosity (Figure 18 A, B, C); also, equant cements commonly fill fractures (Figure 19 D). Monocrystalline overgrowths are single-crystal cements that grow in optical continuity around an underlying grain and occur mainly on echinoderm grains. These syntaxial echinoderm overgrowths are usually restricted to primary pore space, and may grow to engulf neighboring grains (creating a poikilotopic texture). Syntaxial monocrystalline overgrowths can occlude significant porosity in rocks containing abundant echinoderm debris.

Compaction

Compaction manifests itself in the Arab-D in two ways, mechanical compaction and pressure solution. While both processes are active throughout the reservoir and pressure solution has the greatest impact of the two on the reservoir, part of the uniqueness of the Arab-D as a reservoir is that compaction has been minor. Primary depositional porosity has to a large extent been preserved by syndepositional isopachous cements which prevented significant compaction and porosity loss.

Mechanical compaction includes the physical rearrangement, deformation and breakage of grains, and is locally seen in Arab-D limestones (Figure 18 E, F). Elongate grains and molds of leached grains are especially susceptible to crushing and mechanical compaction. Grain squashing, producing elongate grains and grains with plano-convex contacts, is locally common in the muddier rocks. Such compaction is common in peloidal limestones (Figure 19 A) and may imply that squashing occurred while the peloids were still soft and poorly lithified, early in the diagenetic history of the rock.

Pressure solution is present both as grain-to-grain pressure solution and as stylolitization. Grain-to-grain pressure solution (Figure 19 B) results in grains truncated against each other to form a condensed or fitted fabric. The sutures separating these grains are usually marked by a thin accumulation of insoluble residue, or stylocumulate (term from Logan and Semeniuk, 1976), that is composed primarily of clay minerals, residual hydrocarbons and iron sulfides and oxides. Stylolites cross-cut all grains and cements (Figure 19 C) and occur throughout the Arab-D, but are most common in the muddy lower portion of the section. The magnitude of stylolites varies from small-scale, wispy microstylolites containing little stylocumulate to large, high amplitude stylolites with thick stylocumulate. Obviously, the small-scale stylolites suggest that the amount of carbonate material dissovled was probably minor, while the large stylolites result from an appreciable amount of section shortening.

Figure 19. Slabbed core photograph and thin section photomicrographs of grain compaction, stylolitization, and fracturing.

> A. Grain squashing, producing interpenetrated and elongate grains is common in muddy rocks, especially peloidal limestones. Such squashing probably occurred relatively early in the history of the rock, while the peloids were still soft and poorly lithified. Plane-polarized light, 'Uthmaniyah area, Zone 2A.
>
> B. Pyrite (P) replaces grains, matrix and cement and locally occludes porosity. In this example, grains have been compacted to produce sutured, stylolitic grain boundaries (arrows). Plane-polarized light and reflected light, Haradh area, Zone 3B.
>
> C. Stylolites occur as a variety of types. Here, a large-scale stylolite (as denoted by stylolite amplitude and by the amount of stylocumulate present) occurs at the boundary between two rock types, muddy packstone and mudstone. Also, small-scale wispy micro-stylolites (arrows) are present in muddy packstone. Large black area in center of photograph is a plug scar. Slabbed core photograph, Haradh area, Zone 3A.
>
> D. Fractures are generally rare in the Arab-D and are restricted to dolomites and low porosity limestones. In this muddy packstone, a fracture (arrows) is almost completely filled by equant calcite cement. Plane-polarized light, 'Uthmaniyah area, Zone 3B.

Figure 20. Thin section photomicrographs of diagenetic silica, anhydrite, and kaolinite.

> A. Silica is rare overall in the Arab-D, but may occur as a cement or as a replacement mineral. Here, both chalcedony (CH) and authigenic quartz (Q) occur as a cement partially occluding porosity in this dolomite. Plane-polarized light, 'Uthmaniyah area, Zone 3A.
>
> B. Silica also occurs in the Arab-D as chert (arrows), which partially fills both intraparticle and interparticle porosity in this dolomitic _Cladocoropsis_ limestone. Although relatively abundant in this example, chert is generally rare in the Arab-D. Plane-polarized light, 'Uthmaniyah area, Zone 2B.
>
> C. Diagenetic anhydrite (A) occurs as clear, euhedral, lath-shaped crystals that replace grains and occlude porosity. Anydrite of this variety occurs throughout the Arab-D and is volumetrically insignificant. Plane-polarized light, 'Ain Dar area, Zone 3A.
>
> D. Kaolinite (K) occurs as "books" or stacks of clay that may occlude some porosity, but overall have only a minor impact on reservoir quality. Cross-polarized light, Haradh area, Zone 3A.

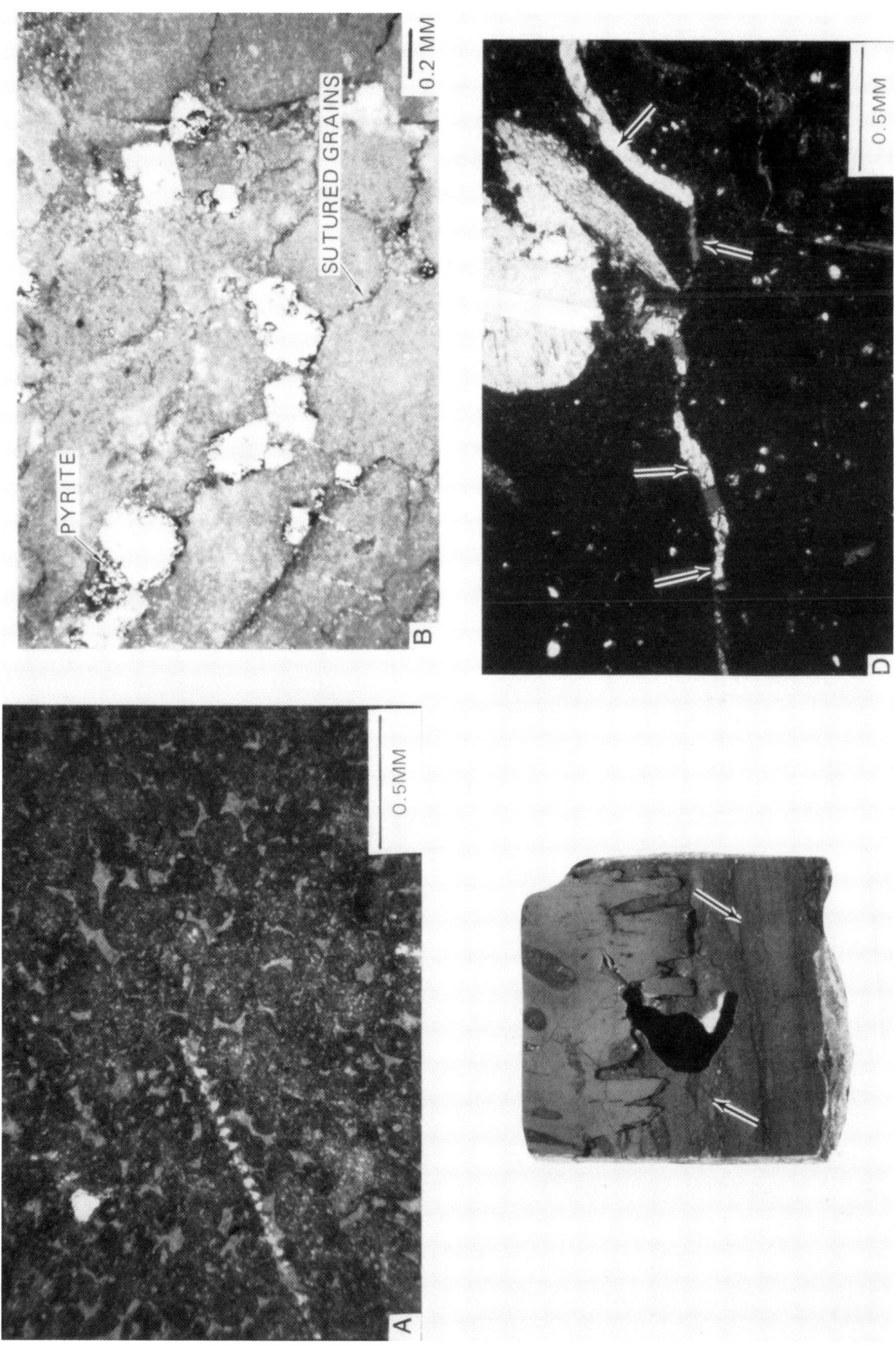

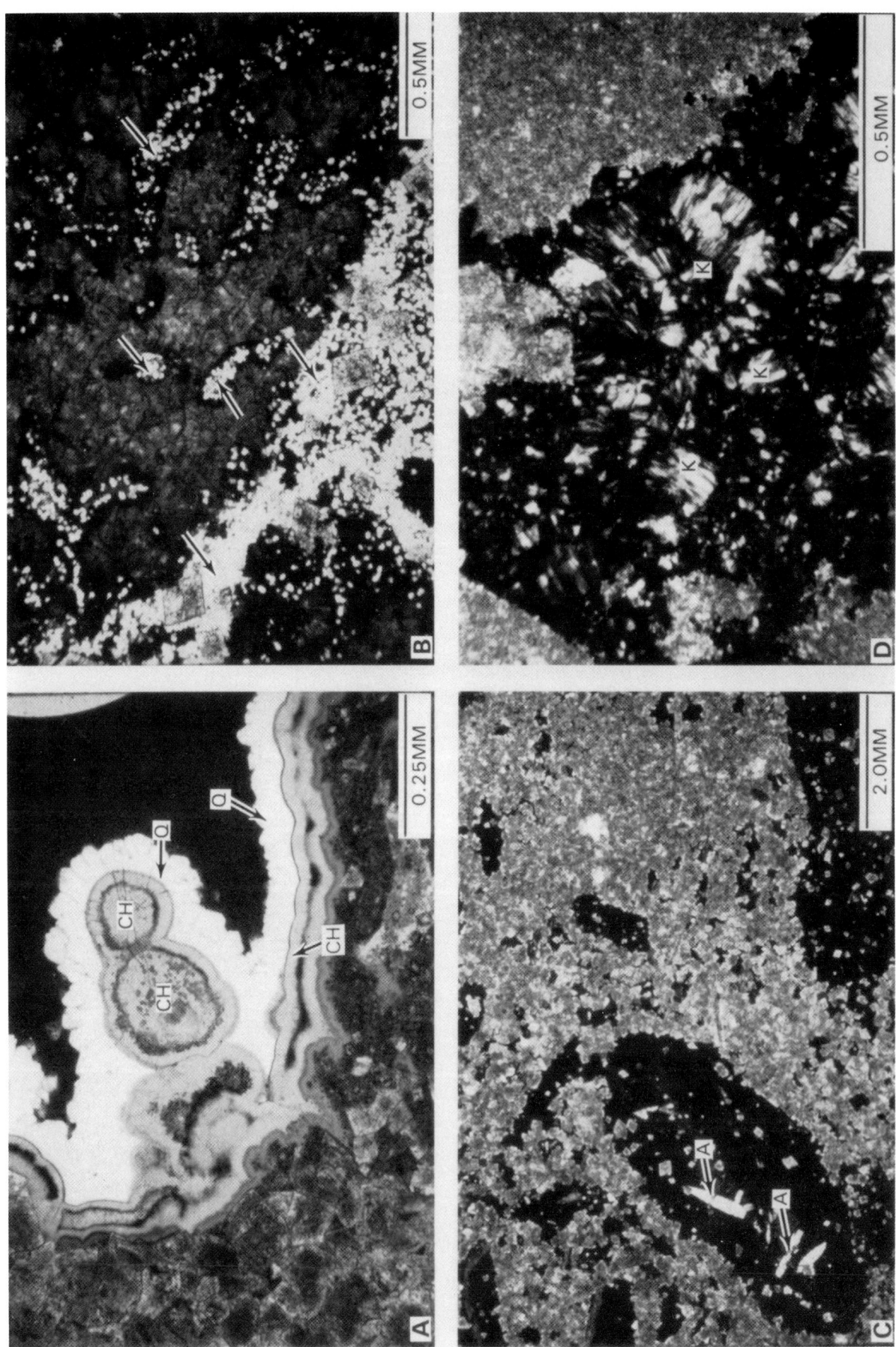

Fracturing

Overall, fracturing is rare in the Arab-D although locally it may be common (Figure 19 D). Fractures are generally limited to dolomites and low porosity limestones. They are typically vertical to subvertical (80° to 90° dip) in orientation and have fracture apertures that range from 0.1 to 2.0 mm. Open fractures are rare, with most fractures filled by equant calcite, dolomite or anhydrite cement.

Auxillary Minerals

Other diagenetic products in the Arab-D include silica, pyrite, diagenetic anhydrite and kaolinite. These products are volumetrically insignificant.

Silica occurs as a replacement of carbonate grains, especially brachipods and echinoderms, and as a void-filling cement (Figure 20 A, B). Powers (1962) noted the existence of chert at several levels in the Arab-D and, in particular, recognized a thin (two to three feet thick), apparently laterally persistent zone of cherty dolomite near the center of the Arab-D. He suggested that silica-filled pores in dolomite implied that silicification post-dated dolomitization. Overall, silica is rare, and does not appreciably affect reservoir quality.

Pyrite (Figure 19 B) occurs throughout the Arab-D, although it is especially common in the muddy, low porosity limestones near the base of the section. Pyrite generally occurs as disseminated framboids and isolated small crystals that replace grains and occlude both primary and secondary porosity. Pyrite is also commonly concentrated along stylolites. While it is volumetrically insignificant and probably has no effect on reservoir quality, pyrite in even minor amounts can influence logging measurements, especially those involving density logs.

Diagenetic anhydrite (Figure 20 C) is also present throughout most of the Arab-D, although it is most abundant in the upper part of the section, near the Arab-D Anhydrite. Anhydrite typically occurs as euhedral, lath-shaped crystals that range in size from 0.5 to 3.0 mm long. Diagenetic anhydrite often partially fills large (especially moldic) pores and replaces grains, matrix and cements. Overall, diagenetic anhydrite is rare and probably does not impact reservoir quality.

Finally, kaolinite is rare in the Arab-D and generally acts as a passive, pore-filling cement (Figure 17 F, 20 D). Kaolinite locally partly fills primary and secondary porosity. Overall, kaolinite has a minimal effect on porosity and permeability.

ACKNOWLEDGEMENT

Appreciation is given to the Saudi Arabian Ministry of Petroleum and Minerals Resources and to the Arabian American Oil Company for permission to publish this paper.

REFERENCES

Aigner, T., 1982, Calcareous Tempestites: Storm-Dominated Stratification in Upper Muschelkalk Limestones (Middle Trais, SW-Germany), in Einsele, G., and Seilacher, A., eds., Cyclic and Event Stratification: Springer-Verlag, p. 180-198.

Champetier, Y. and Fourcade E., 1966, A propos de Cladocoropsis mirabilis Felix dans le Jurassique superieur du Sud-Est de l'Espagne: Estudios Geologicos, v. 22, p. 101-111.

Dravis, J. J., 1979, Rapid and Widespread Generation of Recent Oolitic Hardgrounds on a High Energy Bahamian Platform, Eleuthera Bank, Bahamas: Jour. Sed. Petrology, v. 49, p. 195-207.

Heckel, P. H., 1972, Recognition of Ancient Shallow Marine Environments: in Rigby, J. K., and Hamblin, W. K., (eds.), Recognition of Ancient Sedimentary Environments, Soc. Econ. Paleont. Mineral., Spec. Pub. No. 16, p. 226-286.

Kreisa, R. D., and Bambach, R. K., 1982, The Role of Storm Processes in Generating Shell Beds in Paleozoic Shelf Environments, in Einsele, G., and Sielacher, A., eds., Cyclic and Event Stratification: Springer-Verlag, p. 200-207.

Land, L. S., Salem, M.R.I., and Morrow, D. W., 1975, Paleohydrology of Ancient Dolomites: Geochemical Evidence: Amer. Assoc. Petroleum Geol. Bull., v. 59, p. 1602-1625.

Leeder, M. R., and Zeidan, R., 1977, Giant Late Jurassic Sabkhas of Arabian Tethys: Nature, v. 268, p. 42-44.

Logan, B. W. and Semeniuk, V., 1976, Dynamic Metamorphism; Processes and Products in Devonian Carbonate Rocks, Canning Basin, Western Australia: Spec. Publ. Geol. Soc. Australia no. 6, 138 p.

Loucks, R. G., and Longman, M. W., 1982, Lower Cretaceous Ferry Lake Anhydrite, Fairway Field, East Texas: Product of Shallow-Subtidal Deposition: in Handford, C. R., Loucks, R. G., Davies, G. R., (eds.), Depositional and Diagenetic Spectra of Evaporities - A Core Workshop, Soc. Econ. Paleont. Mineral., Core Workshop No. 3, p. 130-170.

Powers, R. W., 1962, Arabian Upper Jurassic Carbonate Reservoir Rocks: in Ham, W. E., (ed.), Classification of Carbonate Rocks, Amer. Assoc. Petroleum Geol., Memoir 1, p. 122-192.

Sibley, D. F., 1980, Climatic Control of Dolomitization, Seroe Domi Formation (Pliocene), Bonaire, N. A., in Zenger, D. H., Dunham, J. B. and Ethington, R. L., eds., Concepts and Models of Dolomitization: Soc. Econ. Paleont. Mineral., Spec. Publ. No. 28, p. 247-258.

Sibley, D. F., 1982, The Origin of Common Dolomite Fabrics: Clues from the Pliocene: Jour. Sed. Petrology, v. 52, 1087-1100.

Turnsek, D., Buser, S., and Ogorelec, B, 1981, An Upper Jurassic Reef Complex from Slovenia, Yugoslavia, in Toomey D. F., ed., European Fossil Reef Models: Soc. Econ. Paleont. Mineral., Spec. Publ. No. 30, p. 361-370.

Wilson, J. L., 1975, Carbonate Facies in Geologic History, Springer-Verlag, p. 289-293.

PALEOENVIRONMENTAL AND DIAGENETIC RESERVOIR CHARACTERIZATION OF THE SMACKOVER FORMATION JAY FIELD, WEST FLORIDA

DEBORAH M. BLIEFNICK
Chevron USA
P. O. Box 1635
Houston, TX 77251

PHILIP A. MARIOTTI
Chevron Geosciences
2811 Hayes Rd.
Houston, TX 77082

ABSTRACT

The Upper Jurassic Smackover Formation in the Jay Field area, Alabama-Florida, represents an overall transgressive-regressive sequence within which several smaller lithostratigraphic units can be distinguished. The regionally extensive lower Smackover, which constitutes the transgressive phase, is characterized by limestone and dolomite mudstones to packstones with peloids and oncolites. Deposition occurred in a low energy shallow subtidal to intertidal setting. The upper Smackover, which constitutes the regressive phase, is predominantly dolomitic peloid-ooid packstones and grainstones. This reservoir facies interfingers with lower energy subtidal mudstones and wackestones. Intergranular, pore-filling anhydrite becomes common near the top of the Smackover with displacive, nodular growth occurring in some areas. The anhydrite becomes a massive nodular mosaic with minor amounts of dolomudstone near the overlying Buckner.

The Smackover dolomites formed either in a subaerially exposed supratidal setting or during early burial. The replacement is pervasive and original depositional textures are frequently obliterated. Where samples are still limestone, remnants of early marine cements can be identified which are followed by fresh water phreatic pore-filling calcite. Where mudstones and wackestones have been replaced and the dolomite is an anhedral crystal mosaic, porosity remains low. Where packstones and grainstones have been replaced and euhedral dolomite crystals have developed, intergranular and intercrystalline porosity is quite good. Subsequent to dolomitization, hydrocarbons migrated into the reservoir facies and remnants still line open pores giving the rock a black color.

Analysis of pore-throat distributions and capillary pressure curves show that the reservoir quality is a function of the original sediment type and the degree of dolomitization. Micrites range from seals when undolomitized, to intermediate in quality when extensively dolomitized. Reservoirs developed from packstone/grainstone precursors vary from intermediate in quality when slightly dolomitized to very good quality when extensively dolomitized. In general, the more extensively dolomitized the rock and the more coarsely crystalline the dolomite, the better the reservoir quality.

INTRODUCTION

The Upper Jurassic Smackover Formation is a major hydrocarbon source and reservoir unit in the U.S. Gulf Coast. Jay Field is the largest Smackover field with recoverable reserves in excess of 450 million barrels oil equivalent. Other significant Jurassic fields in the Jay field area of Escambia County, Alabama, and Escambia and Santa Rosa Counties, Florida include Big Escambia Creek, Flomaton and Blackjack Creek fields. Jay and Blackjack Creek produce sour oil and Big Escambia Creek produces sour gas and condensate from the Smackover whereas Flomaton produces sour gas from the Norphlet Formation (Sigsby, 1976).

Smackover reservoirs along the Jay trend are all dolomite. In fact, massive dolomite replacement is responsible for making the formation a major hydrocarbon producer not only along this trend but also in the rest of southwest Alabama-southeast Mississippi and other areas around the Gulf Coast. Traps tend to be subtle structural - stratigraphic combinations. The key for exploration is the delineation of low relief structures which occur in conjunction with favorable facies. Since the development of reservoir - grade porosity is facies selective, an understanding of the depositional environments as well as the varying responses of the original carbonate material to dolomitization is needed to effectively interpret the textures and porosity/permeability relationships observed in the area.

For this study eight wells with continuous core or core chips through the Smackover were selected. Some of the cores included rock from the Buckner and Norphlet. Thin sections and SEM samples from all wells (a total of 1500 feet) were examined to determine lithology, texture, diagenetic sequence and depositional environment. Mercury injection analyses were performed on core from six wells to determine porosity and reservoir characteristics. The integration of these data help define depositional and diagenetic controls on reservoir development.

GEOLOGIC SETTING

The Jay trend is located between the pre-Jurassic Conecuh Ridge and Pensacola Arch (Figure 1), where the Smackover Formation is buried to between 15,000 and 16,000 feet (Figure 2). Structural mapping shows gentle south to southwest dip except where the Pickens-Gilbertown fault system is associated with local aberrations including rollover into the fault at Jay and Blackjack Creek fields. The Smackover varies in thickness between 300 and 370 feet reflecting fairly uniform subsidence in the study area.

The Conecuh and Pensacola ridges served as source areas for terrigenous clastics deposited before and during the Smackover transgression and influenced the distribution of carbonate sediments as well (Ottmann, et al, 1973). During the Smackover transgression and subsequent regression, shallow water carbonates were deposited on a broad, gently sloping shelf. The Smackover producing facies in Mississippi and further west is a high energy ooid grainstone whereas in the Jay area, the reservoir is a lower energy peloid packstone to mixed peloid-ooid grainstone (Ottmann, et al, 1973; Sigsby, 1976). The presence of grain shoals instead of shelf-margin reefal facies is typical of a ramp setting (Figure 3). The dominance of packstone textures indicates that the ramp-like embayment between the ridges was sufficiently restricted so that higher energy sediments did not form. Local variations in energy conditions and paleotopography resulted in the textural variations observed in the study wells.

STRATIGRAPHY AND LITHOFACIES

The stratigraphic section present in the Jay Field area is shown in Figure 4. The contact between the Smackover and underlying Norphlet Formation is conformable, grading upward over a short distance from dolomitic sandstone to sandy dolomite (Figure 8). The Norphlet is a porous quartz sandstone that is frequently cross-bedded (Figure 5A) and was deposited as a coastal dune complex. It produces sour gas to the north of Jay Field in Flomaton Field. The Smackover conformably underlies the Buckner member of the Haynesville

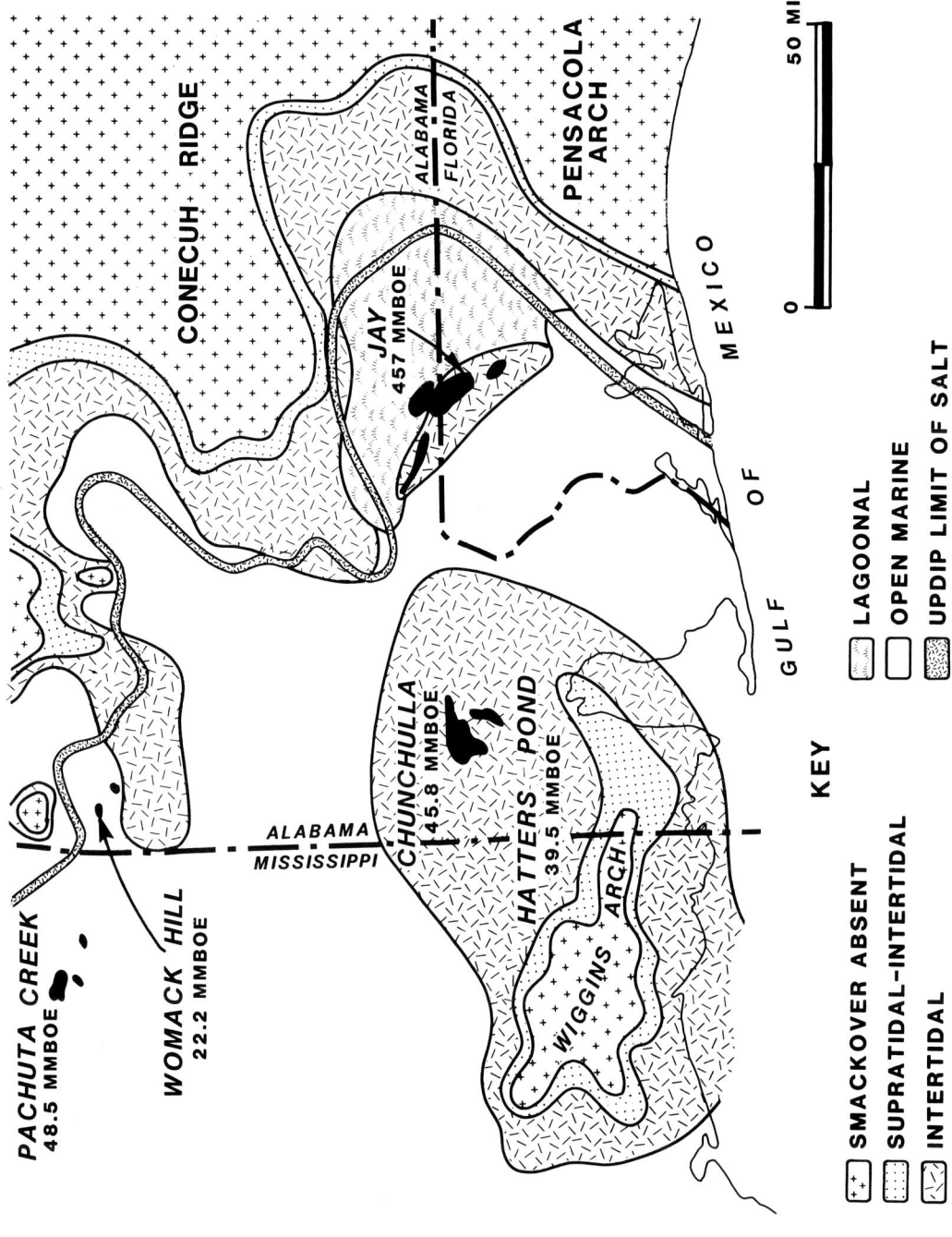

Figure 1 Paleoenvironmental setting of the Jay Field area during the Late Jurassic, specifically during Upper Smackover deposition. Sedimentation was influenced by the Conecuh Ridge and Pensacola Arch as well as subsurface features such as basement irregularities and salt swells.

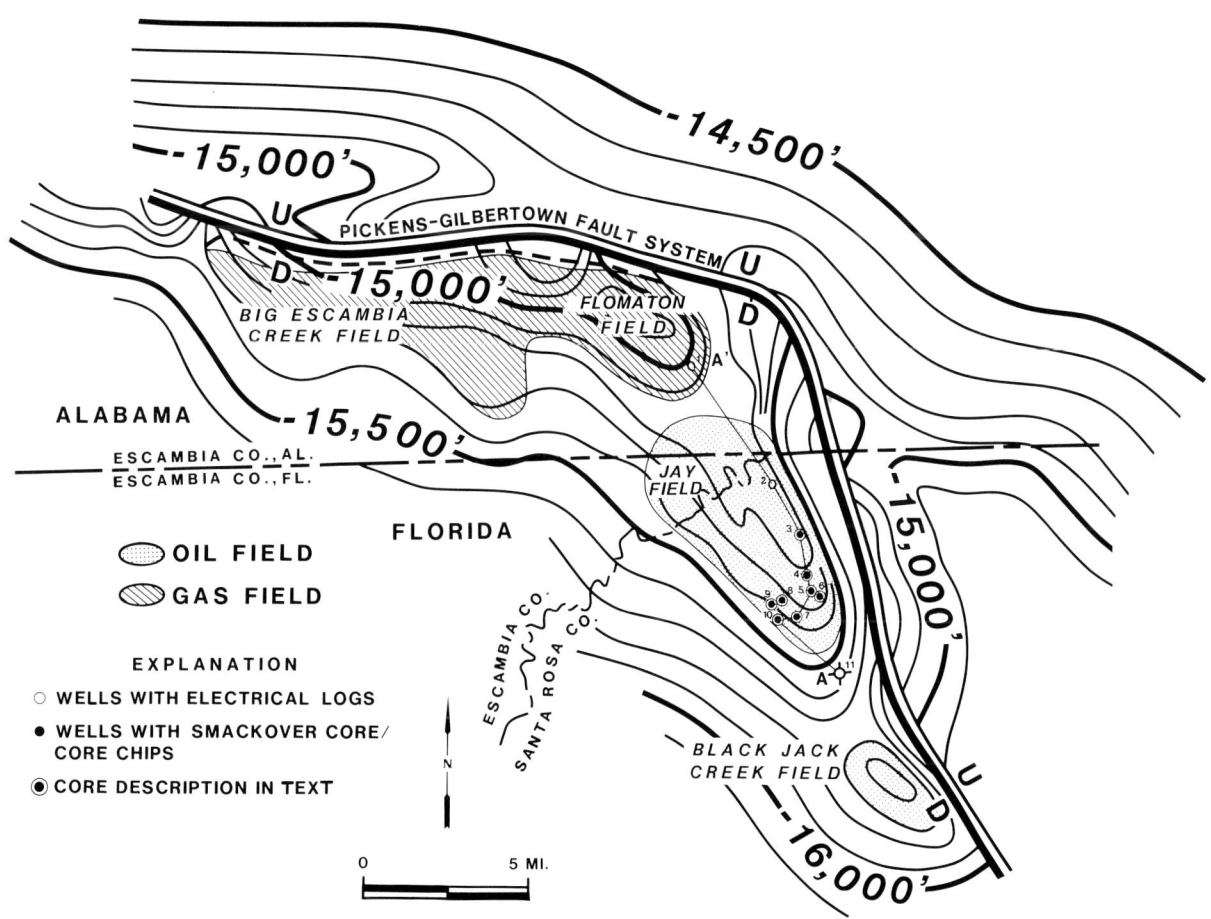

Figure 2 Structure map on top Smackover in Jay trend area with distribution of fields (modified after Sigsby, 1976). Wells with core or core chips include: (3) Chevron No. 1 C. E. Hayes 40-4; (4) Chevron No. 1 Ruby Kent 18-2; (5) Chevron No. 1 Ruby Kent 23-2; (6) Chevron No. 1 Golden 23-1; (7) Chevron No. 1 B. D. Hendricks 23-3; (8) Chevron No. 1 Bragg 22-1; (9) Chevron No. 1 D. Boutwell 22-2; and (10) Chevron No. 1 D. Boutwell 22-3. Other wells with electric logs used on cross-section A-A' shown in Figure 6 include: (1) Humble #1 Martin Powell Gas Unit 19; (2) Exxon #1 St. Regis Paper Co. 6-2; and (11) Exxon #1 St. Regis Paper Co. 36-3. Core from well numbers 3, 4 and 10 were photographed and displayed for the core workshop.

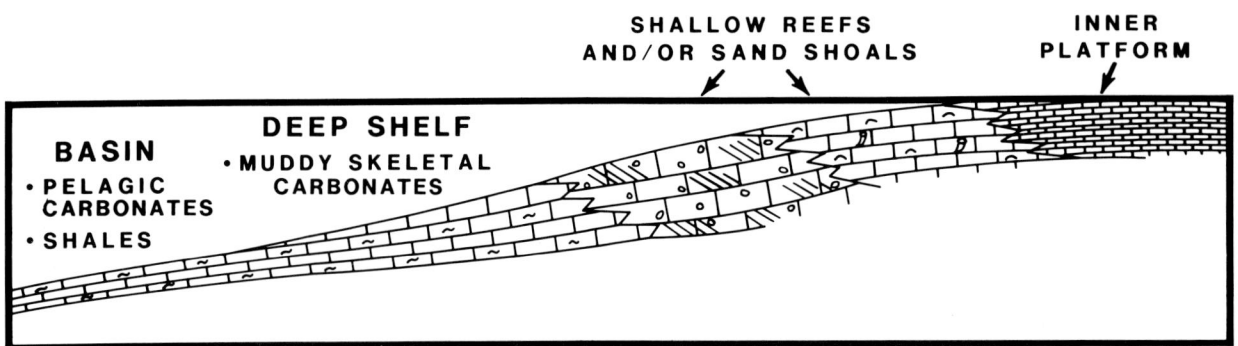

Figure 3   An idealized model for carbonate deposition in a ramp setting with generalized environments of deposition and rock types accumulating in the various environments.

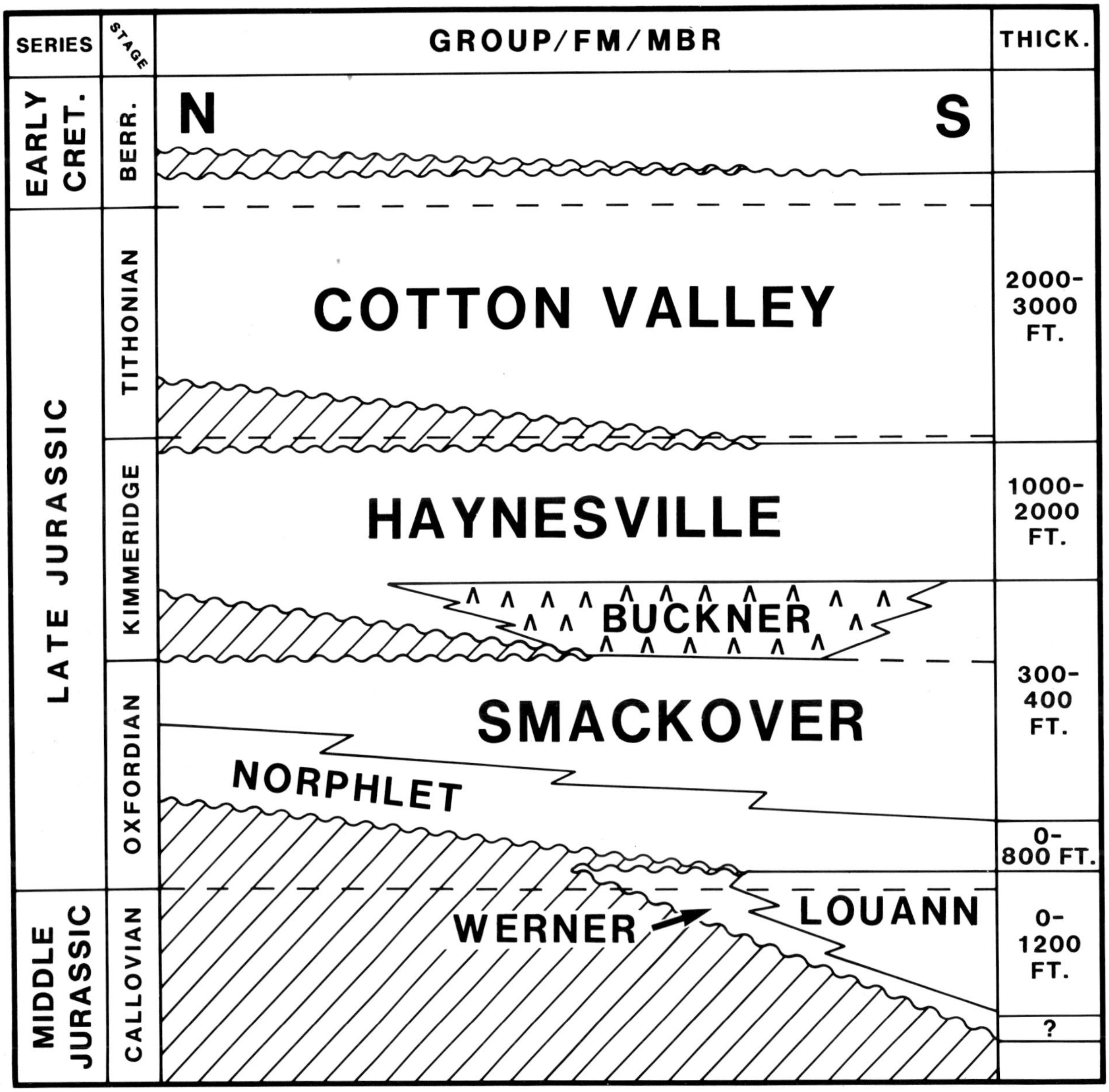

Figure 4   Stratigraphic nomenclature in the Alabama-Florida area.

Figure 5 A. Porous quartz sandstone of the Norphlet Formation with high angle cross-bedding outlined by heavy minerals. Chevron No. 1 Hayes 40-4, 15771-15777 ft., slabbed core, scale in cm.

B. Nodular mosaic anhydrite interbedded with anhydritic dolomudstone, Buckner Member, Haynesville Formation. Chevron No. 1 Boutwell 22-3, 15720-15726 ft., slabbed core, scale in cm.

Formation. The contact grades from dolomite through interbedded dolomite and anhydrite to nodular mosaic anhydrite (Figures 11, 13 and 5B). The anhydritic Buckner serves as a regional seal for the Smackover reservoirs.

An examination of core from eight wells and electric logs from numerous other wells indicates that different facies can be delineated within the Smackover carbonates (Figure 6). Lithofacies in the eight cored wells is similar in correlative portions of the Smackover. Cored intervals from six of these wells were selected to illustrate various facies and their vertical succession (Figures 7-13); three of these were displayed at the core workshop. All cores are extensively dolomitized and most are heavily oil stained making delineation of sedimentary structures difficult. Thin sections and SEM samples were required to identify grains, textures, and diagenetic overprints. The lower portion of the Smackover includes:

1) lime mudstone: partially to completely dolomitized, occasionally bioturbated, contains argillaceous laminae and stylolites, and has very few skeletal fragments; laminated mudstones have been reported from other areas (Mancini, 1979); (Figures 14A and B)

2) peloid wackestone: both limestone and dolomite; frequently bioturbated; contains pellets, including <u>Favreina</u> pellets, and sparse skeletal (predominantly echinoderm) material (Figure 14C); low intercrystalline porosity is present in dolomitized portions with an irregular distribution reflecting the bioturbated fabric (Figure 14D);

3) peloid-algal packstone: both limestone and dolomite; similar to the peloid wackestone but pellets are more abundant and large oncolites are common (Figures 14E and F); skeletal material is rare; porosity is generally low.

The lithofacies of the regionally extensive lower Smackover represent intertidal to shallow subtidal deposition in a transgressive sequence (Ottmann, et al, 1973; Sigsby, 1976; Mancini, 1979). Laminated mudstones are frequently deposited during the initial transgressive phase (Barrett, 1986). As water depths increase and subenvironments differentiate, peloid wackestones and packstones become common.

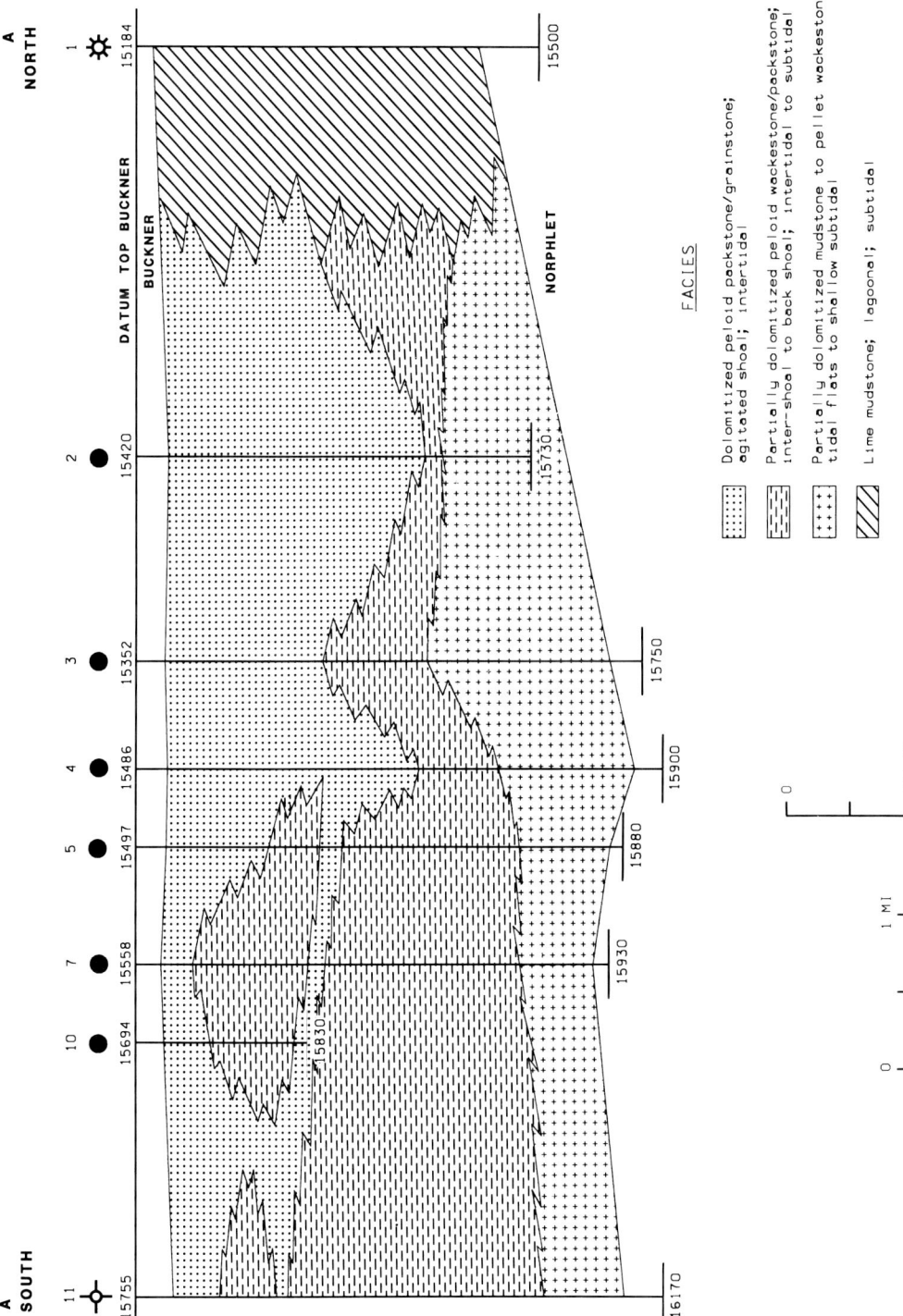

Figure 6   Facies cross-section A-A'. Initial tidal flat deposition on the Norphlet gives way to subtidal peloid wackestones and packstones which become more open marine to the south. This facies differentiates into back-shoal and inter-shoal settings as the peloid packstone/grainstone shoals develop during the upper Smackover regression. The high energy packstones pinch out up dip into tight lagoonal mudstones.

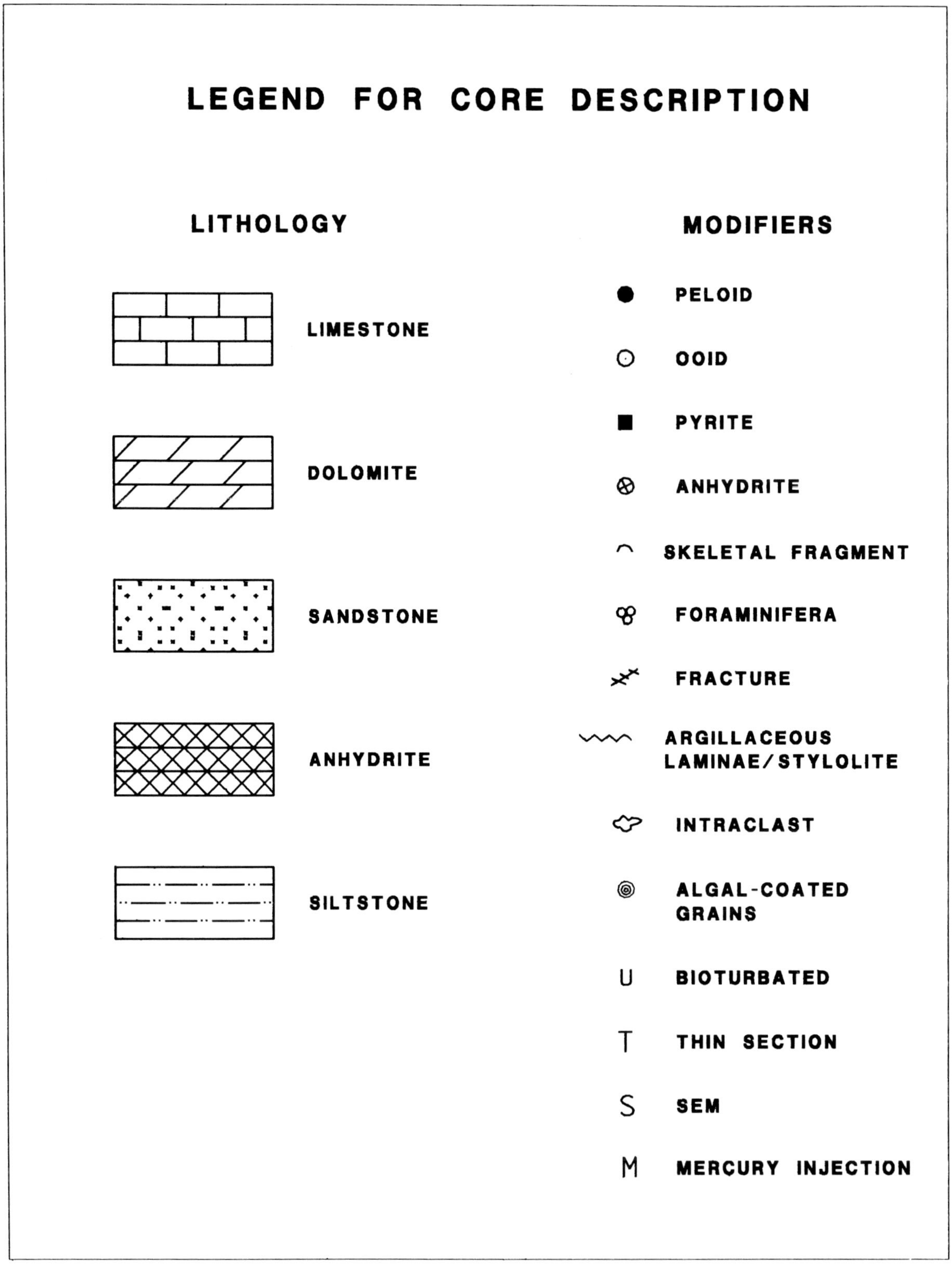

LEGEND FOR CORE DESCRIPTION

LITHOLOGY

- LIMESTONE
- DOLOMITE
- SANDSTONE
- ANHYDRITE
- SILTSTONE

MODIFIERS

- ● PELOID
- ⊙ OOID
- ■ PYRITE
- ⊗ ANHYDRITE
- ⌒ SKELETAL FRAGMENT
- ⚭ FORAMINIFERA
- ⤨ FRACTURE
- ∿ ARGILLACEOUS LAMINAE/STYLOLITE
- ⌬ INTRACLAST
- ◎ ALGAL-COATED GRAINS
- U BIOTURBATED
- T THIN SECTION
- S SEM
- M MERCURY INJECTION

Figure 7 Legend for core descriptions (Figs. 8 to 13).

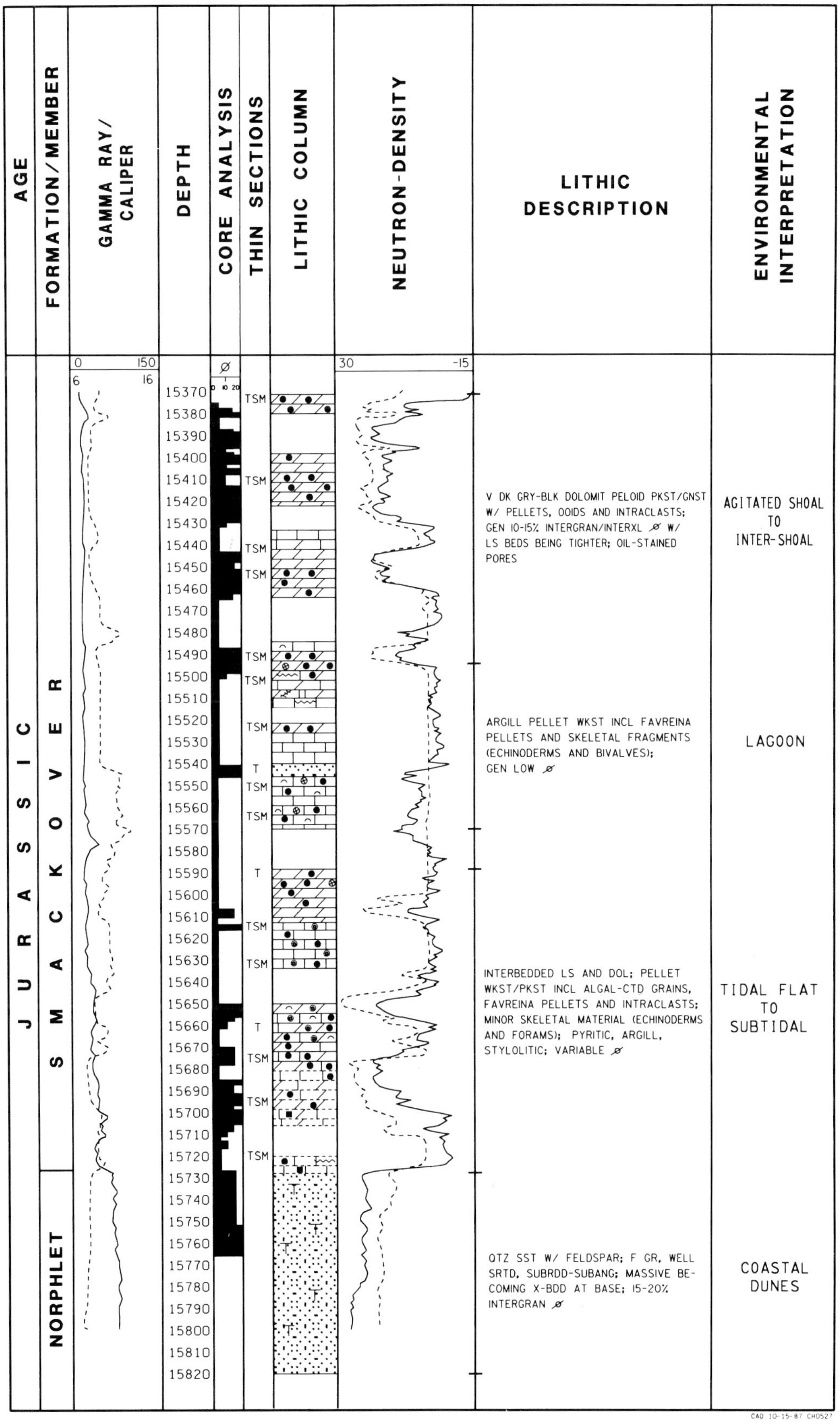

Figure 8 Electric log response and core description for the Chevron No. 1 C. E. Hayes 40-4 (well number 3 in Figure 2).

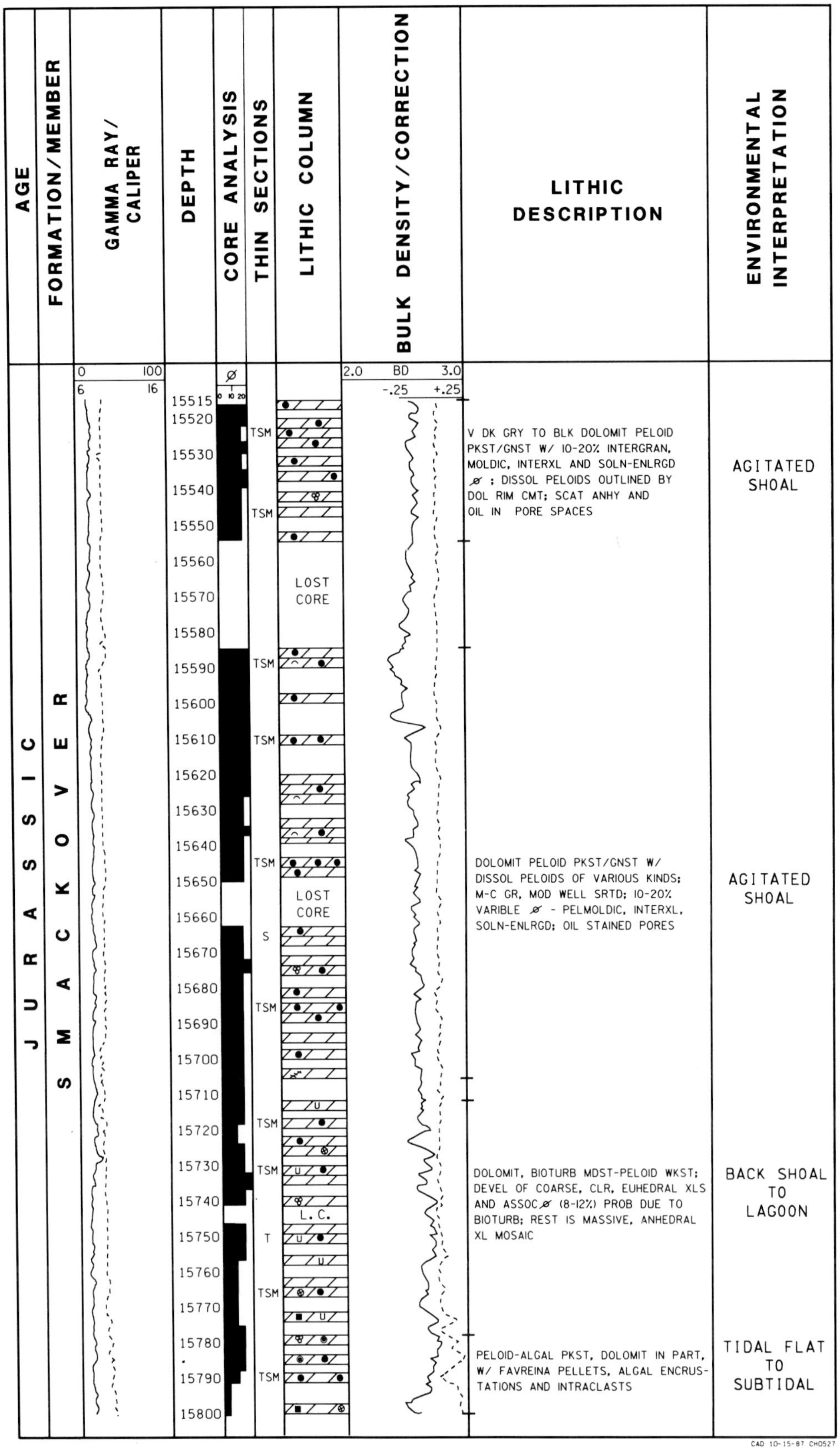

Figure 9 Electric log response and core description of the Chevron No. 1 Ruby Kent 18-2 (well number 4 in Figure 2).

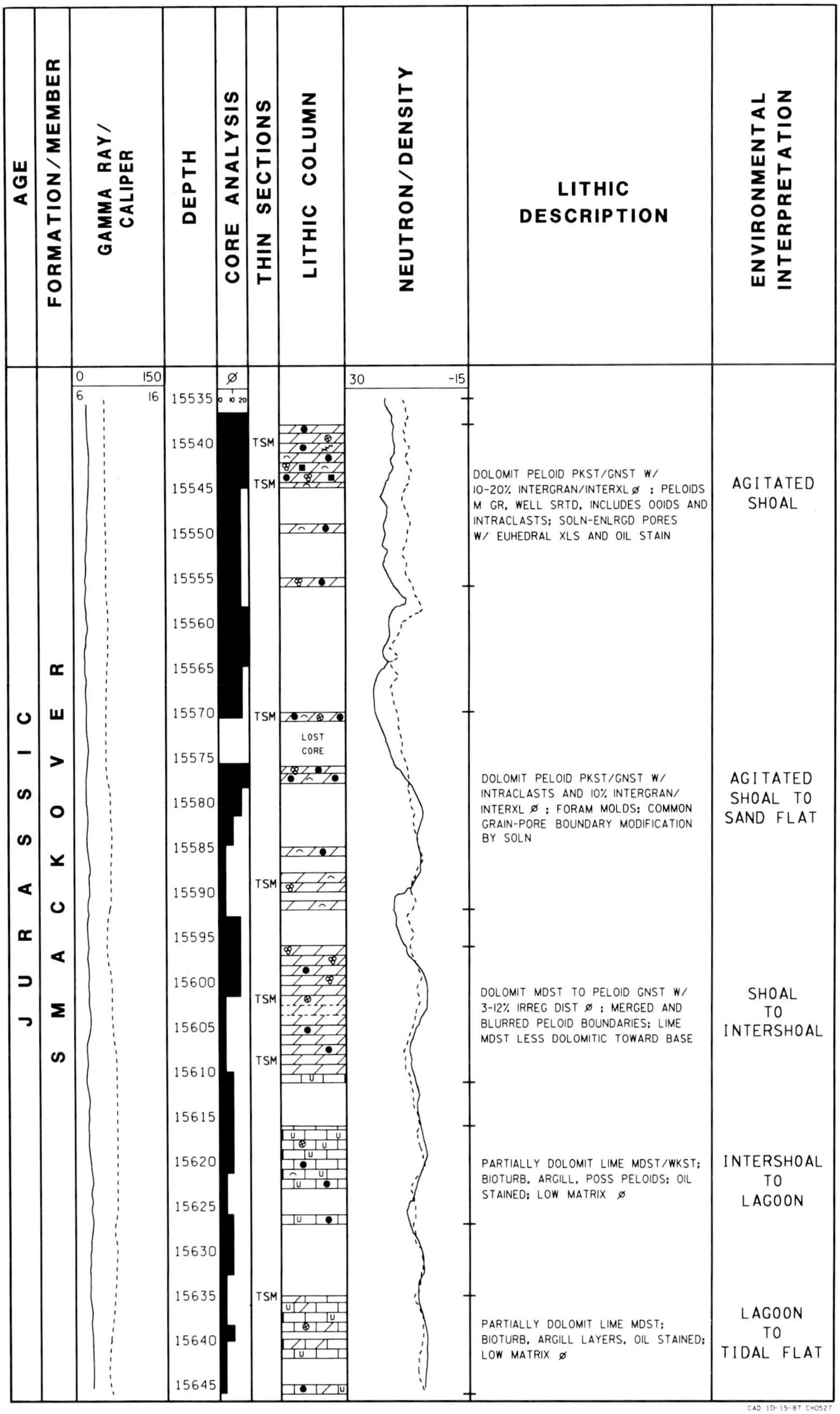

Figure 10 Electric log response and core description for the Chevron No. 1 Golden 23-1 (well number 6 in Figure 2).

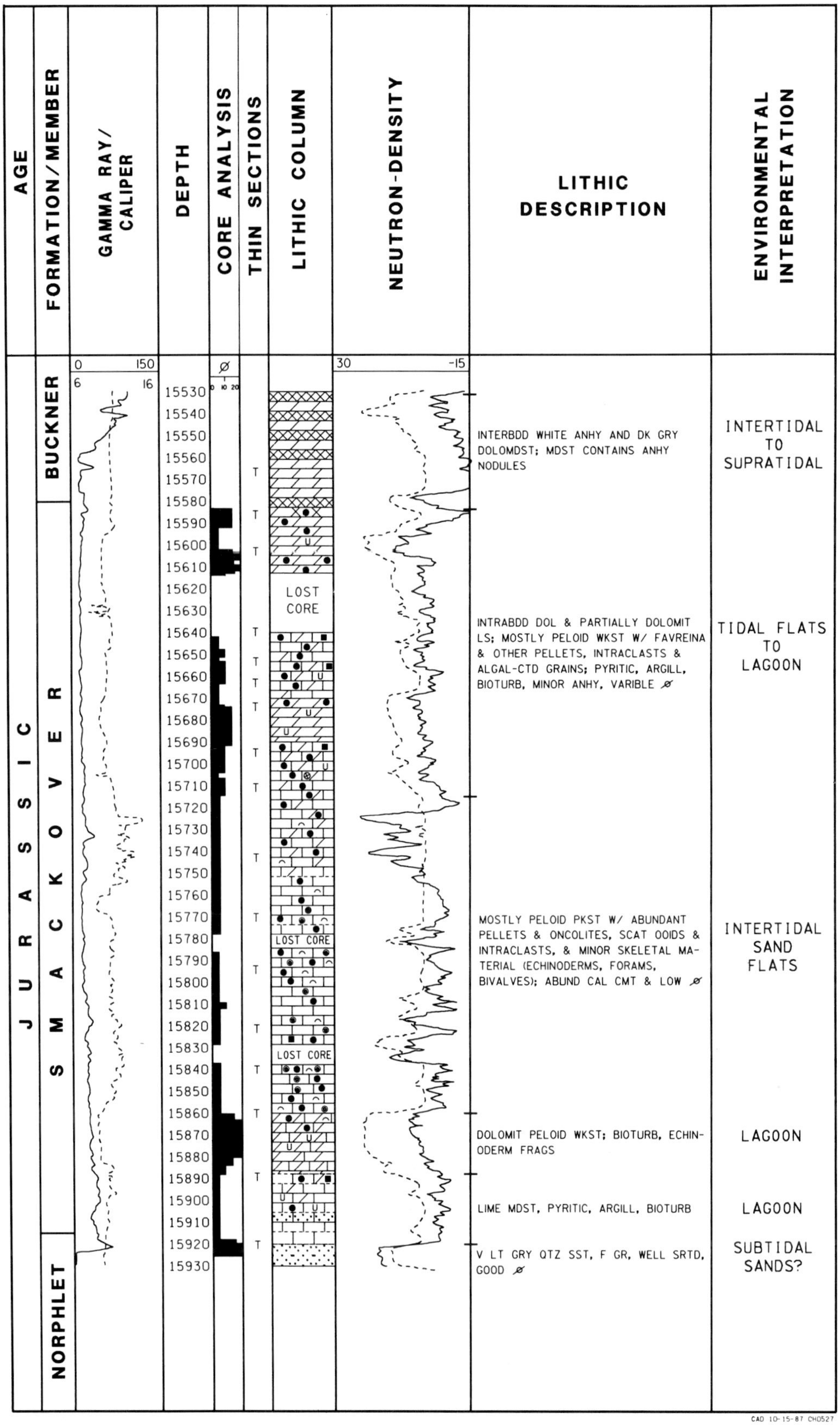

Figure 11 Electric log response and core description for the Chevron No. 1 B. D. Hendricks 23-3 (well number 7 in Figure 2).

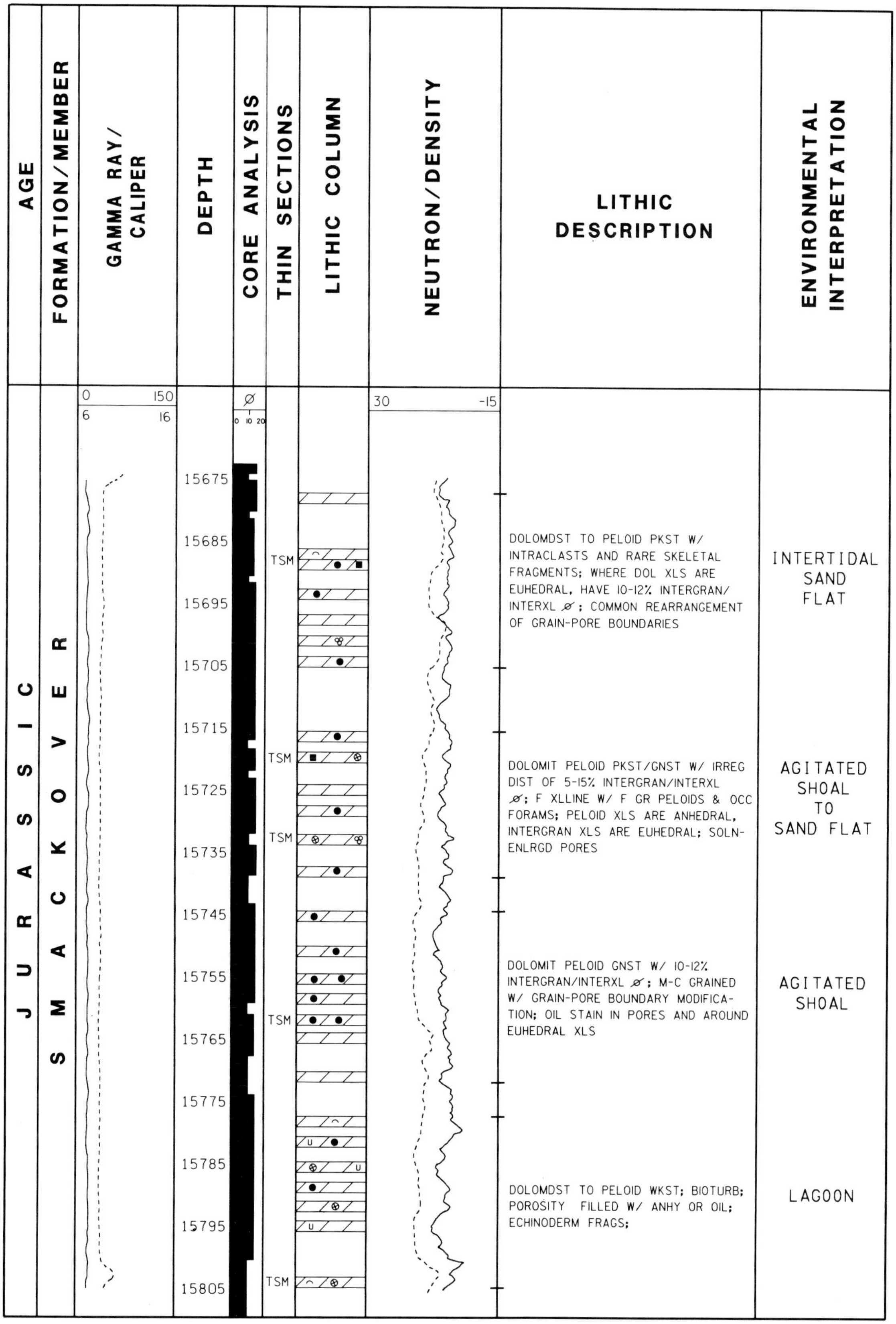

Figure 12 Electric log response and core description for the Chevron No. 1 D. Boutwell 22-2 (well number 9 in Figure 2).

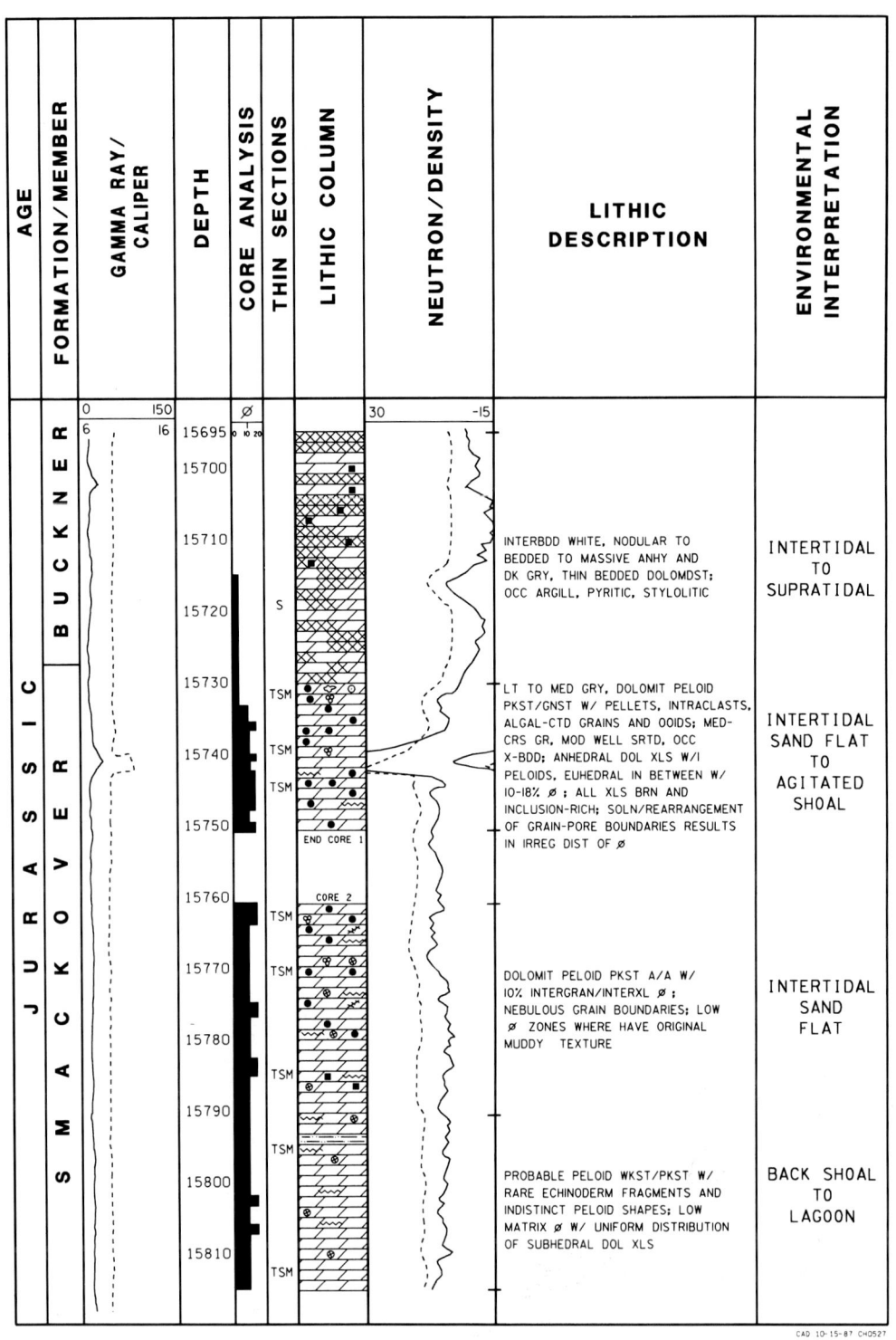

Figure 13 Electric log response and core description for the Chevron No. 1 D. Boutwell 22-3 (well number 10 in Figure 2).

Figure 14   A. Partially to completely dolomitized lime mudstone; bioturbated with argillaceous laminae, stylolites and oil staining. Chevron No. 1 Kent 18-2, 15731- 15747 ft., slabbed core, scale in cm.

B. Pyritic lime mudstone with argillaceous laminae. Chevron No. 1 Hendricks 23-3, 15916 ft., plane light.

C. Peloid wackestone with Favreina (F) and other pellets. Porosity is low due to calcite cementation (white areas). Chevron No. 1 Hendricks 23-3, 15823 ft., plane light.

D. Dolomitized peloid wackestone; bioturbation results in a patchy distribution of increased porosity and precipitation of coarser dolomite crystals (upper left and lower center). Chevron No. 1 Bragg 22-1, 15843 ft., plane light.

E. Slightly dolomitized peloid-algal packstone. Pellets, intraclasts and skeletal fragments are algally coated. S - echinoderm fragment; M - algal coating around multiple grains. Chevron No. 1 Hendricks 23-3, 15797 ft., plane light.

F. Peloid-algal packstone with abundant pellets (upper right and lower right) and a large oncolite (pores filled with anhydrite). Chevron No. 1 Hendricks 23-3, 15843 ft., plane light.

Figure 15   A. Dolomitized peloid grainstone with pellets, intraclasts, ooids and algal-coated grains. Grains are medium to coarse and reflect high energy deposition. Chevron No. 1 Boutwell 22-3, 15731-15738 ft., slabbed core, scale in cm.

B. Thin section of A, 15732 ft., plane light. Dark areas are open pores.

C. Massive, dolomitized grainstone similar to A but heavily oil stained. Chevron No. 1 Kent 18-2, 15544-15547 ft., slabbed core, scale in cm.

D. Dolomitized peloid grainstone with fine crystalline dolomite rim cements (R) and pelmoldic porosity (dark areas). Chevron No. 1 Kent 23-2, 15526 ft., plane light.

E. Dolomitized peloid grainstone with solution-enlarged pores (S) and resulting indistinct grain boundaries. Chevron No. 1 Hayes 40-4, 15404 ft., plane light.

F. Partially dolomitized pellet wackestone with very low porosity; argillaceous with minor pyrite and anhydrite. Chevron No. 1 Hendricks 23-3, 15643 ft., plane light.

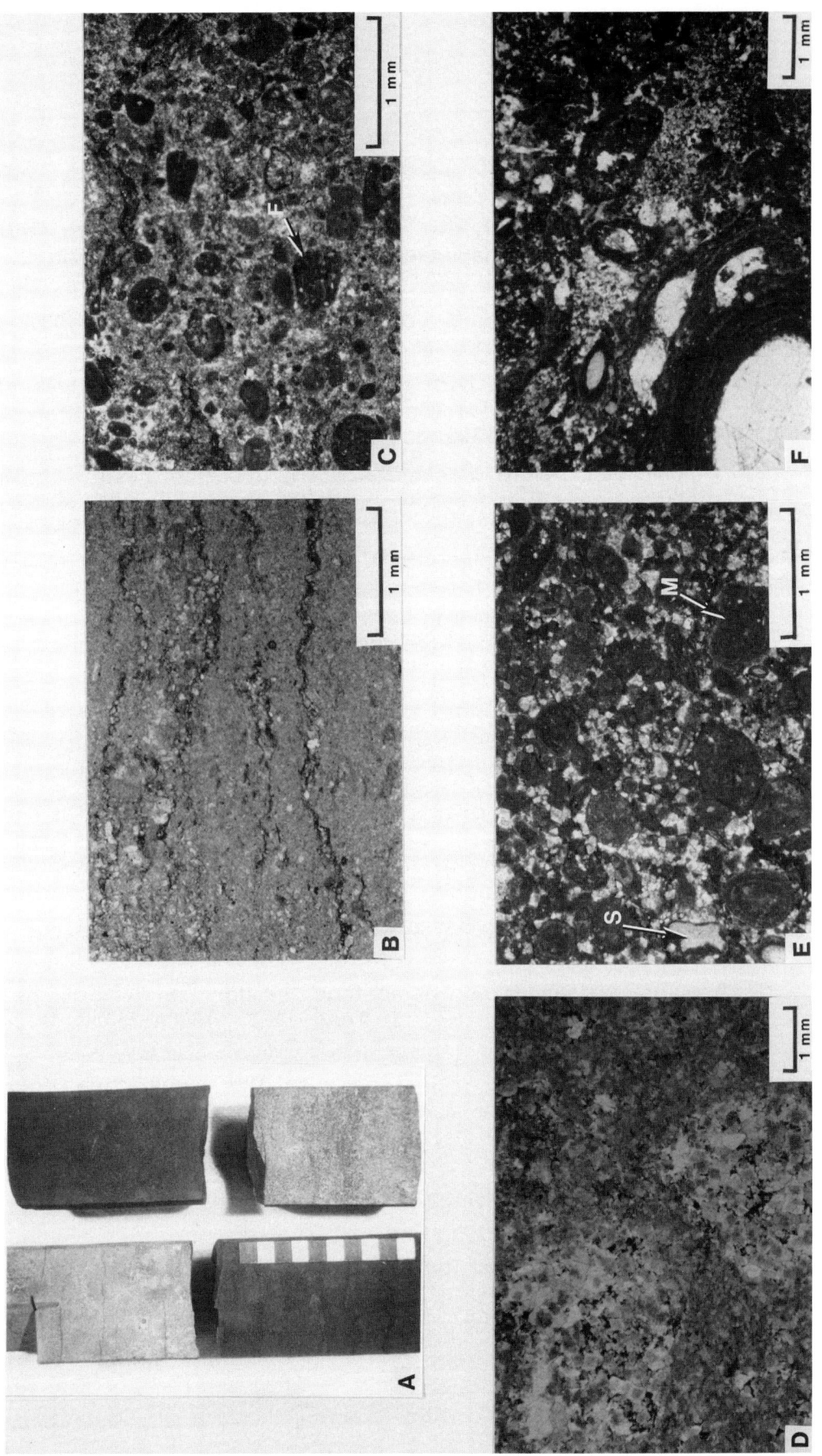

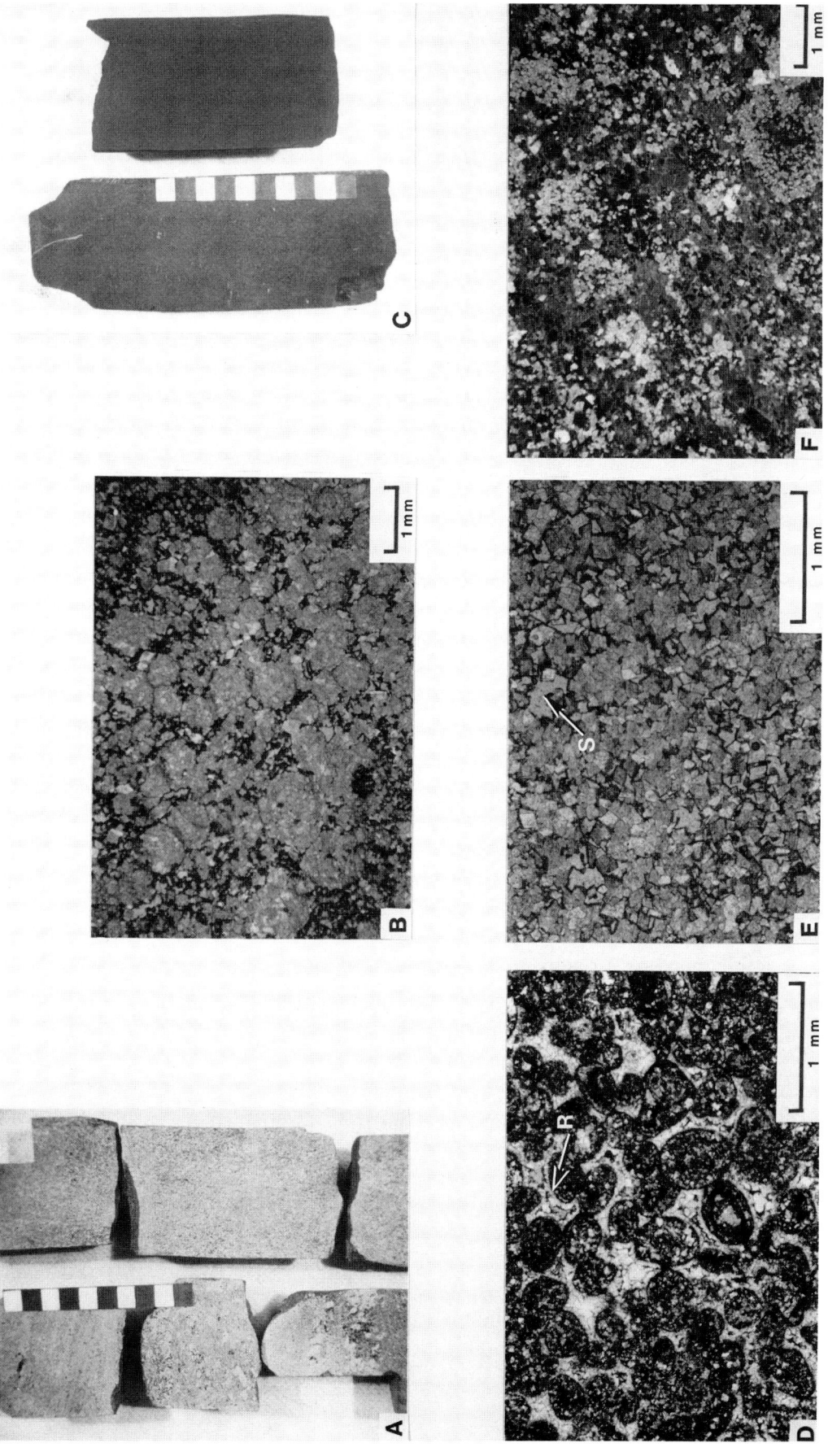

These sediments represent a shallow shelf to lagoonal environment characterized by low energy, subtidal conditions where muddy carbonates accumulated. Extensive bioturbated fabrics indicate relatively low sedimentation rates, with open marine conditions reflected by the appearance of oncolites and sparse skeletal material derived from echinoderms, bivalves, and forams.

While the upper Smackover contains some facies similar to those in the lower portion, it is the high energy, grain-rich facies which make it distinctive. These facies include:

1) peloid packstone/grainstone: dolomite; contains pellets, ooids, intraclasts, occasional algal-coated grains and rare skeletal fragments (Figure 15A-C); peloids are medium to coarse grained, moderately well sorted and are frequently outlined by fine crystalline dolomite rim cements which represent early isopachous calcite cements (Figure 15D); porosity tends to be high (15-20%) and occurs as pelmoldic, (Figure 15D), intercrystalline, and intergranular types (Figure 15B); solution-enlarged pores are also a common feature (Figure 15E);

2) mudstones and pellet wackestones: dolomite, rarely limestone; may be bioturbated, argillaceous, pyritic, stylolitic and anhydritic (Figure 15F); anhydrite increases upward. This facies was not common in the cores used for this study, but according to Vinet (1984), it may dominate the upper member between producing fields.

The upper member comprises approximately the upper 100 feet of the Smackover and represents subtidal to supratidal deposition in a shoaling or regressive sequence capped by the supratidal, nodular mosaic anhydrites of the Buckner. The packstones and grainstones vary little in texture with minor variations in grain type. These sediments represent a shoaling environment characterized by moderate to high energy conditions. In the study area, these conditions resulted in virtually the same type of grain shoal environment, although local low areas in the Jay trend accumulated mudstones and wackestones (Vinet, 1984).

FACIES DISTRIBUTION

The lateral distribution of facies in the upper Smackover indicates that a linear northwest-to-southeast trending shoal complex extended more than 25 miles in a belt 2 to 4 miles wide (Figure 16). The higher energy packstones and grainstones may have accumulated over subtle paleotopographic highs created by basement features, faults or salt movement. Waves impinging on these features winnowed out the lime mud and deposited sand-size grains in a belt of shallow bars or banks. The paucity of skeletal material reflects the relatively inhospitable conditions of the upper Smackover seas in the study area.

The dominant component in the reservoir facies of the Smackover fields in Arkansas and Louisiana is ooids (Moore, 1984). The Jay trend area is somewhat unique in that pellets are more prevalent than ooids. As mentioned previously, the embayment between the Conecuh and Pensacola ridges has been interpreted as being sufficiently restricted such that high energy ooids did not form in abundance. They are present in minor amounts in Jay Field (this study) and have been reported from Blackjack Creek Field (Sigsby, 1976). An ooid shoal seaward of the Jay trend has been described in the LL and E McDavid Land - St. Regis No. 1 well (Lomando, 1981). The presence of a belt of ooid shoals parallel to the Jay trend may help explain the presence (albeit in low quantities) of ooids in the dominantly hardened pellet facies of Jay and Blackjack Creek fields. This complex of ooid shoals may also have influenced the formation of the lower energy pellets behind it which were subsequently wave and current-swept into high energy shoals.

A protected lagoonal area where lower energy mudstones and wackestones accumulated, formed shoreward of the shoal-water deposits (Figures 6 and 16). Lower energy, open marine wackestones were deposited seaward of the shoals. The resulting encasement of the higher-energy grainy facies within the lower-energy muddy facies, in conjunction with facies-selective dolomitization and development of secondary porosity, forms the basis of the hydrocarbon trapping mechanism along the Jay Field trend. Pay thickness corresponds to the distribution and thickness of the peloid packstone/grainstone facies. Productive

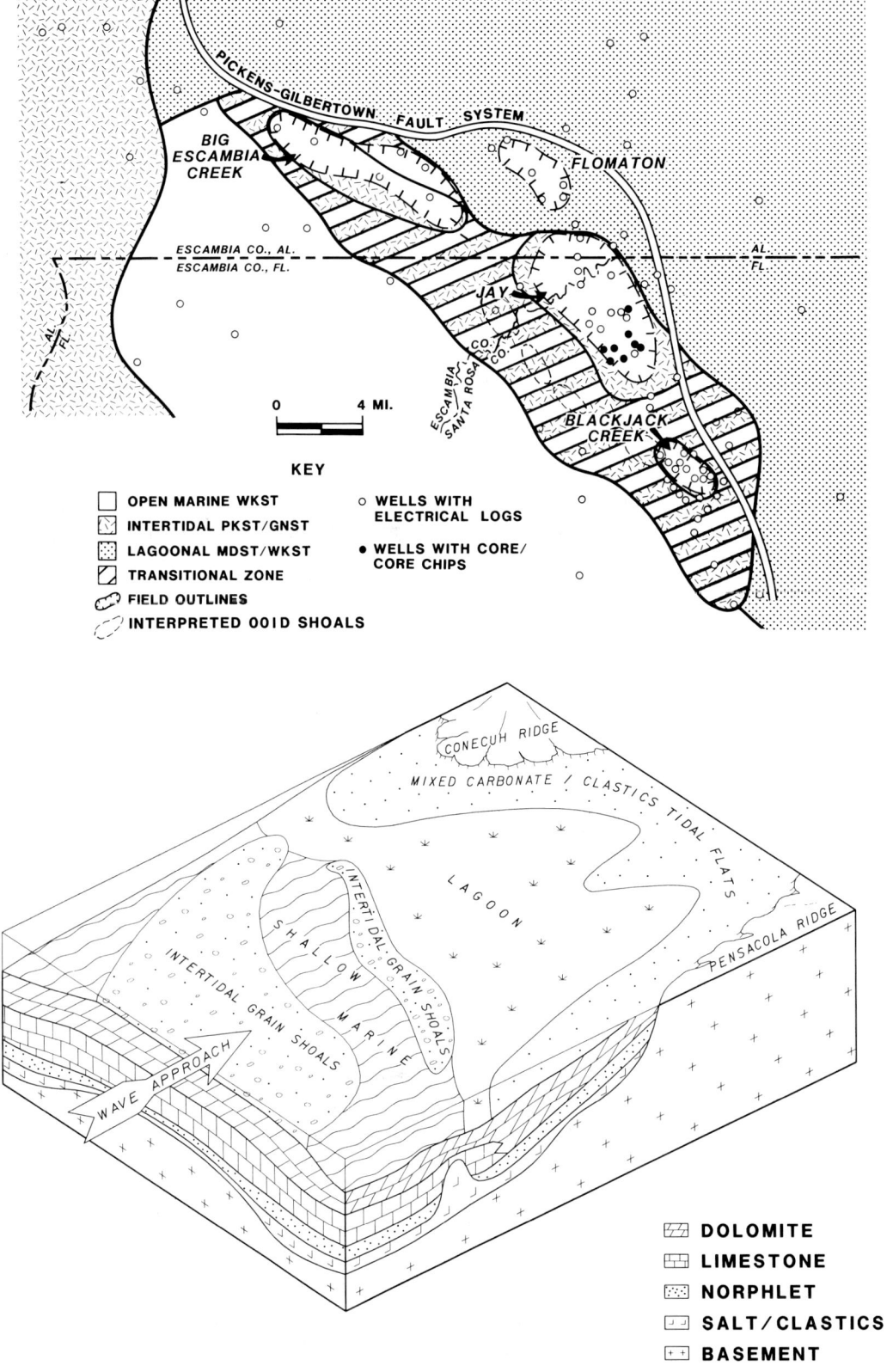

Figure 16 Top: Facies distribution along the Jay trend during deposition of the upper Smackover. Bottom: Depositional model for the facies of the upper Smackover shoaling complex. The cores displayed at the workshop are located in intertidal grain shoals. These shoals are distributed as a discontinuous belt along the trend with intervening lagoonal/shallow marine facies.

limits of the field correspond to pinch-outs of secondary porosity developed in that facies (Ottmann, and others, 1973).

DIAGENESIS

Upper Smackover sediments have undergone syndepositional and early diagenetic alterations typical of carbonate lithologies. These include grain micritization, calcite cementation, dissolution of allochems (peloids, ooids, skeletal grains), general stabilization of mineralogies, compaction and local fracturing. Following early diagenesis was a major dolomite replacement phase which in turn was followed by events of iron-rich dolomite cement, baroque dolomite cement and anhydrite/calcite cement (Barret, 1986).

There are several types of dolomite in the Smackover (Vinet, 1984) with a wide range in replacement percentage and degree of original fabric preservation. In calcite-dolomite transitions, it is a well-documented pattern that increased replacement and recrystallization results in increased primary textural alteration and loss. Peloid wackestones vary from limestone to partial to complete dolomite with concomitant textural loss (Figure 17). Some of the original fabric and pore geometry may be preserved when a fabric-selective replacement occurs, as in early-cemented grainstones (Figure 18A and B). At the other extreme, the same dolomitizing fluid in the same rock type destroys the original fabric, totally rearranging the original pore geometry and leaving only ghosts of peloidal grains (Figure 18C) (Barrett, 1986). The highest intercrystalline porosity and permeability in Smackover dolomite occurs in this type of fabric where original grain outlines have been completely obliterated by the growth of coarse rhombic dolomite crystals in loose contact (Sigsby, 1976) (Figure 18D).

Reservoir development is closely connected to the dolomite replacement process with most of the reservoir-grade porosity being secondary and facies-selective (Benson and Mancini, 1982). It is not our intent to discuss possible dolomitization models here. Recent work utilizing light stable isotopes, fluid inclusion data and hydrologic models seems to favor a combination of penecontemporaneous dolomitization and a mixed marine-meteoric hydrologic

Figure 17 A. Partially dolomitized pellet wackestone with some original fabric preservation. Chevron No. 1 Hayes 40-4, 15669, plane light.

B. SEM micrograph of partial dolomitization of a pellet wackestone. Note the incorporation of the micrite crystals into the dolomite rhombs.

C. SEM micrograph of nearly complete dolomitization with remnant micrite crystals in the center. Very little of the original textural relationships has been retained.

D. SEM micrograph showing the limestone-dolomite transition and the incorporation of micrite crystals into the dolomite rhomb.

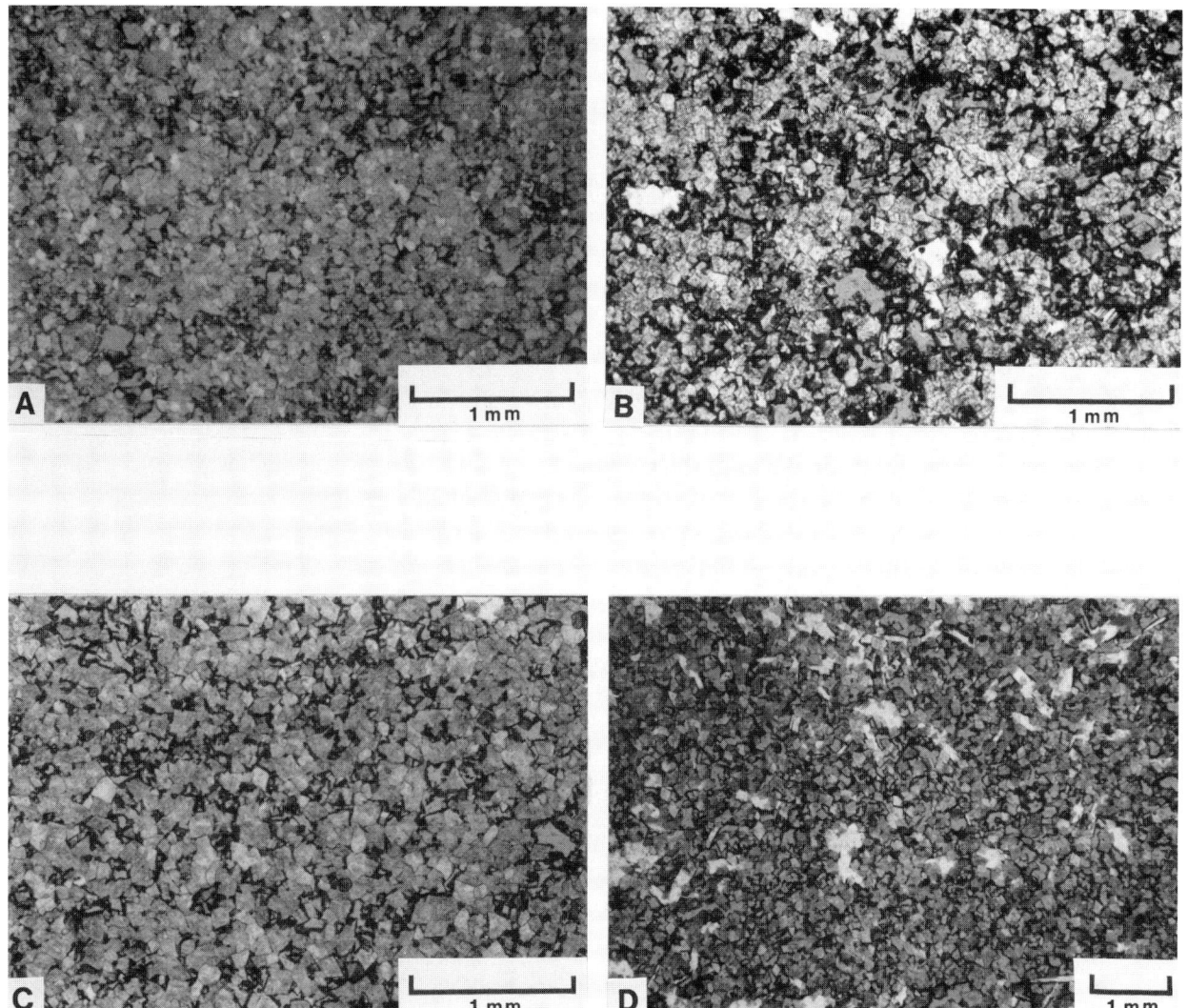

Figure 18 A. Dolomitized peloid grainstone with preservation of original rock fabric. Dark areas are intergranular pore spaces. Chevron No. 1 Bragg 22-1, 15660 ft., plane light.

B. Dolomitized peloid grainstone with partial retention of original fabric. Peloid outlines have become blurred and rearrangement of grain-pore boundaries has begun. Chevron No. 1 Golden 23-1, 15540 ft., plane light.

C. Only vague outlines or ghosts of peloids can be seen in this dolomitized peloid grainstone. Pores (dark areas) have been rearranged and no longer exhibit original textural relationships. Chevron No. 1 Boutwell 22-3, 15761 ft., plane light.

D. Dolomite rhombs in loose contact result in the highest porosity found in the Smackover reservoir. White areas are anhydrite crystals. Chevron No. 1 Hendricks 23-3, 15604 ft., plane light.

system (Vinet, 1984; Moore, 1984; Barrett, 1986). Within this framework of dolomitization processes, the formation of Smackover reservoirs was controlled by original sediment type, environmental setting, and by the varying responses of these sediment packages to diagenetic alteration.

An empirical observation made by many researchers working with the Smackover in Alabama and Florida is that the main porosity development is associated with the dolomitized peloid packstone/grainstone facies. The dolomitization of the grain-rich sediments resulted in a range of dolomite crystal sizes. Very fine, anhedral crystals replaced the peloids while coarse, euhedral crystals occur as pore-filling cement (Figure 19A). The coarse crystalline dolomite and attendant good porosity are generally restricted to the packstone/grainstone facies (Vinet, 1984). The importance of original lithologies is further emphasized by the noted decrease in reservoir quality associated with vertical and lateral changes to finer grained carbonates (Figures 8-13 and 16).

Aragonite and magnesian calcite allochems are more susceptible to dolomitization and leaching than calcite allochems or micritic matrix (Sibley, 1980; Gaines, 1980). Thus the pellets, ooids, oncolites and other grains in the packstone/grainstone facies form moldic and dolomite intercrystalline porosity (Figure 19B and C). The original grain sizes and grain/pore distribution are interpreted to be controlling factors in the final preservation of porosity (Barrett and Hardie, 1986b). Thus a grainstone with an irregular distribution of pores would result in retention of greater porosity. Conversely a finer grained material with similar porosity but a more uniform distribution of smaller pores would result in a more tightly cemented final product (Figure 19D).

It became apparent that the interaction of depositional and diagenetic factors controlled the development of the reservoir facies at Jay Field. Muddy sediments and grain-rich sediments reacted differently to diagenesis (mainly dolomitization) and the resulting reservoir quality of the pore systems also varied. In an effort to better understand the importance of original sediment composition on the development of diagenetic pore networks, mercury injection analyses were performed on fifty samples from six wells.

Figure 19   A. SEM micrograph showing the fine, anhedral dolomite crystals comprising the replaced peloids and the coarse, euhedral crystals which precipitated in open pores as cement.

B. Moldic porosity (dark areas) from dissolution of pellets, including Favreina (F), and other originally aragonitic allochems. Chevron No. 1 Kent 18-2, 15588 ft., plane light.

C. Intercrystalline porosity within dolomitized peloids. Very fine crystalline matrix is darkened with hydrocarbons. Chevron No. 1 Hayes 40-4, 15480 ft. plane light.

D. SEM micrograph of a dolomitized mudstone/wackestone with very low porosity.

MERCURY INTRUSION THEORY

Mercury does not wet, i.e., is not spontaneously absorbed into, the pore network of a rock. It must be forced in under pressure. The relationship most commonly used to model the penetration of mercury into a porous substance was proposed by Washburn (1921). It is given by:

$$r = \frac{-2G\cos(CA)}{P}$$

where r is the radius of a capillary tube being penetrated at pressure, P, G is the surface tension of the fluid and CA is the contact angle between fluid and the substance. In the oil industry, the surface tension of mercury, G, is typically given the value of 480 dynes/cm and 140 degrees is used for the contact angle, CA. The pore throat radius, r, is reported in microns and the pressure, P, is measured in psia.

The model assumes the pore network to be a bundle of capillary tubes of varying radii. Since the relationship between pressure and pore throat radius is inverse, the higher the intrusion pressure, the smaller the pore throat being penetrated.

The Micromeritics Autopore 9210 Mercury Intrusion Porosimeter used in this study allows intrusion pressures over the range of 1.5 psia to 60,000 psia which correspond to Washburn radii of 71 microns to 0.0018 microns, respectively. The pressure scan used is a pre-programmed sequence of increasing pressures. Once a programmed pressure step is attained, a pre-programmed equilibration time is allowed to transpire. At the conclusion of this time period, the pressure is again measured. A decrease indicates that equilibrium has not been attained and the machine re-pressures to the desired pressure and again waits. Once equilibrium is obtained, the volume of mercury in the sample is measured by an internal capacitance bridge and the machine proceeds to the next pre-programmed intrusion pressure. The data are stored digitally during the run. Experience with a wide range of geological materials indicates data reproducibility equal to the degree that a rock body can be assumed to be homogenous.

MODELLING HYDROCARBON/BRINE CAPILLARY PRESSURE BEHAVIOR

If it is assumed that a typical reservoir is water-wet prior to the entry of hydrocarbon, the capillary behavior of mercury can be used to model the capillary behavior of hydrocarbon, since hydrocarbon is non-wetting if the pore network is first water-wet just as mercury is non-wetting in dry rock. The Washburn equation relates pore-throat radius to entry pressure by means of the surface tension and the cosine of the contact angle. In any system, these will evaluate to a constant. If the surface tension and contact angle of the hydrocarbon/brine system are known or can be estimated with some confidence, the numerator, $-2G\cos(CA)$ can be evaluated. The ratio of the mercury/rock numerator to that of hydrocarbon/brine can be used to relate a given mercury/rock capillary pressure to an equivalent hydrocarbon/brine capillary pressure. This allows mercury-based capillary pressure data to be used as a basis for the calculation of estimates of air/brine, oil/brine or gas/brine capillary behavior. Comparisons of converted mercury-based data indicate reasonable agreement with data obtained in these other systems (Slider, 1974).

INTERPRETATION OF RESULTS WITH GEOLOGICAL MATERIALS

Clearly, the pore network of a rock is not so straight forward as a bundle of capillary tubes with smooth, straight walls and uniform radii. The capillary bundle model is excessively simple. However, examination of rocks of known reservoir character can make it possible to interpret the reservoir properties of other rocks of unknown properties, even if semi-quantitatively.

Figure 20 shows the results of mercury intrusion for a shale sample taken from full core in a well in West Bay, Louisiana at a depth of 9,515 feet. This particular sample was part of a 25-foot thick continuous sequence of shale immediately overlying an oil and gas producing horizon. It is probable that this shale is part of a sealing sequence and therefore has extremely low permeability. The mercury-based porosity estimate is 15%. The pore volume distribution curve (Figure 20A) is derived by dividing each intrusion volume increment by the

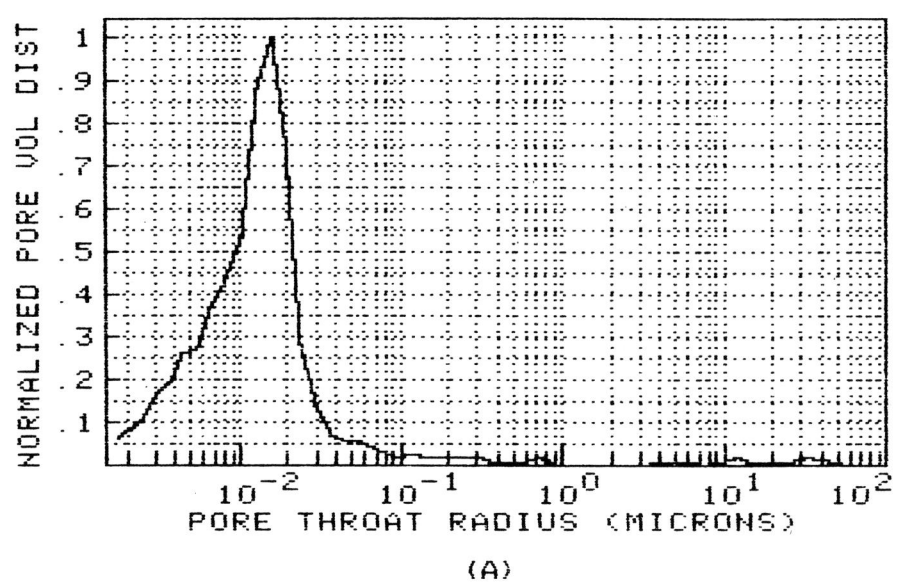

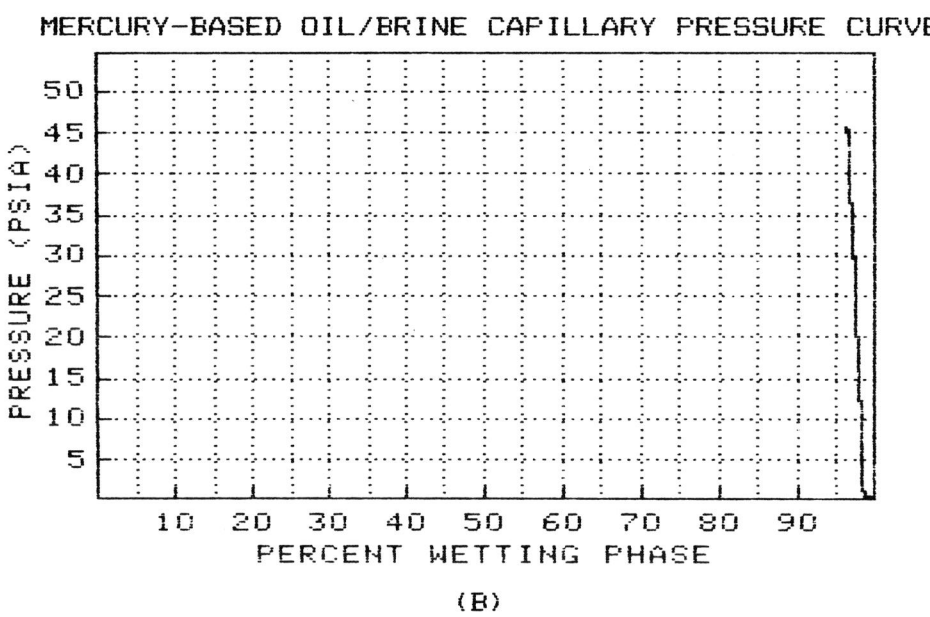

Figure 20   Shale from West Bay, Louisiana core at 9515 feet.

    A. Normalized pore volume distribution curve. The modal pore throat size is nearly 0.02 microns. The pore volume is lognormally distributed with slight skewness toward smaller radii.

    B. Mercury-based Oil/Brine Capillary Pressure Curve. Wetting phase (water) saturation shows virtually no change with increasing pressure and is 96 percent at 45 psia.

maximum intrusion volume increment and plotting the result as a function of pore throat radius. For this shale, the pore volume distribution is lognormally distributed around a modal pore throat size of 0.02 microns (Figure 20A) and the capillary pressure curve (Figure 20B) shows wetting-phase saturations of 95 percent. The hydrocarbon saturation would be correspondingly low, less than 5 percent. Both the pore throat size distribution and the capillary pressure curve indicate a very tight rock.

The pore volume distribution of a friable, open sand also from a Gulf of Mexico core is illustrated in Figure 21. The permeability is 2100 millidarcies and the porosity estimate from mercury porosimetry is 28%. The pore volume distribution (Figure 21A) is, like the shale, lognormally distributed but the modal pore throat radius is 10 microns, nearly four orders of magnitude larger than that of the shale. The capillary pressure curve (Figure 21B) indicates very low wetting-phase saturations, or, conversely, very high hydrocarbon saturations at low capillary pressures.

Using these two samples as end members of a continuum of reservoir quality, it is possible to qualitatively assess the reservoir character of a given sample. In general, the larger the modal pore throat size for a unimodal sample, the greater the permeability and the better the reservoir character. With the capillary pressure curve, the lower the wetting phase saturation, the better the reservoir, both from an oil storage standpoint and an oil recovery standpoint.

IMPACT OF ORIGINAL SEDIMENT CHARACTER ON DOLOMITIZATION AND DEVELOPMENT OF RESERVOIR QUALITY

Wackestone (Micrite) Precursor

A typical undolomitized micrite has a porosity of 2.2 percent as determined by mercury porosimetry. The pore volume distribution curve (Figure 22A) and capillary pressure curve (Figure 22B) are very similar to those of the West Bay Shale (Figure 20A). The intercrystalline pore network of this micrite has the properties of a seal – low permeability and low storage capacity (Figure 17).

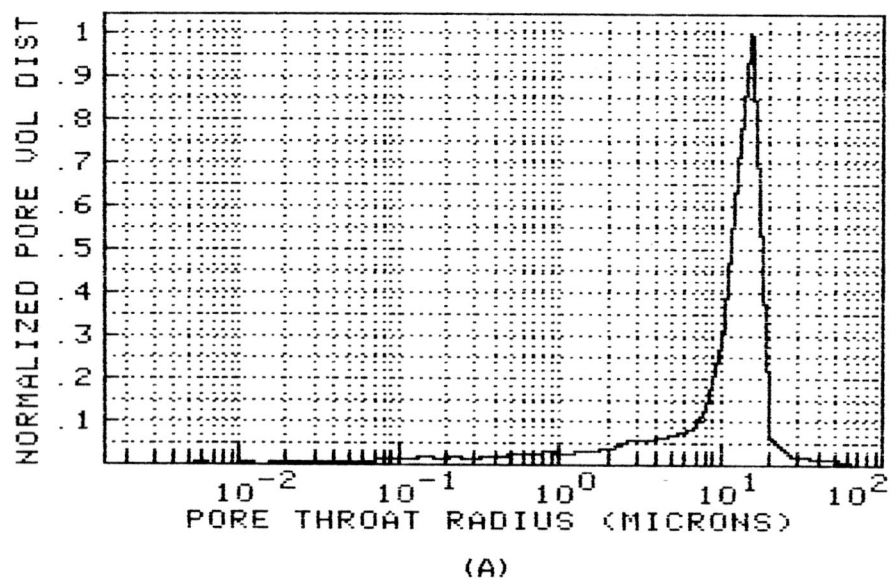

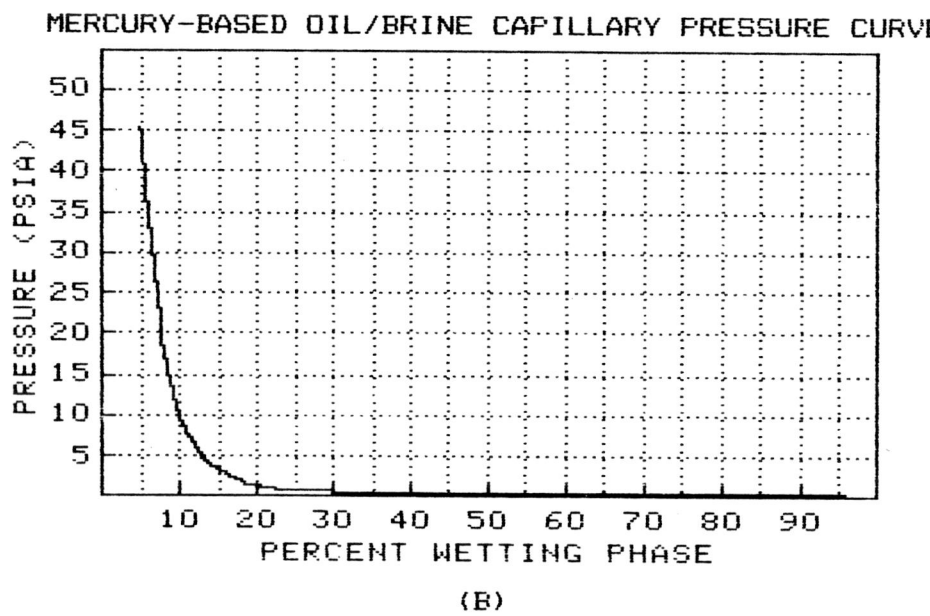

Figure 21 Sand from Bastian Bay Field, Louisiana core at 12,497 feet.

- A. Normalized Pore Volume Distribution Curve. Modal pore throat radius is 15 microns. The pore volume is lognormally distributed with slight skewness toward smaller radii. If this were a reservoir, the transition from the water table, where the water saturation would be 100 percent, to near maximum oil saturation would occur over a pressure range of 15 psia.

- B. Mercury-based Oil/Brine Capillary Pressure Curve. Wetting phase saturation is 5 percent at 45 psia. This reservoir would show a very sharp transition of hydrocarbon saturation from zero at the water table to near maximum a few feet within the oil zone.

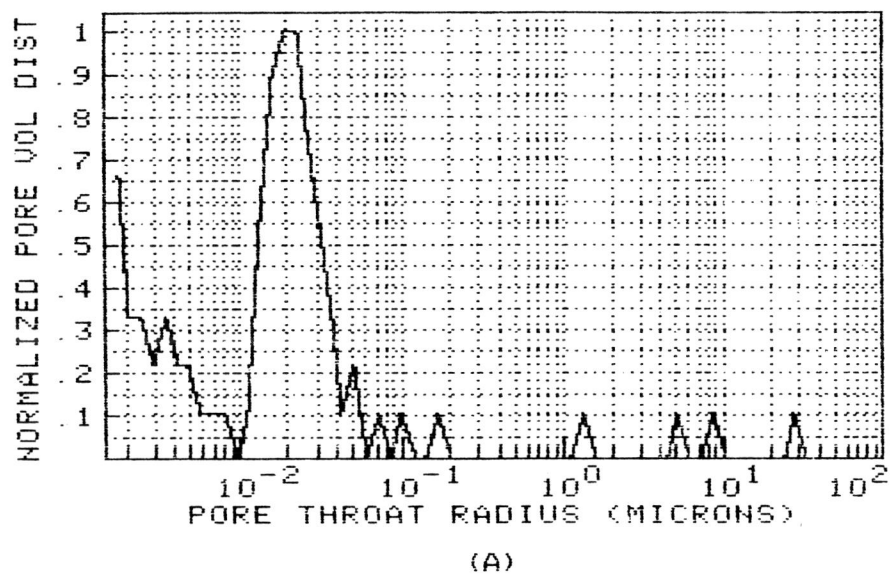

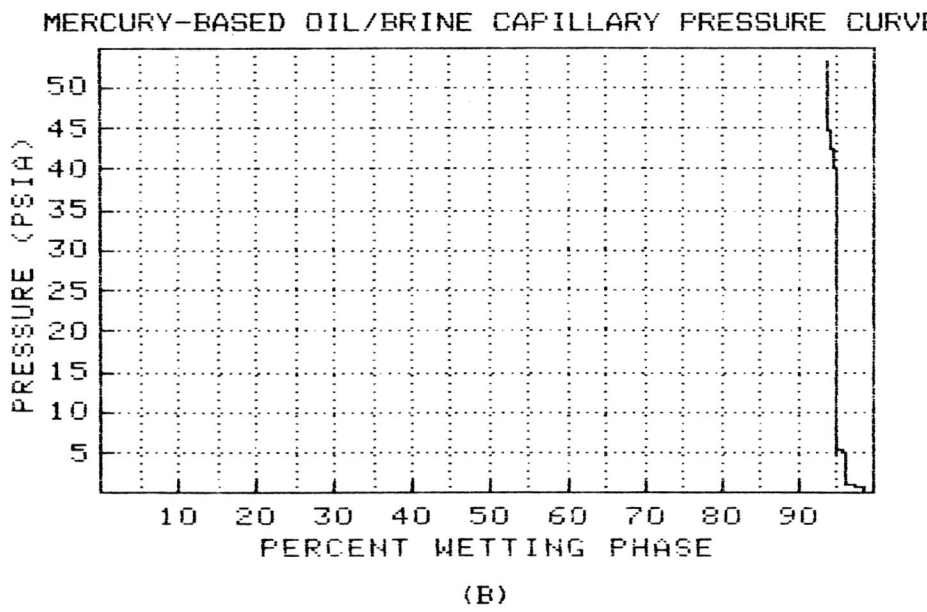

Figure 22 Undolomitized micrite from Hayes 40-4 at 15,379 feet.

    A. Normalized Pore Volume Distribution Curve. The modal pore throat size is 0.02 microns. Jagged appearance of the curve is due to the low porosity of the sample. Many of the intrusion increments were small enough to be near the porosimeter's detection limit.

    B. Mercury-based Oil/Brine Capillary Pressure Curve. The wetting phase saturation is 94 percent at 45 psia.

As dolomitization of micrite proceeds, the modal pore throat size increases by approximately one order of magnitude (Figure 23A) and the porosity increases to around 4 percent. However, there is virtually no change in the capillary pressure curve (Figure 23B) indicating that in spite of having double the porosity, under normal drainage pressures, there would be little increase in the storage capacity of the rock.

More complete dolomitization (Figure 24) yields porosities of nearly 15 percent and yet another order of magnitude increase in modal pore throat size (Figure 24A). The capillary pressure curve shows a concomitant increase in the reservoir quality (Figure 24B).

Packstone/Grainstone Precursor

All of the packstones are at least partially dolomitized. In the slightly dolomitized examples, porosities are typically 8 percent. Figure 25A shows that the modal pore throat size and pore volume distribution of a minimally dolomitized packstone are very similar to that of the most heavily dolomitized wackestone (Figure 24A). While the porosity of the dolomitized packstone is about one-half that of the wackestone, the reservoir quality is approximately the same (Figures 24B and 25B).

Increasing dolomitization produces a relatively open pore network distribution (Figure 26A) with a modal pore throat size of 2 microns. The capillary pressure curve (Figure 26B) indicates high hydrocarbon saturation at low capillary pressures. Reservoir quality is thus improving with increasing dolomitization.

Complete dolomitization produces porosities in the range of 20 percent and modal pore throat sizes of 9 microns (Figure 27A). A typical capillary pressure curve (Figure 27B) shows very high non-wetting phase (hydrocarbon) saturations at relatively low drainage pressures. This combination results in the best reservoir development observed in the Jay Field samples (Figure 18).

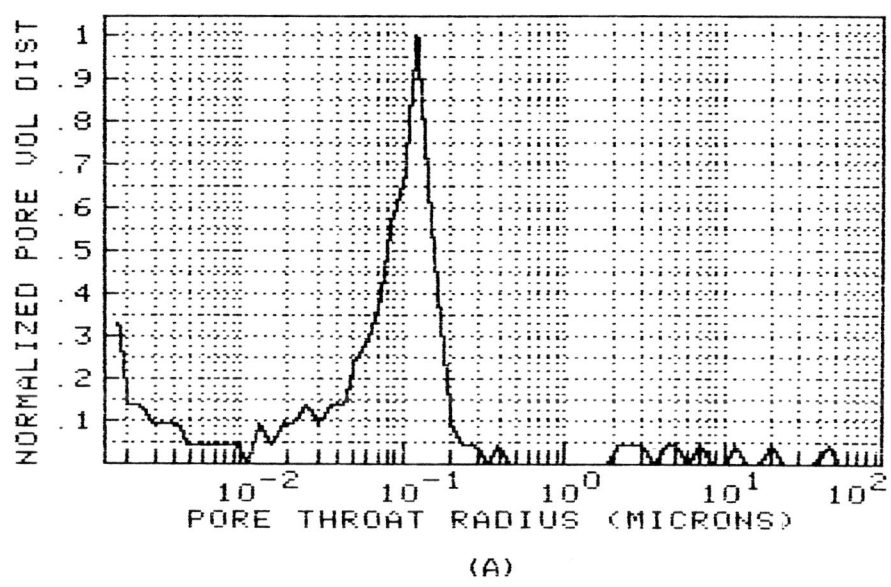

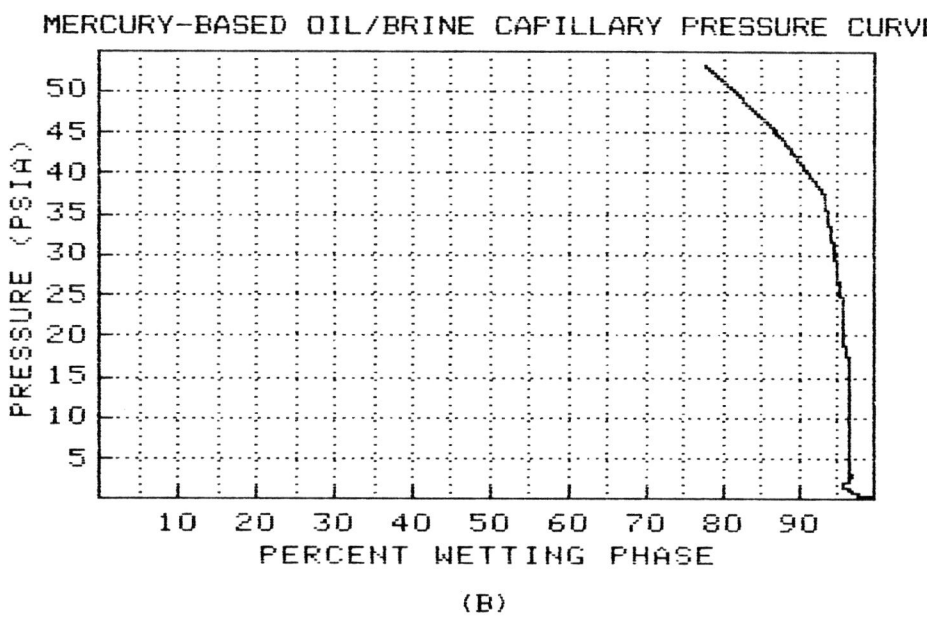

Figure 23 Slightly dolomitized micrite from Hayes 40-4 at 15,669 feet.

    A. Normalized Pore Volume Distribution Curve. The modal pore throat size is slightly larger than 0.1 microns. The jagged appearance stems from the low porosity of the sample. Many of the intrusion increments were small enough to be near the porosimeter's detection limit.

    B. Mercury-based Oil/Brine Capillary Pressure Curve. The wetting phase saturation at 45 psia is 87 percent. Note that the saturation abruptly changes at around 35 psia. If this rock were a reservoir, it would have a long transition zone.

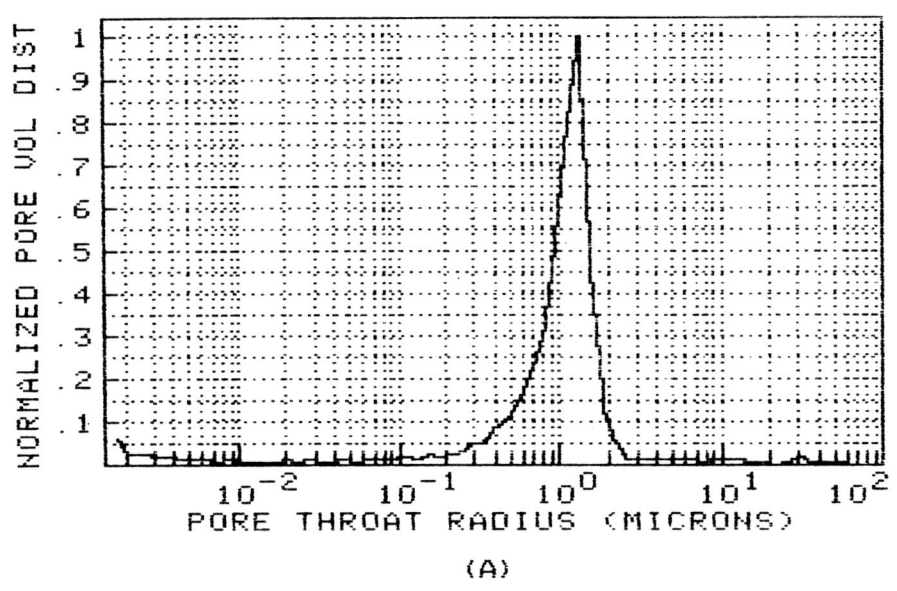

(A)

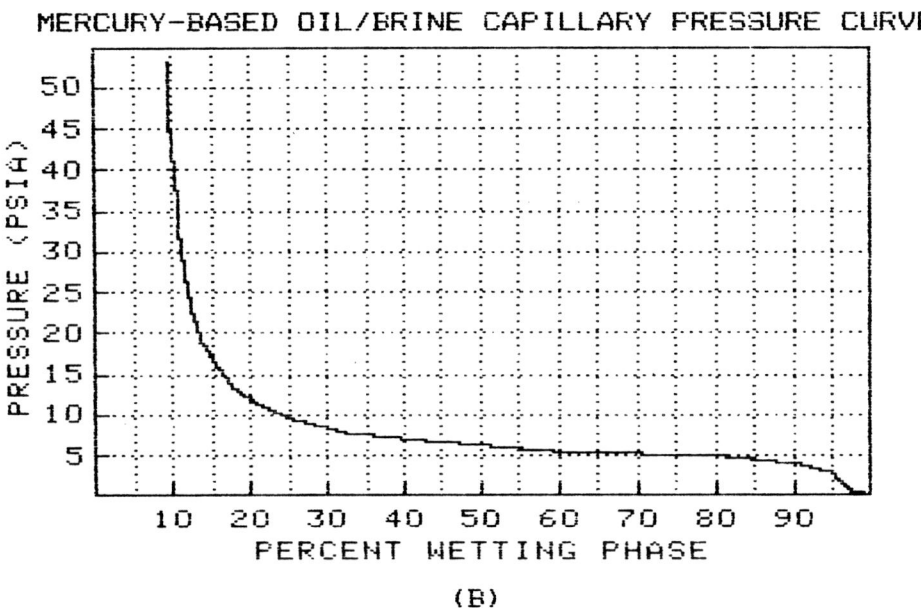

(B)

Figure 24 Extensively dolomitized micrite from Hayes 40-4 at 15,490 feet.

    A. Normalized Pore Volume Distribution Curve. Modal pore throat size is 1.5 microns.

    B. Mercury-based Oil/Brine Capillary Pressure Curve. The transition zone occurs over a 20 psia interval. The wetting phase saturation at 45 psia is 10 percent.

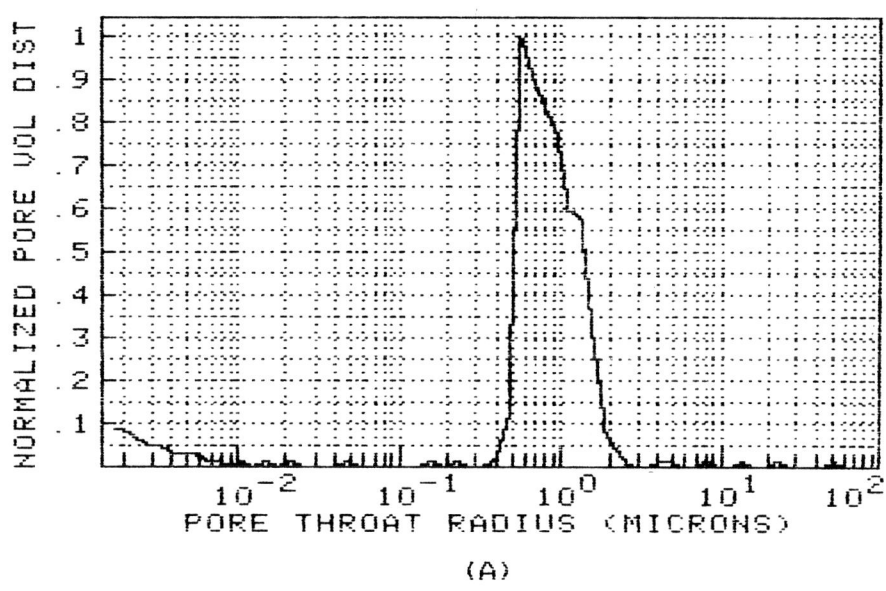

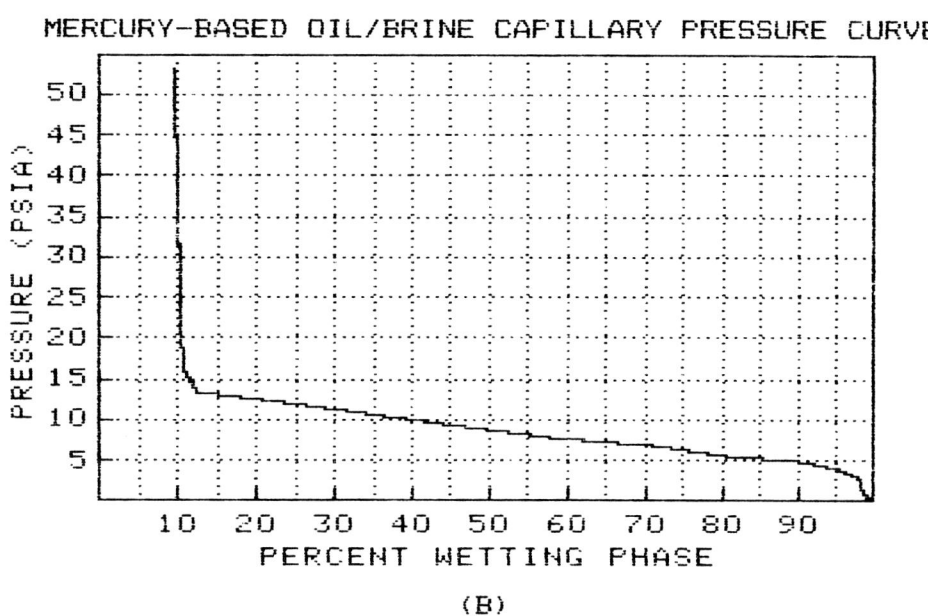

Figure 25   Slightly dolomitized packstone/grainstone from Hayes 40-4 at 15,638 feet.

    A. Normalized Pore Volume Distribution Curve. The modal pore throat size is 0.55 microns.

    B. Mercury-based Oil/Brine Capillary Pressure Curve. The transition zone occurs over a 15 psia interval. The wetting phase saturation at 45 psia is 10 percent.

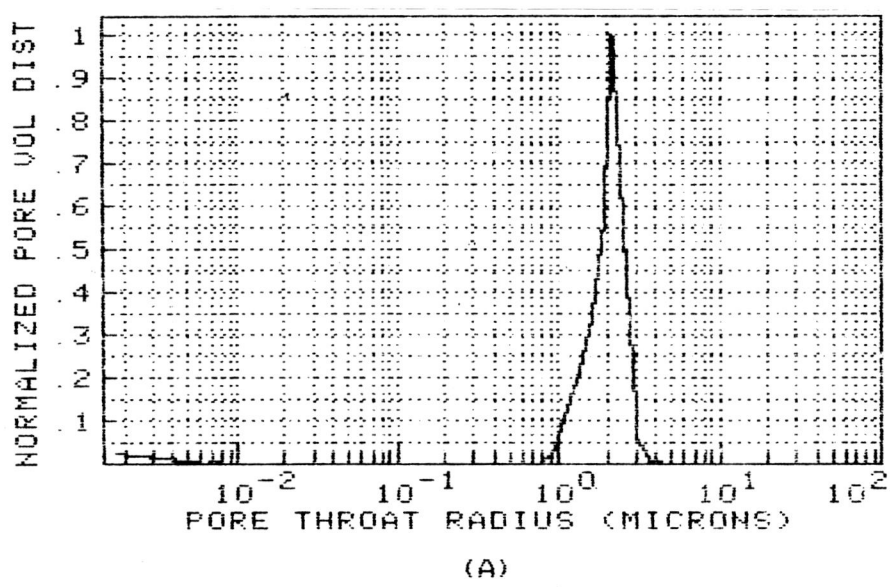

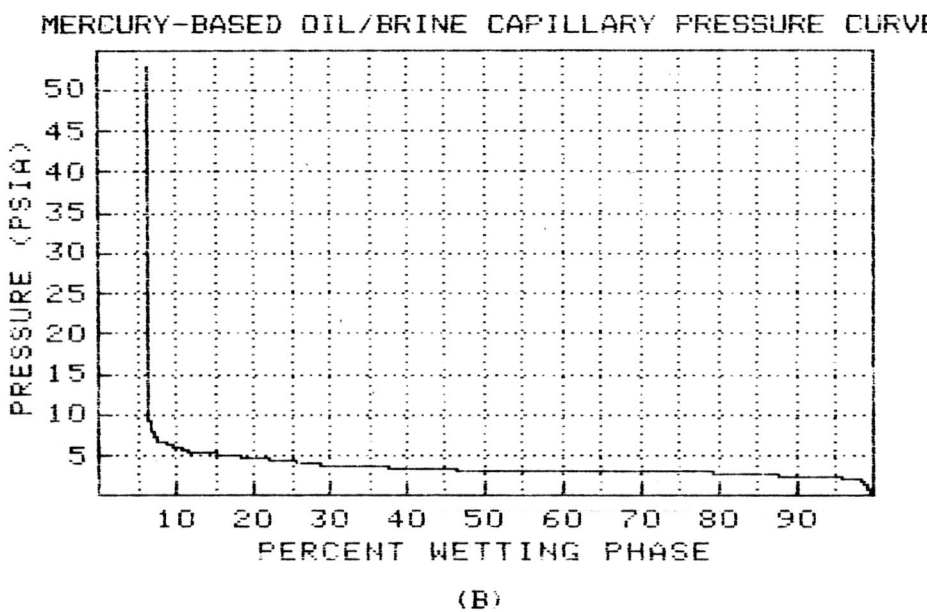

Figure 26 Moderately dolomitized packstone/grainstone from Kent 18-2 at 15,523 feet.

  A. Normalized Pore Volume Distribution Curve. The modal pore throat size is 2 microns.

  B. Mercury-based Oil/Brine Capillary Pressure Curve. The transition zone occurs over a 7 psia pressure range. The wetting phase saturation is 6 percent at 45 psia.

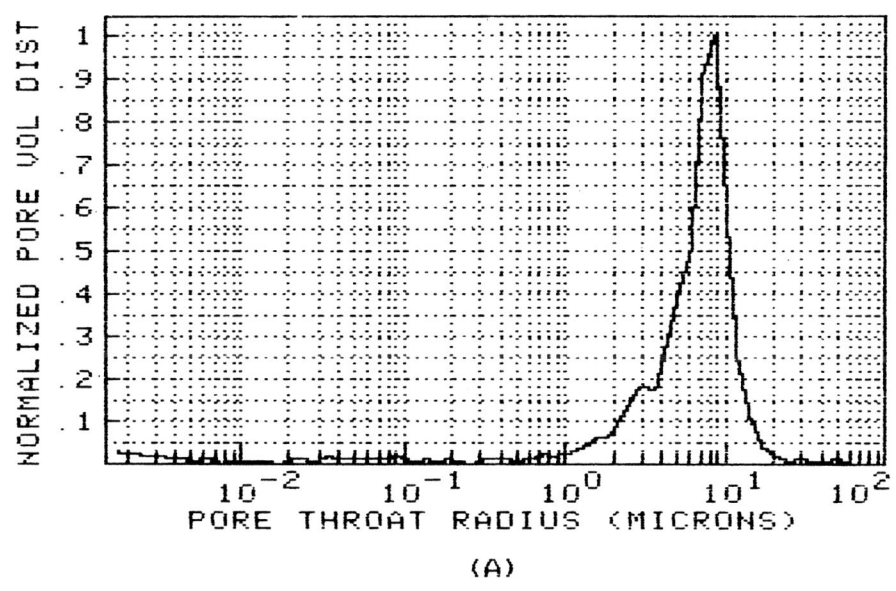

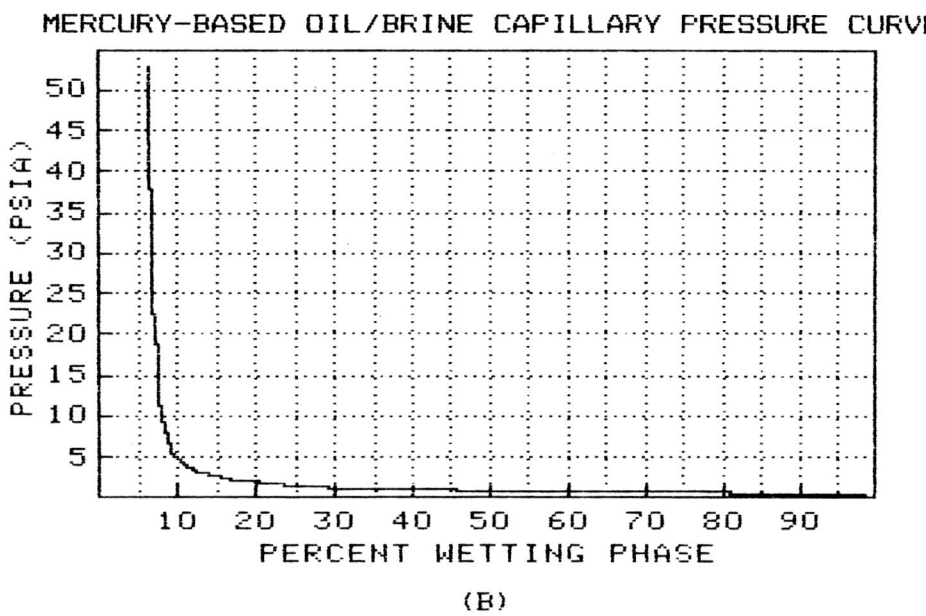

Figure 27   Completely dolomitized coarsely crystalline packstone/grainstone from Golden 23-1 at 15,570 feet.

    A. Normalized Pore Volume Distribution Curve. The modal pore throat size is 9 microns.

    B. Mercury-based Oil/Brine Capillary Pressure Curve. The transition zone occurs over a 5 psia pressure range. The wetting phase saturation is 6 percent at 45 psia.

Coarseness of Dolomite

In completely dolomitized packstone/grainstone samples, the porosity and non-wetting phase saturation profiles increase directly with an increase in the crystal size of the dolomite. Figure 28A is typical of a medium crystalline dolomite. The porosity is around 13 percent and the modal pore throat size is 0.8 microns. Where diagenesis has resulted in a more coarsely crystalline dolomite, the modal pore throat size is 9 microns (Figure 27A), one order of magnitude larger than the medium crystalline dolomite. Likewise, the capillary pressure curve (Figure 28B) of the medium crystalline dolomite indicates relatively low non-wetting phase saturations over the same pressure range that the coarse dolomite (Figure 27B) would have over 90 percent non-wetting phase saturation.

LATE DIAGENESIS

Even though dolomitization destroyed many of the original petrologic textures and modified pore configurations, diagenesis continued to modify the rock fabric. SEM analysis shows further modification by late dissolution which created vuggy, solution-enlarged pores and etched dolomite rhombs (Figure 29 A-C). This dissolution event post-dates both the inclusion-rich early dolomite and the clear iron-rich late dolomite, but precedes hydrocarbon migration as indicated by the oil staining of pore-facing edges of crystals (Figure 29 D and E). A minor loss of porosity can be seen in the bridging of pore space by clays (Figure 29F). The importance of original fabrics is again seen in the interpretation of the dissolution mechanism. Greater volumes of corrosive brines, migrating vertically and laterally, passed through the coarsely crystalline dolomite (originally porous grainstone) than through the finely crystalline dolomite (originally mudstones and wackestones) thus enhancing porosity already present (Benson and Mancini, 1982).

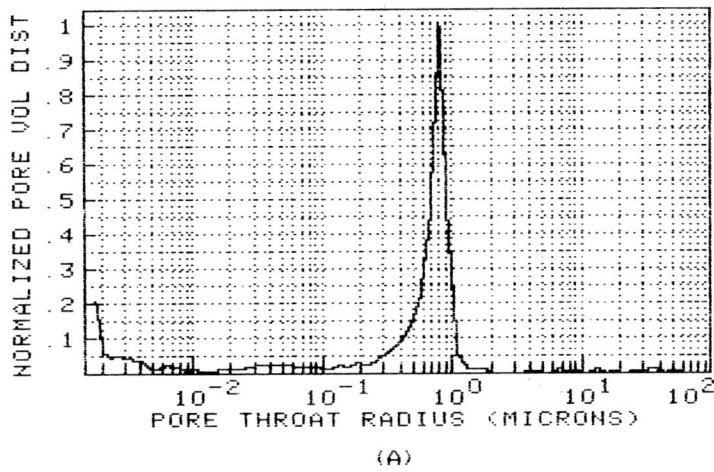

(A)

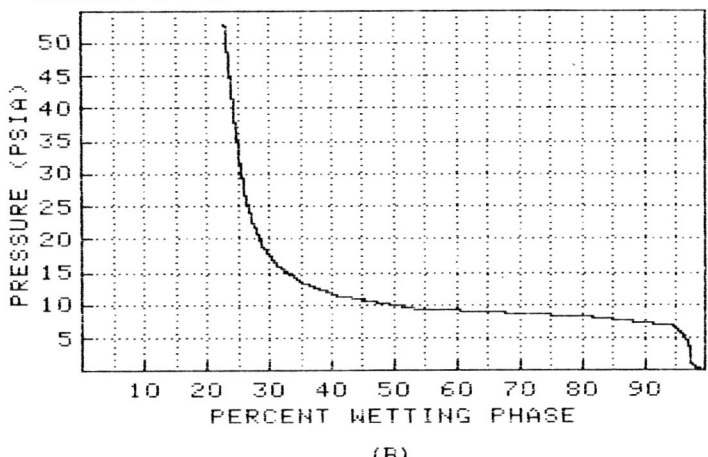

(B)

Figure 28 Medium crystalline dolomite from Kent 23-2 at 15,641 feet.

    A. Normalized Pore Volume Distribution Curve. The modal pore throat size is 0.8 microns.

    B. Mercury-based Oil/Brine Capillary Pressure Curve. The wetting phase saturation is 28 percent at 45 psia. Transition is abrupt but occurs at 10 psia.

Figure 29 A. Vuggy, solution-enlarged pores in dolomitized peloid grainstone with variable amounts of hydrocarbon staining. Chevron No. 1 Kent 18-2, 15766-15797 ft., slabbed core, scale in cm.

    B. SEM micrograph of etched dolomite rhombs with dissolution along crystal lattices.

    C. SEM micrograph showing extensive solution modification of dolomite rhombs.

    D. Solution-enlarged pores in a dolomitized wackestone. The matrix is composed of inclusion-rich, anhedral dolomite crystals. Pores were partially filled with clear, euhedral dolomite crystals which have been modified by dissolution. Open pores are now filled with hydrocarbons. Chevron No. 1 Kent 18-2, 15752 ft., plane light.

    E. SEM micrograph showing similar dolomite crystals in a solution-enlarged pore.

    F. SEM micrograph showing the bridging of intercrystalline pore spaces by mixed layer illite/smectite clays.

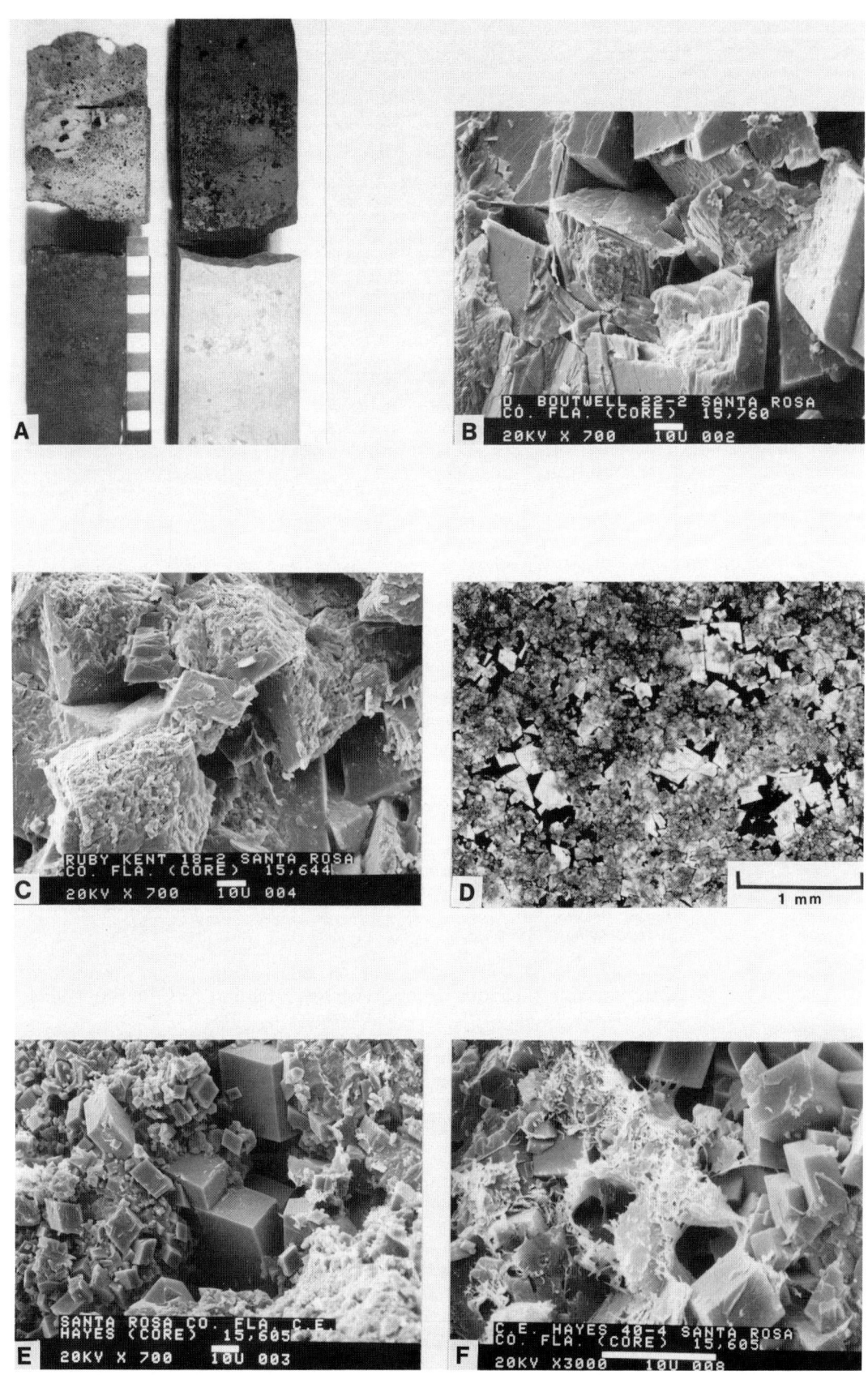

IMPACT OF DEPOSITIONAL FACIES AND DOLOMITIZATION ON RESERVOIR QUALITY

Local variations in depositional settings, the carbonate sediments produced, and their response to diagenetic events resulted in rapidly changing porosity and permeability values over short distances across the field and a very heterogeneous distribution of reservoir-quality dolomite (Figure 6). The low energy lime mudstones commonly are incompletely dolomitized and could qualify as a seal. The most extensively dolomitized examples of micritic precursors are of intermediate reservoir quality, not unlike moderately dolomitized packstones. In general, lime mudstones led to poor-to-intermediate reservoir quality.

Dolomitization of higher energy packstones and grainstones led to the development of reservoir rocks with high non-wetting phase saturation profiles at low capillary pressure, the highest measured porosities and the largest pore throats. All are characteristics of a high quality reservoir facies. Furthermore, the reservoir quality of these dolomitized packstones and grainstones varies directly with the size of the dolomite crystals. The coarser the dolomite, the higher the quality of the reservoir.

It is this depositional/diagenetic facies change which produced the stratigraphic component of the trap at Jay Field where dense lime mudstone provides the lateral seal for porous packstones and grainstone (Figure 16). The environmentally controlled distribution of high energy peloid packstones and grainstones and the secondary porosity produced within them are reflected in the variable pay thicknesses and the abrupt productive limit of the field (Ottmann and others, 1973). Understanding and predicting their occurrence is vital in generating successful Smackover exploration plays.

ACKNOWLEDGEMENTS

The authors thank Chevron U.S.A., Inc. and Chevron Geosciences for permission to publish this study and for providing assistance in preparation. J. R. Kirk and W. T. Dees reviewed the manuscript and provided editing suggestions that improved the paper.

REFERENCES

BARRETT, M. L., 1986, Replacement geometry and fabrics of the Smackover (Jurassic) dolomite, southern Alabama: Gulf Coast Assoc. Geol. Soc. Trans., v. 36, p. 9-18.

BARRETT, M. L. and HARDIE, L. A., 1986b, The development of permeable dolomite by void coalescence: Smackover dolomite of southern Alabama: AAPG Ann, Mtg., Atlanta, Georgia.

BENSON, D. J. and MANCINI, E. A., 1982, Petrology and reservoir characteristics of the Smackover Formation, Hatter's Pond Field: implications for Smackover exploration in southwestern Alabama: Gulf Coast Assoc. Geol. Soc. Trans., v. 32, p. 67-75.

GAINES, A. M., 1980, Dolomitization kinetics: recent experimental studies: in Zenger, D. H., Dunham, J. B. and Ethington, R. L., eds., Concepts and Models of Dolomitization, SEPM Spec. Pub. 28, p. 81-86.

LOMANDO, A. J., 1981, Deposition and diagenesis of the Smackover Formation, Jay Field area, western Florida: unpublished Masters thesis, City Univ. of New York, Queens College.

MANCINI, E. A., 1979, Upper Jurassic Smackover petroleum geology of southwest Alabama: in Arden, D. D., Beck, B. F. and Morrow, E., eds., Proceedings Second Symposium on the Geology of the Southeastern Coastal Plain, Georgia Geol. Survey, Information Circular 53, p. 76-87.

MOORE, C. H., 1984, The upper Smackover of the Gulf rim: depositional systems, diagenesis, porosity evolution and hydrocarbon production: in Ventress, W.P.S., Bebout, D. G., Perkins, B. F. and Moore, C. H., eds., The Jurassic of the Gulf Rim, Proceedings Third Annual Research Conference, Gulf Coast Section, SEPM, p. 283-307.

OTTMANN, R. D., KEYES, P. L. and ZIEGLER, M. A., 1973, Jay Field - a Jurassic stratigraphic trap: Gulf Coast Assoc. Geol. Soc. Trans., v. 23, p. 146-157.

SIBLEY, D. F., 1980, Climatic control of dolomitization, Seroe Domi Formation (Pliocene), Bonaire, Netherlands Antilles: in Zenger, D. H., Dunham, J. B. and Ethington, R. L., eds., Concepts and Models of Dolomitization, SEPM Spec. Pub. 28, p. 247-258.

SIGSBY, R. J., 1976, Paleoenvironmental analysis of the Big Escambia Creek - Jay - Blackjack Creek Field areas: Gulf Coast Assoc. Geol. Soc. Trans., v. 26, p. 258-278.

SLIDER, H. C., 1976, Practical Petroleum Reservoir Engineering Methods: Petroleum Publishing Co., Tulsa, OK, 272 p.

VINET, M. J., 1984, Geochemistry and origin of Smackover and Buckner dolomites (Upper Jurassic), Jay Field area, Alabama: in Ventress, W. P. S., Bebout, D. G., Perkins, B. F. and Moore, C. H., eds., The Jurassic of the Gulf Rim, Proceedings Third Annual Research Conference, Gulf Coast Section, SEPM, p. 365-374.

WASHBURN, E. W., 1921, Note on a method of determining the distribution of pore sizes in a porous material: Proc. Nat. Acad. Sci., v. 7, p. 115-116.

SEDIMENTOLOGY AND DEPOSITIONAL ENVIRONMENTS OF THE IVISHAK SANDSTONE, PRUDHOE BAY FIELD, NORTH SLOPE, ALASKA

CHRISTOPHER D. ATKINSON,[1] PHILIP N. TRUMBLY,[2] and MEG C. KREMER[2]

[1] ARCO Oil and Gas Company, Plano, Texas 75075
[2] ARCO Alaska, Inc., Anchorage, Alaska 99510

ABSTRACT

The Prudhoe Bay Field is the largest oil and gas field in North America with original reserves in place of over 22 billion stock tank barrels of oil and 40 trillion standard cubic feet of gas. The field was discovered in 1968 and the main reservoir interval is the Triassic Ivishak Sandstone of the Sadlerochit Group.

The Ivishak Sandstone is a fluvio-deltaic deposit comprising initially a coarsening-, and then a fining-upward sequence of sandstones, conglomerates and shales. On the basis of petrophysical parameters, the reservoir has been subdivided into four stratigraphic zones (1-4) arranged in ascending order. Reservoir quality generally increases upwards from zone 1 (the lowest) through zone 2, reaches a maximum at the top of zone 2 and in zone 3 and then diminishes upwards through zone 4.

Identified facies and facies associations can be grouped into the following main depositional environments: sub-aerial, braided river-dominated coastal plain, transitional fluvio-marine deltaic and marine reworked transgressive. These environments are arranged in such a way that at any location in the field a vertical passage from fluvio-marine deltaic deposits (zone 1) up into more proximal, then more distal, braided river dominated, coastal plain sediments (zones 2-4) characterizes the succession. At the top of zone 4, especially in the western and southwestern parts of the field, marine reworking is associated with a transgressive phase which deposited the Shublik Formation.

Within the field the more proximal sediments are located in the north and the more distal deposits to the south. Less extreme environmental transitions occur in a lateral west-east sense across the field. Such lateral transitions suggest that the Ivishak coastal plain was characterized by several major fluvial axes separated by areas of less active river flow. The active channel areas, here termed "core" deposits, tend to be dominated by multistorey, coarser-grained channel bodies. In contrast, the more inactive regions, here termed "lateral" deposits, are relatively shale-prone, and exhibit more isolated and finer-grained channel bodies. Further complicating this picture is evidence, approximately mid-way through the succession at the zone 3 level, of a major phase of river incision which affected the coastal plain. This incision was most probably induced by a base-level fall of either tectonic (subsidence?) and/or eustatic origin.

The facies characteristics of the coastal plain component of the Ivishak Sandstone most closely resemble the deposits of modern day fluvioglacial outwash/coastal braidplains (sandurs). For this reason, we agree with recent studies which have interpreted the Ivishak in terms of a "braid delta" depositional model. Development of the Ivishak braid deltas was promoted by a high latitudinal setting (45-60° north), a prevailing wet, cool temperate climate, limited vegetation cover and the nearby presence of an uplifted source terrain.

INTRODUCTION

The Prudhoe Bay Field is the largest oil and gas field in North America with original reserves in place of over 22 billion stock tank barrels (3.5 billion m^3) of oil and 40 trillion standard cubic feet (1.13 trillion m^3) of gas. The discovery well, Prudhoe Bay State #1, was drilled in 1968 by ARCO-Humble (Exxon) and completed in the Triassic Ivishak Sandstone of the Sadlerochit Group. The Field covers an area of over 300 square miles (770 km^2) and the reservoir attains a maximum of 425 ft (130 m) of light oil column in the main field region. Production commenced in 1977 and of the total recoverable reserves estimated at 10 billion barrels (1.6 billion m^3), production exceeded 5 billion (0.8 billion m^3) barrels in March of 1987.

The earliest studies on the field concentrated on describing its general geologic setting and the overall lithologic appearance of the Ivishak (Detterman, 1970; Rickwood, 1970; Morgridge and Smith 1972; Jones and Speers, 1976; Eckelmann, Dewitt and Fisher, 1976 and Jamison, Brockett and McIntosh, 1980). Recently, several more specific studies have dealt with detailed aspects of the sedimentology, inferred depositional environments and diagenesis of the Ivishak deposits (Melvin and Knight, 1984; McGowen and Bloch, 1985; Lawton, Geehan and Vorhees,1987). The purpose of this paper is to illustrate and describe the typical appearance in core of the various facies and facies associations which comprise the succession and to develop, using this information, a modified depositional model for the interpretation of the Ivishak Sandstone in the Prudhoe Bay Field.

GEOLOGIC SETTING

The Prudhoe Bay Field lies on the southern flank of the subsurface Barrow Arch, a major uplift which generally parallels the present coastline and plunges to the southeast (Figs. 1a, b). The Arch had a history of recurrent movement from the Pennsylvanian through the Lower Cretaceous (Jones and Speers, 1976) and several unconformities are associated with its sedimentary cover (Fig.1b). The most important of these is the Lower Cretaceous Unconformity (LCU) which truncates all the previous stratigraphic sequences and forms the major seal over the eastern flank of the Prudhoe Bay Field (Fig. 1b). A structure map of the field (Fig. 2) illustrates that the top of the Ivishak ranges in subsea depth from -8000 ft (-2440

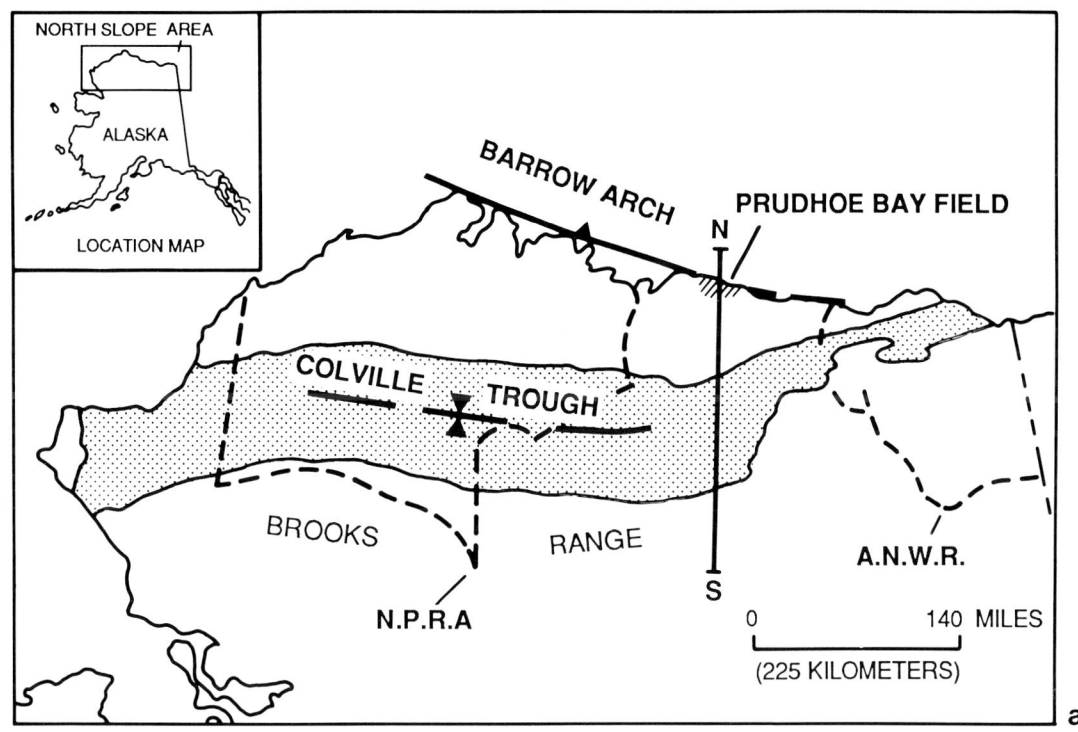

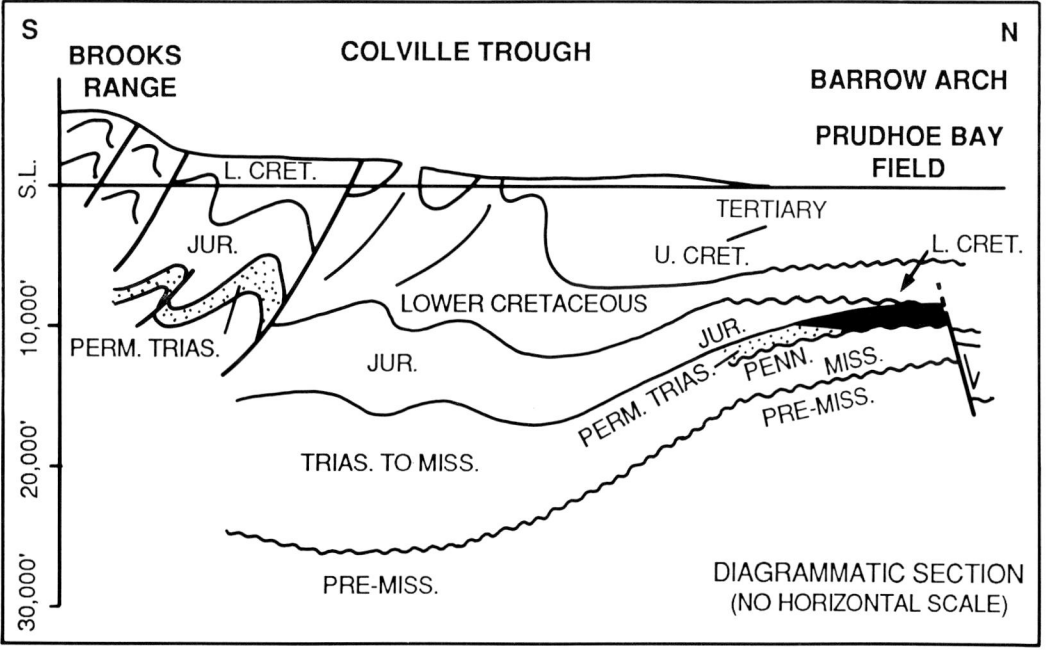

Figure 1 A. Location of the Prudhoe Bay Field.

B. North-to-south generalized cross-section from the Brooks Range to Prudhoe Bay (modified after Jamison, Brockett, and McIntosh, 1980).

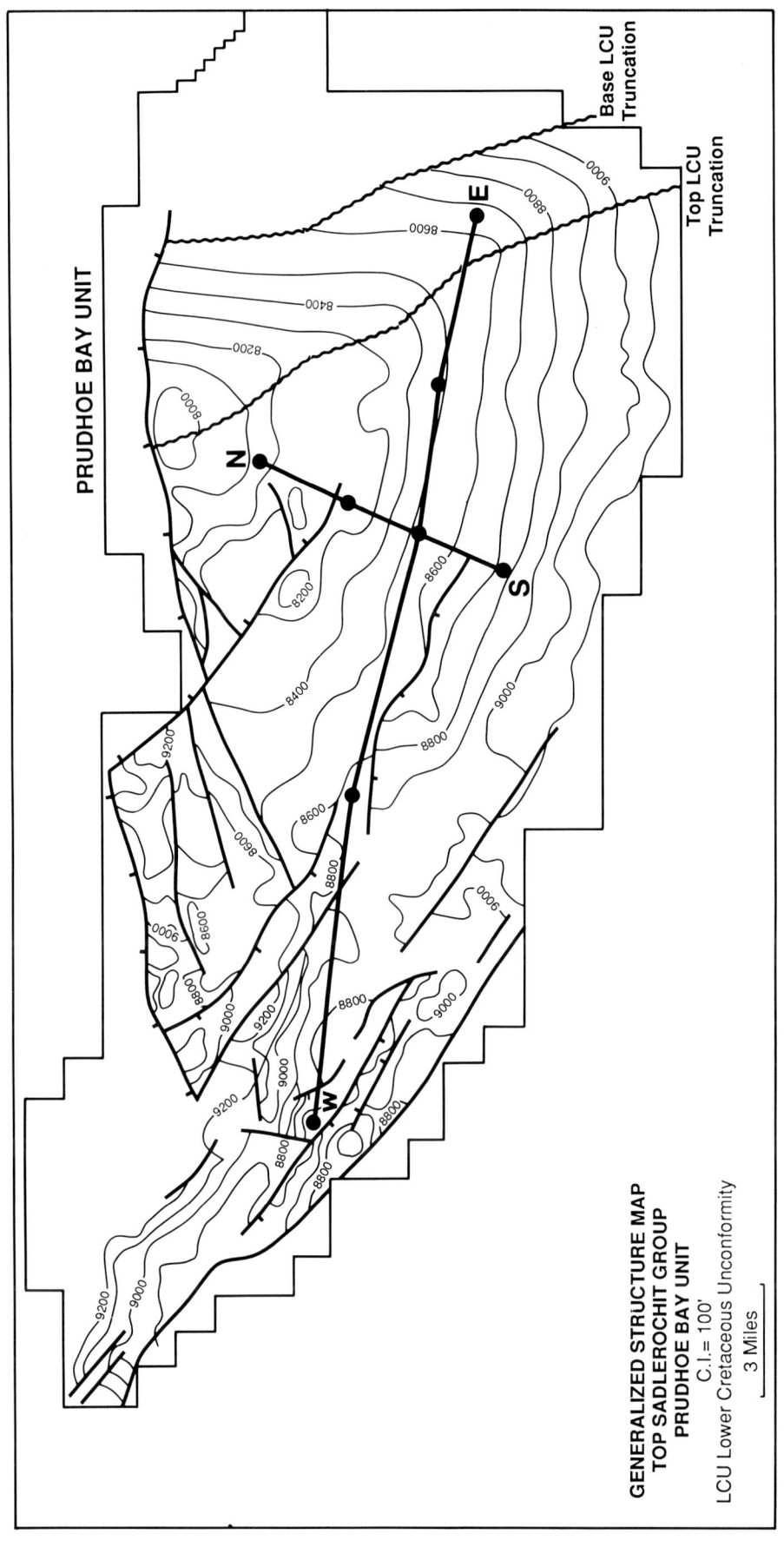

Figure 2 Generalized structure map of the Top Sadlerochit Group in the Prudhoe Bay Field. Location of the N-S and W-E cross-sections for figures 20-23 are indicated.

m) to -9200 ft (-2800 m) and dips to the south and west at approximately 1-2°. The Ivishak reservoir is bounded to the north by normal faults; on the east by the LCU and overlying Cretaceous shales; on the south by the oil/water contact (OWC); and on the west by another series of normal faults.

In the Prudhoe Bay area the Ivishak Sandstone forms part of the Sadlerochit Group which also includes the underlying Echooka Formation and Kavik Shale (Jones and Speers, 1976). This group possesses an unconformable contact with the underlying Mississippian/Pennsylvanian Lisburne Group and is overlain by the Triassic Shublik Formation (Fig. 3). The Ivishak Sandstone is conformable and locally gradational with the Kavik Shale (Fig 4). In contrast, the upper contact with the Shublik Formation is sharp and commonly associated with a phosphatic and pyritic pebbly "lag" which is interpreted to represent a transgressive disconformity. This Shublik transgression eroded the upper portion of the Ivishak, particularly in the northeast portion of the field. Overlying the Shublik Formation are the the Sag River Formation and Kingak Shale, respectively (Figs. 3 & 4).

The Ivishak Sandstone ranges from 400 to nearly 700 ft (120-215 m) in thickness at Prudhoe Bay (Fig. 4). The succession consists of an initial coarsening-upward "mega-cycle" involving a vertical transition from predominantly interbedded sandstones and shales to more amalgamated sandstones and conglomerates. This is overlain abruptly by a fining-upward interval of more distal-to-source sandstones and occasional interbedded shales (Fig. 4). Various authors have suggested that the Ivishak was deposited by a northerly-sourced, southwards-prograding fluvio-deltaic complex (Jamison et al., 1980; Melvin and Knight,1984; McGowen and Bloch, 1985; Lawton et al., 1987). The overall coarsening-/fining-upward character of the succession is thus generally interpreted as reflecting an initial phase of major fluvial progradation followed by a later phase of fluvial retreat.

On the basis of petrophysical parameters the succession in the field has been sub-divided into four stratigraphic zones (1-4) arranged in upward ascending order (Fig. 4). As will be discussed later, these zones tend to correlate to changes in lithologic composition and inferred depositional sub-environments within the Ivishak succession.

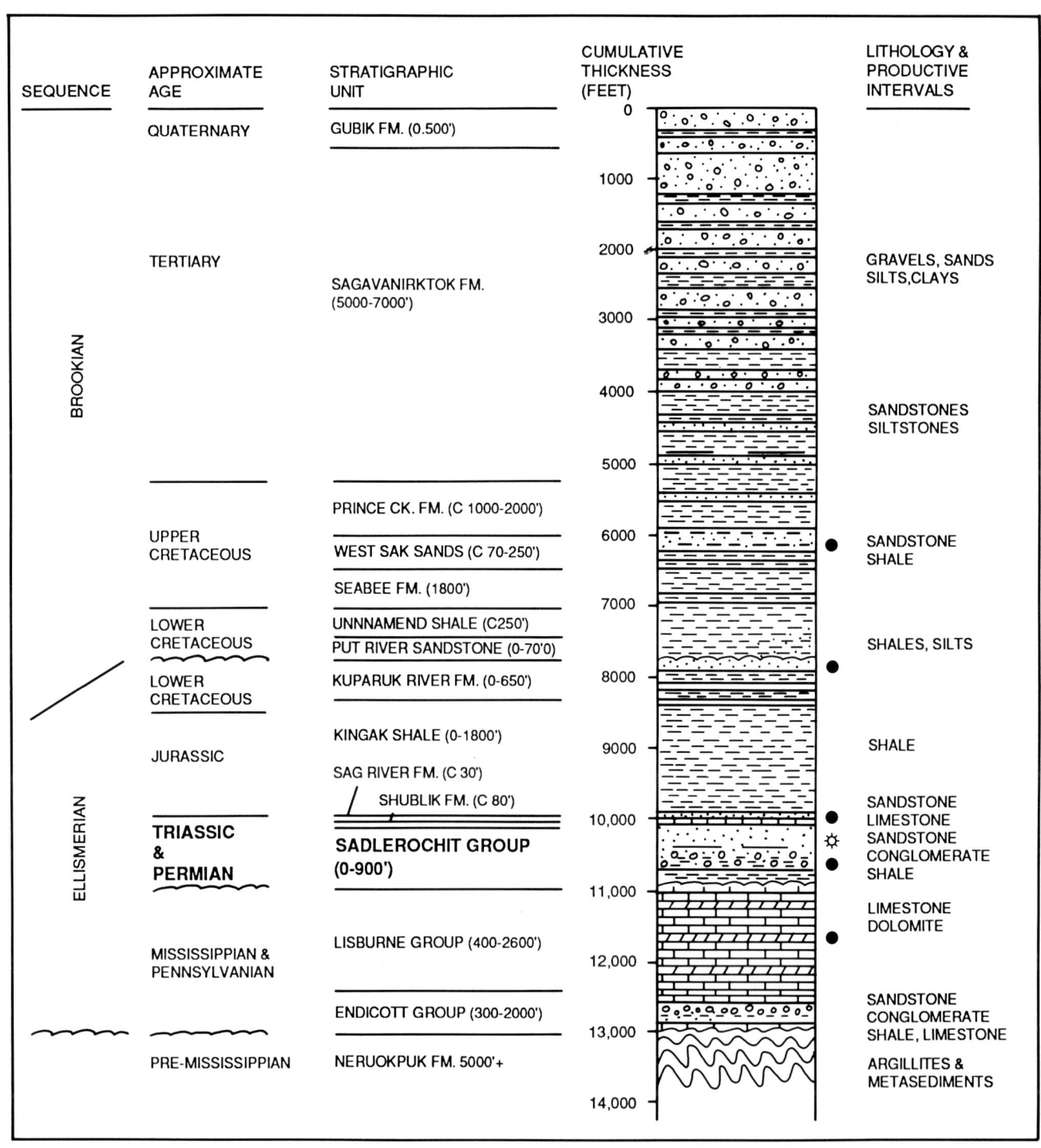

Figure 3 Stratigraphic succession of the Prudhoe Bay area indicating the position of the Permo-Triassic Sadlerochit Group and main hydrocarbon bearing intervals.

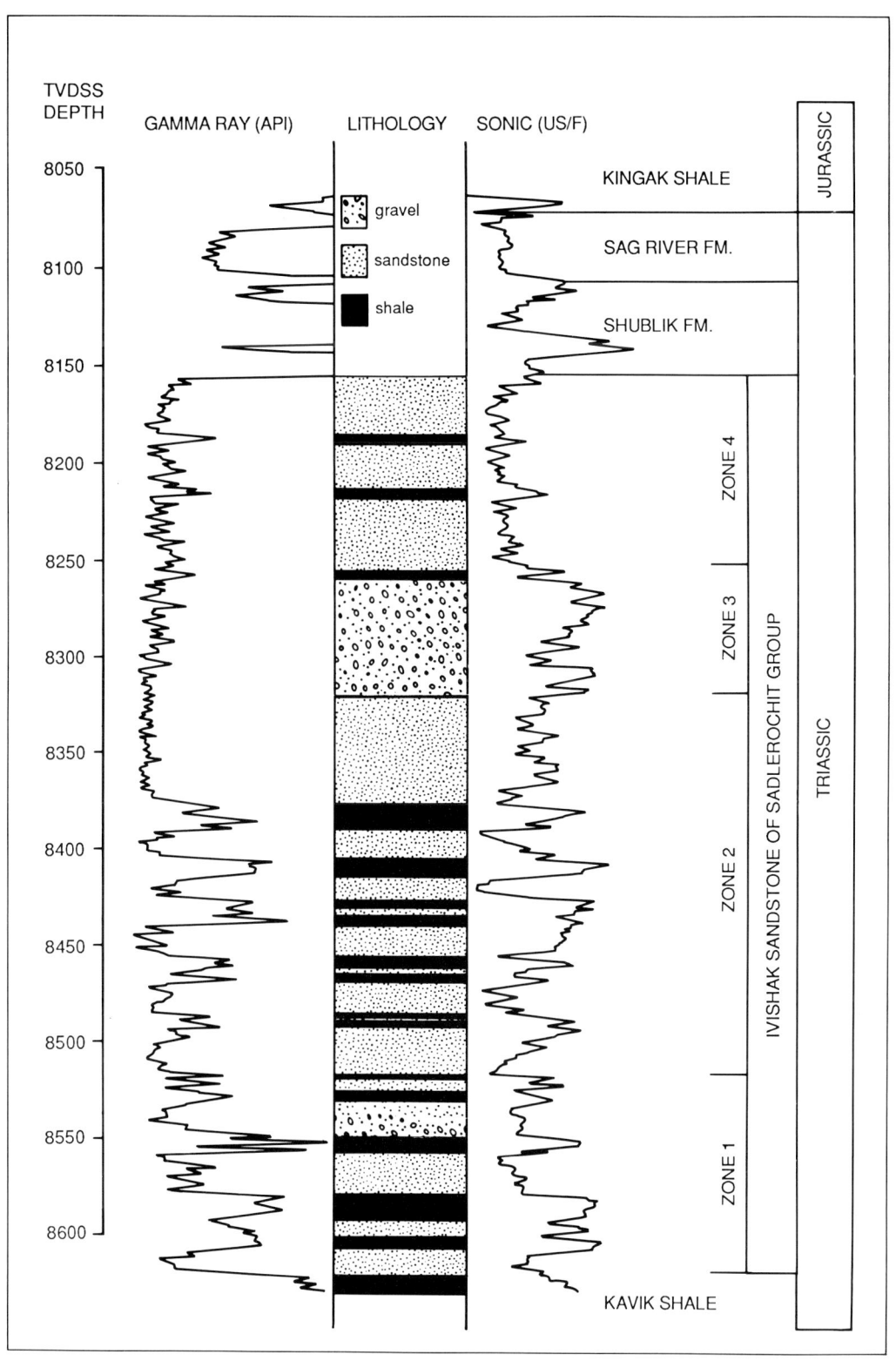

Figure 4 Typical log response and lithology of the Ivishak Sandstone reservoir as seen in the ARCO-Exxon Prudhoe Bay State #1 discovery well. Also indicated is the petrophysically-based zonation of the reservoir (modified after Jamison, Brockett, and McIntosh, 1980).

FACIES AND DEPOSITIONAL ENVIRONMENTS

Continuous cores throughout the Ivishak interval from 21 wells in the field form the main data base for this study. Nineteen sedimentological facies* have been identified and Table I summarizes the salient characteristics of each facies type. A modified form of the facies description/abbreviation code developed by Miall (1977) is employed. The various facies are seen to be grouped into distinctive vertical sequences or "facies associations,"* each of which represents deposition in a particular type of sub-environment. Facies associations belonging to three main types of environment have been recognized: "subaerial, fluvially-dominated coastal plain," "transitional fluvio-marine, deltaic," and "marine-reworked, transgressive shelf." Table II provides a summary description of each facies association together with information concerning its well log response and reservoir properties.

Subaerial, Fluvially-Dominated Coastal Plain

The subaerial coastal plain represents a complex environment comprising multiple facies associations each possessing a complex, intergradational suite of grain sizes and stratification types. Both fluvial channel-fill and marginal overbank floodplain deposits are recognized. On the basis of grain size and maximum clast diameter, the various channel-fills can be sub-divided into those of either the mid- or distal coastal plain. Tentatively these "end-member" classes can be further sub-divided into facies associations representing various segments of that particular setting. This transitional classification scheme allows one to make the distinction between slightly coarser (better reservoir quality) or finer (poorer reservoir quality) grained units within the same overall coastal plain setting. Facies associations of lateral overbank origin are dominated by fine-grained

---

* The term facies is used in the sense of Reading (1978) and Walker (1979) to define "a sedimentary rock of similar color, bedding, composition, texture, fossils and sedimentary structures." A facies association is thus "a group of facies that occur together and are considered to be genetically related." Facies thus provide the building blocks from which facies associations are constructed. Facies associations generally reflect a depositional setting or environment and as such usually form genetic reservoir bodies.deposits and comprise two main groups distinguished by the presence or absence of paleosol (ancient soil) modification.

Table I: Ivishak Sandstone Depositional Facies (facies code modified from Miall, 1977)

Lithology	Facies	Description	Sedimentary Structures	Depositional Process
Gravel (G)	Massive, gravel (Gm)	Poor-moderately sorted, granule to pebble-grade gravel with less than 30% sand/silt. Dominant chert extraclasts, subordinate shale intraclasts. Maximum diameter at least 90 mm.	Massive to crudely bedded. Weak imbrication, and occasional spaced stratification.	High energy. Traction current or wave winnowed lag.
Gravel-Sand (GS)	Massive, gravel-sand (GSm)	Poor-moderately sorted fine to granule grade sands containing 5-70% pebbles of both extra- (chert) and intraformational origin. Maximum clast diameter usually less than 90 mm.	Massive to crudely bedded. Occasional imbrication.	High energy, traction current deposit.
Gravel-Sand (GS)	Cross-stratified, gravel-sand (GSx)	Poor-moderately sorted, fine to granule grade sands containing 5-50% pebbles of both extra-and intraformational origin. Maximum clast diameter usually less than 50 mm.	Trough and planar cross-stratification (undifferentiated). Foreset angle dip usually >10°.	Moderate-high energy, traction current deposit
Gravel-Sand (GS)	Horizontal to low angle stratified, gravel-sand (GSl)	Poor-moderately sorted, fine to granule grade sands containing 5-50% pebbles of both extra- and intraformational origin. Maximum clast diameter usually less than 50 mm.	Horizontal to low-angle stratification, frequently spaced.	Moderate to high energy, traction current deposit.
Sand (S)	Massive, sand (Sm)	Moderately sorted very fine to coarse grained sand. Generally non-pebbly.	Massive (partly reflects uniform grain size, core condition, etc.).	Low-high energy, traction current, rapidly deposited, liquified (?) sands - originally Sl or Sx ?
Sand (S)	Cross-stratified, sand (Sx)	Moderately sorted very fine to coarse grained sand. Generally non-pebbly.	High angle (>10°) cross-stratification. Trough or planar form indistinguishable.	Low-moderate energy, traction current deposit.
Sand (S)	Horizontal to low angle laminated, sand (Sl)	Moderate to well sorted very fine to medium grained sand. Non-pebbly.	Horizontal to low angle laminations (mm-cm scale). In zone 1 of succession laminations highlighted by carbonaceous debris.	Low-high energy, traction current/wave deposit.
Sand (S)	Ripple cross-laminated, sand (Sr)	Moderate to well sorted very fine to medium grained sand. Non-pebbly and frequently draped by finer grained argillaceous material.	Both asymmetric and symmetric ripple laminations recognized.	Low energy, traction current/wave deposit.
Sand (S)	Deformed, sand (Sd)	Poor-moderately sorted, very fine to coarse grained sands. Occasional floating pebbles and disrupted mud drapes.	Disturbed bedding, load and flame structures.	Post depositional sediment compaction, dewatering and/or slumping.

Table I: Ivishak Sandstone Depositional Facies (facies code modified from Miall, 1977)

Lithology	Facies	Description	Sedimentary Structures	Depositional Process
Sand (S)	Bioturbated, sand (Sb)	Moderate-well sorted, very-fine to medium grained sands. Non-pebbly.	Burrow structures, Planolites sp. Teichichnus sp. Chondrites sp.	Low energy deposit, sediment reworking.
Mud-Sand (MS)	Horizontal laminated/bedded, mud stone (MSl)	Interbedded (cm) interlaminated (mm) sand and clay/silt. Generally sands are fine grained or less and display micro-fining upward trend.	Horizontal, spaced bedding/laminae.	Alternating moderate and low energy, traction current/wave deposit.
Mud-Sand (MS)	Ripple cross-laminated, mud-sand (MSr)	Interbedded (cm)/interlaminated (mm) sand and clay/silt. Generally sands are fine-grained or less.	Ripple cross-laminated, both asymmetric and symmetric forms recognized.	Low energy, traction current/wave deposit.
Mud-Sand (MS)	Massive (bioturbated, rootletted), mud-sand (MSm)	Massive, churned mixture of very fine to fine sands and clay-silt. Occasionally color mottled (grey-brown-red).	Massive, except for occasional burrow/rootlet structures.	Reworked, low energy, traction current/wave deposit (soil formation?).
Mud-Sand (MS)	Deformed, mud-sand (MSd)	Poor-moderately sorted, interbedded (cm)/interlaminated sand and clay silt.	Disturbed bedding, micro-faulting, load and frame structures.	Post depositional sediment compaction, dewatering and/or slumping.
Mud (M)	Laminated, mud (Ml)	Finely interlaminated (mm) silt and clay. Occasionally characterized by macerated carbonaceous debris.	Weak, parallel to low angle laminations.	Low energy, suspension deposit of sub-aqueous origin.
Mud (M)	Rippled, mud (Mr)	Finely laminated (mm) silt and clay. Occasionally, characterized by macerated carbonaceous debris.	Small-scale (cm), asymmetric and symmetric, ripple cross-lamination.	Low energy, traction current/wave deposit.
Mud (M)	Massive-unmottled (bioturbated), mud (Mb)	Homogeneous mixture of clay-silt, variable colors ranging from dark grey to light grey/green. No visible color mottling.	Massive, except for rare burrow trace.	Low energy, suspension deposit. Sub-aqueous deposition suggested.
Mud (M)	Massive-mottled (rootletted) mud (Mp)	Homogeneous mixture of clay-silt, displaying diagnostic color mottling (grey-brown-red).	Massive, except for color mottled rootlet traces of pedogenic origin.	Low energy, suspension deposit of floodplain origin.
Mud (M)	Deformed, mud (Md)	Complex mixture of churned silt and clay.	Disturbed laminations, load and flame structures.	Post-depositional sediment compaction, dewatering and/or slumping.

Table II. Facies Associations - Description, Interpretation, Well Log Response and Reservoir Properties

Environment		Facies Associations	Dominant Facies Types	Core Characteristics and Interpretation	Well-Log Response	Reservoir Properties		
						θ (range)	Kh (range)	Permeability Profiles
Sub-Aerial, Fluvially Dominated Coastal Plain		I. UPPER MID-COASTAL PLAIN	Gm, GSm, GSx, GSl	Gravel content >50%. Max. clast size >50 mm. No apparent vertical grain size or facies trends. Interpreted as the deposits of gravelly braid bars and sandy side-channel fills.	Blocky, uniform gamma-ray. Relatively large separation between overlaid density and sonic logs.	12-21%	200-5000 mD*	Overall high, but irregular. Tendency to form isolate zones of very high K (>2000 mD) surrounded by lower K (500 mD).
		II. MIDDLE TO LOWER MID-COASTAL PLAIN	Gm, GSm, GSx, GSl	Gravel content >50%. Max. clast size <50 mm. No constant vertical grain size/facies trends - however tendency for abrupt fining from gravel to sandstone in some beds. Interpretation as I.	Blocky, uniform gamma-ray. Separation between overlaid density and sonic logs (not as great as in I).	15-25%	200-3000 mD*	Overall high but irregular. Similar tendency to that seen in I to isolate higher K zones.
		III. LOWER MID- TO UPPER DISTAL COASTAL PLAIN	Sx, Sm, GSm	Approximately 80% F-VC sandstone, 20% gravel. Occasional siltstone beds. Max. clast size <40 mm >5 mm. Poorly developed FUS 5-20 ft (1.5 - 6 m) thick. Interpreted as the deposits of sandy bars and interbar gravelly channels. Either braided/meandering pattern.	Blocky, uniform gamma-ray, if siltstones present log becomes more serrated (rare). Little separation between overlaid density and sonic logs.	24-30%	200-1500 mD	Vertical decreasing K profiles rare. Uniform to irregular trends more common.
		IV. UPPER TO MIDDLE DISTAL COASTAL PLAINS	Sx, Sl, Sm	F-C sandstone content >90%. Weakly developed FUS 5-15 ft (1.5-4.5 m) thick. Gravel restricted to channel lag, vertical facies trend GSm-Sx-Sm-Sl. Interpretation as III.	Blocky, weakly serrated gamma-ray. No separation between overlaid density and sonic logs.	22-27%	100-1000 mD	Vertically decreasing K profiles expected though not pronounced. K range from 500-1000 mD at channel base to 100-500 mD at top.
		V. MIDDLE TO LOWER DISTAL COASTAL PLAIN	Sx, Sm, Sl	Predominantly F-M sandstone, gravel <2% (intraformational). Well developed FUS 2-15 ft (0.6-2.5 m) thick. Vertical facies trend Sx/Sm-Sl-Sr. Interpreted as the deposits of sandy braided/meandering streams.	Serrated gamma-ray, occasional bell-shapes seen. Highly serrated density/sonic, no separation.	20-25%	50-700 mD	Well developed vertically decreasing K profiles. Channel base, K = 300-700 mD, channel top K = 50-100 mD.
		VI. ABANDONED CHANNEL/FLOOD PLAIN	Sm, Sl, Sr, MSm, MSl, MSx	Interbedded VF-F sandstones and siltstones. Occur at top of FU channel sequences -overlain by floodplain channel abandonment deposits	Sharp bell-shape gamma-ray response above blocky channel fill. Density log displays better developed bell shape.	Sand poor θ range K range		Sand-rich θ range 5-15% K range 1-50 mD
		VII. FLOODPLAIN/POND	Sr, Sm, Ml, Md, Mp	Sand-rich (up to 70% VF sandstone) and sand-poor (<20% VF sandstone). FUS and CUS common, both often overprinted by pedogenic mottling. Interpreted as crevasse splay, pond, floodplain deposits. Often contain gley, pseudo-gley ancient soil profiles.	High gamma-ray values, both bell-and funnel-shapes common. Similar shapes typify density and sonic logs.	<5%	<1 mD	Associations possess poor reservoir properties - mainly seal lithologies

FUS Fining-upward sequence
CUS Coarsening-upward sequence

*Poor core recovery, may be higher values.

Table II. Facies Associations - Description, Interpretation, Well Log Response and Reservoir Properties

Environment	Facies Associations	Dominant Facies Types	Core Characteristics and Interpretation	Well-Log Response	Reservoir Properties		
					θ (range)	K_h (range)	Permeability Profiles
Transitional, Fluvio - Marine Deltaic	VIII. DISTAL COASTAL PLAIN/RIVER MOUTH	Sl, Sx	VF-M sandstone (up to 95%), siltstone and gravel (<5%). Carbonaceous debris common. Organized into FUS 2-15 ft (0.6-4.5 m) thick, intraformational gravel at base, facies trend Sx-Sm-Sl/Sr. Interpreted as representing distal fluvial-river mouth deposits.	Weakly developed, blocky to funnel-shaped serrated gamma-ray, density and sonic log response.	20-29%	100-1500 mD	Very irregular and grain size dependent. Strong tendency to isolate high K zones (500-1500 mD) surrounded by lower K (100 mD).
	IX. CHANNEL MOUTH BAR (DELTA FRONT)	Sm, Sl, Sx, Sd	Up to 100% VF-M sandstone, minor shale rare intraformational lags. Well developed CUS. Carbonaceous debris common. Soft sediment deformation features seen. Interpreted as CU mouth-bar deposits.	Well developed, blocky to funnel-shaped serrated gamma-ray, density and sonic log response.	15-25%	10-500 mD	Strong tendency to develop vertically increasing K profiles from K = 10-50 mD at base to K = 100-500 mD at top.
	X. DELTA FRONT	Sr, Sl, Ml, Md, Mb	VF-F sandstone (up to 25%), shale (up to 75%). Well developed CUS, 30-70 ft (9-21 m) thick. Carbonaceous debris very common. Interpreted as sediments accumulating largely from suspension in a delta front setting.	Serrated gamma-ray, density and sonic log response	5-20%	<100 mD	Overall values low (<100 mD). Tendency to develop vertically increasing profiles from K = <1 mD at base to K = 50-100 mD at top.
Marine Reworked	XI. TRANSGRESSIVE LAG/MARINE EROSION SURFACE	GSm, GSl, Sm, Sb	Thin interval of F-C sandstones and gravels. Characterized by erosive base and presence of glauconite, phosphate and diagenetic pyrite nodules. Interpreted as a transgressive lag deposit.	Where glauconite percentages are high see very high gamma-ray values ("shale" response). Sonic/density logs often affected by pyrite presence.	15-25%	100-800 mD	Thin and variable high K zone associated with transgressive surface.

FUS Fining-upward sequence
CUS Coarsening-upward sequence

*Poor core recovery, may be higher values.

The seven facies associations assigned to the subaerial fluvially dominated coastal plain will be discussed in order of decreasing coarse clastic content and not in any implied stratigraphic sense.

Upper Mid-Coastal Plain Association (I; Fig. 5)

This association comprises an interbedded mixture of gravel and sandstone lithologies in which gravel content is dominant and usually exceeds 50% of the total rock volume (Figs. 5 & 6). The gravel is poorly sorted, sandy, granule to cobble grade with maximum clast diameters generally >50 mm. Dominant clast types include white ("tripolitic"), gray and black chert, vein and red quartz and quartzite (Fig. 7a). Gravel facies Gm (Fig. 7b), GSm, GSx and GSl (Table I) predominate. The average thickness of foreset beds is approximately 1.5 ft (0.4 m). The sandstone is poorly sorted, gravelly and fine- to coarse-grained (Fig. 5). Sandstone facies include GSx (Fig. 7c), GSl (Fig. 7d), Sx, Sl and Sm. The average thickness of the foreset beds ranges from 0.75 to 1.4 ft (0.2 to 0.4 m). In the field there is a decrease in overall gravel clast size and bed thickness within this association from north to south. No apparent vertical grain size or facies trends characterize the beds (Fig. 5).

The association is interpreted to represent the deposits of migrating shallow, longitudinal gravel bars and sandy side channel fills within restricted, and probably partially incised, upper mid-coastal plain braided streams.

Middle to Lower Mid-Coastal Plain Association (II; Fig. 5)

This association possesses a similar gravel:sandstone ratio to the previous one but is characterized by somewhat smaller clast diameters, generally <50 mm but >30mm (Fig. 5). The gravel is poorly sorted, sandy and granule to pebble grade. Clast composition is identical to that described in the previous association. Gravel facies consist primarily of Gm, GSm, GSx (Fig. 8a) and GSl. Foreset height averages 2.4 ft (0.6 m). The sandstone is moderately sorted, and medium- to coarse-grained (Fig. 5). Primary sandstone facies include GSx, GSm, Sx and Sm. The average thickness of the foreset strata ranges from 0.75 to 1.5 ft (0.2 to 0.4 m). Across the field the thickness of both the gravel and sandstone beds decreases to the south but overall gravel clast size exhibits no variation (McGowen, pers.

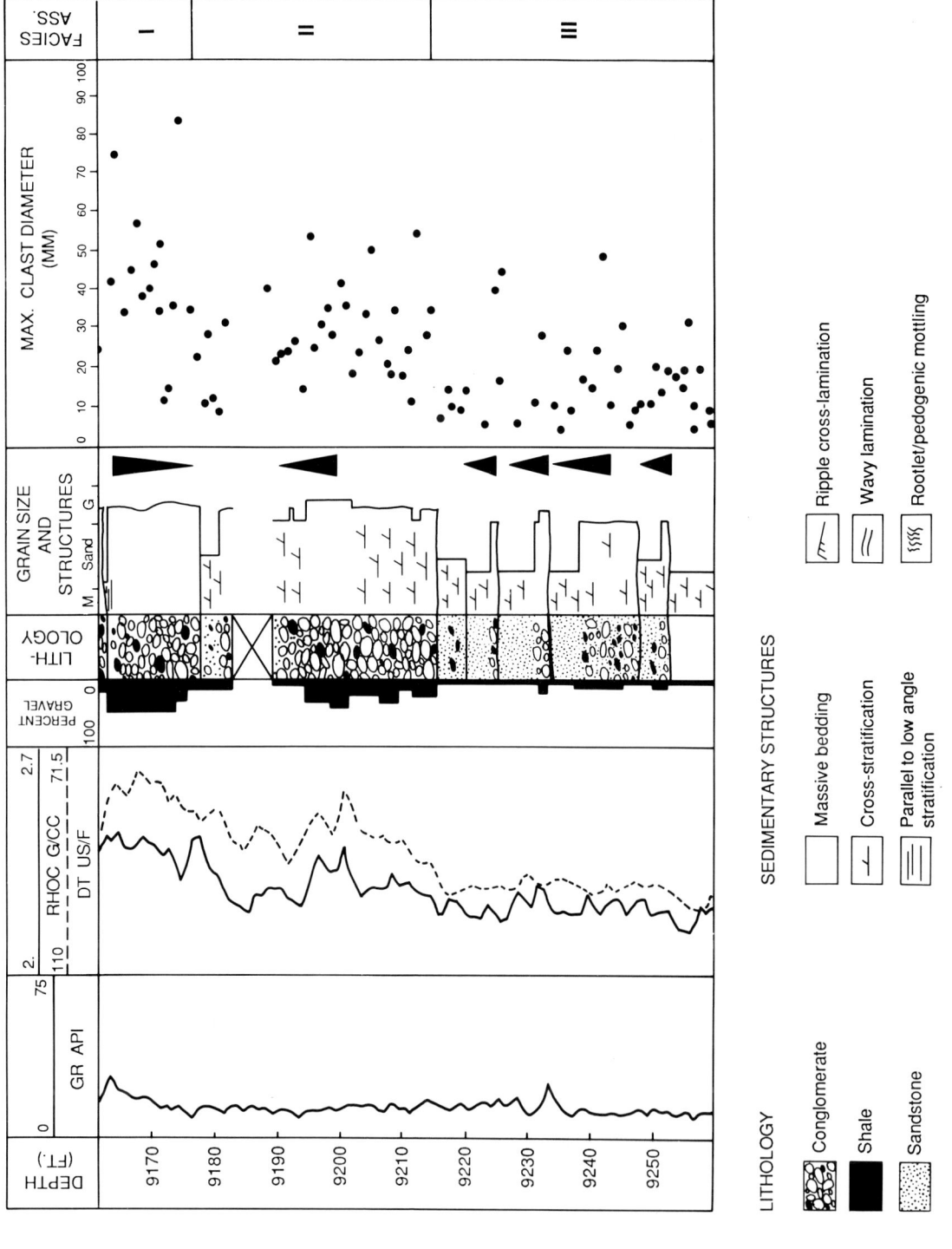

Figure 5. Sedimentological characteristics and log response of (I) upper mid-coastal plain, (II) middle to lower mid-coastal plain, and, (III) lower mid- to upper distal coastal plain facies associations.

Figure 6   Core representing facies associations I, upper mid-coastal plain (right, 3 boxes) and II, middle to lower mid-coastal plain (left, 3 boxes). Newspaper indicates missing core and testifies to poor core recovery commonly associated with these conglomerate-rich deposits. Scale in tenths of a foot.

Figure 7   Upper mid-coastal plain association (I).

　　　　　　A. Typical appearance and clast composition of the conglomerates, w = white chert, b = black chert, and g = gray chert.

　　　　　　B. Massive gravel, facies Gm (Table I).

　　　　　　C. Cross-stratified gravel-sand, facies GSx (Table I). Note well developed grading on foreset laminae.

　　　　　　D. Parallel to low angle laminated gravel-sand, facies GSl (Table I).

Figure 8   A and B Middle to lower mid-coastal plain association (II). A. Cross-stratified to massive gravel-sand, facies GSx to GSm (Table I). B. Abrupt grain size reduction and evidence of winnowing (arrow) at the top of a gravelly bed.

　　　　　　C. Lower mid- to upper distal coastal plain association (III). Cross-stratified medium to very coarse grained sandstones, facies Sx (Table I).

　　　　　　D. Upper to middle distal coastal plain association (IV). Gravel forming lag accumulation above basal erosion surface (arrows).

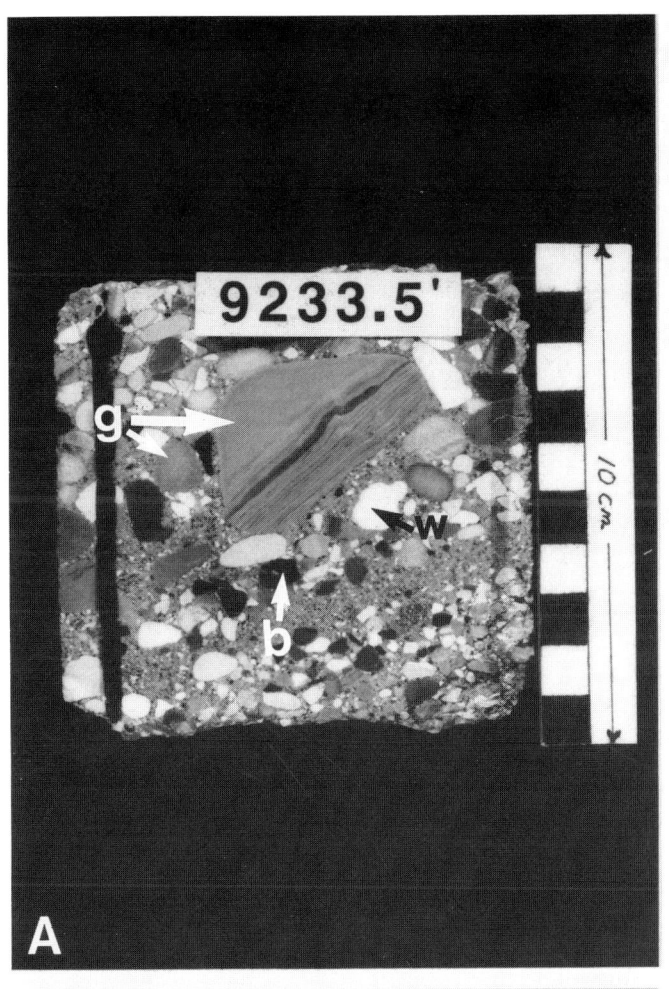

comm.). No repeated vertical grain size or facies trends are seen, although the tops of some of the gravel beds fine abruptly into more sand-rich deposits (Figs. 5 & 8b).

The association is interpreted as representing the deposits of shallow gravel bars and sandy side channel fills within laterally restricted, middle to lower mid-coastal plain braided streams.

Lower Mid- to Upper Distal Coastal Plain Association (III; Fig. 5)

This association marks the change from predominantly gravel-rich to gravel-poor deposits (Figs. 5 & 9). It comprises approximately 80% sandstone and 20% gravel. Maximum clast diameter in the gravel range from 5 mm to 40 mm. The sandstones are mostly moderately sorted, dominantly medium-grained but can range from fine- to very coarse-grained. Sandstone facies, in order of decreasing abundance, are Sx (Fig. 8c), Sm and Sl. Foreset thickness ranges from 0.7 to 2.2 ft (0.2 to 0.6 m). The subordinate gravels are poorly sorted, sandy and predominantly granule in grade. Clast types are identical to those described previously but usually include more significant amounts of intraformational shale "rip-up" clasts. The main gravel facies is GSm followed by GSx and GSl. Foreset beds average 1.1 feet (0.35 m) in thickness. Generally the sandstone within this association is coarser in the north, contains more pebble size clasts and is thicker bedded than to the south. The sandstones and gravels exist in fining-upward sequences ranging in thickness from 5 to 20 ft (1.5 to 6 m). Each sequence comprises a basal interval of gravel overlain by a succession of gravel-sands which become more sand-rich upwards (Fig. 5). The size of the foreset beds within each sequence shows a systematic vertical decrease from base to top (Fig. 5). Mudstone and siltstone interbeds (Fig. 10) are minor components of the association and occur in massive to thinly laminated units 0.1 to 0.2 ft (0.03 to 0.06 m) in thickness.

The association is interpreted to represent the deposits of finer-grained, more laterally extensive, lower mid- to upper distal coastal plain river channels. Sediment accumulation within the channels occurred via two mechanisms; (i) aggradation and migration of transverse, low amplitude, sandy bedforms (linguoid bars ?), and (ii) deposition of coarse-grained lag deposits at the base of interbar

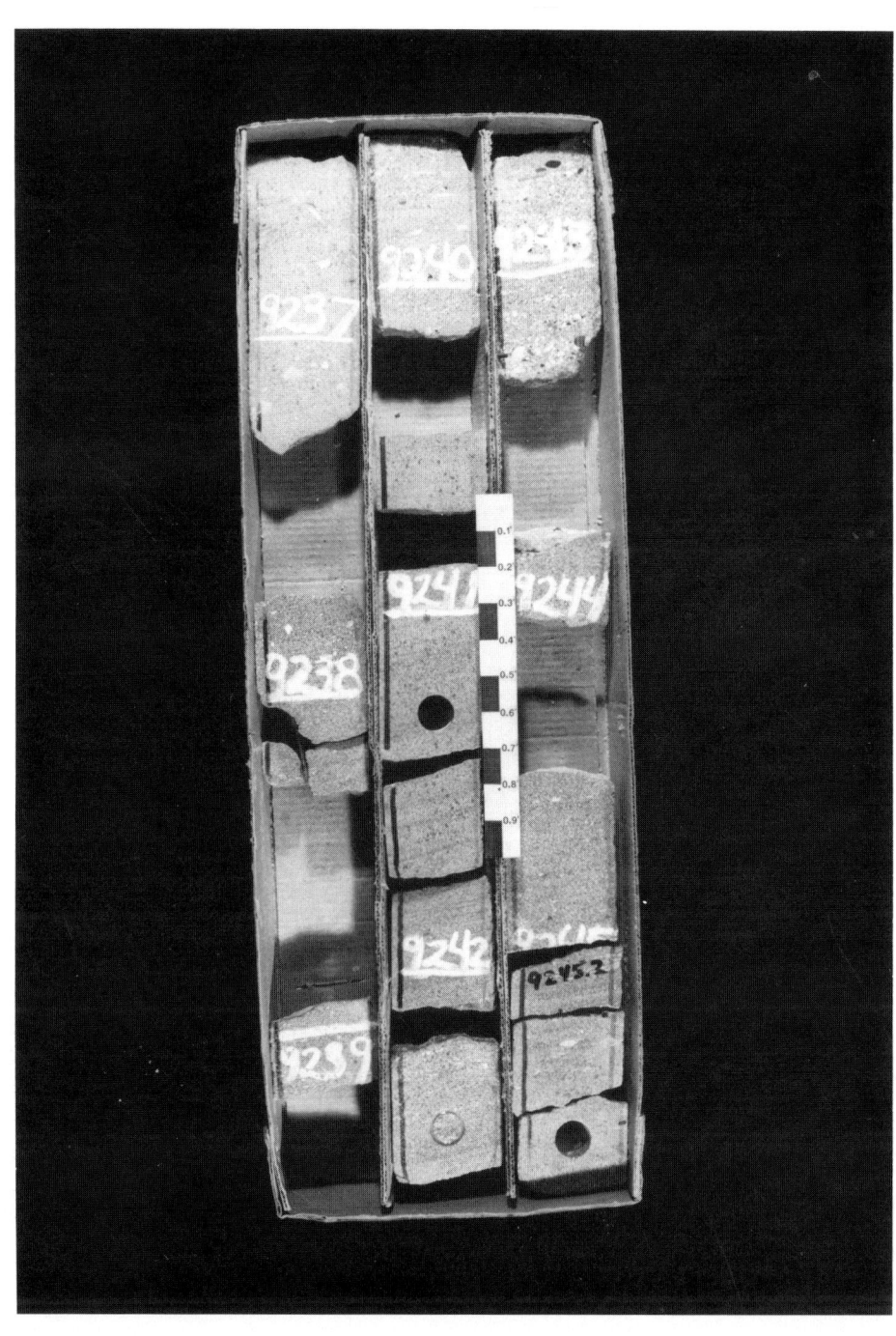

Figure 9 Core representing facies association (III), lower mid- to upper distal coastal plain. Note general absence of gravel and domination by coarse-grained sandstones.   Scale in tenths of a foot.

Figure 10 Interval from 9092' to 9103' represents the lower mid- to upper distal coastal plain association (III). Arrows indicate presence of thin (< 6 in. thick) mud/siltstone drapes. Scale in tenths of a foot.

channels. Given the available data, paleochannel pattern is difficult to assess and may have been braided, meandering or a combination of both.

Upper to Middle Distal Coastal Plain Association (IV; Fig. 12)

This association consists of predominantly sandstone with minor amounts of gravel (usually < 10%). The sandstones are moderately sorted, fine to occasionally coarse-grained and often occur in well developed fining-upward sequences ranging in thickness from 5 to 15 ft (1.5 to 4.5 m). Primary sandstone facies, in decreasing order of abundance are; Sx, Sl and Sm. Foreset bedding in the sandstones averages between 0.5 to 1 ft (0.15 to 0.3 m) in thickness. The gravel occurs at the base of the fining-upward sequences where it forms a lag accumulation above a basal erosion surface (Fig. 8d). It is poorly sorted, sandy and ranges from granule to pebble grade. Maximum clast diameter is less than 10 mm and clast composition is identical to that described from the previous association. The dominant gravelly facies are GSx and GSm. Above the gravel lag the sandstones record an upward decrease in grain size and change in facies character from Sx to Sm and eventually to Sl. The average grain size of the association and the thickness of foreset bedding in the sandstones shows a systematic decrease from north to south across the field.

The association is interpreted to represent the deposits of sand-rich, laterally extensive, upper to middle distal coastal plain rivers. The main sedimentation in these streams took place on sandy, mid-channel, transverse/linguoid bars or lateral point bars. Winnowing and accumulation of coarser gravel debris was restricted to the inter-bar channels where more persistent flows operated even at times of reduced discharge. Given the available data, paleochannel pattern is difficult to assess and may have been braided, meandering or a combination of both.

Middle to Lower Distal Coastal Plain Association (V; Fig. 12)

This association is composed predominantly of sandstone (Figs.11 &12) with only a minor amount of patchy gravel (usually < 2%). Distinct differences typify sandstones of this association in the north and south of the field. In the north the sandstones tend to be, at best, only moderately sorted and range from fine- to coarse-grained. Dominant sandstone facies, in decreasing order of abundance, are

Figure 11 Core representing facies association (V), middle to lower distal coastal plain. Parallel laminations (facies Sl) and cross-stratification (Sx) are highlighted by differential oil staining. Scale in tenths of a foot.

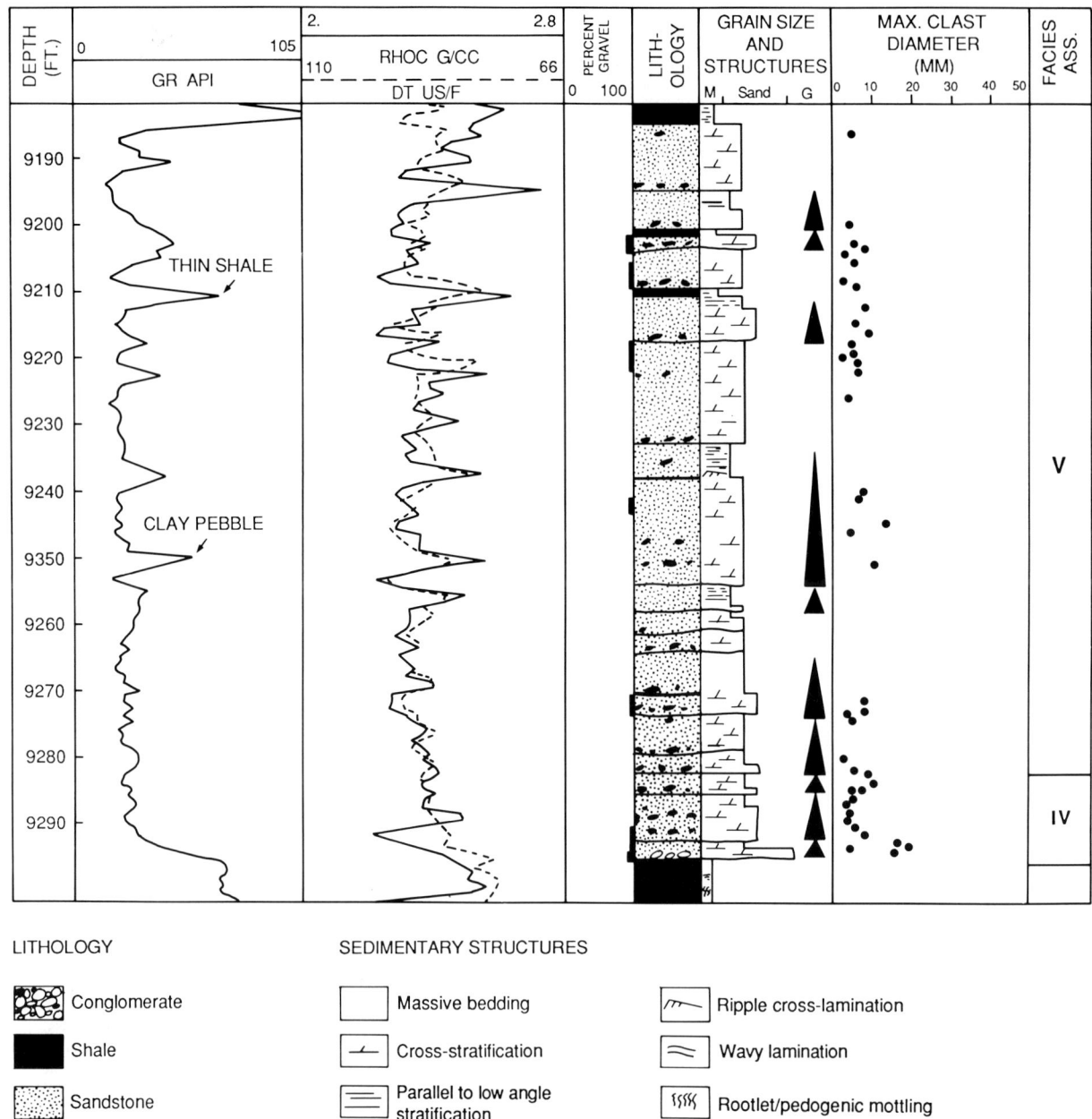

Figure 12   Sedimentological characteristics and log response of facies association (V), middle to lower distal coastal plain. Note difficulty when using gamma ray log response to differentiate between thin shale beds and intraformational shale pebbles.

Figure 13   A and B. Middle to lower distal coastal plain association (V). A. Cross-stratified medium to coarse grained sandstone, facies Sx (Table I). B. Intraformational mudstone pebbles (arrowed) at the base of a channel-fill unit.

C and D. Floodplain/pond association (VII). C. Ripple laminated (weakly climbing) fine sandstone, facies Sr (Table I). D. Vertical transition from sandstone channel top to sand-poor, floodplain mudstones exhibiting weak pedogenic mottling.

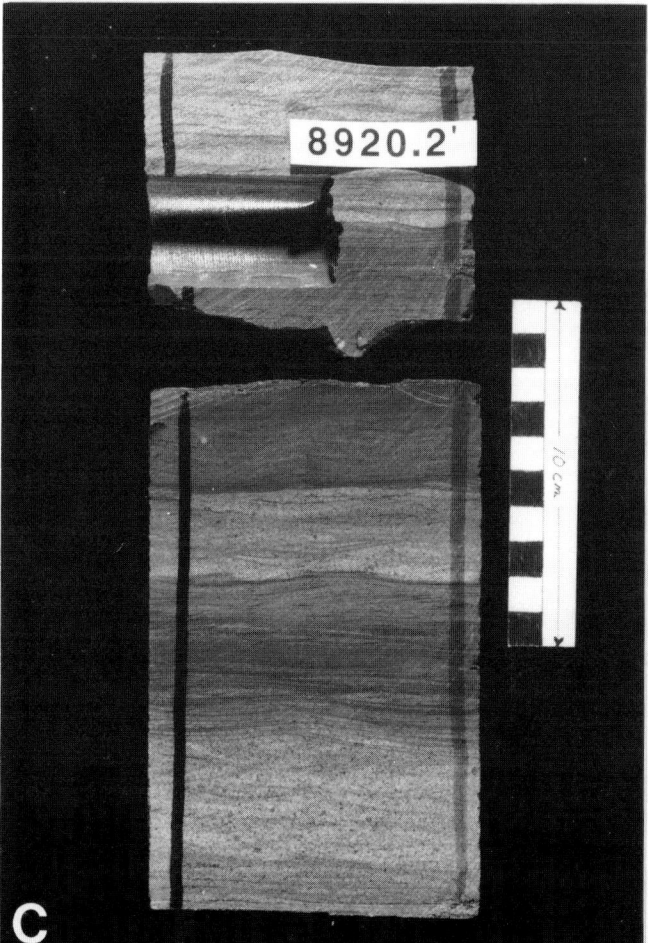
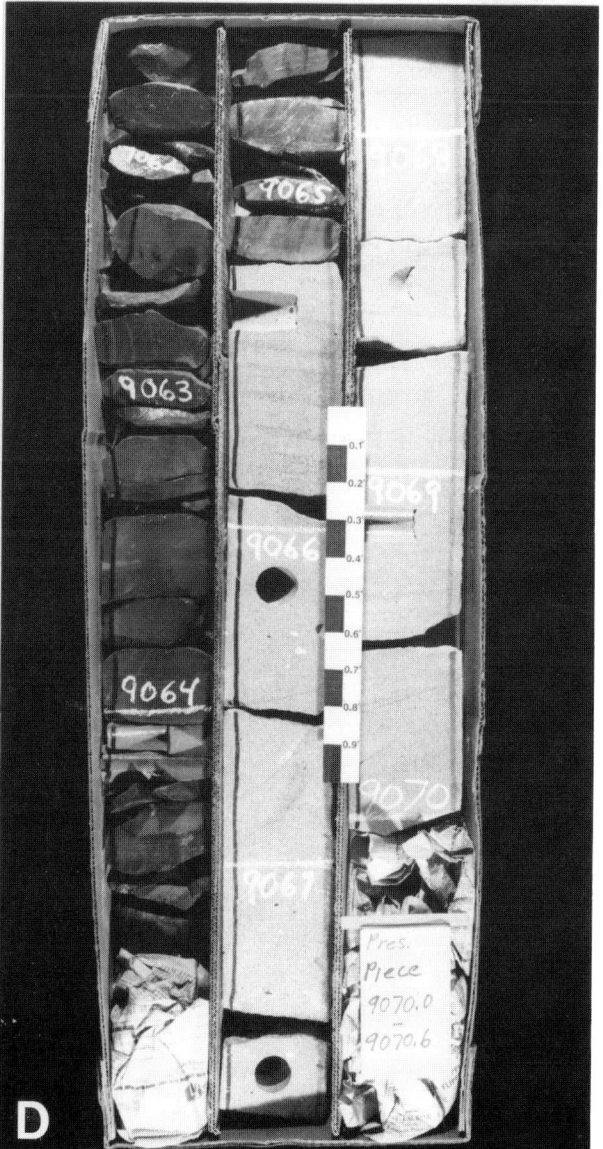

Sx (Fig. 13a), Sm and Sl. Average foreset thickness is approximately 0.8 ft (0.25 m). In the south the sandstones are better sorted (moderate to well) and finer-grained (very fine to coarse, average medium-grained). The major facies, in decreasing order of abundance, are; Sm, Sl, Sx and Sr. Average thickness of foreset beds is the same as in the north.

In both the north and south, but more evident in the south, is the tendency for this association to be organized into larger scale fining-upward sequences ranging from 2 to 15 ft (0.6 to 4.5 m) in thickness (Fig. 12). Each sequence begins with a basal scour overlain by a thin, intra- (shale rip-up clasts) and/or extraformational lag gravel (Figs. 12 and 13b). This lag is succeeded by a sequence of sands whose average grain size decreases upwards. Accompanying the upward-fining is a vertical transition from facies Sx and Sm to facies Sl and Sr. The top of the sequence, where not eroded by the overlying unit, is capped by very fine sandstones and siltstones that are frequently highly convoluted with evidence of loading structures and other traces of soft sediment deformation (Sd, MSd).

The association is interpreted as the deposits of an extensive, sandy distal coastal plain generated by the frequent lateral migration of a series of shallow river channels. The dominant mode of sand accumulation would have been as shallow linguoid or lateral point-bars within the channels. Patchy lag gravels would have been deposited in and along the channel axes and silt accumulation would have occurred out of suspension as flow strengths waned and areas of the channel became abandoned. At times of maximum water discharge local water-tables would have been high and thixotropic deformation of previously deposited sediment would have taken place. Given the available data, paleochannel pattern is difficult to assess and may have been braided, meandering or a combination of both.

Abandoned Channel/Floodplain Association (VI; Fig. 14)

This association, where preserved, occurs at the top of fining-upward sand-gravel sequences in mid- and distal coastal plain settings (Fig. 14). It comprises a rather complex alternation of interbedded very fine to fine sandstone and siltstone which exhibit an upward reduction in overall grain size and increase in clay content. Dominant lithofacies include Sm, Sl, Sr, MSm, MSl and MSr.

The association was deposited under the influence of alternating traction and suspension and is interpreted to represent the final stages in a channel abandonment fill. The uppermost part of this association is frequently gradational into overlying paleosol (ancient soils) mottled floodplain deposits.

Floodplain/Pond Association (VII; Fig. 14)

This association can be divided into two end-members: a sand-rich type and a sand-poor type (Fig. 14). The sand-rich type comprises up to 70% sandstone, approximately 20% siltstone and 10% shale. The sandstones are moderately sorted and are usually very fine grained. They tend to occur in thin (0.5-2 ft [0.15-0.6 m] thick), fining-upward beds with sharp to erosional bases. Internally they are dominated by facies Sr (Fig. 13c), Sm, Sl and Sd. The siltstones exhibit similar types of sedimentary structures and are typified by facies Ml, Md and Mr. Both the siltstones and the shales often possess a distinct color mottling (Fig. 14) attributed to soil formation (Mp). These ancient soil profiles or "paleosols" display a variety of color mottling ranging from drab, box-work patterns of dark brown and grey (Fig. 15a) to irregular-shaped patches of red-brown limonite and hematite set within a pale brown to cream colored matrix (Fig. 15b-d). In some cases the degree of color mottling is so intense that the whole interval may display a characteristic deep red appearance.

The sand-poor type consists of up to 70% siltstone with sandstone percentages rarely exceeding 20% (Fig. 13d). Both the siltstones and the sandstones are similar to those described above, in terms of grain size, lithofacies, and paleosol development. However, the siltstones often show more abundant evidence of burrowing (?) and/or rootlet disturbance.

Both end-members comprising this association were deposited in low physical energy environments as testified by their overall high percentages of fine grained deposits. Such environments would have existed marginal to active stream courses in floodplain settings. The presence of interbedded sandstones within the shales (Fig. 14) suggests periodic influxes by higher energy flows onto the floodplain, presumably as a result of overbank flooding from the adjacent rivers. The types of paleosols so far identified are dominated by hydromorphic profiles of "gley" and "pseudo-gley" affinity - such immature soils are typical of modern day floodplain

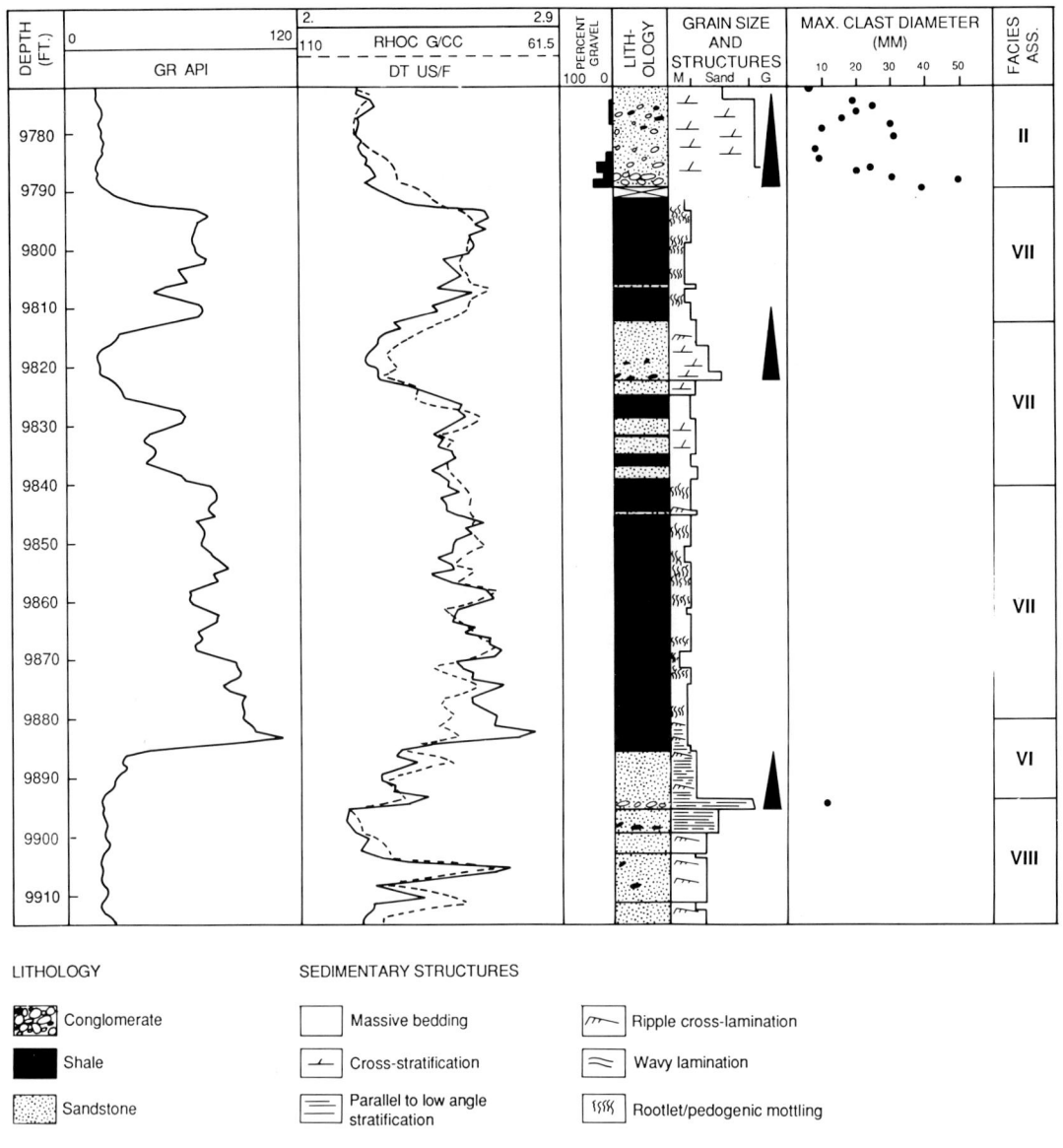

Figure 14 Sedimentological characteristics and log response of (VI) abandoned channel/floodplain and (VII) floodplain/pond facies associations.

Figure 15 Floodplain/pond association (VII). Scale in inches.

    A. Weakly developed paleosol color mottling. Arrows point to gray (reduced) rootlet traces set within a drab, dark brown mudstone matrix.

    B. Better developed paleosol color mottling. Arrow points to red (oxidized) iron concentration at rim of rootlet trace. Center of rootlet is gray (reduced) and matrix color is light to dark brown.

    C. Vertical transition from poor to well developed (arrow) mottling within the same paleosol profile.

    D. Close-up detail of well developed paleosol color mottling, g = gray (reduced) and r = red (oxidized). Note the typical "rubbly" and "pitted" appearance of the core a characteristic of pedogenic modification to mudstones in the Ivishak.

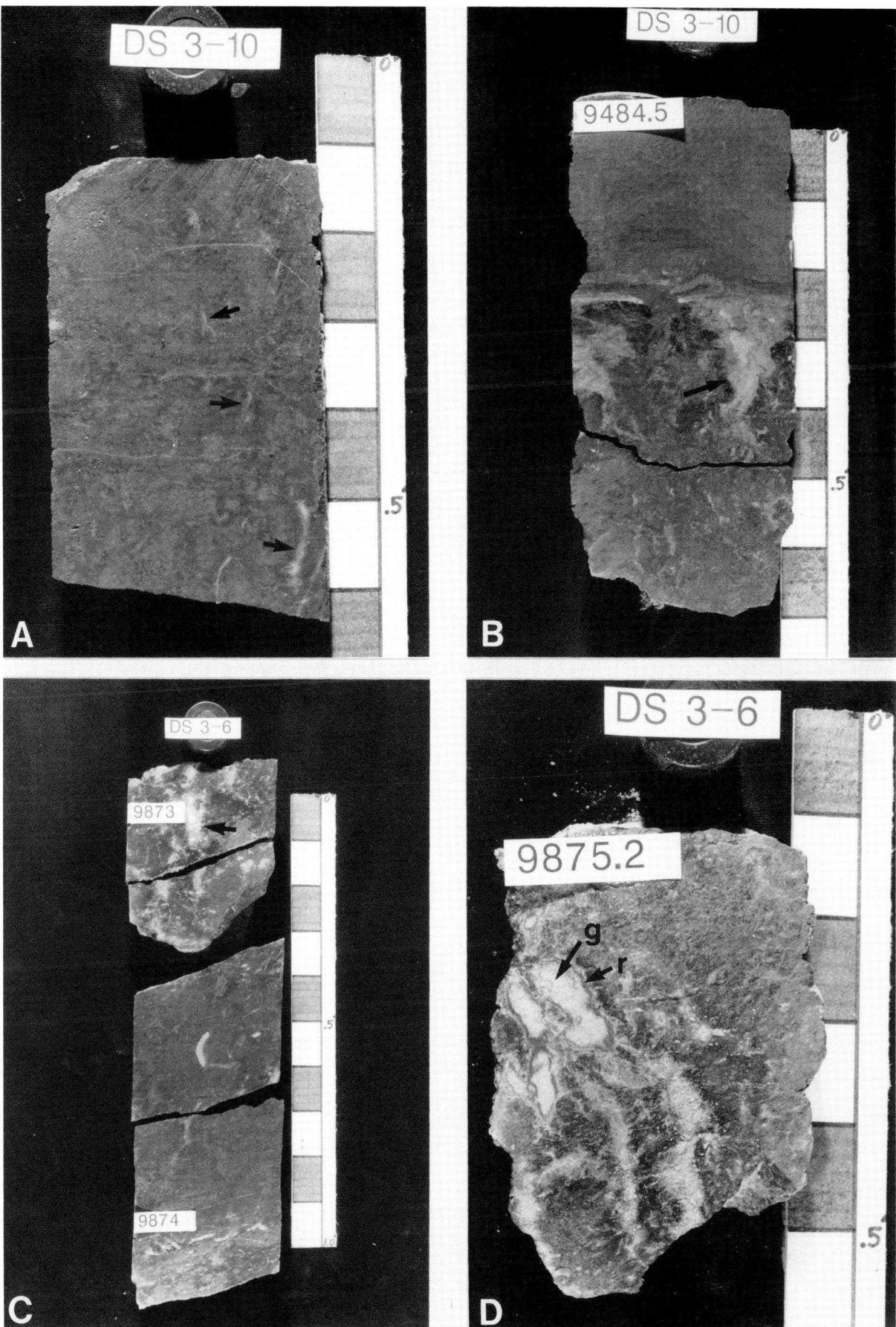

settings where sedimentation rates are relatively high and groundwater levels fluctuate on a seasonal basis (Buurman 1980, Atkinson 1986). The only index of paleosol maturity is color and degree of mottling. More mature soils are characterized by increased levels of mottling and more intense red colors. In some cases the presence of burrow structures (?) and obvious sets of small-scale wave ripples in the siltstones argues for the establishment of shallow ponds and lakes on the floodplain. Such features would most probably have arisen following periods of maximum river discharge when flood waters would have expanded over large areas of the floodplain.

Transitional, Fluvio-Marine Deltaic

The distal portion of the fluvially-dominated coastal plain grades seaward into coastal and nearshore deposits. These deposits accumulated in the area where rivers discharged directly into the sea. They thus exhibit properties inherent to both fluvial and marine processes and three component associations (depending upon the relative importance of fluvial and marine processes during deposition) have been recognized; an alternating distal coastal plain/river-mouth association, a channel mouth-bar association and a delta front association. The three facies associations are presented in a proximal to distal sequence and generally occur in the vertical succession as described.

Distal Coastal Plain/River-Mouth Association (VIII; Fig. 16)

This association (Fig. 16) comprises predominantly sandstone (up to 95%) with subordinate amounts of gravel (< 5%) and siltstone (< 5%). The sandstones are moderately sorted and range from very fine- to medium-grained (dominantly fine-grained). Sandstone facies Sl (Fig. 17a) and Sx predominate, although Sr and Sm also occur in varying degrees. River-mouth sandstones tend to be somewhat finer-grained than the fluvially-derived deposits and are clearly the dominant constituent of the association. They are distinguished from true distal coastal plain sands by the fact that they contain carbonized plant debris. The gravel, a minor component of the overall association, is poorly sorted and consists of thin beds of predominantly intraformational shale clasts (Fig. 17b) which lie above sharp erosive surfaces at the base of sandstone units (GSm). The siltstone occurs as thin interbeds at various levels within the sands. There is a tendency for the

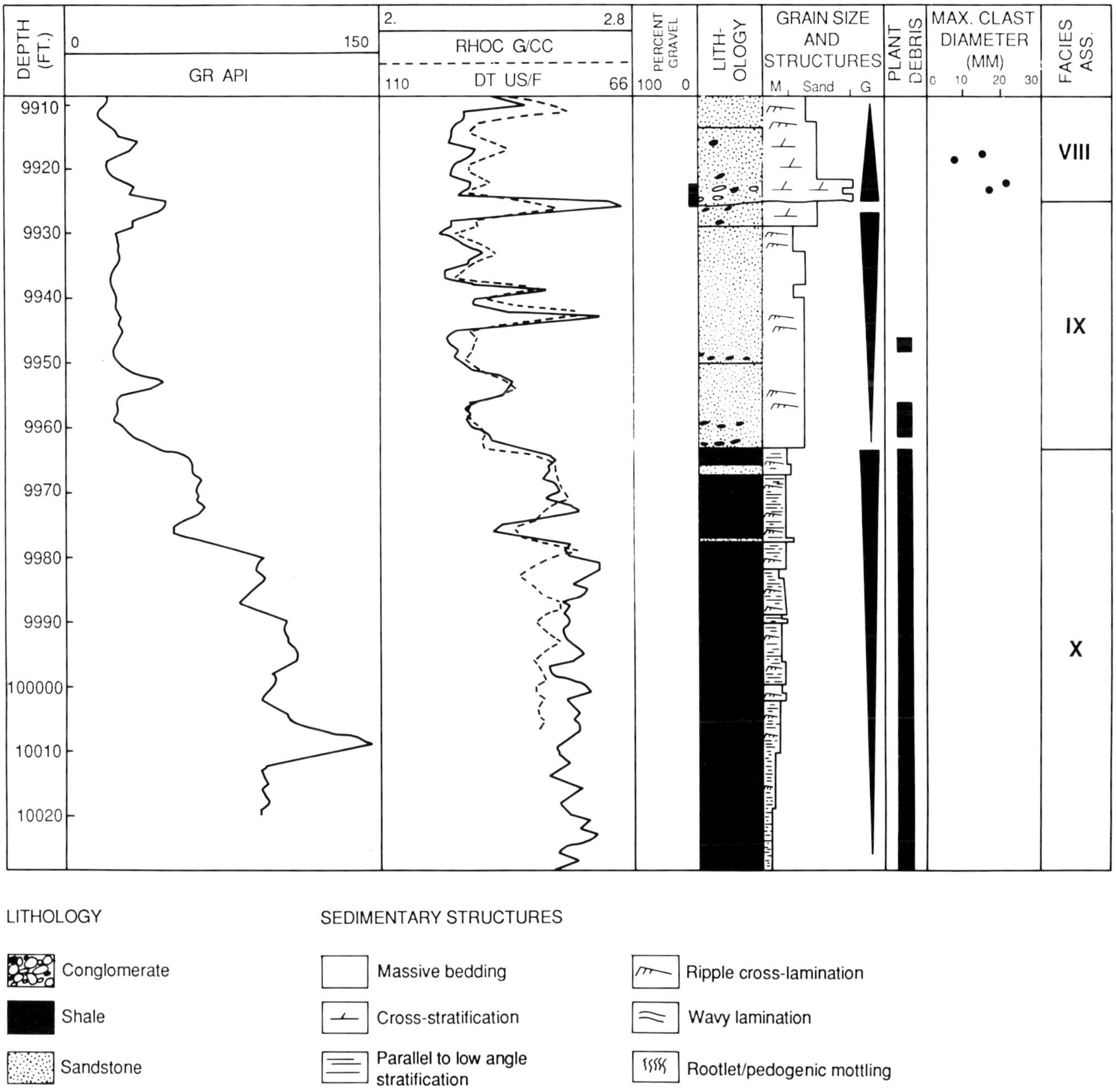

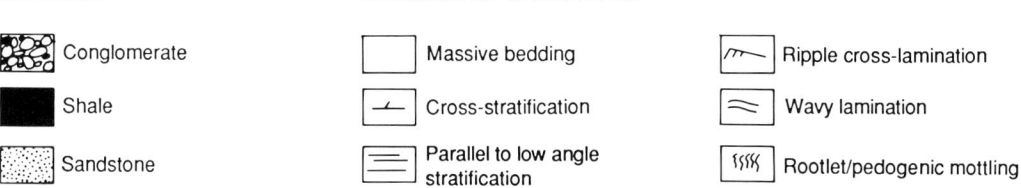

Figure 16 Sedimentological characteristics and log response of (VIII) distal coastal plain/river-mouth, (IX) channel mouthbar (delta front) and (X) delta front facies associations.

Figure 17 A and B. Distal coastal plain/river-mouth association (VIII). A. Scoured, low-angle laminated very fine to fine sandstone, facies Sl (Table I). B. Intraformational mudstone clasts at the base of a sandstone unit. Facies Sl is present above.

C and D. Channel mouthbar (delta front) association (IX). C. Cross-stratified, fine to medium-grained sandstone, facies Sx (Table I). D. Soft sediment deformation and associated microscale faulting (arrow).

591

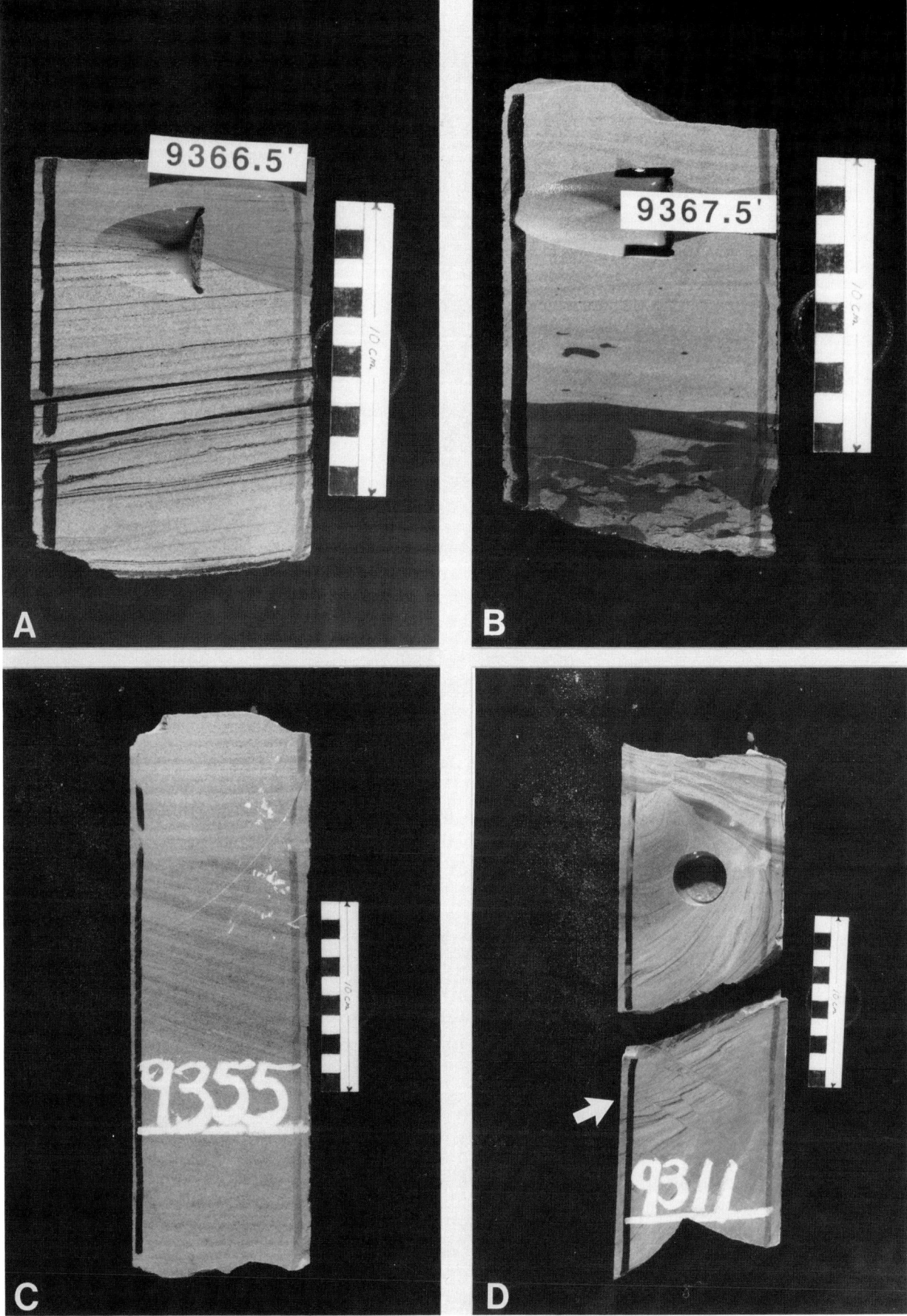

association to be organized into vertical sequences characterized by a thin (< 2 ft [0.6 m] thick) lag gravel at the base overlain by a fining-upward sequence of sands ranging from medium to very fine grained (Fig. 16). Accompanying the upward-fining is a vertical transition in facies from GSm to Sx, Sm to Sl and Sr.

The association is interpreted as representing fluvial-marine interface deposits comprising alternations of distal coastal plain (distributary-like) river channels and more marine-influenced delta front sediments. In such a setting the coarser gravels and sands represent the periodic introduction, into an otherwise low energy, marine-dominated environment, of higher energy fluvially derived debris during river flood events. The association possesses a laterally variable character across the field owing to changes in the degree of preservation of either the fluvial or delta front component.

Channel Mouth Bar (Delta Front) Association (IX; Fig. 16)

This association is entirely composed of coarsening-upward sandstone sequences which range from moderately sorted, very fine-grained at the base to moderately sorted, fine- to medium-grained at the top (Figs. 16 & 18a). Dominant facies within the sandstones (in order of decreasing abundance) are; Sm, Sl, Sx (Fig. 17c) and Sr. Foreset thickness in the cross-bedded sandstones average 0.4 ft (0.12 m). Carbonized plant debris (macerated) is ubiquitous throughout the association. Evidence of soft sediment deformation, e.g. slumping, folding, microscale faulting, etc. is common in this association (Fig. 17d).

The association is interpreted as a mouth-bar type of deposit owing to its characteristic coarsening-upward nature and the alternating presence of traction-generated cross-beds (fluvial outflow) and wave generated ripple cross-lamination (marine reworking).

Delta Front Association (X; Fig. 16)

This association consists of approximately 25% sandstone and 75% siltstone/mudstone (Figs. 16 & 18b). In gross terms the association coarsens-upwards. However, more variable sequences, some of which even fine-upwards, are often seen. The sands range from moderately sorted, very fine- to fine-grained and are typified by facies

 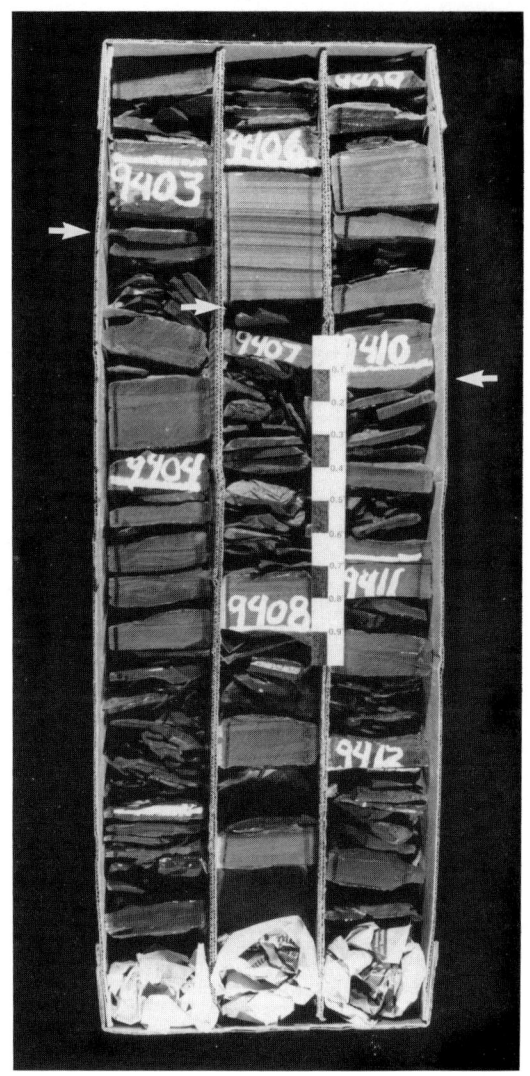

Figure 18   A. Core representing association (IX), channel mouthbar (delta front). Note domination by massive (Sm) to parallel laminated (Sl) sandstones and abrupt, vertical grain size change indicated above arrow. Scale in tenths of a foot.

B. Core representing association (X), delta front. Arrows indicate position of thin (< 1 ft thick), sharp-based, parallel laminated (Sl) fine sandstone beds. Scale in tenths of a foot.

Figure 19   A, B and C. Delta front association (X). A. Wavy to parallel laminated very fine grained sandstone, facies Sr (Table I). Concentration of carbonaceous debris along laminae indicated by arrow. B. Parallel to low-angle laminated very fine sandstones, facies Sl (Table I). Erosion surface lined by mudstone intraclasts indicated by arrow. C. Planolites burrows (arrows) within delta front mudstone interval.
D. Transgressive lag/marine erosion surface association (XI). Pebbly, phosphate- and pyrite-rich lag (arrow) overlain by intensely bioturbated glauconitic sandstones.

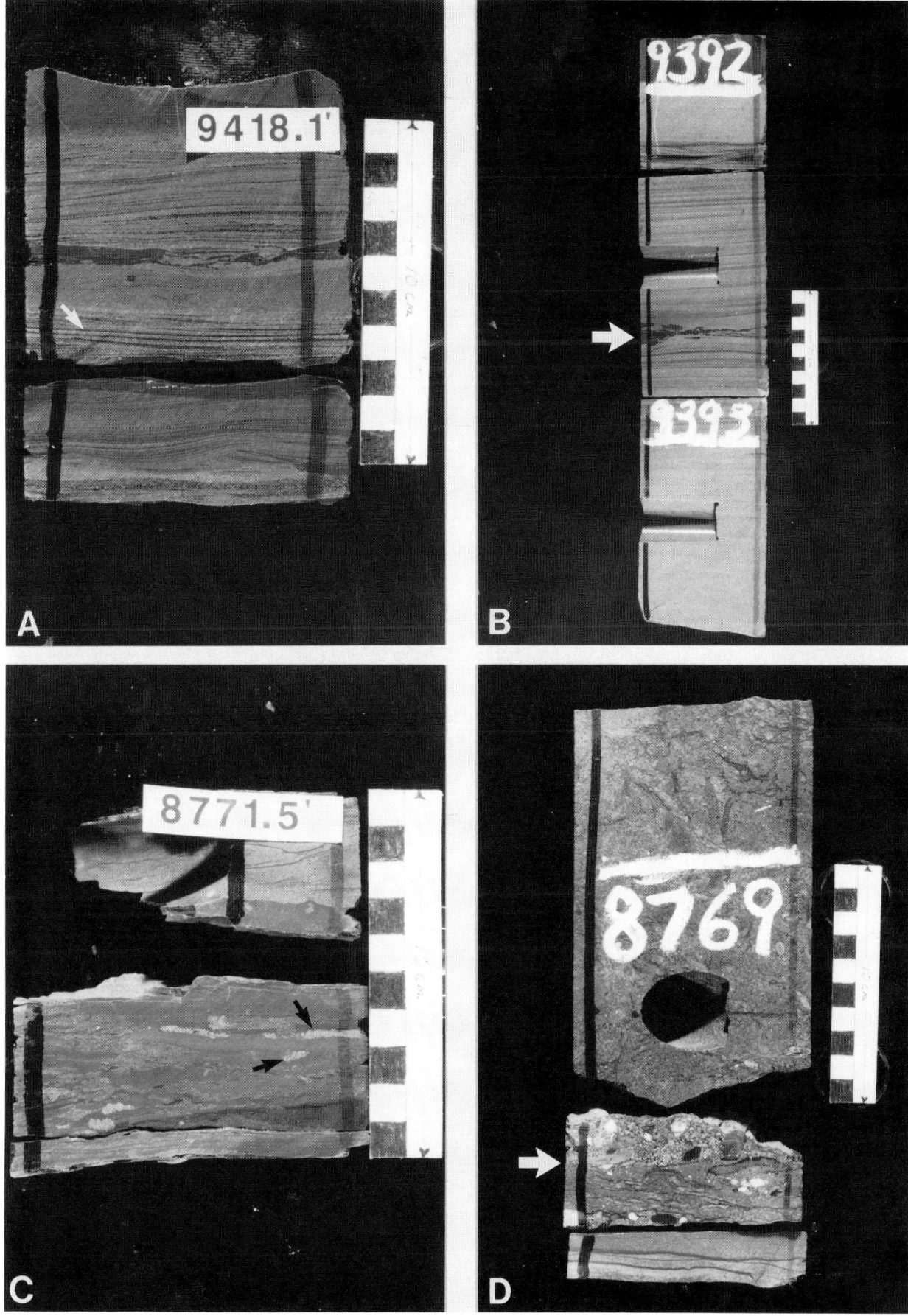

Sr (Fig. 19a) and Sl (Fig. 19b). Less important sandstone facies include Sm, Sx, and Sb. Climbing or ripple drift cross-lamination is particularly noticeable in some zones dominated by lithofacies Sr. Carbonized and macerated plant debris is common within the sands and is concentrated along laminae comprising the Sl lithofacies (Fig. 19a). The silts are restricted primarily to the base of the coarsening upward cycles and are dominated by lithofacies Ml, Md and Mb (Fig. 19c).

The primary sedimentary structures of this association are small-scale and imply deposition from low physical energy flows. The types of stratification present indicate that these flows were non-channelized and that most sedimentation was probably achieved via "raining out" from suspension. This conclusion is backed up by the presence of climbing ripple laminations (Sr) and escape-type burrows in some of the sands (Sb). The association is interpreted as deposits which accumulated at some distance from the active channel mouth in a delta front setting characterized by unconfined, expansive flows which deposited sediment largely from suspension.

Marine Reworked Transgressive Shelf

Directly above the uppermost portion of the Ivishak succession proper, and at the base of the Shublik Formation, is a thin, reworked, transgressive shelf deposit. Although the association is not truly a part of the Ivishak Formation, it is composed predominantly of reworked Ivishak material and occurs in close association with the underlying fluvial deposits.

Transgressive Lag/Marine Erosion Surface Association (XI; Fig. 19d)

Throughout most of the field, this association comprises a thin (< 5 ft [1.75 m]) interval of sandstones and gravels characterized by the presence of glauconite and phosphatic pebbles and later diagenetic nodules of pyrite. The sandstones are moderate to well sorted, range from fine- to coarse-grained and often exhibit intense bioturbation (Fig. 19d). The gravel clasts are well rounded and are composed of both extraformational and intraformational debris. Dominant facies include GSm, GSl, Sm and Sl.

The association is interpreted to represent sediments generated by the reworking of previously deposited fluvial material during the Shublik transgression. Throughout most of the field it can best be thought of as a type of "lag" deposit which was produced by post transgressive winnowing on a submarine ravinement surface. In the western part of the field the transgressive surface is less distinct and is replaced by deposits locally termed the Eileen Sandstone. These sandstones range up to 40 ft in thickness and presumably represent more extensive, reworked shelfal accumulations.

VERTICAL AND LATERAL DISTRIBUTION OF FACIES ASSOCIATIONS

A series of dip- and strike-oriented cross-sections across the field (Fig. 2) illustrate the lateral facies association relationships and variations with depth in the four petrophysical zones comprising the Ivishak Sandstone (Figs. 20-23). All the wells depicted on the cross-sections are cored throughout the interval. In a general sense each petrophysical zone is characterized by different groups of facies associations and depositional sub-environments. Since grain size and sorting are largely controlled by these factors, the various zones also tend to possess contrasting reservoir properties.

Zone 1

Zone 1 is distinguished petrophysically across the field by a cleaning-upwards, "funnel-shaped" gamma-ray character (Figs. 20 & 22). In core it comprises a coarsening-upward succession of interbedded sandstones and shales ranging from 40 to 130 ft (12-40 m) in thickness (Figs. 21 & 23). It is conformably transitional into the underlying marine deposits of the Kavik Shale Formation. The top of zone 1 is picked based on gamma-ray response where a shift of 5-10 API units occurs above the "clean" sand baseline established in zone 2. Internally, it is characterized by a vertical sequence of facies associations (Figs. 21 & 23) belonging mainly to transitional fluvio-marine, coastal to nearshore environments (associations VIII, IX and X). These associations are arranged in a typical progradational form in which the lowermost delta front deposits grade upwards into channel mouth-bar sediments and eventually, in some instances, into erosive-based distal coastal plain, fluvially-dominated channel units. These associations are not found in zones 2, 3, and 4.

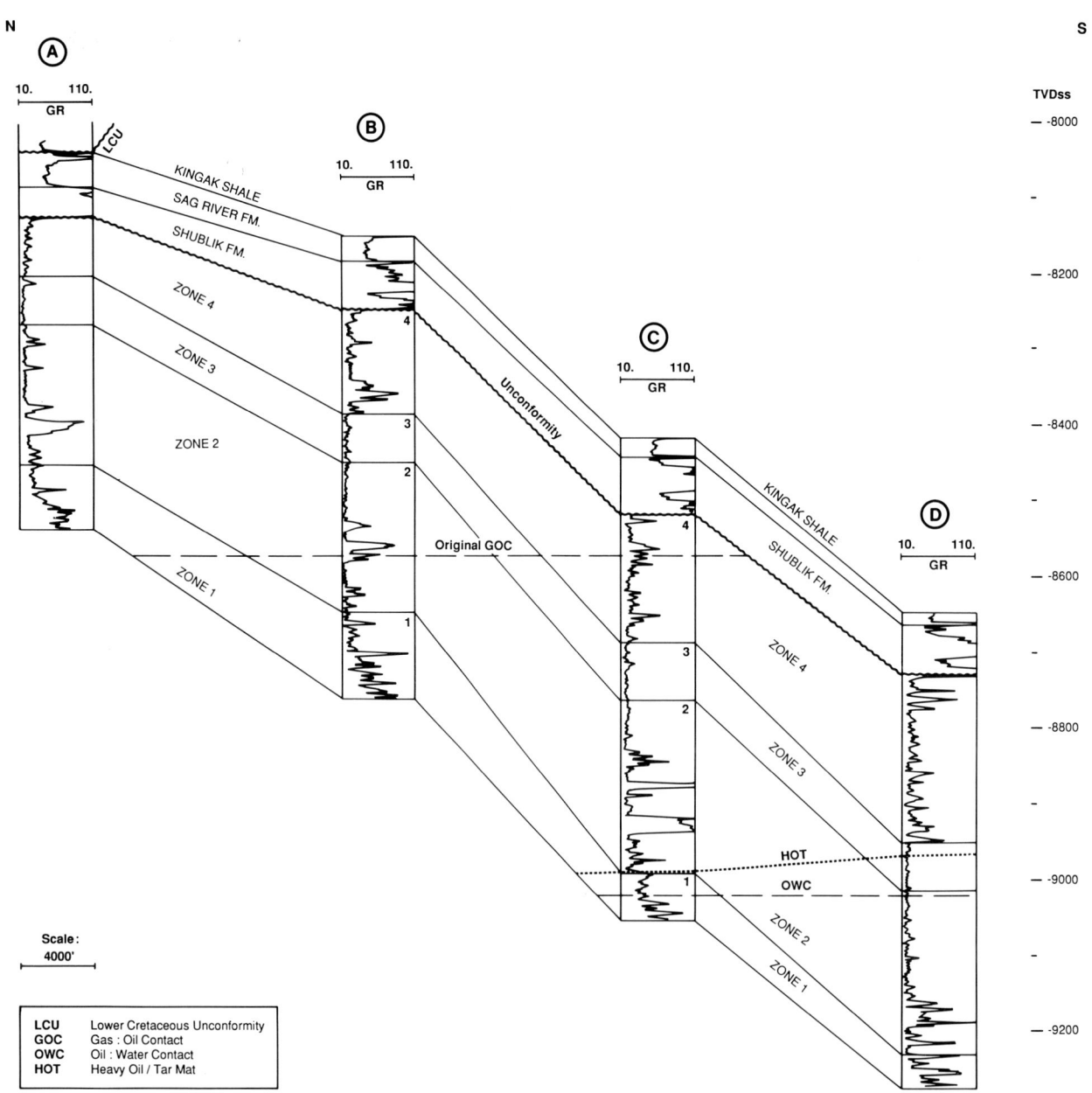

Figure 20 North-south structural cross-section through the field illustrating petrophysical zonation within the reservoir.

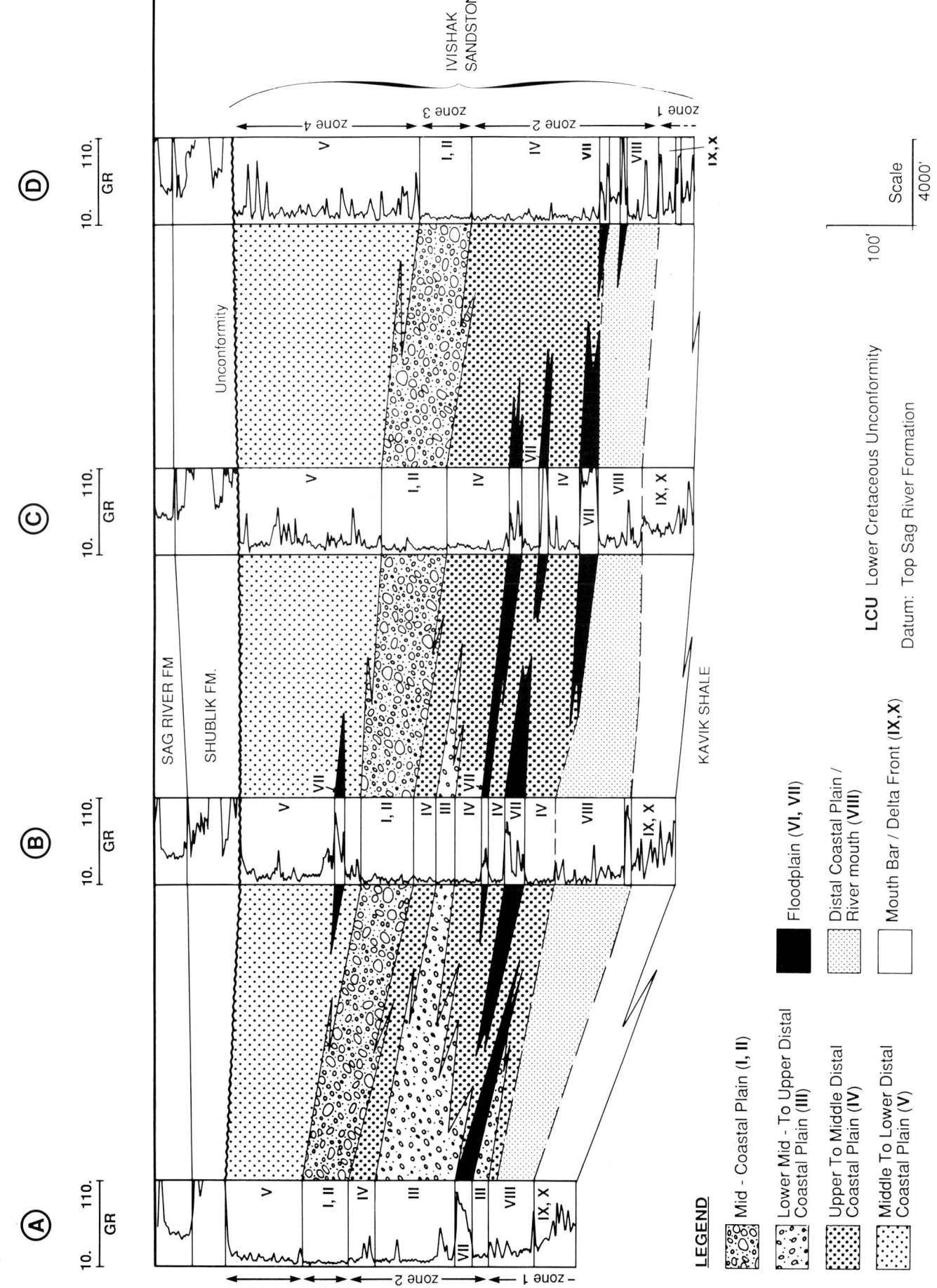

Figure 21 North-south stratigraphic cross-section through the field illustrating facies association distribution within the reservoir.

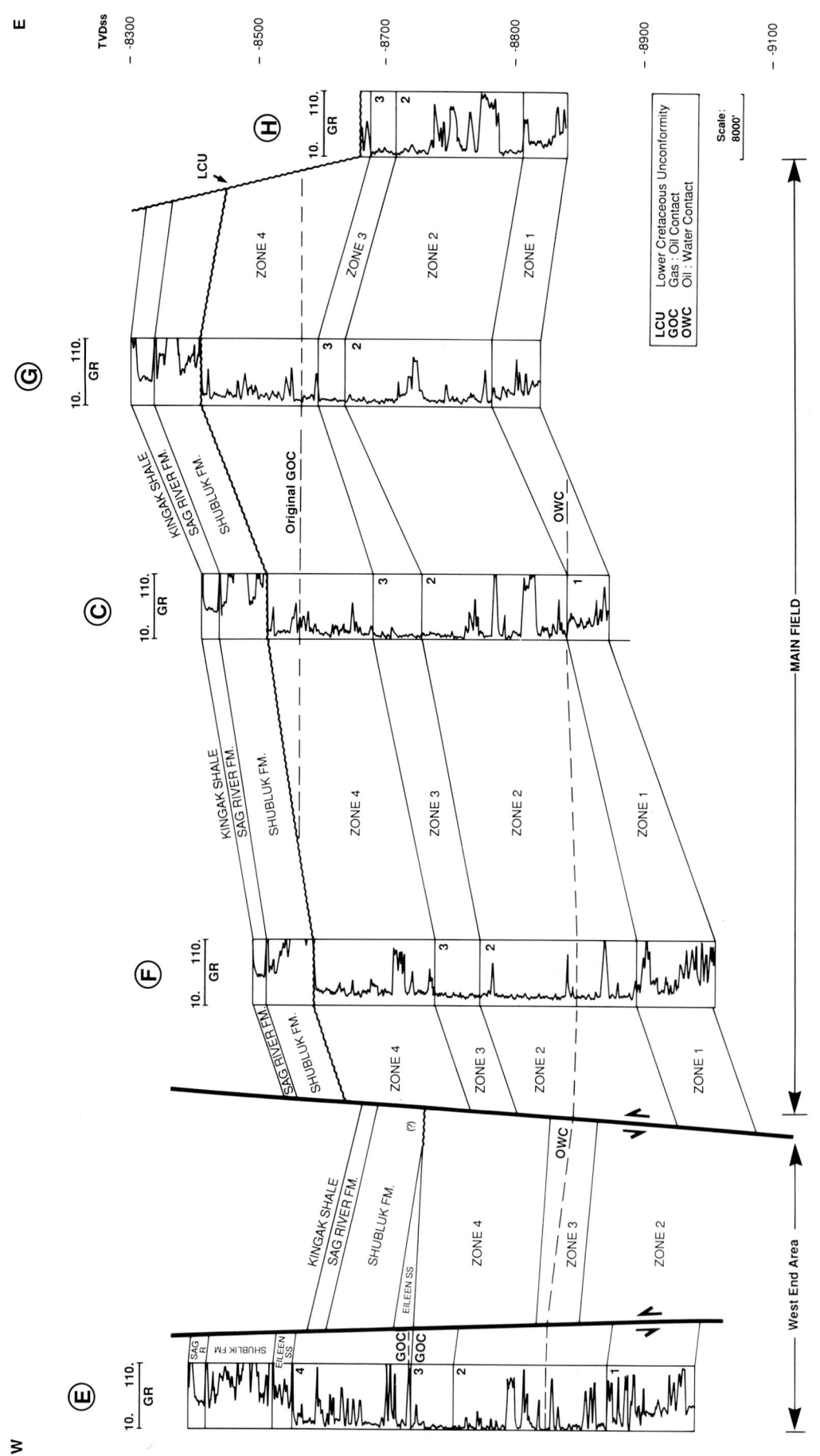

Figure 22 West-east structural cross-section through the field illustrating petrophysical zonation within the reservoir.

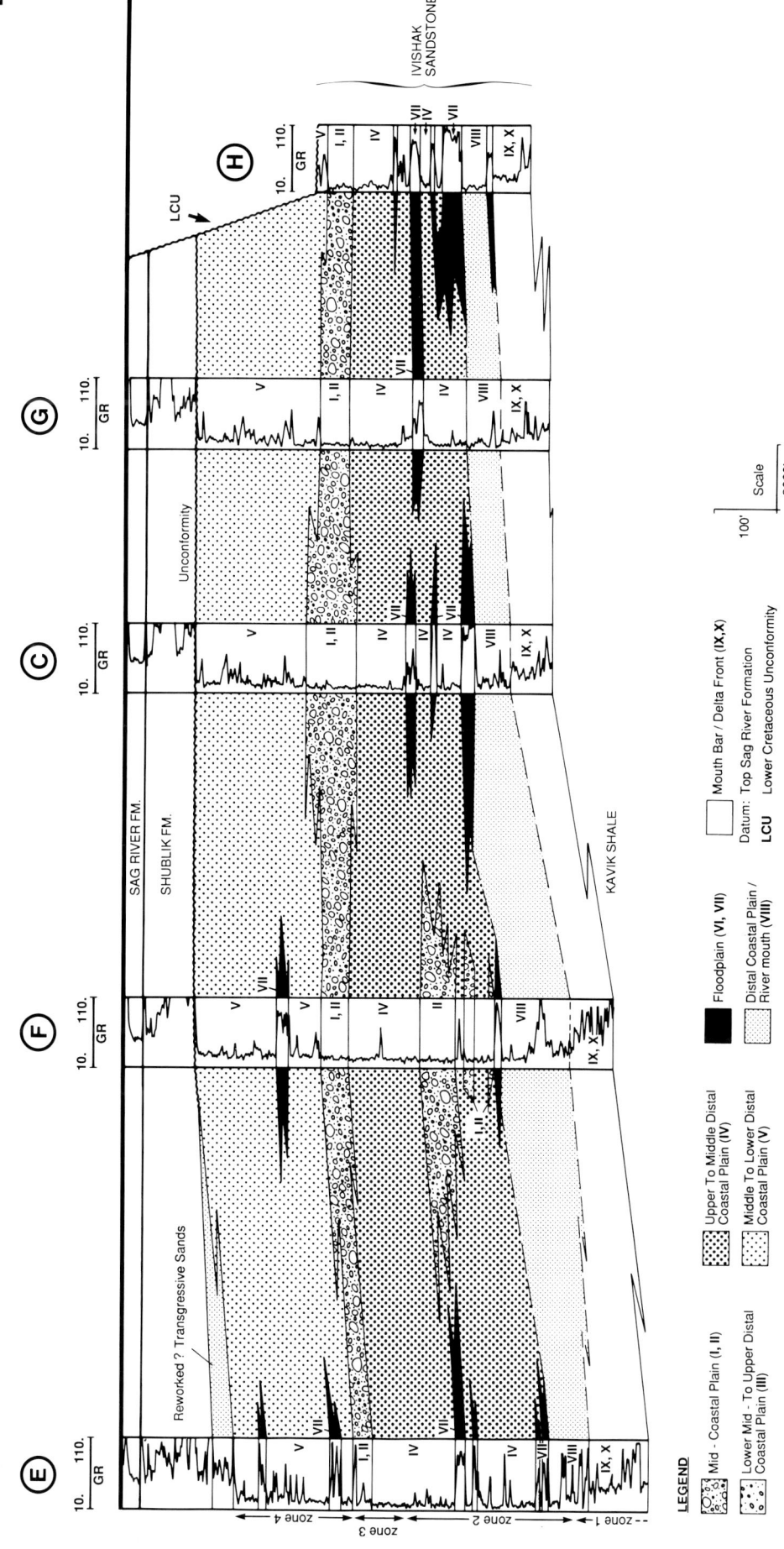

Figure 23 West-east stratigraphic cross-section through the field illustrating facies association distribution within the reservoir.

The average porosity of this zone across the field is 21% and the average horizontal air permeability is 140 mD.

Zone 2

The log character of zone 2 is more variable across the field than zone 1. The gamma-ray log displays a range of values at the base of zone 2 and becomes relatively uniform and blocky toward the top of zone 2 (Figs. 20 & 22). The top of the zone is defined by the base of the overlying zone 3, where the sonic interval transit time exceeds 85 microseconds. The zone ranges from 200-250 ft (60 to 76 m) in thickness and is dominated by a laterally variable, interbedded arrangement of conglomerates, sandstones and shales (Figs. 21 & 23) belonging to facies associations of the subaerial, fluvially-dominated coastal plain (associations III, IV and V). Occasionally, more proximal, facies association II deposits are seen in the zone. Generally, both average grain size and maximum clast diameter increase upwards through the zone (especially in its upper part) and an overall vertical transition takes place from lower distal- to lower mid-coastal plain environments (Figs. 21 & 23). On the dip-oriented cross-section (Fig. 21) it is apparent that at the same stratigraphic level within zone 2 more- distal coastal plain deposits (association IV) in the "B" and "C" wells grade northwards into more proximal, mid-coastal plain sediments (association III) in the "A" well. A similar, but less pronounced, transition typifies the zone on the strike-oriented cross-section (Fig. 23) between the "E", "F" and "C" wells. Knowing that the Ivishak was sourced from the north, the former north-south transition is to be expected. However, the presence of similar transitions in a lateral sense suggests that at certain times, major north-south oriented fluvial axes, in which river flow was concentrated, were present on the coastal plain.

Important, relatively thick (5 to 65 ft [1.5 to 20 m]) floodplain shale complexes dominate zone 2 in its lower part throughout the field (Figs. 21 & 23, association VII). In the western and eastern portions of the field these shale complexes are fairly extensive and form continuous, intra-reservoir vertical seals over relatively large areas. Often they contain numerous paleosol horizons, some of which are clearly correlatable between wells at the same stratigraphic level. Within these shale intervals isolated, fining-upwards, channel-fill units are occasionally seen. These channels are up to 20 ft (6 m) in thickness with estimated widths on the order of 2000 to 3000 ft (600 to 900 m). Towards the central portion of the field the number of channel sandbodies within the shales increases and eventually the shales are replaced entirely by a stacked sequence of multistorey channel bodies. Above the shale-rich lower portion of the zone, complete fining-upwards sequences capped by abandonment deposits are rare and vertical stacking of channel units becomes the norm throughout the field (Figs. 21 & 23). Accompanying this trend is an overall increase in the grain size and maximum clast diameter of the channel-fills suggesting a progradational transition into more upstream coastal plain river systems.

The average porosity of this zone remains constant across the field at 24%. In contrast, owing to the overall upward increase in grain size, values of horizontal air permeability increase from an average of 480 mD at the base of the zone to an average of 770 mD at the top.

Zone 3

Zone 3 is the most uniform of the four zones in terms of lithology and log response across the field (Figs. 20-23) and varies in thickness from 30 ft (9 m) in the east to in excess of 100 ft (30 m) in the center of the field (Figs. 22 & 23). It is composed of an interbedded sequence of conglomerates and conglomeratic sandstones with a characteristic "blocky" gamma-ray response (Figs. 20 & 22). The top and base of zone 3 are picked where sonic interval transit times of less than 85 microseconds per foot for oil and water and 95 microseconds per foot for gas occur. Conglomerate percentages observed in core vary from 1-2% to >50%. In uncored wells the degree of cross-over separation between the sonic and bulk density logs (plotted at the scale of 110 to 50 microseconds and 2 to 3 g/cm^3, respectively) can provide a useful approximation of pebble (conglomerate) content (Lawton et al., 1987). The zone is characterized by the presence of fluvially-dominated facies associations (Figs. 21 & 23) of mid-coastal plain affinity (associations I, II, and infrequently III). In contrast to the zone 2 sequences below, erosive-based channel bodies occur as thin (<10 ft [3 m]), strongly multistorey, stacked units with little or no preservation of their presumed fining-upwards tops. Maximum clast diameters in the conglomerates are usually greater than 50 mm and a distinct clast size "jump" typifies the transition from the underlying zone 2 bodies (Fig. 24). If one equates maximum clast size with maximum stream energy and competence to carry bedload then both these factors imply that a major phase of river incision

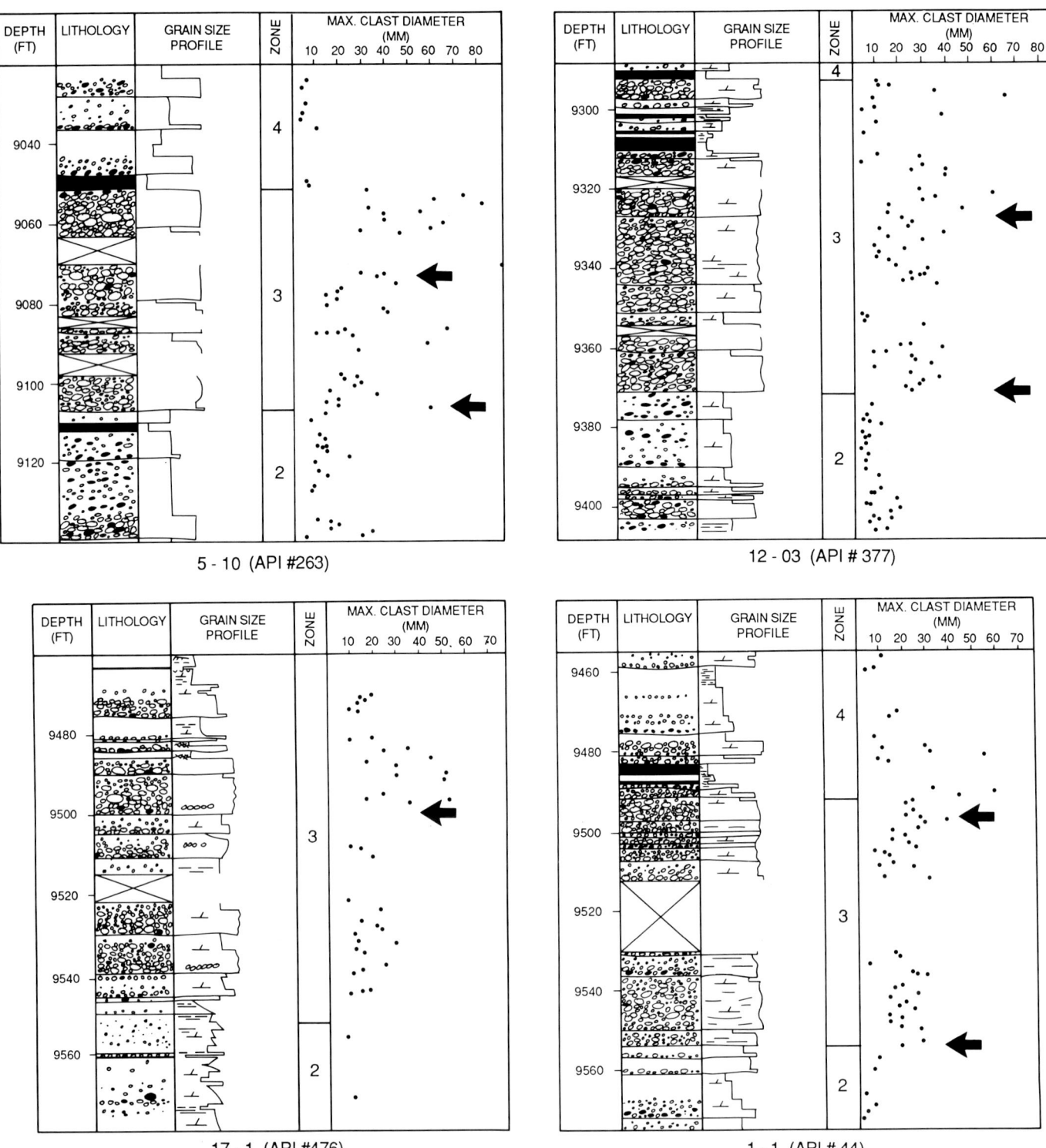

Figure 24 Abrupt clast size increases (arrowed) seen at the base and within the zone 3 mid-coastal plain conglomerate deposits. Note also the rapid change in fluvial architecture at the zone 2/3 boundary whereby fairly thick, well developed fining-upward sequences in zone 2 are replaced by thinner, more amalgamated and irregular conglomeratic sequences in zone 3.

probably occurred prior to the deposition of the zone 3 sequence. Interestingly, several other abrupt clast size increases occur within the zone 3 succession (Fig. 24) suggesting that several episodes of incision may have taken place. This incision phase may explain why the zone 3 conglomeratic interval is continuous and remains so uniform in character throughout the field (Figs. 21 & 23).

The poorly sorted texture and bimodal grain size distribution in the conglomerates generates a relatively low average porosity for this zone (19%). However, its predominantly coarse grain size favors the development of large, interconnected pore throats which promotes high, average horizontal air permeability values (>700 mD). Owing to the poor core recovery in most of the conglomeratic-rich horizons and problems associated with measurements of whole core properties in very coarse grained rocks, permeability values in this zone are probably larger than the present averages suggest and could be as high as 10 darcies locally.

Zone 4

Zone 4 possess a similar varied log character to that of zone 2 (Figs. 20 & 22). The zone ranges up to 230 ft (70 m) in thickness and comprises an interbedded succession of sandstones (rare conglomerates) and shales in which the percentage of sandstone becomes dominant upwards (Figs 21 & 23). Internally, both the mean grain size and facies association distribution of zone 4 record a rapid cessation in coastal plain progradation following the previous zone 3 maximum. Conglomerate percentages and maximum clast diameters fall off rapidly upwards through the interval and at the same time mid-coastal plain fluvial channel associations are replaced abruptly by associations representing the distal coastal plain (Figs. 21 & 23, associations III, IV and V). Despite this domination by distal coastal plain deposits, overall grain sizes tend to be coarser in wells located further to the north (e.g. well "A", Fig. 21). In both the western and eastern parts of the field relatively thick (up to 50 ft [20 m.]) floodplain shales, similar to those of zone 2, typify the base of the interval (Figs. 21 & 23). The shales again contain evidence of soil formation, however the paleosol horizons are generally both thinner and less mature than in zone 2 (association VII). Laterally towards the center of the field the shales die out owing to the increased presence of channel bodies at that stratigraphic level. Occasionally, within the thicker shales, isolated fining-upwards sequences (5 to 10 feet [1.5 to 3 m] in thickness) interpreted to represent complete channel-fill units are seen. Above this lower shale prone interval, the zone is generally characterized by channel stacking and multistorey sand body development (Fig. 21 & 23). At the top of the zone, especially in the western portion of the field (Fig. 23), the distal coastal plain sediments have been modified to varying degrees by marine reworking associated with the advancing Shublik transgression (McGowen, pers. comm.).

The average porosity of this zone remains fairly uniform across the field at 23-24%. However, the general tendency towards an upward reduction in mean grain size results in horizontal air permeability decreasing from an average of 400 mD at the base of the zone to an average of 220 mD at the top.

DEPOSITIONAL MODEL FOR THE IVISHAK SANDSTONE

Three main depositional models have been put forward in the past to explain the Ivishak succession at Prudhoe Bay:

(i) a multi-point source, braided river dominated, coastal plain (Eckelmann et al., 1976; Lawton et al., 1987),

(ii) a multi-point source, braided fluvial plain/fan-delta, fed by coarse grained alluvial fans (Melvin and Knight, 1984), and,

(iii) a braided river fed, fan-delta (McGowen and Bloch, 1985).

Faced with the fact that no environmentally equivalent Ivishak deposits outcrop in the North Slope area and that the only direct geological data is widely spaced core information, it remains difficult to prove conclusively if any one of the above models is the "ground truth" interpretation for the Ivishak. In fact, depending upon how one personally interprets the intricacies of the internal facies and facies association characteristics of the succession evidence could be found to support any of the proposed models. However, all the models have one factor in common and that is that the Ivishak appears to have been deposited on a coarse grained, braided river-dominated, coastal plain.

Taking into account this fact, the closest modern day depositional analogue for the Ivishak appears to be the facies models developed by Boothroyd and Ashley (1975) and Boothroyd and Nummedal (1978) for fluvioglacial outwash or coastal sandurs. Their work was based upon the modern day outwash deposits along the southern Alaskan and Icelandic coasts. In these environments large volumes of relatively coarse-grained material are transported rapidly to sea mainly via numerous, fairly extensive, shallow braided streams. These rivers are characterized by a downstream transition from pebble to sand dominated bedload, relatively rapid lateral shifting of courses, a tendency to generate stacked, multistorey channel-fills and frequent evidence of channel incisement. All these characteristics, as described above, are well developed in the Ivishak deposits. Importantly, Boothroyd and Nummedal (1978) distinguished two main areas of sedimentation on the outwash plain, a so-called "core" region occupied by the main active river system and a "lateral" region outside the area of river flow dominated by floodplain-type deposits (see their figure 12). A similar distinction between core and lateral deposits exists in zones 2 and 4 of the Ivishak succession where stacked channel complexes in certain areas of the field are correlatable laterally into floodplain shales which may, or may not, contain isolated channel units (Fig. 23).

Bearing in mind the similarity between the Ivishak and modern day fluvioglacial outwash it appears that of the three proposed models, the one first suggested by Eckelmann et al. (1976), and more recently expanded upon by Lawton et al. (1987), provides the best analogy for the Ivishak succession. Interestingly, the guiding modern analogue for this model are the proglacial deposits of the Canterbury Plains of South Island, New Zealand (Carson, 1984). In their Ivishak model Lawton et al.(1987) distinguished between both braided and meandering channel deposits on the ancient coastal plain. The main criterion for producing this distinction is that meandering channels tend to have better developed fining-upwards grain size trends than those of braided origin. We feel such a distinction may be rather arbitrary since it is becoming increasingly clear in the sedimentological literature that fining-upwards trends are by no means ubiquitous to, nor diagnostic of, meandering stream point bar deposits (McGowen and Garner, 1970; Jackson, 1975, 1978; Jordan and Pryor 1987). For this reason we prefer not to subdivide into meandering or braided river types. Instead, we would rather limit our interpretation and be guided by modern day analogies where similar coastal plains are "dominated by braided streams" but may have subordinate meandering courses.

McPherson, Shanmugam and Moiola (1987 p.338) have recently presented a similar conclusion to that expressed above, but stressed the need to introduce a new, and technically more correct term, "braid delta" to describe sequences of this type. In the case of the Ivishak, an important addition to their braid delta model (see their Fig. 1) would be the need to distinguish both a sand/conglomerate-rich, multistorey channel "core" and a shale-dominated, isolated channel-body "lateral" facies component (Fig. 25). In such a model, the relative preservation of each component at any given location would be a function of both the rate of lateral river migration and floodplain aggradation, coastal plain width, number of river entrants and the presence of any pre-existing topography on the coastal plain along which rivers may have a tendency to concentrate their flow.

As McPherson et al. (1987) point out, modern braid deltas occur in middle to high latitude regions where precipitation is high, vegetation is minimal and a ready supply of coarse-grained sediment exists. In the case of the Ivishak, available evidence suggests that all four of these prerequisites were met. Paleomagnetic reconstructions place the Alaskan North Slope at a latitude of between 45 to 60 degrees north during the late Permian to late Triassic (Smith and Briden, 1977). According to Frakes (1979, p.155) Triassic paleoclimates were similar to the present day, if somewhat slightly cooler and more humid during early Triassic times. Precipitation pattern maps show that during the early Triassic the North Slope area lay within the middle to high latitude humid belt and most probably possessed a wet seasonal climate of temperate affinity (Robinson, 1973). Under these conditions extensive plant growth is expected to have been retarded. The high percentage of conglomerate within the Ivishak succession supports the idea that an uplifted source terrain, most probably a much larger predecessor of the Barrow Arch, lay close by.

CONCLUSIONS

1. Facies and facies associations within the Ivishak Sandstone can be grouped into three main types of depositional setting: (i) sub-aerial, fluvially-dominated, coastal plain; (ii) transitional fluvio-marine deltaic; and (iii) marine reworked, transgressive shelf. Within each depositional setting recognized facies associations are highly intergradational both vertically and laterally.

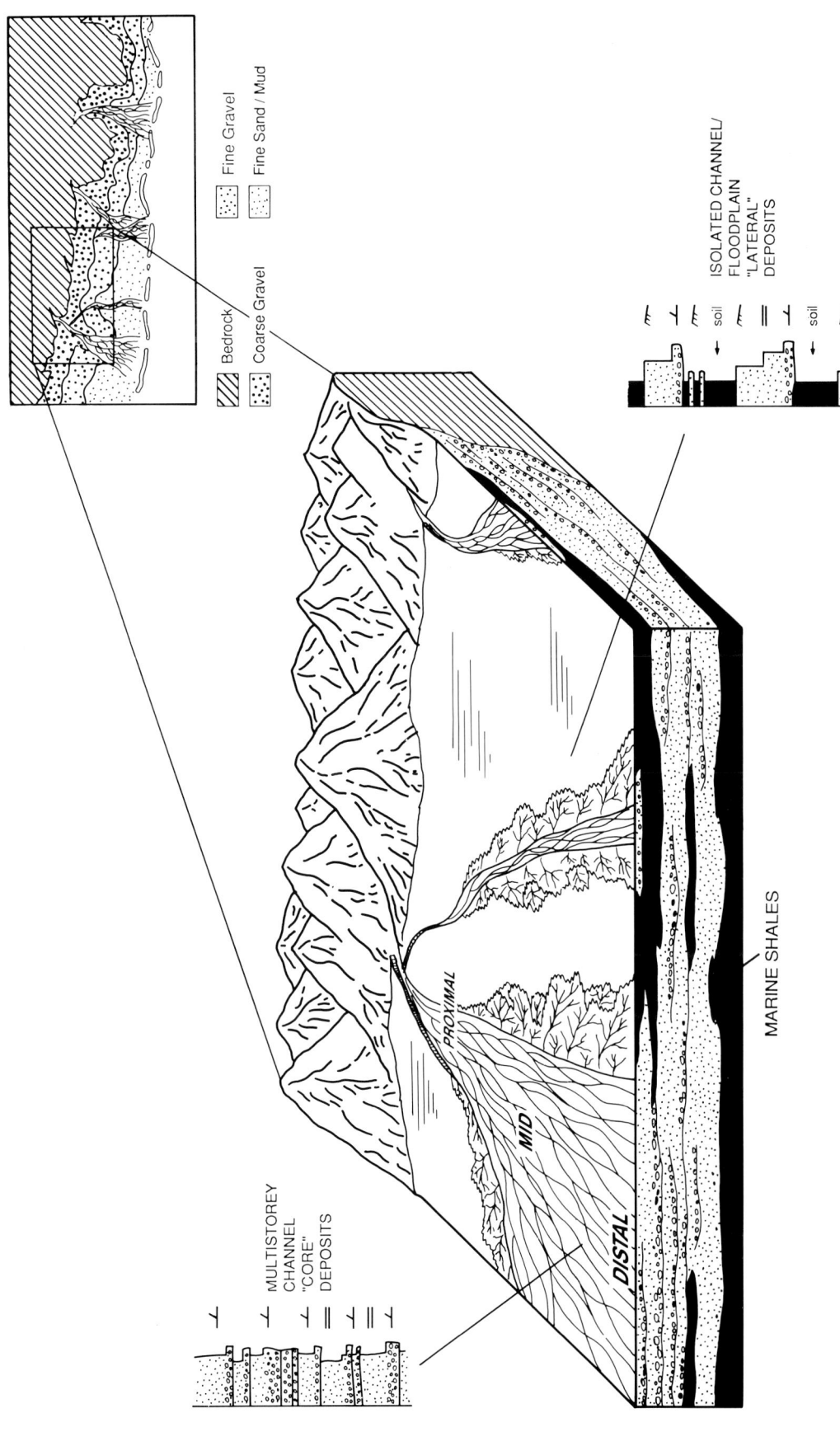

Figure 25 Proposed braid-delta depositional model for the Ivishak Sandstone as suggested by Lawton et al. (1987) and McPherson et al. (1987) but modified to include "core" and "lateral" facies tract components. The model implies that the Ivishak accumulated as a laterally extensive, clastic wedge which paralleled the trend of its uplifted source.

2. Reservoir quality is highest in the coarse-grained (conglomeratic) facies associations (I-V) belonging to fluvially-dominated coastal plain environments. Poorer reservoir qualities characterize the finer-grained facies associations (VI-X) of the floodplain and transitional deltaic environments.

3. At any location within the field a vertical transition characterizes the Ivishak succession from fluvio-marine deltaic deposits up into at first, increasingly more proximal and then later, increasingly more distal braided river dominated, coastal plain sediments. Later reworking and eventual truncating of these distal deposits took place during the transgressive phase associated with the deposition of the Shublik Formation.

4. At any particular stratigraphic level within the braided river-dominated, coastal plain portion of the succession, more proximal channel deposits are always developed to the north and more distal deposits to the south. A similar transition is often apparent in a lateral west-east sense. It involves the lateral gradation from strongly multistorey channel deposits of more proximal affinity into either more isolated and complete distal channel bodies and/or floodplain shale complexes. These lateral transitions imply that the Ivishak coastal plain was probably characterized by several major, approximately north-south oriented, fluvial axes.

5. The presence of a laterally extensive, field wide, conglomerate-rich horizon (zone 3) characterized by a major "jump" in clast size and abrupt change in fluvial style and architecture is interpreted to reflect an important phase of river incision which affected the coastal plain during deposition. This incision was most probably a result of base-level fall promoted by either tectonics and/or eustasy.

6. The Ivishak succession is best interpreted by a braid delta depositional model in which a distinction is made between multistorey, coarse member-rich "core" deposits and finer-grained, more floodplain prone "lateral" deposits. Adopting this model, the relative preservation of each component at any given location would be a function of both the rate of lateral river migration and floodplain aggradation, coastal plain width, number of river entrants, and the presence of any

pre-existing floodplain topography along which the rivers would have had a tendency to flow.

7. Development of the Ivishak braid deltas was promoted by: a high latitude setting (45-60° north), a wet, seasonal, cool temperate climate; limited vegetation cover and the nearby presence of an uplifted source terrain.

ACKNOWLEDGEMENTS

The authors are indebted to ARCO Alaska, Inc. and the other co-owners of the Prudhoe Bay Field for permission to publish this paper. We further acknowledge the use of additional information obtained from previous in-house studies conducted by Joe McGowen and Sal Bloch although we wish to state that our ideas and interpretations do not necessarily agree with their thinking. Naresh Kumar, Wayne Zeck, Dave Hite, Roger Slatt, and Mark Scheihing kindly criticized and improved the original manuscript. Special thanks are also due to Jim Bishop for preparing the photographs and Katherine Vaughan for formatting and organizing the text. Finally, we are indebted to Ms. Romy de Groot Heupner for her speedy work in drafting the published figures.

REFERENCES

ATKINSON, C. D., 1986, Tectonic control on alluvial sedimentation as revealed by an ancient catena in the Capella Formation (Eocene) of northern Spain: in Wright, V. P. (ed.), Paleosols: Their Recognition and Interpretation, Blackwell Scientific Publications, Oxford, U.K., pp. 139-179.

BOOTHROYD, J. C. and ASHLEY, G. M., 1975, Process, bar morphology and sedimentary structures on braided outwash fans, northeastern Gulf of Alaska: in Jopling, A. V. and McDonald, B. C. (eds.), Glaciofluvial and Glaciolacustrine Sedimentation, Soc. Econ. Paleont. Mineral. Spec. Pub. 23, pp. 193-222.

BOOTHROYD, J.C. and NUMMEDAL, D., 1978, Proglacial braided outwash: a model for humid alluvial-fan deposits: in Miall, A. D. (ed.), Fluvial Sedimentology, Can. Soc. Petrol. Geol. Memoir 5, pp. 641-668.

BUURMAN, P., 1980, Paleosols in the Reading Beds (Paleocene) of Alum Bay, Isle of Wight, U.K., Sedimentology, v. 22, pp. 593-606.

CARSON, M. A., 1984, Observations on the meandering-braided river transition, the Canterbury Plains, New Zealand: Part one, New Zealand Geographer, v. 40, pp. 12-17.

DETTERMAN, R. L., 1970, Sedimentary history of the Sadlerochit and Shublik Formations in northeastern Alaska: in Proceedings of the Geological Seminar on the North Slope of Alaska, Pacific Section Am. Assoc. Petrol. Geol., pp. O1-O13.

ECKELMANN, W. R., DEWITT, R. J. and FISHER, W. L., 1976, Prediction of fluvial-deltaic reservoir geometry, Prudhoe Bay field, Alaska, Proc. 9th World Petroleum Congress, v.2, pp. 223-227.

FRAKES, L. A., 1979, Climates Throughout Geologic Time, Elsevier Publ., Amsterdam, 310p.

JACKSON, R. G., II., 1975, Velocity - bed-form - texture patterns of meander bends in the lower Wabash River of Illinois and Indiana, Geol. Soc. Am. Bull., v. 86, pp. 1511-1522.

JACKSON, R. G., II., 1978, Preliminary evaluation of lithofacies models for meandering alluvial streams: in Miall, A. D. (ed.), Fluvial Sedimentology, Can. Soc. Petrol. Geol. Memoir 5, pp. 543-576.

JAMISON, H. C., BROCKETT, L.D. and McINTOSH, R. A., 1980, Prudhoe Bay - a 10-year perspective: in Halbouty, M. T. (ed.), Giant Oil and Gas Fields of the Decade, Am. Ass. Petrol. Geol. Memoir 30, pp. 289-310.

JONES, H. P. and SPEERS, R. G., 1976, Permo-Triassic reservoirs of Prudhoe Bay field, North Slope, Alaska: in Braunstein, J. (ed.), North American Oil and Gas Fields, Am. Ass. Petrol. Geol. Memoir 24, pp. 23-50.

JORDAN, D.W. and PRYOR, W. A., 1987, Sedimentology and depositional environments associated with a Mississippi river bar sand: Bryant Bar near Dorena, Missouri, Abst. SEPM Annual Mid-Year Meeting, Austin, Texas, p.40.

LAWTON, T. F., GEEHAN, G. W. and VOORHEES, B.J. 1987, Lithofacies and depositional environments of the Ivishak Formation, Prudhoe Bay Field, in Tailleur, I. and Weimer, P. (ed.) Alaskan North Slope Geology, SEPM Pacific Section and Alaskan Geol. Soc., v. 50 in press.

McGOWEN, J. H.and BLOCH, S., 1985, Depositional facies, diagenesis and reservoir quality of Ivishak Sandstone (Sadlerochit Group), Prudhoe Bay field, Abstr. Am. Ass. Petrol. Geol. Bull., v. 69, p. 286.

McGOWEN, J. H. and GARNER, L. E., 1970, Physiographic features and stratification types of coarse grained point bars: Modern and ancient examples, Sedimentology, v. 14, 77-111.

McPHERSON, J. G., SHANMUGAM, G. and MOIOLA, R. J., 1987, Fan-deltas and braid deltas: varieties of coarse-grained deltas, Geol. Soc. Amer. Bull. v. 99, pp. 331-340.

MELVIN, J. and KNIGHT, A.S., 1984 Lithofacies, diagenesis and porosity of the Ivishak Formation, Prudhoe Bay area, Alaska: in McDonald, D.A. and Surdam, R. C. (eds.), Clastic Diagenesis, Am. Ass. Petrol. Geol. Memoir 37, pp.347-365.

MIALL, A.D. 1977, A review of the braided river depositional environment, Earth Science Reviews, v. 13, pp. 1-62.

MORGRIDGE, D. L. and SMITH, W. B., Jr., 1972, Geology and discovery of Prudhoe Bay field, eastern Arctic Slope, Alaska: in King, R. E. (ed.), Stratigraphic Oil and Gas Fields - Classification, Exploration Methods and Case Histories, Am. Ass. Petrol. Geol. Memoir 16, pp. 499-501.

READING, H. G., 1978, Facies: in Reading, H. G. (ed.), Sedimentary Environments and Facies, Blackwell, Oxford, U.K., pp. 4-14.

RICKWOOD, F.K. 1970, The Prudhoe Bay field: in Proc. Geological Seminar on the North Slope of Alaska, Pacific Section, Am. Ass. Petrol. Geol., pp. L1-L11.

ROBINSON, P.L. 1973, Palaeoclimatology and continental drift: in Tarling, D. H. and Runcorn, S. K. (eds.), Implications of Continental Drift to the Earth Sciences, 1., Academic Press, London, pp. 451-476.

SMITH, A.G. and BRIDEN, J. C. 1977, Mesozoic and Cenozoic Paleocontinental Maps, Cambridge University Press, U. K., 63p.

WALKER, R. G., 1979, Facies and Facies Modles. General Introduction: in Walker, R. G., (ed.), Facies Models, Geosciences Canada Reprint Series 1, Ainsworth Press, Ontario, pp. 1-7.

DEPOSITIONAL FACIES AND POROSITY DISTRIBUTION, PERMIAN (GUADALUPIAN) SAN ANDRES AND GRAYBURG FORMATIONS, P.J.W.D.M. FIELD COMPLEX, CENTRAL BASIN PLATFORM, WEST TEXAS

R. P. MAJOR, D. G. BEBOUT, and F. JERRY LUCIA
Bureau of Economic Geology
The University of Texas at Austin
Austin, Texas 78713-7508

ABSTRACT

The Penwell, Jordan, Waddell, Dune, McElroy (P.J.W.D.M.) Field Complex of Ector, Crane, and Upton Counties, Texas, is a single giant oil field which, because of separate initial field discoveries, is divided into five fields. Production is from Guadalupian San Andres and Grayburg carbonate reservoirs. The original oil in place in this giant field complex is calculated to be 4,100 million barrels, cumulative production is more than 800 million barrels, and 900 million barrels of mobile oil will remain after conventional primary and secondary recovery.

The P.J.W.D.M. Field Complex is located on the Central Basin Platform in the Permian Basin of West Texas and Southeast New Mexico. San Andres and Grayburg sediments were deposited as an upward-shoaling sequence of low-energy ramp sediments. The lower, marine portions of both formations contain pellet and skeletal grainstone and skeletal wackestone with normal marine fauna. The upper, peritidal portions of both formations contain mudstone and pisolite packstone/grainstone characterized by desiccation features and sulfate cements. The section has been thoroughly dolomitized and contains anhydrite partially hydrated to gypsum.

The trap is a low, broad anticline formed by compaction deformation over a buried Pennsylvanian fault system. The reservoir zones are primarily dolomitized subtidal grainstones, and the seals are fine-grained supratidal and peritidal dolomite plugged with sulfate cements. Linear grainstone trends impart lateral heterogenity to these reservoirs, and because they contain a concentration of remaining mobile oil, these features are targets for infill drilling.

INTRODUCTION

The P.J.W.D.M. Field Complex is located on the east side of the Central Basin Platform of the Permian Basin, southwestern United States (Fig. 1). The field complex contains five fields: Penwell, Jordan, Waddell, Dune, and McElroy, which were discovered in the 1920's and 1930's and initially developed separately. Subsequent expansion of field boundaries has demonstrated that these five fields are part of a single giant feature and that production is from the Grayburg Formation in the southeastern part of the field and from the San Andres Formation in the northwestern part of the field (Figs. 2, 3).

The five fields in this giant complex may be arranged in size on the basis of original oil in place from the largest, McElroy Field with more that 2,500 million barrels, to the smallest, Jordan Field with more than 200 million barrels. The total estimated oil in place for the entire giant field complex is approximately 4.1 billion barrels, nearly 800 million barrels of which have been produced. Large parts of the field complex have been waterflooded since the late 1960's and 1970's. Projected ultimate recovery is nearly 940 million barrels (Galloway and others, 1983).

The projected ultimate recovery of 940 million barrels is based on current well-spacing and development practices. However, reservoir facies of the San Andres/Grayburg exhibit heterogeneous porosity and permeability distribution, and it is estimated that an additional 900 million barrels of mobile oil can be recovered using improved conventional oil-field methods, such as recompletions, workovers, and geologically targeted infill-drilling based on integrated geological/engineering studies. This substantial mobile-oil target is the incentive for detailed field studies and was the impetus for combined geological and engineering studies by the Bureau of Economic Geology. Part of this B.E.G. research is reported in this paper; a more complete account of the Dune Field study is available in Bebout and others (1987).

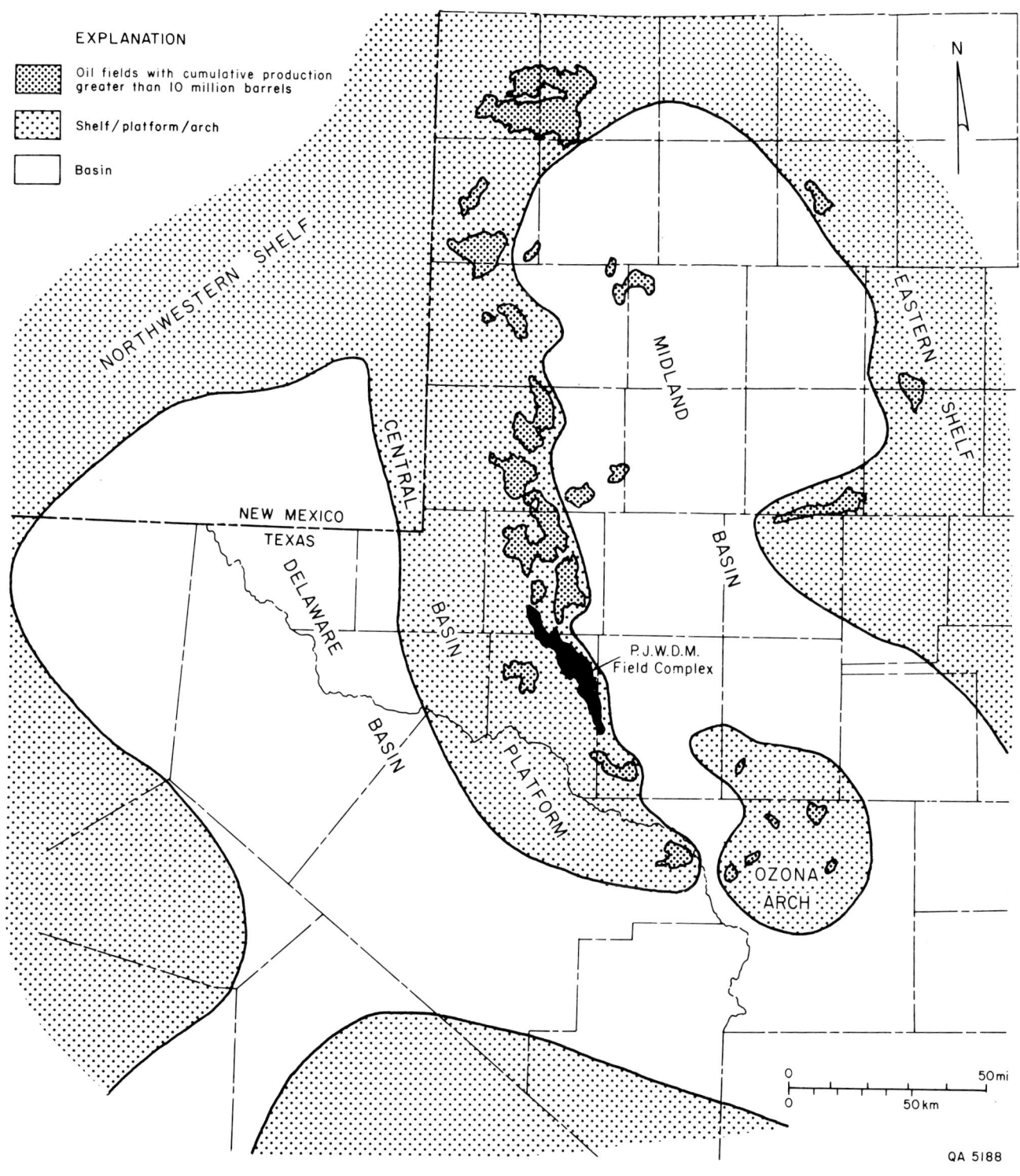

Figure 1. Location of the P.J.W.D.M. Field Complex relative to the paleogeography of the Permian Basin. The highlighted fields produce oil from the San Andres and Grayburg Formations.

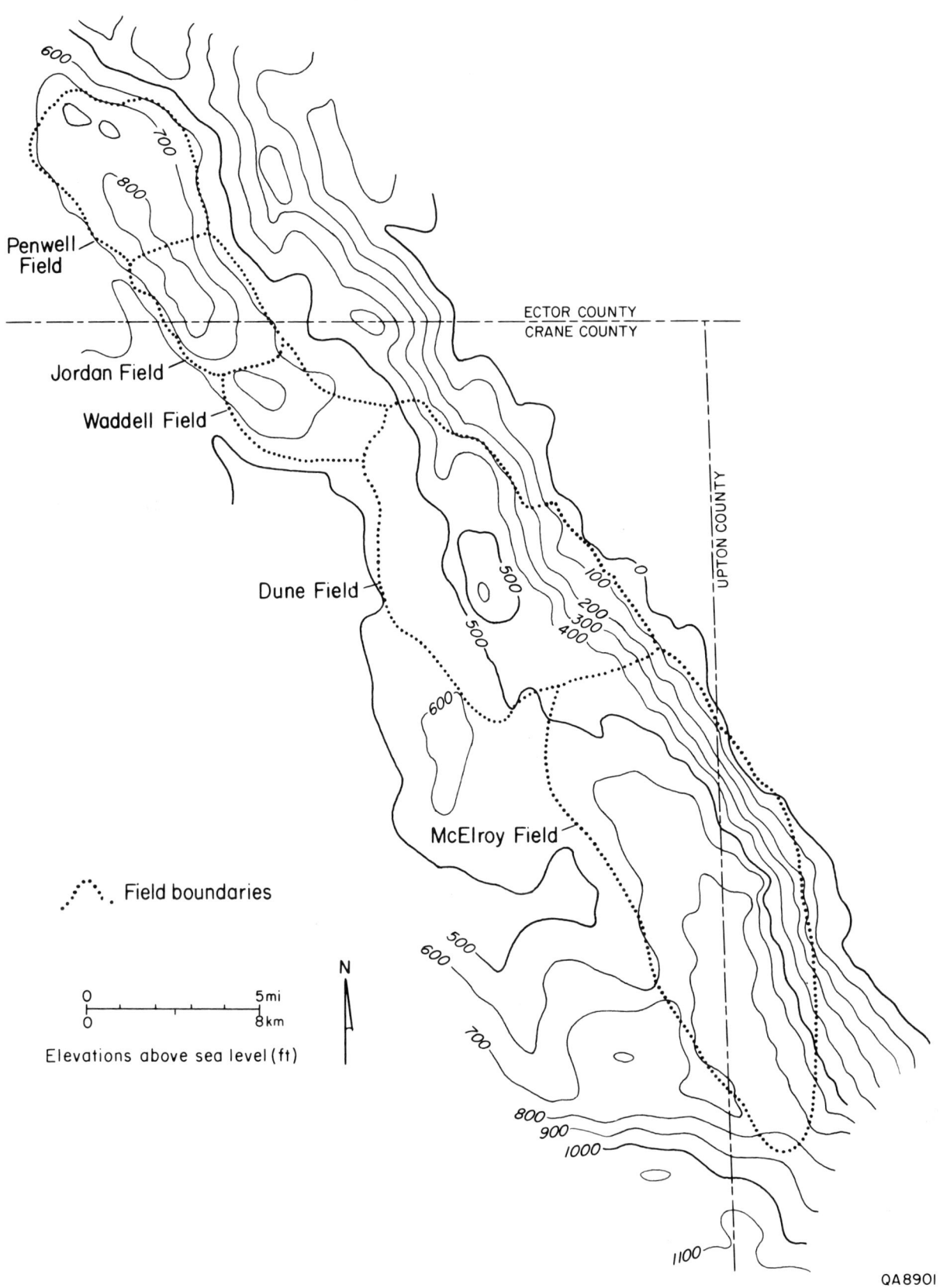

Figure 2. Yates structure map of the P.J.W.D.M. Field Complex. The Yates is identified on well logs in Figure 3 (modified from a map authored by L. F. Long, on file at the University Lands Office, Midland, Texas).

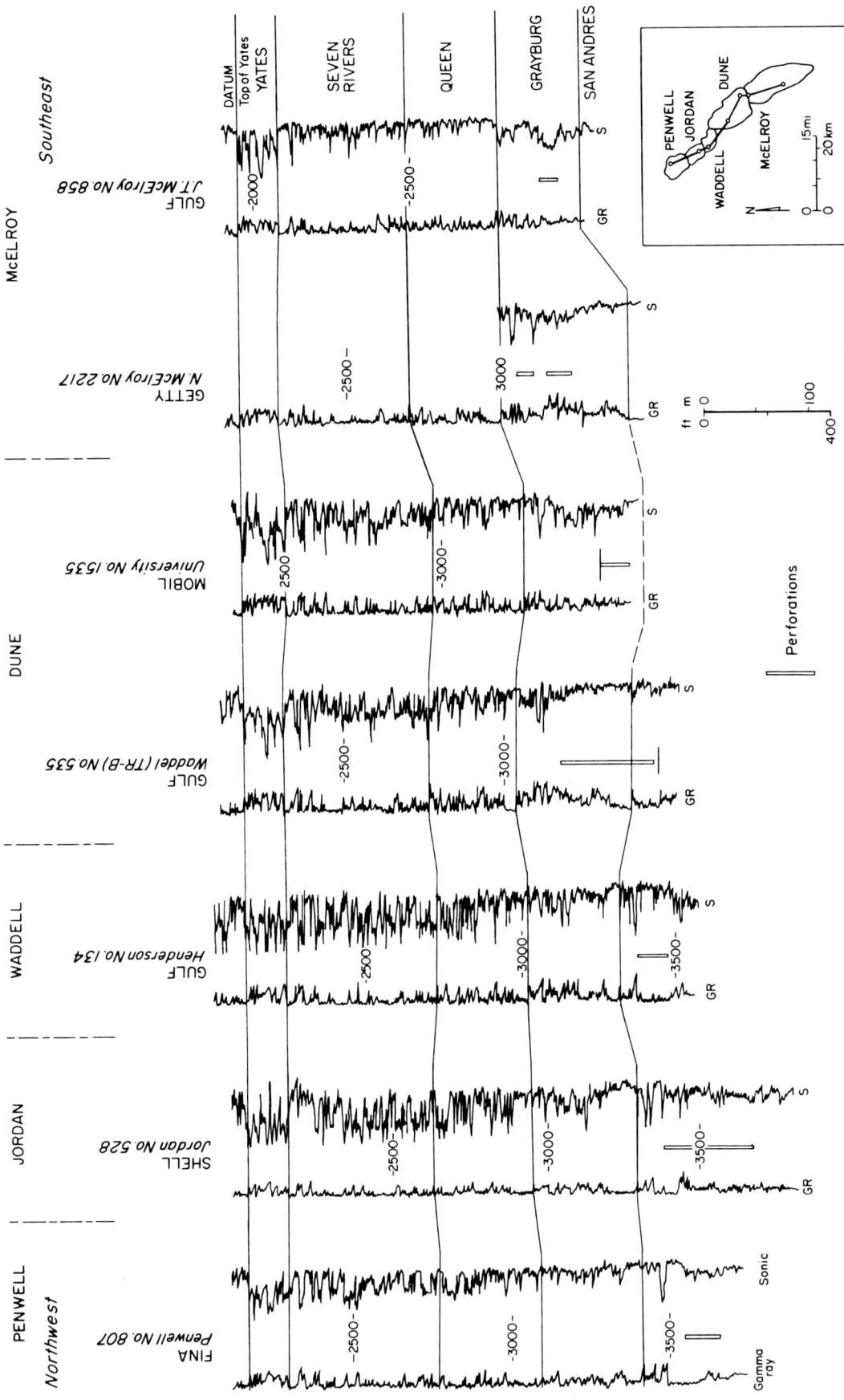

Figure 3. Northwest-southeast stratigraphic cross section through the P.J.W.D.M. Field Complex. The datum is the Yates Formation. Solid bars mark perforation intervals. Note that perforated intervals are in the San Andres Formation in the northwest and in the Grayburg Formation in the southeast (log correlations after Garrett, 1986).

GEOLOGIC SETTING

The paleogeography of the Permian Basin was established by Pennsylvanian tectonism that deformed the Precambrian basement. During Permian time the region was divided into two basins, the relatively deep Delaware Basin on the west and the relatively shallow Midland Basin on the east, separated by the south-southeast trending Central Basin Platform (Fig. 1). The Central Basin Platform was the site of shallow-water ramp margin and ramp interior carbonate sedimentation, whereas the central parts of the Delaware and Midland Basins were the sites of siliciclastic deposition (Ward and others, 1986).

The Permian stratigraphic section on the Central Basin Platform is dominated by Wolfcampian, Leonardian, and Guadalupian shallow-water carbonate sediments, much of these now thoroughly dolomitized, and by relatively minor siliciclastic-rich carbonates. Guadalupian carbonates are in gradational contact with overlying Ochoan evaporites and siliciclastic red beds that were deposited during increasingly restricted basin conditions. The Permian carbonates of the Central Basin Platform form numerous hydrocarbon reservoirs. The Guadalupian San Andres and Grayburg Formations contain the majority of oil production (Fig. 4). These Guadalupian rocks contained an estimated 32.6 billion barrels of original oil in place and still contain 8.7 billion barrels of remaining mobile oil in fields with cumulative production of more than 10 million barrels (Tyler and others, 1984).

The P.J.W.D.M. Field Complex is on the eastern side of the Central Basin Platform. This low-relief, broad anticlinal structure was formed by drape of Permian sediments over buried Pennsylvanian faults that trend oblique to the margin of the basin. Thus, the southern part of the field complex, McElroy Field, occurs at the approximate margin of the Central Basin Platform, and the northern part of the field complex, Penwell Field, occurs approximately 15 miles west of the platform margin (Figs. 1, 2). In general, the San Andres/Grayburg reservoir rocks in the P.J.W.D.M. Field Complex exhibit depositional textures indicative

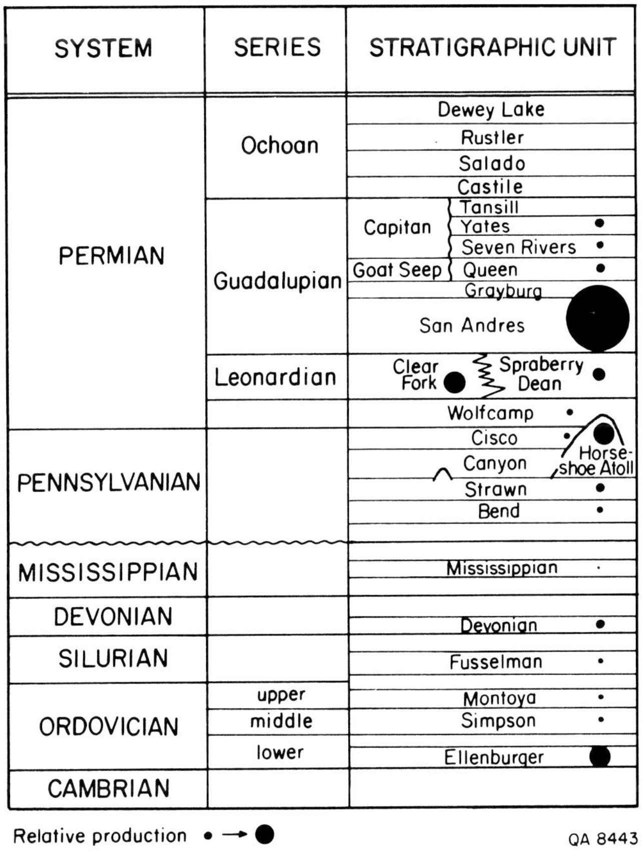

SYSTEM	SERIES	STRATIGRAPHIC UNIT	
PERMIAN	Ochoan	Dewey Lake	
		Rustler	
		Salado	
		Castile	
	Guadalupian	Capitan { Tansill	
		Yates	•
		Seven Rivers	•
	Goat Seep { Queen	•
		Grayburg	
		San Andres	● (large)
	Leonardian	Clear Fork ● ⁓ Spraberry • / Dean	
		Wolfcamp	•
PENNSYLVANIAN		Cisco	● Horse-
		Canyon	shoe Atoll
		Strawn	•
		Bend	•
MISSISSIPPIAN		Mississippian	·
DEVONIAN		Devonian	•
SILURIAN		Fusselman	•
ORDOVICIAN	upper	Montoya	•
	middle	Simpson	•
	lower	Ellenburger	●
CAMBRIAN			

Relative production • → ●

QA 8443

Figure 4. Simplified Paleozoic stratigraphic column for Texas. The size of the black circles is proportional to the amount of hydrocarbon production from the various stratigraphic units (modified from Galloway and others, 1983).

of relatively low-energy environments where they occur in platform interior locations and contain some facies that exhibit relatively high-energy environments where they occur at platform margin locations. The facies in both the San Andres and Grayburg Formations form upward-shoaling sequences in which subtidal grainstones are the primary reservoir rocks and superjacent peritidal and supratidal rocks are the seal. These facies prograded from west to east across the platform, and the peritidal section thickens westward. This westward thickening of low-porosity and low-permeability peritidal rocks provides an updip seal, and the P.J.W.D.M. Field Complex produces mainly from the eastern flank of this broad anticline (Fig. 5).

SAN ANDRES FORMATION

Facies and Depositional Environments

The following facies descriptions and interpretations are based on examination of cores and well logs from the San Andres Formation in the Penwell and Jordan Fields, in the northern platform interior part of the P.J.W.D.M. Field Complex.

Pellet Grainstone

The volumetrically dominant subtidal facies in the Penwell San Andres section is thoroughly dolomitized grainstone composed of spherical to ovoid faecal pellets. Common accessory skeletal grains are fusulinids and mollusks, rarely preserved and most commonly evident as molds. Where fusulinids or mollusks compose 10 percent or more of grains, this facies is described as pellet-fusulinid grainstone or pellet-mollusk grainstone. Burrow structures are rare, but a complete lack of bedding suggests this sediment was thoroughly bioturbated.

Pellets were deposited as soft mud, as is common in modern low-energy subtidal settings (Wanless and others, 1981). The preservation of pellets in these thoroughly dolomitized rocks is variable, and at Penwell and Jordan Fields they are commonly not visible on slabbed core surfaces. Thus, these rocks may be

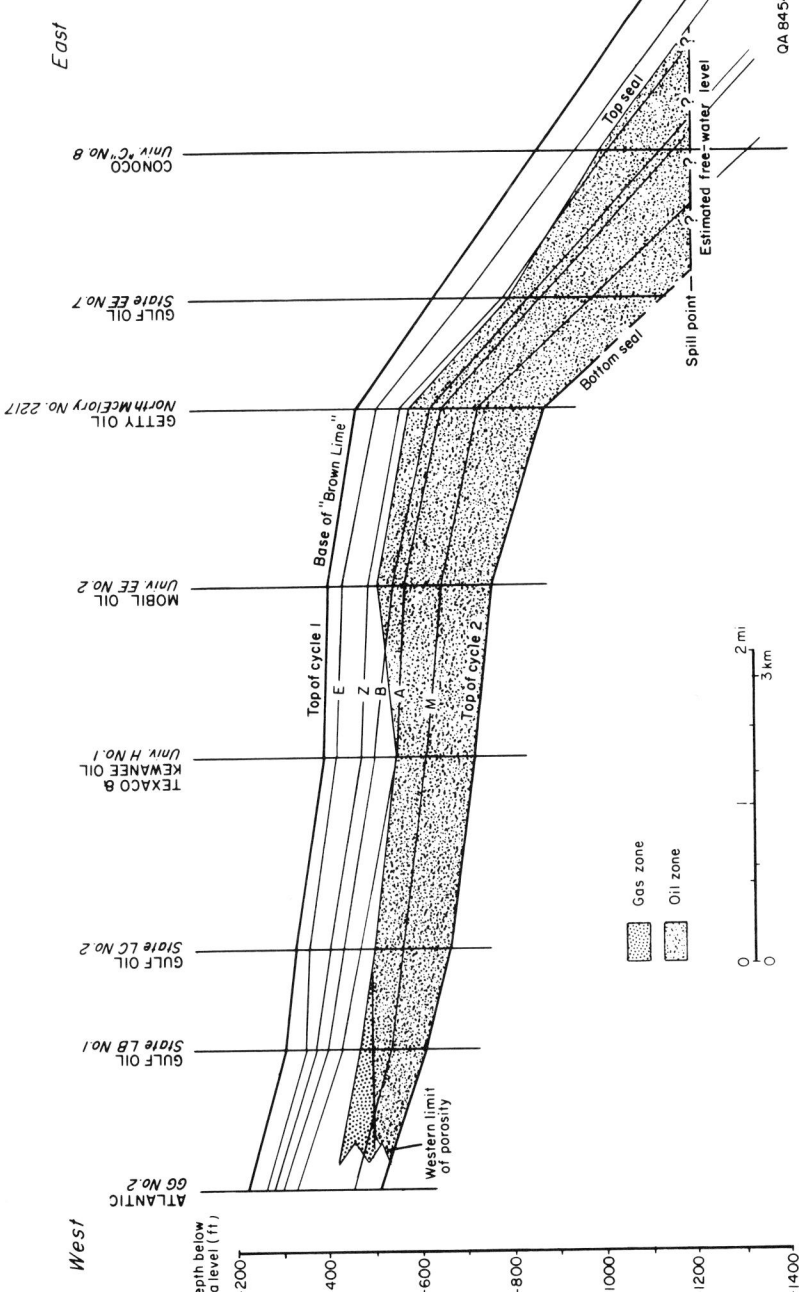

Figure 5. Structural cross section across Dune Field, illustrating the top seal of the reservoir, the oil-producing zone, the bottom seal, and the free-water level. Updip loss of porosity and permeability cause the trapping of the hydrocarbons on the eastern flank of the anticline.

incorrectly described as mudstone or as fusulinid or mollusk wackestone where fusulinid and mollusk grains compose 10 percent or more of the rock volume (Fig. 6a, b). Pellets can, in many cases, be identified on dry-sanded surfaces with a low-magnification microscope, in thin section using a standard petrographic microscope (Fig. 6c), or on broken surfaces using a scanning electron microscope. Thin sections observed under cathodoluminescent illumination may be required to identify pellets where syndepositional or postdepositional compaction and/or pervasive dolomitization have nearly destroyed the pellet fabric (Fig. 6d).

Pellet grainstone is the primary San Andres reservoir rock at Penwell and Jordan Fields. Interparticle porosity is commonly well preserved and results in a relatively high permeability rock. Fusulinid and mollusk molds contribute somewhat to reservoir volume but have little or no effect on permeability.

Algal Grainstone

Algal grainstone, with both micritized and well-preserved dascycladacean algae grains, occur in thin and discontinuous beds. Some algal grainstones are bedded, and some are crossbedded. Pervasive dolomitization has somewhat obscured the depositional texture, but algal grainstone is generally free of muddy matrix (Fig. 6e, f). This facies is interpreted as having been deposited in a relatively high energy ramp-interior setting, probably as tidal-channel deposits. Where not thoroughly cemented by sulfates, this facies has high porosity and permeability.

Sponge-Algal Boundstone

Thin zones of sponge-algal boundstone are a rare subtidal facies in the San Andres (Fig. 7a). These zones are only a foot or two thick and, although they contain some interparticle porosity, are not of sufficient volume to be considered a significant part of the reservoir. This facies is interpreted to have formed as isolated ramp-margin reef mounds, and their occurrence is apparently restricted to the downdip (eastern) side of the structure.

Figure 6. a) Slabbed-core view of pellet grainstone. Note that the grainstone texture is nearly invisible at this scale of resolution (E.P.S.A.U. # 1110, 3,499 ft). b) Slabbed-core view of pellet-fusulinid grainstone. Fusulinids are preserved as molds. Note that because the pellets are difficult to see in slabbed core this rock might be incorrectly described as a fusulinid wackestone (E.P.S.A.U. # 1313, 3,900 ft). c) Plane-light photomicrograph illustrating pellet texture (E.P.S.A.U. #1110, 3,499 ft). d) Cathodoluminescence photomicrograph of the same field of view as c. Note that pellets are easier to discern under cathodoluminescent illumination. e) Slabbed-core view of crossbedded algal grainstone (E.P.S.A.U. #207, 3,807 ft). f) Plane-light photomicrograph illustrating dascycladacean algae grains, dolomite cement, and porosity in the algal grainstone facies (same sample as e).

Figure 7. a) Slabbed-core view of sponge-algal boundstone (E.P.S.A.U. #1313, 3,947 ft). b) Slabbed-core view of crinoid grainstone (E.P.S.A.U. #2307, 3,659 ft). c) Slabbed-core view of faintly laminated mudstone (E.P.S.A.U. #1313, 3,772 ft). d) Slabbed-core view of pisolite packstone/grainstone, illustrating fenestrae and sheet cracks characteristic of this facies (E.P.S.A.U. #1505, 3,467 ft). e) Slabbed-core view of brecciated pisolite grainstone. Broken pieces of pisolite packstone/grainstone found at the high angle illustrated here are interpreted as tepee structures (E.P.S.A.U. #1505, 3,464 ft).

Figure 8. a) Slabbed-core view of pisolite grainstone in which fenestrae are incompletely plugged by sulfate cements. This rock may act as a floodwater thief zone (E.P.S.A.U. #1505, 3,487 ft). b) Slabbed-core view of karst dissolution of pisolite grainstone and infilling by greenish-gray siltstone (E.P.S.A.U. #1914, 3,745 ft). c) Slabbed-core view of laminated siltstone (E.P.S.A.U. #1914, 3,730 ft). d) Cross-nicols photomicrograph illustrating sulfate cement. The large cement zone is gypsum; the higher-birefringence zone in the center of the field of view is anhydrite (E.P.S.A.U. #1505, 3,562 ft). e) Plane-light photomicrograph of a vug containing sulfate cement (bottom) with subsequently deposited residual hydrocarbons (top)(E.P.S.A.U. #1110, 3,679 ft).

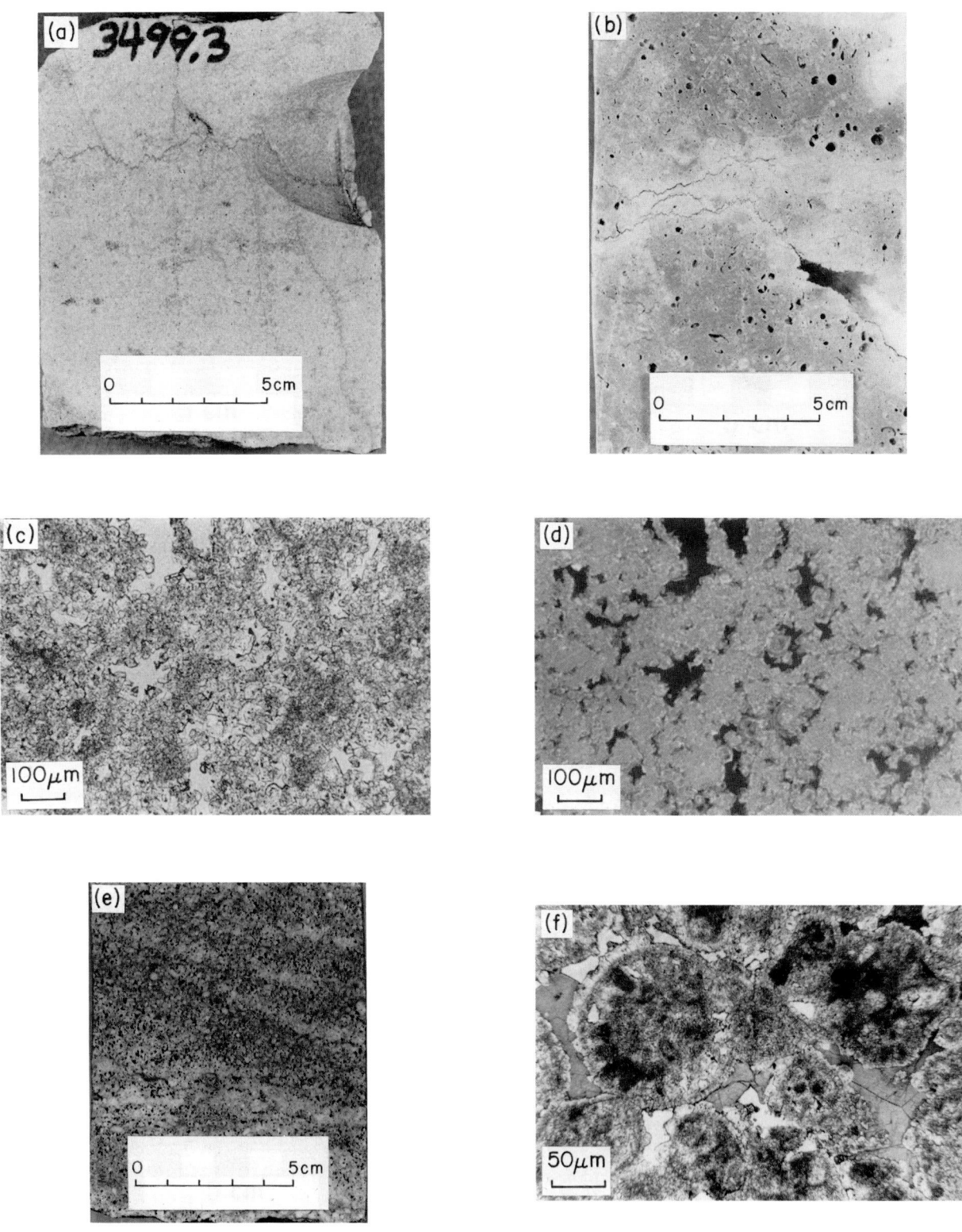

Figure 6

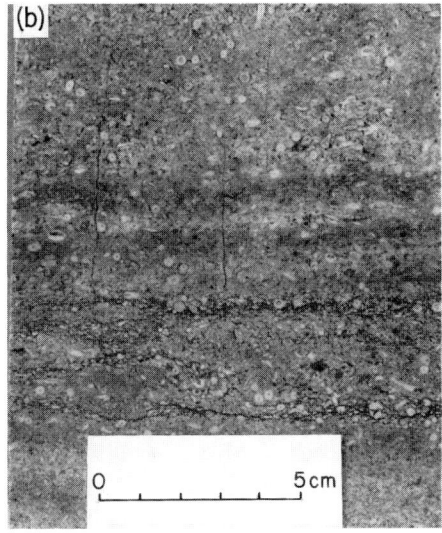
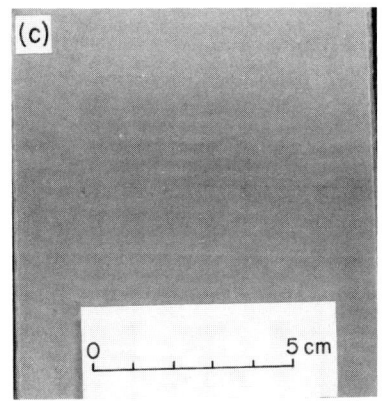
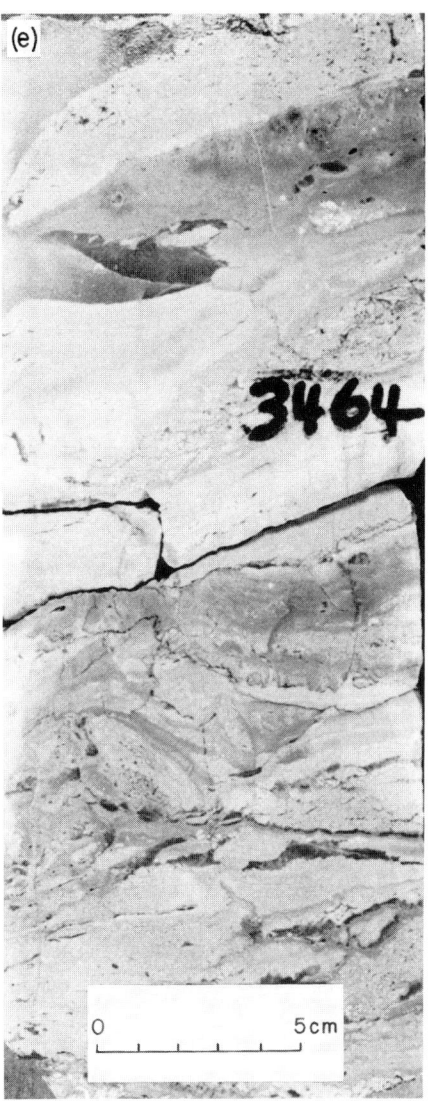
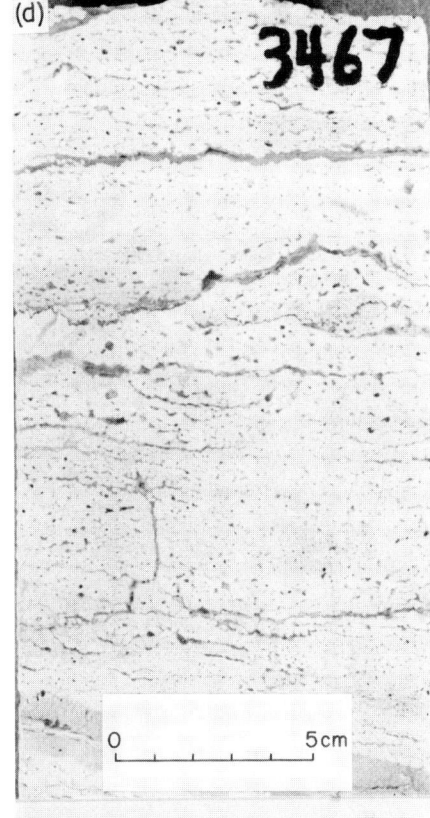

Figure 7

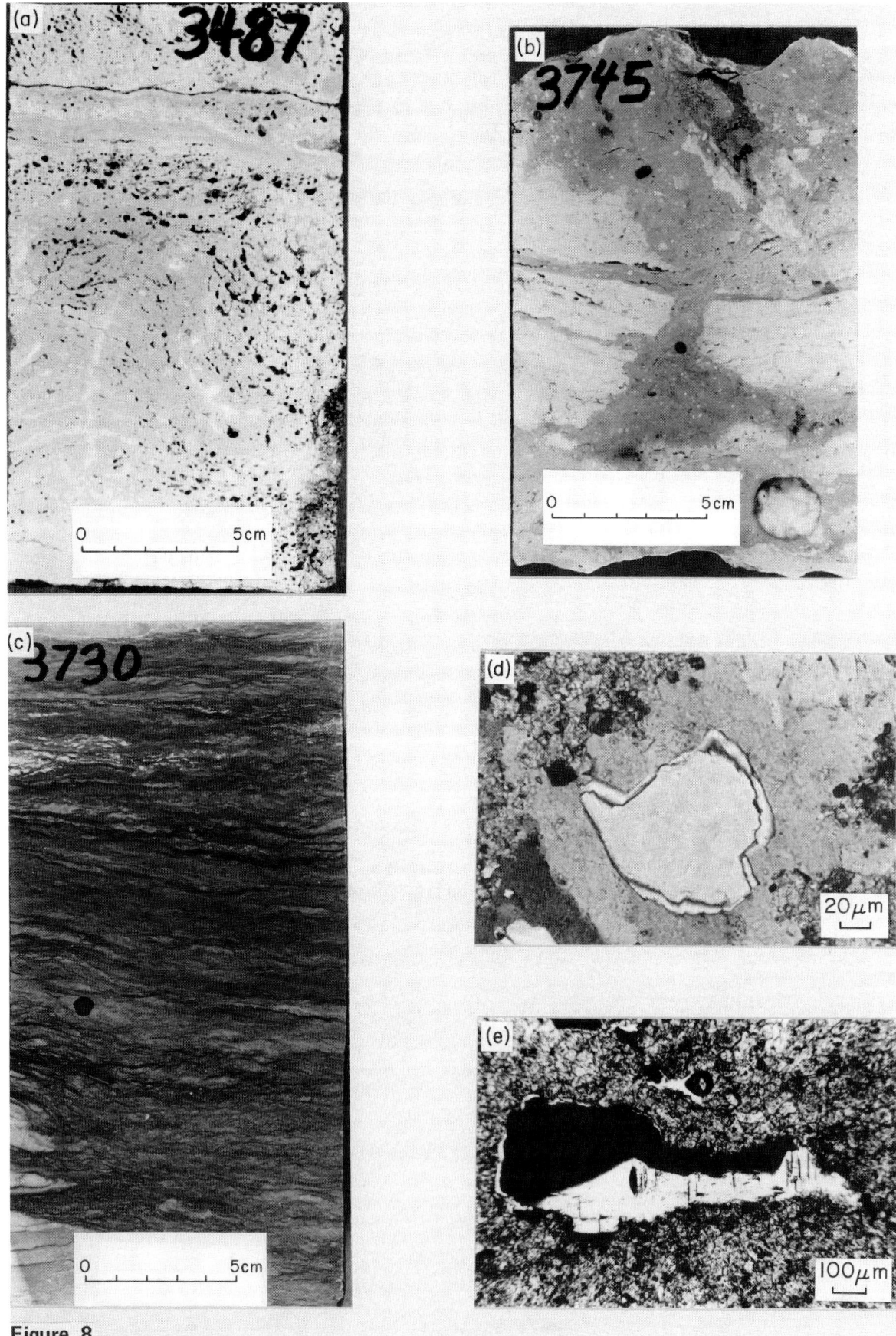

Figure 8

Crinoid Grainstone

Crinoid fragments occur as rare accessory grains in pellet grainstone but are observed in sufficient quantities to warrant designation of a separate facies in only one downdip core. Crinoid grainstone is poorly sorted and apparently free of muddy matrix, although it is thoroughly cemented by dolomite and is not part of the reservoir (Fig. 7b). This facies is intepreted to represent crinoid meadows, which were a deep-water, subtidal ramp-margin deposit.

Mudstone

Much of the reservoir seal in the San Andres is dolomitic mudstone deposited in a peritidal environment (Fig. 7c). These sediments were deposited as carbonate mud, in some cases finely laminated and generally not pelleted, presumably because rapid salinity fluctuations in the restricted peritidal environment were too stressful for organisms that produce pellets and bioturbate the sediment in deeper-water, open-marine environments. The mudstone facies is generally barren of fossils, although algal laminites and rare fusulinids and mollusks have been observed. The fusulinids and mollusks were probably grains transported by storms from deeper-water, open-marine environments.

Pisolite Packstone/Grainstone

The facies in the San Andres section exhibiting the most evidence for syndepositional subaerial exposure is pisolite packstone/grainstone. These rocks are composed of poorly sorted and fitted-fabric pisolites and are characterized by sheet cracks, fenestrae, and desiccation cracks (Fig. 7d, e). Pisolite packstone/grainstone is commonly interbedded with mudstone and is characteristically the same cream color. This facies is also commonly barren of skeletal grains. Intergranular porosity in pisolite packstone/grainstone is commonly thoroughly cemented by anhydrite. However, thin partially cemented zones are rare and only a foot or two thick, but they may be high-permeability floodwater thief zones (Fig. 8a). Locally, minor karst dissolution is indicated by severe brecciation and infilling by greenish-gray siltstone (Fig. 8b).

Siltstone

Siliciclastic siltstone beds occur within the peritidal part of the San Andres at the Penwell and Jordan Fields. Some of these siltstones are finely laminated, but most are massive (Fig. 8c). These rocks are often carbonaceous and in transitional contact with peritidal mudstone and pisolite packstone/grainstone. Siltstone is interpreted to be windblown detritus deposited in shallow water and reworked in the shallow-water marine environment.

Summary of Depositional Environments

The San Andres Formation at the Penwell and Jordan Fields is an upward-shoaling sequence of ramp sediments (Figs. 9 and 10). The deeper-water marine section is characterized by pelleted mud and open-marine fauna, mostly fusulinids and mollusks, with minor occurrences of sponges, algae, and crinoids. The subtidal section contains rare, locally developed sponge-algal boundstone interpreted as reef mounds. The shoreward tidal-flat environment was characterized by tranquil high-salinity waters in which environmental stress excluded most fauna, resulting in deposition of barren carbonate mud. High-exposure tidal flats were sites of pisolite formation and desiccation features such as sheet cracks and fenestrae. The tidal-flat and open-marine parts of the ramp are locally cut by dip-oriented relatively high energy tidal channels. These deposits are characterized by skeletal grainstones in which the grains are dominantly dascycladacean algae (Fig. 11).

Porosity Distribution and Production Characteristics

The main reservoir facies are subtidal pellet and skeletal grainstones in the lower part of the upward-shoaling sediment sequence (Figs. 9 and 10). Porosities are as high as 10 to 15 percent and permeabilities attain values of 10 md or greater. Net-pay zones are as thick as 100 ft or more (Fig. 10). Peritidal pisolite packstone/grainstone and mudstone in the upper part of the upward-shoaling sequence are generally impermeable and serve as a seal. Locally, the pisolite packstone/grainstone facies is incompletely cemented by sulfates, resulting in thin permeable zones that may be floodwater thief zones.

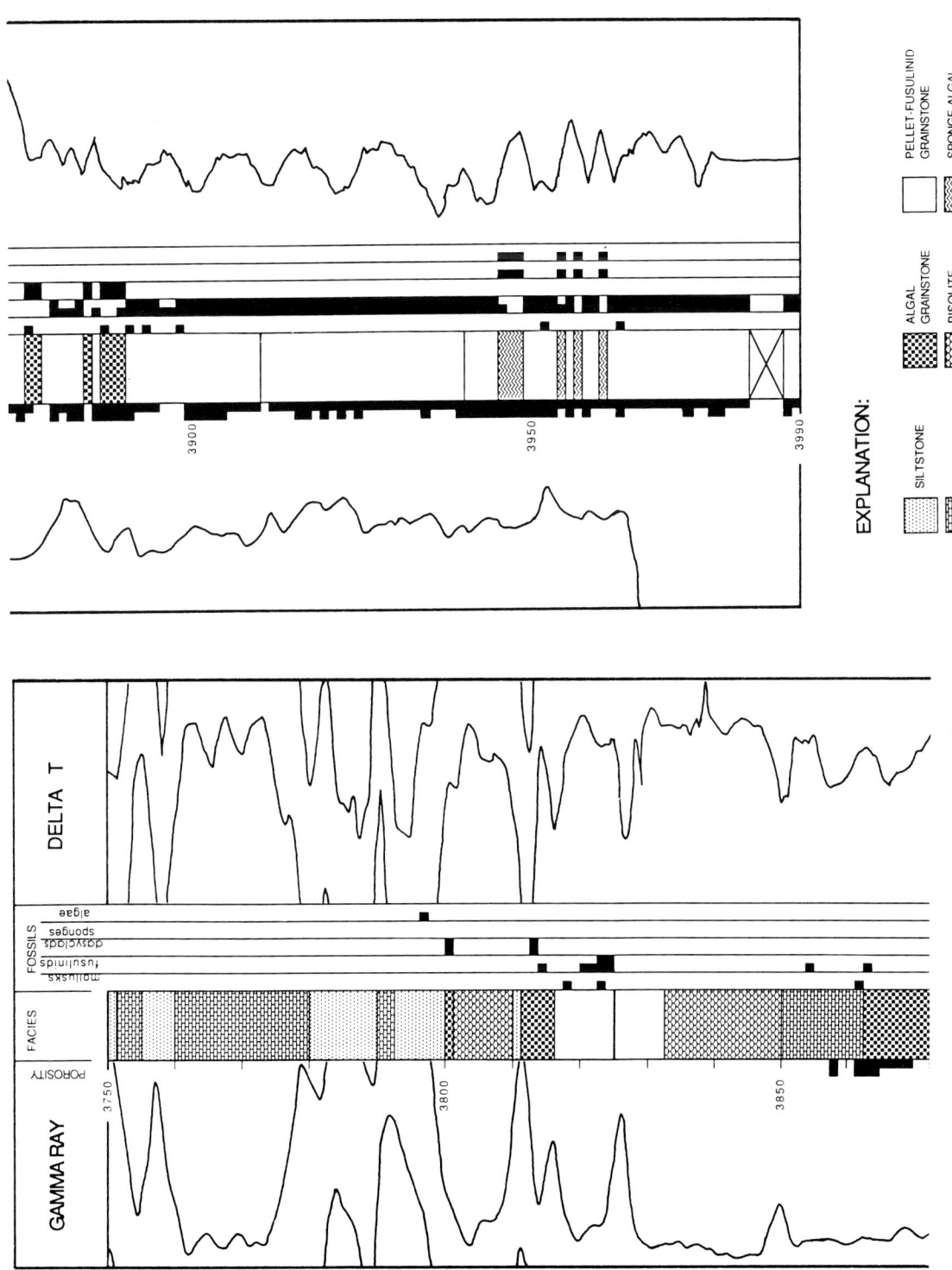

Figure 9. Wire-line log and core description data for the upper San Andres interval in the East Penwell San Andres Unit No. 1313.

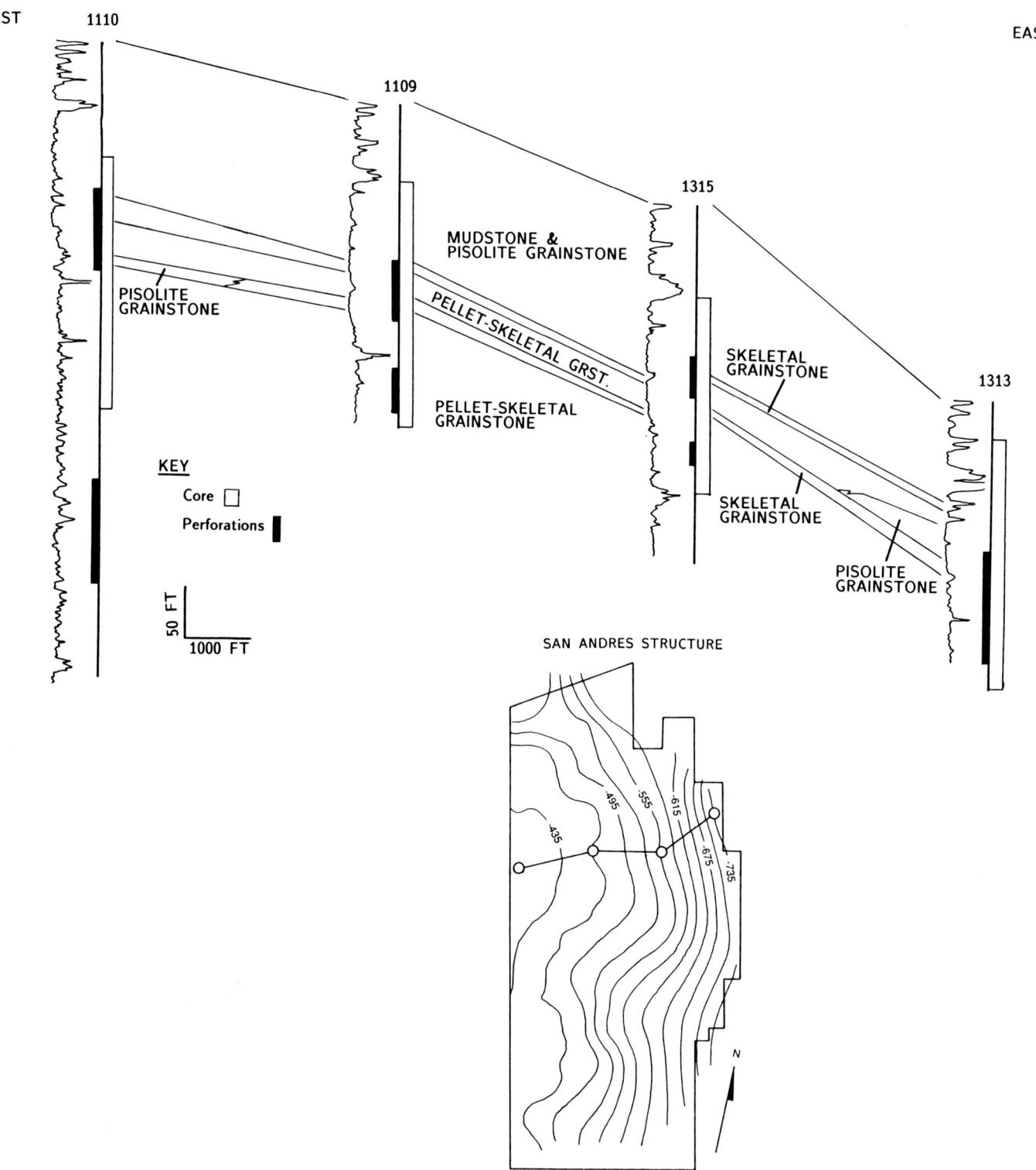

Figure 10. West-east cross section illustrating gamma-ray logs and generalized depositional facies, and San Andres structure map indicating the trace of the cross section, East Penwell San Andres Unit. The facies constitute a low-energy, upward-shoaling sediment sequence.

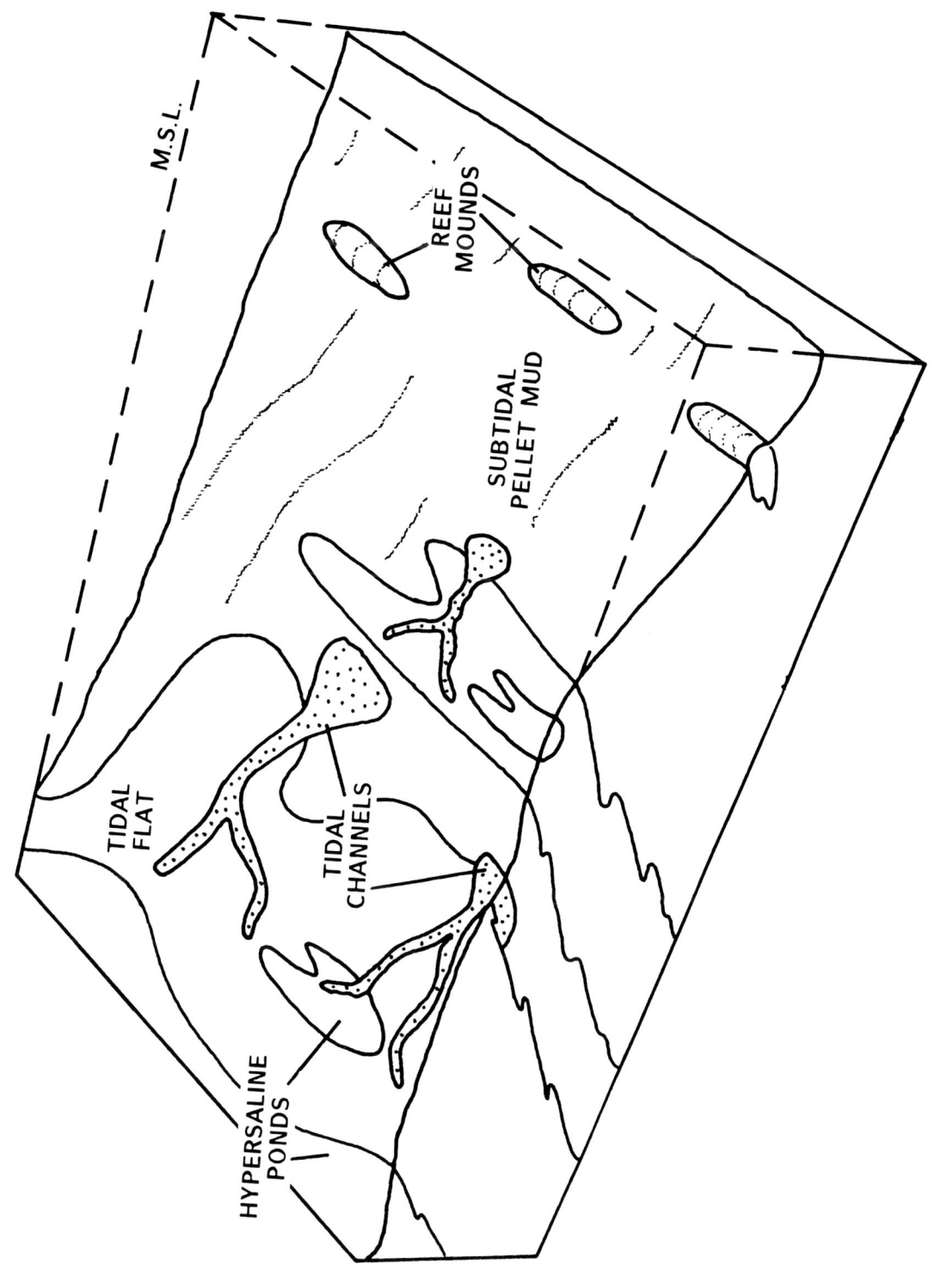

Figure 11. Schematic presentation of interpreted depositional environments in the San Andres Formation at Penwell Field. Deposition was on a low-energy carbonate ramp.

Facies isopachous maps of the Penwell Field indicate that peritidal sediments are thick in the southwestern part of the East Penwell San Andres Unit and thin to the north and east. Cumulative production maps generally mimic this pattern, although there are some anomalously high production wells. Highest cumulative production and highest remaining mobile oil occur in a northeast-southwest trend in the northern part of the field, and this feature is attributed to the influence of tidal channels that are approximately parallel to dip.

GRAYBURG FORMATION

Facies and Depositional Environments

The following facies descriptions and interpretations are based on examination of cores and well logs from the Grayburg Formation in Dune Field, in the southern part of the P.J.W.D.M. Field Complex.

Mollusk Wackestone

This facies is dominantly mollusk wackestone, although it contains thin zones of mollusk grainstone. Mollusks are commonly preserved as molds only, and most of these are plugged with anhydrite cement. This facies is most clearly distinguished by an absence of fusulinids. Crinoid and sponge fragments are present as rare accessory grains; burrows and wavy laminations are common. The mollusk wackestone facies is interpreted as a shallow-water shelf deposit.

Compacted-Pellet Facies

This facies is composed primarily of grainstones composed of unidentifiable grains. These grains are interpreted to have been pellets that have been compacted and deformed somewhat during burial. This facies is also distinguished from other subtidal facies by the lack of skeletal grains. The compacted-pellet facies is interpreted as having been deposited on a protected shallow-water shelf.

Crinoid Packstone/Grainstone

The crinoid packstone/grainstone facies is most readily characterized by the presence of crinoids, which occur throughout but are never abundant, and the generally high grain content (Fig. 12a, b). This facies is actually wackestone in the west side of Dune Field, where it grades into the vertically structured facies. To the east, toward the shelf edge, thick packstone and grainstone beds alternate, and porosity is best developed in the grainstone sections. Bryozoan and sponge fragments are common, but unidentifiable grains, probably faecal pellets, compose much of the rock on the eastern side of the field. The crinoid packstone/grainstone facies is interpreted to be a shallow-water shelf and slope deposit.

Vertically Structured Facies

This facies is composed of wackestone and packstone containing abundant bryozoan, mollusk, and sponge skeletal grains and is characterized by vertically aligned structures with sharp to diffuse boundaries (Fig. 12c, d). The structures are from a few to several centimeters wide, reach 20 to 30 centimeters in length, and are commonly cylindrical to oblate in cross section. Across the boundaries of these vertical structures there is commonly a difference in color, dolomite crystal size, packing and orientation of skeletal grains, porosity, and anhydrite content. There are a few examples of sponge pore systems preserved at the edges of these structures, but more commonly these structures contain faint laminations suggesting algal origin. Burrows, brecciation, fractures, stylolites, and anhydrite nodules are also common. The vertically structured facies is interpreted as a low-energy shelf edge deposit.

Fusulinid Wackestone

The fusulinid wackestone facies is represented by a mud-supported carbonate with common to abundant fusulinids (Fig. 13a, b). The fusulinids exhibit a range of preservation from complete to partially leached with geopetal structure of test remains at the base of the mold to completely leached and filled

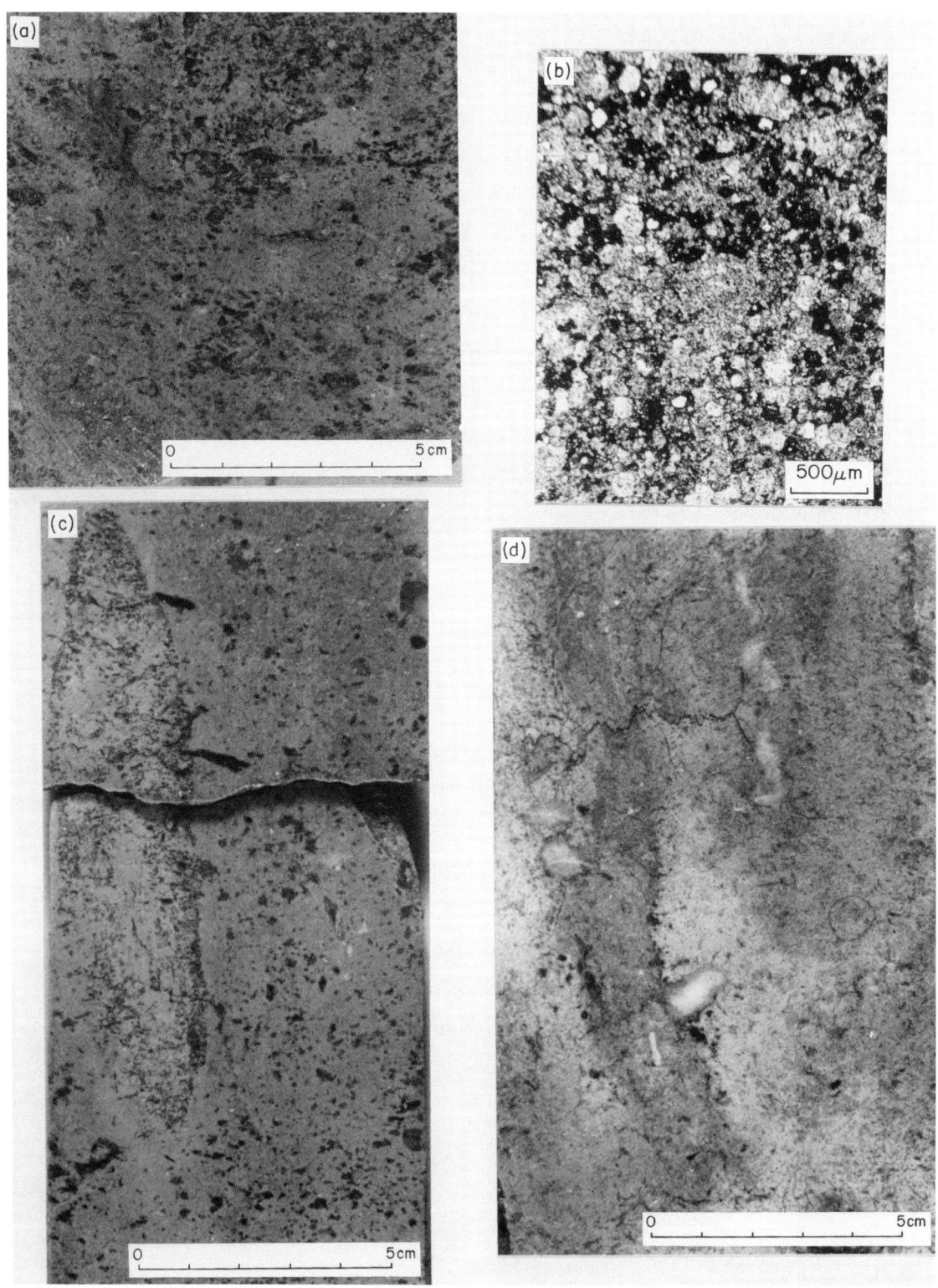

Figure 12. a) Slabbed-core view of crinoid packstone/grainstone (Mobil Univ. #1559, 3,491 ft). b) Plane-light photomicrograph of crinoid packstone/grainstone (Mobil Univ. #1559, 3,502 ft). c,d) Slabbed-core views of the vertically structured facies (Mobil Univ. Sec. 1/2 28, 3,303 ft; Gulf State EB #9, 3,766 ft).

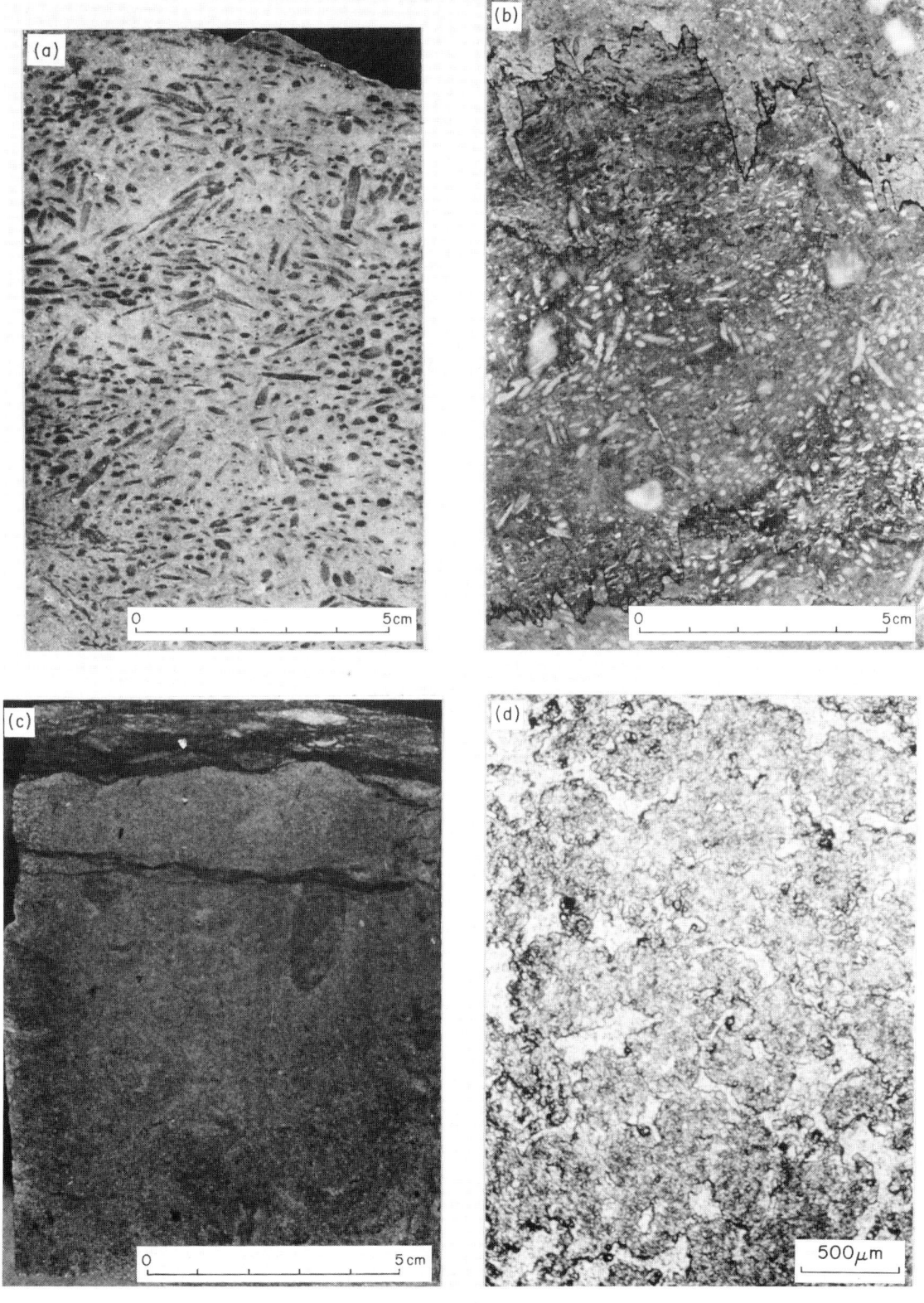

Figure 13. a,b) Slabbed-core views of fusulinid wackestone (Mobil Univ. #1635, 3,438 ft; Gulf State EB #9, 3,884 ft). c) Slabbed-core view of pellet grainstone (Mobil Univ. #1539, 3,301 ft). d) Plane-light photomicrograph of pellet grainstone (Mobil Univ. #1559, 3,341 ft).

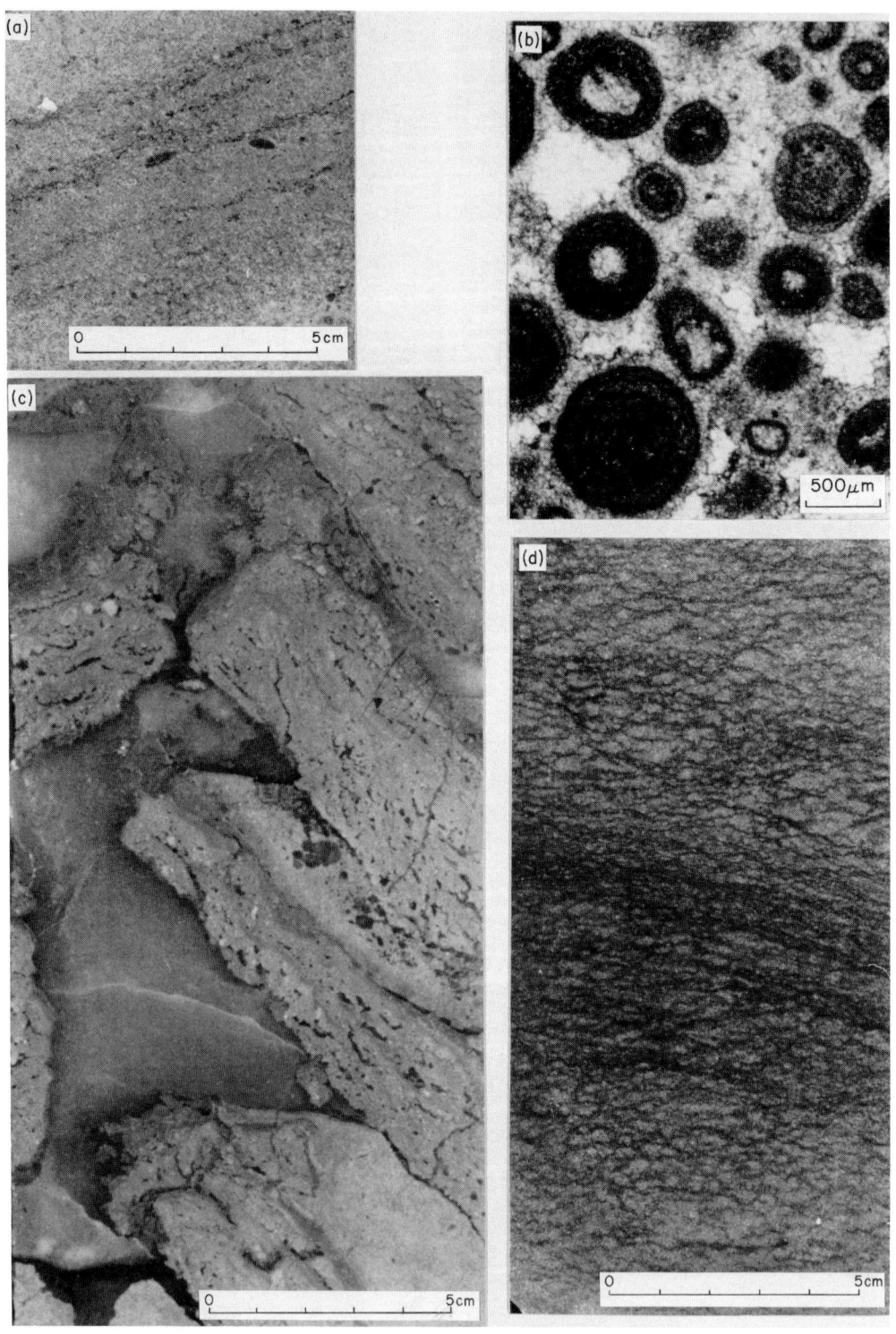

Figure 14. a) Slabbed-core view of crossbedded ooid grainstone (Mobil Univ. #1539, 3,291 ft). b) Plane-light photomicrograph of ooid grainstone (Mobil Univ. CC #3, 3,139 ft). c) Slabbed-core view of a brecciated tepee structure in the pisolite facies (Gulf State #13, 3,684 ft). d) Slabbed-core view of laminated siltstone (Mobil Univ. #1540, 3,550 ft).

with anhydrite cement. Total fusulinid preservation is rare. Although molds may constitute a significant amount of porosity in this facies, the greatest volume of porosity is in the form of intercrystalline matrix and vuggy porosity. The fusulinid wackestone facies is interpreted to be a shallow-water platform or shelf deposit that formed in water of varying depth.

Pellet Grainstone

This facies is dominated by round featureless grains interpreted as faecal pellets, but it also contains intraclasts, mollusks, and rare crinoids and fusulinids (Fig. 13c, d). Burrows are the most common structures, and laminations occur locally. This facies contains both interparticle and moldic porosity. The pellet grainstone facies is interpreted as a shallow-water burrowed grain flat, possibly partially exposed at occasional very low tide.

Ooid Grainstone

This grainstone facies is characterized by parallel laminations and crossbedding, and the dominant grains are ooids, although accessory pellets, mollusks, dascycladacean algae, and intraclasts are also present (Fig. 14a, b). Porosity is low because of abundant anhydrite cement. The ooid grainstone facies is interpreted as a high-energy shoal deposit formed adjacent to tidal flats and offshore bars on the basinward side of pellet flats.

Pisolite Facies

This facies comprises a wide range of carbonate fabrics from pisolite grainstone/packstone to mudstone and laminated mudstone. Fenestrae, tepee structures, laminated crusts, and sheet cracks are common and indicate syndepositional desiccation (Fig. 14c). Polygonal-fitted fabrics and inverse grading are also common in this facies. Crinkly (algal?) laminations occur in some of the mudstone parts of this facies. The pisolite facies contains abundant anhydrite cement and is generally characterized by low porosity and permeability. The pisolite facies was deposited on an arid tidal flat where conditions ranged from supratidal flats to restricted subtidal ponds.

Siltstone

The siltstone facies occurs as thin beds containing 20 to 80 percent siliciclastic silt interbedded with the dolostone units. Small burrows are the most common structures, although laminations occur locally (Fig. 14d). The siltstone is interpreted to have been transported to the Dune Field area by eolian processes and was deposited over the entire facies tract. Thus, siltstone formed discrete marker beds across all carbonate facies, and correlative siltstone zones are interpreted to approximate time lines. The high gamma-ray response associated with this facies allows reliable correlation over the entire field using well logs.

Summary of Depositional Environments

The Grayburg Formation at Dune Field is an upward-shoaling sequence of ramp sediments (Fig. 15), somewhat similar to the upward-shoaling sequence in the San Andres Formation. Correlation of siltstone markers allows the Grayburg Formation in Dune Field to be subdivided into three units (Fig. 16). The Lower Unit is entirely composed of fusulinid wackestone and is interpreted to be a tranquil-water normal-marine sediment deposited below wave base. Rare, thin pellet grainstone zones probably represent local high-energy shoals.

The Middle Unit is dominated by the vertically structured and crinoid packstone/grainstone facies. The origin of the vertical stuctures is problematic. One interpretation is that the structures are the result of poorly preserved framework constructed by blue-green algae and calcareous sponges. The common occurrence of fragments of cellular skeletal material and laminations preserved in some of the structures support this interpretation. Alternatively, the structures may be the result of burrowing in exposed-flat sediments, as has been proposed by Longacre (1980). Both of these interpretations indicate the presence of a northwest-trending bank that ranged at times from a few feet in water depth to subaerially exposed. The lack of fusulinids, which are common in most other subtidal Grayburg environments, and the occurrence of sponge and bryozoan fragments indicate a unique environment different from adjacent facies (Fig. 17a).

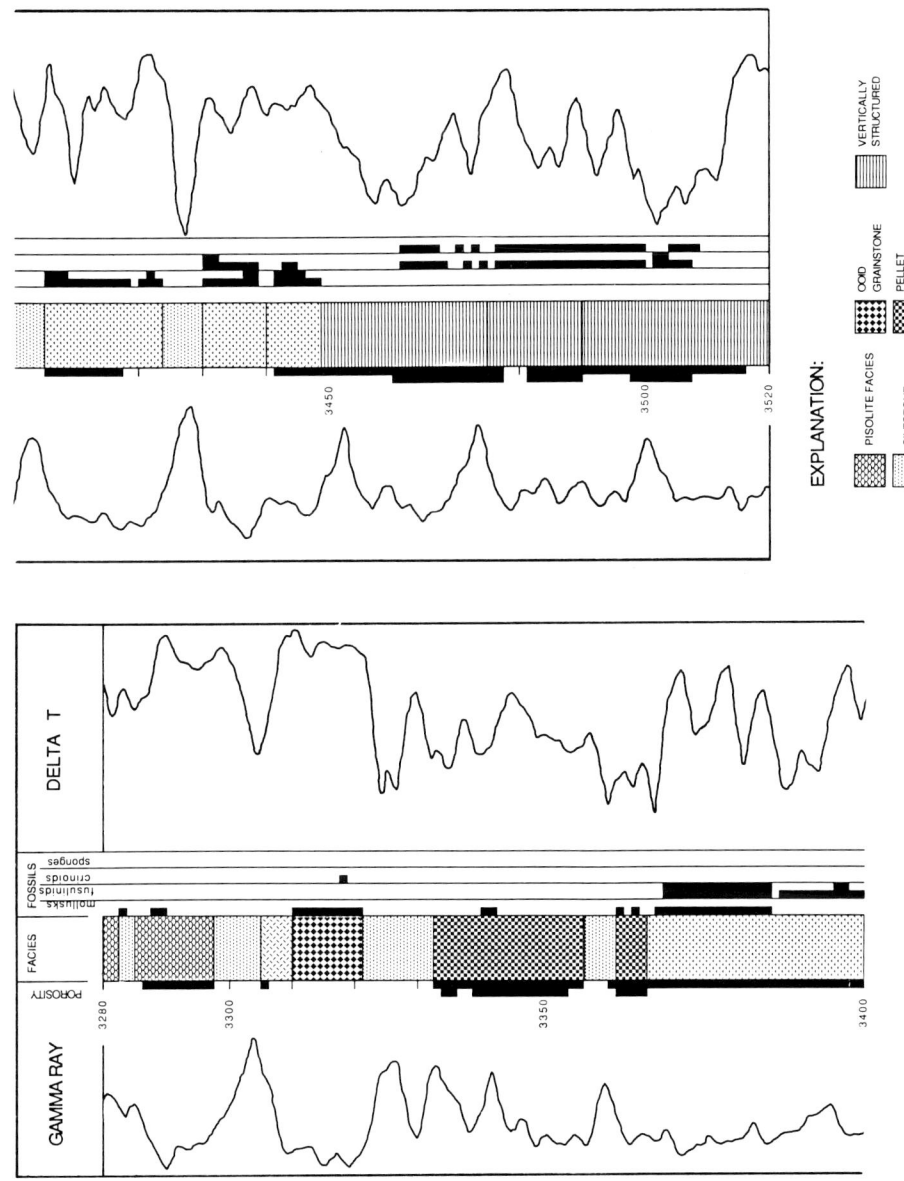

Figure 15. Wire-line log and core description data for the Grayburg interval in the Dune University No. 1560.

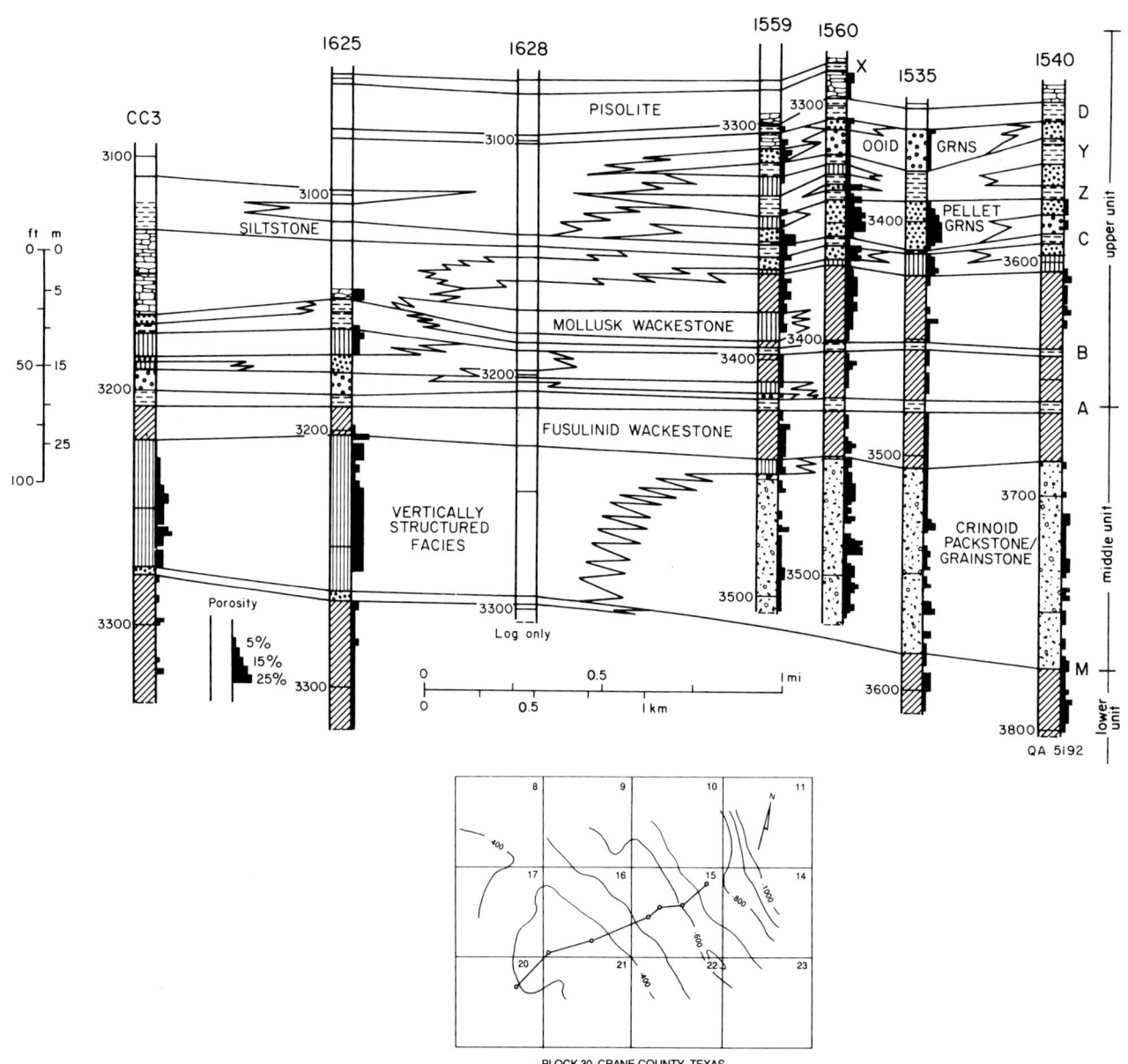

Figure 16. Facies distribution along a section across the Mobil Unit 15/16 as determined from detailed core descriptions. The core control is indicated by the patterns within the wellbores. Porosity as determined from visual examination of the cores is indicated by black bars on right side of columns.

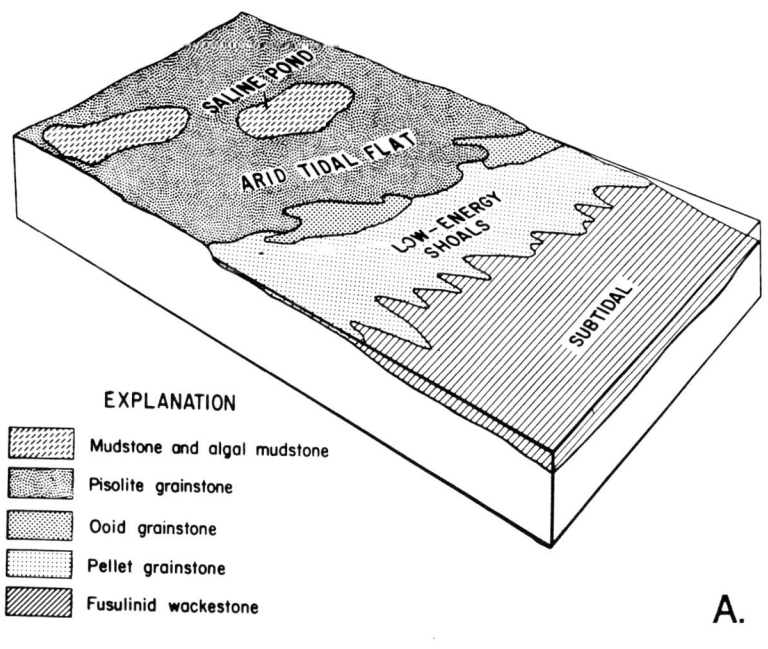
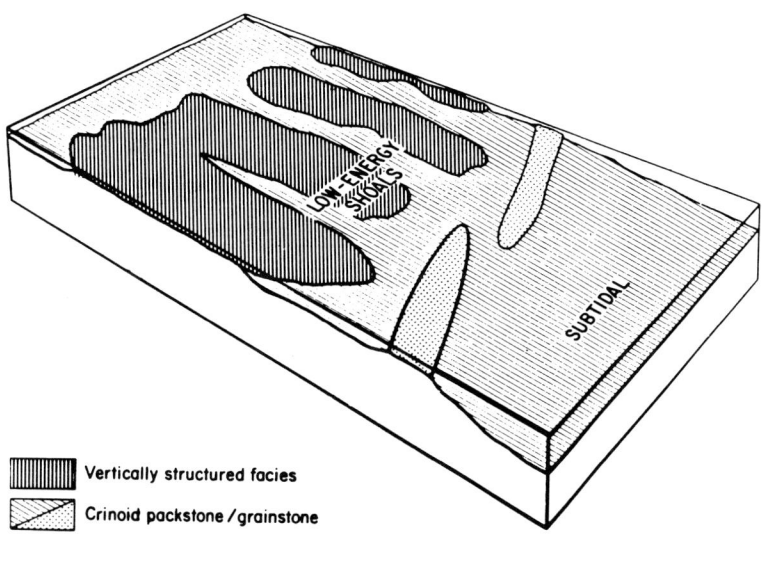

Figure 17. Interpreted depositional environments of the upper Grayburg unit (A) and the middle Grayburg unit (B) at Dune Field.

The Upper Unit is characterized by a vertical succession in ascending order of fusulinid wackestone, pellet grainstone, ooid grainstone, and pisolite facies. This sequence is interpreted as a progradational and upward-shoaling sequence from shallow-water subtidal to arid tidal-flat environments. Highest energy occurred along the edges of these tidal flats where crossbedded and laminated ooids accumulated as fringing bars and beaches. The pellet grainstone, basinward of the ooid facies, represents a broad area of low-energy, burrowed, stable grain flat with water depths generally below normal wave base. Farther offshore was the extensive shallow-water subtidal shelf where fusulinid wackestone was deposited (Fig. 17b).

This vertical sequence of facies in the Dune Field is very similar to that of the adjacent McElroy Field, described by Harris and others (1984) and by Longacre (1980, 1983).

Porosity Distribution and Production Characteristics

Intercrystalline dolomite porosity is widely distributed in the subtidal and intertidal wackestone, packstone, and grainstone facies; however, porosity values are highest in the packstone and grainstone facies where the intercrystalline porosity is supplemented by high interparticle porosity. Permeability is highest in the grainstone facies where both porosity types occur. Low permeability zones, largely represented by the wackestone facies, occur both laterally and vertically and result in a heterogeneous reservoir. Because of the progradational nature of the upper part of the Grayburg, the shelfward-occurring low-porosity supratidal pisolite facies migrated basinward over the porous and permeable intertidal and subtidal facies and provided a superposed and marginal seal for the reservoir (Fig. 16).

The majority of remaining mobile oil in the Grayburg Formation at Dune Field is in packstones and grainstones of the Middle Unit. In particular, northwest-southeast trending crinoid grainstone bars, interpreted as offshore marine grainstone bars deposited subparallel to depositional strike, are characterized by excellent porosity and permeability. These features contain the greatest concentration of remaining mobile oil (Bebout and others, 1987).

SAN ANDRES/GRAYBURG DIAGENESIS

Diagenesis in the San Andres and Grayburg Formations began with induration of soft pelleted mud. Where induration resulted in good pellet preservation, interparticle porosity is preserved. Where pellets were compacted, most of the porosity was destroyed.

The entire section has been pervasively dolomitized, and dolomitization of the original carbonate sediment was the major diagenetic event. Strontium-isotope values (Leary and Vogt, 1986) indicate that dolomitization took place during Guadalupian time. Oxygen and carbon isotopic data (Bein and Land, 1982; Leary and Vogt, 1986; Naiman, 1982) indicate dolomitization by hypersaline waters that originated through evaporation of sea water. Therefore, these San Andres and Grayburg carbonates probably were dolomitized by hypersaline water that originated on arid tidal flats and percolated through the shallow subsurface during the Guadalupian. This hypersaline brine was also probably the source of the anhydrite and gypsum, which are common in the San Andres/Grayburg section.

Petrographic evidence suggests that sulfates were probably entirely anhydrite at some time during the diagenetic sequence and are now partly hydrated to gypsum (Fig. 8d). Presence of gypsum in these formations is especially noteworthy because the bound water in this mineral affects interpretation of core-analysis data and wireline logs. Specifically, the bound water in gypsum may be driven off during high-temperature core-cleaning procedures, thus artificially increasing core porosity, and the bound water is recorded as porosity by neutron logs (Tilly and others, 1982; Bebout and others, 1987). Rare oil staining postdates sulfate leaching, suggesting that leaching may be associated with hydrocarbon migration (Fig. 8e).

Calcite and native sulfur replaced some of the sulfate minerals. The light carbon values from the calcite (Leary, 1984) suggest that these minerals are the result of reduction of sulfate by bacteria, introduced by meteoric water, in the presence of hydrocarbons.

SUMMARY

The San Andres and Grayburg Formations comprise two upward-shoaling cycles of normal marine through peritidal carbonate rocks. Both formations contain similar depositional facies and were deposited in a carbonate ramp environment.

Oil production is primarily from subtidal grainstones. Linear grainstone trends are the locus of highest porosity and highest remaining mobile oil. These grainstone bodies result in lateral reservoir heterogeneity and are the target for geologically sited infill wells.

ACKNOWLEDGMENTS

This research was funded by the University of Texas System and is part of an ongoing study of hydrocarbon reservoirs on University Lands. We are grateful for access to data through the courtesy of The University of Texas Lands Office, Fina Oil and Chemical Company, and Mobil Oil Corporation. In the course of this work we have benefited from critical discussion with several colleagues, notably G. E. Fogg, C. M. Garrett, Jr., E. Guevara, C. Hocott, M. Holtz, C. Kerans, S. C. Ruppel, N. Tyler, and G. W. Vander Stoep.

Publication authorized by the Director, Bureau of Economic Geology, The University of Texas at Austin.

REFERENCES

BEBOUT, D. G., LUCIA, F. J., HOCOTT, C. R., FOGG, G. E., and VANDER STOEP, G. W., 1987, Characterization of the Grayburg Reservoir, University Lands Dune Field, Crane County, Texas: The University of Texas at Austin, Bureau of Economic Geology Report of Investigations No. 168, 104 p.

BEIN, A., and LAND, L. S., 1982, San Andres carbonates in the Texas Panhandle: The University of Texas at Austin, Bureau of Economic Geology Report of Investigations No. 121, 48 p.

GALLOWAY, W. E., EWING, T. E., GARRETT, C. M., TYLER, N., and BEBOUT, D. G., 1983, Atlas of major Texas oil reservoirs: The University of Texas at Austin, Bureau of Economic Geology Special Publication, 139 p.

GARRETT, C. M., JR., 1986, Correlation of San Andres and Grayburg (Guadalupian) reservoirs - Central Basin Platform, in: Bebout, D. G., and Harris, P. M., eds., Hydrocarbon Reservoir Studies, San Andres/Grayburg Formations, Permian Basin: Permian Basin Section, Society of Economic Paleontologists and Mineralogists Publication No. 86-26, p. 97-98.

HARRIS, P. M., DODMAN, C. A., and BLIEFNICK, D. M., 1984, Permian (Guadalupian) reservoir facies, McElroy Field, West Texas, in: Harris, P. M., ed., Carbonate Sands - A Core Workshop, Society of Economic Paleontologists and Mineralogists, Core Workshop No. 5, p. 136-174.

LEARY, D. A., 1984, Diagenesis of the Permian (Guadalupian) San Andres and Grayburg Formations, Central Basin Platform, West Texas: M.S. Thesis, The University of Texas at Austin, 129 p.

LEARY, D. A., and VOGT, J. N., 1986, Diagenesis of the Permian (Guadalupian) San Andres Formation, Central Basin Platform, West Texas, in: Bebout, D. G., and Harris, P. M., eds., Hydrocarbon Reservoir Studies - San Andres/Grayburg Formations, Permain Basin: Permian Basin Section, Society of Economic Paleontologists and Mineralogists Publication No. 86-26, p. 67-68.

LONGACRE, S. A., 1980, Dolomite reservoirs from Permian biomicrites, in: Halley, R. B., and Loucks, R. G., eds., Carbonate Reservoir Rocks, Society of Economic Paleontologists and Mineralogists, Core Workshop No. 1, p. 105-117.

LONGACRE, S. A., 1983, A subsurface example of a dolomitized Middle Guadalupian (Permian) reef from West Texas, in: Harris, P. M., ed., Carbonate Buildups - A Core Workshop, Society of Economic Paleontologists and Mineralogists, Core Workshop No. 4, p. 304-326.

NAIMAN, E. R., 1982, Sedimentation and diagenesis of a shallow marine carbonate and siliciclastic shelf sequence - the Permian (Guadalupian) Grayburg Formation, southeastern New Mexico: M.A. Thesis, The University of Texas at Austin, 197 p.

TILLY, H. P., GALLAGHER, B. J., and TAYLOR, T. D., 1982, Methods for correcting porosity data in a gypsum-bearing carbonate reservoir: Journal of Petroleum Technology, October, p. 2449-2454.

TYLER, N., GALLOWAY, W. E., GARRETT, C. M., JR., and EWING, T. E., 1984, Oil accumulation, production characteristics, and targets for additional recovery in major oil reservoirs of Texas: The University of Texas at Austin, Bureau of Economic Geology Geological Circular 84-2, 31 p.

WANLESS, H. R., BURTON, E. A., and DRAVIS, J., 1981, Hydrodynamics of carbonate fecal pellets: Journal of Sedimentary Petrology, v. 51, p. 27-36.

WARD, R. F., KENDALL, C. G. St. C., and HARRIS, P. M., 1986, Upper Permian (Guadalupian) facies and their association with hydrocarbons - Permian Basin, West Texas and New Mexico: American Association of Petroleum Geologists Bulletin, v. 70, p. 239-262.

STRATIGRAPHY AND LITHOFACIES OF THE SAN ANDRES FORMATION, C. S. DEAN "A," XIT, AND SW LEVELLAND UNITS OF LEVELLAND-SLAUGHTER FIELD, PERMIAN BASIN

PAUL M. HARRIS
Chevron Oil Field Research Company
La Habra, California 90633

EMILY L. STOUDT
Texaco Research Center
Houston, Texas 77215

ABSTRACT

A core study in the C. S. Dean "A," XIT, and SW Levelland Units of the giant Levelland-Slaughter Field in Cochran and Hockley Counties, Texas, helped define the geologic factors which affect enhanced recovery operations. The general nature of the Permian San Andres Formation in the area is one of widespread stratigraphic zones (both porous and tight intervals) that can be recognized on electric logs, confirmed by core description, and correlated between wells. Zones 7 through 3 recognized in the C. S. Dean "A," XIT, and SW Levelland Units conform uniformly to regional dip from the northwest to the southeast at approximately 30 feet per mile.

Producing intervals of the San Andres Formation occur within a large-scale regressive sequence. Within the C. S. Dean "A," XIT, and SW Levelland Units, this sequence is divided into zones that record, from bottom to top, open marine deposits (lying below Zone 7) overlain by a thick package of restricted shelf and tidal-flat deposits (Zones 7, 6, and 5) and capped by supratidal-flat deposits (Zones 4 and 3). There appears to be a good correlation between porous reservoir intervals and subtidal deposits, which are cyclical in nature. Porosity and permeability patterns within each porous interval reflect variations in the original deposits. The most common porous lithofacies are skeletal and peloidal dolopackstones-wackestones with moldic and intercrystalline porosities.

GEOLOGIC SETTING

The San Andres Formation (Permian, Early Guadalupian) is the most prolific oil-bearing horizon of the Permian Basin. The extent of San Andres production on the Northwest Shelf is evidenced particularly by the extensive Levelland-Slaughter Field of Cochran and Hockley Counties, Texas (Fig. 1). Since its discovery in 1936, the Levelland-Slaughter Field has produced more than 1.2 billion barrels of oil. The field is now undergoing enhanced recovery and contains more than 6,000 wells, including producers, shut-in producers, injectors, and water-supply and disposal wells. Most San Andres fields on the Northwest Shelf are primarily stratigraphic traps controlled by a combination of extensive dolomitization and anhydrite plugging. However, subtle structural nosing and changes in dip are important in localizing hydrocarbons (Gratton and LeMay, 1969; Ramondetta, 1982; Cowan and Harris, 1986).

In Cochran and Hockley Counties, Texas, the San Andres Formation consists of interbedded dolomites, anhydrites, and minor limestone. The San Andres is overlain by evaporitic dolomites of the Artesia Group (in ascending order: the Grayburg-Queen, Seven Rivers, Yates, and Tansill formations) and underlain by sandy anhydritic dolomites of the Glorieta and Yeso formations (Fig. 2). The San Andres is equivalent to basinal deposits of the Bone Spring, Brushy Canyon, and Cherry Canyon formations. The exact geometric configuration of the San Andres shelf is not well understood, although there are indications that some depositional topography existed within the shelf and the shelf-to-basin relief was not great, at least in the Midland Basin (Ramondetta, 1982). Carbonate sediments of the San Andres Formation were deposited mostly during "normal" sea-level stands, whereas terrigenous clastic material was introduced into the basin during low sea-level stands (Silver and Todd, 1969). As a result of this pattern of sedimentation, most of the carbonate rocks in the basin are distributed around its periphery and a comparatively thin "starved-basin" sequence is equivalent to a considerably thicker sequence of shelf-margin beds.

The San Andres Formation over much of the Northwest Shelf can be divided into upper and lower parts based on the occurrence of a regionally correlatable marker bed, a 5- to 10-foot siltstone, known as the "Pi" marker, which typically occurs 400 to 650 feet below the formation top (Fig. 2). The lower part of the San Andres, in which the reservoir zones occur, is an upward-shoaling

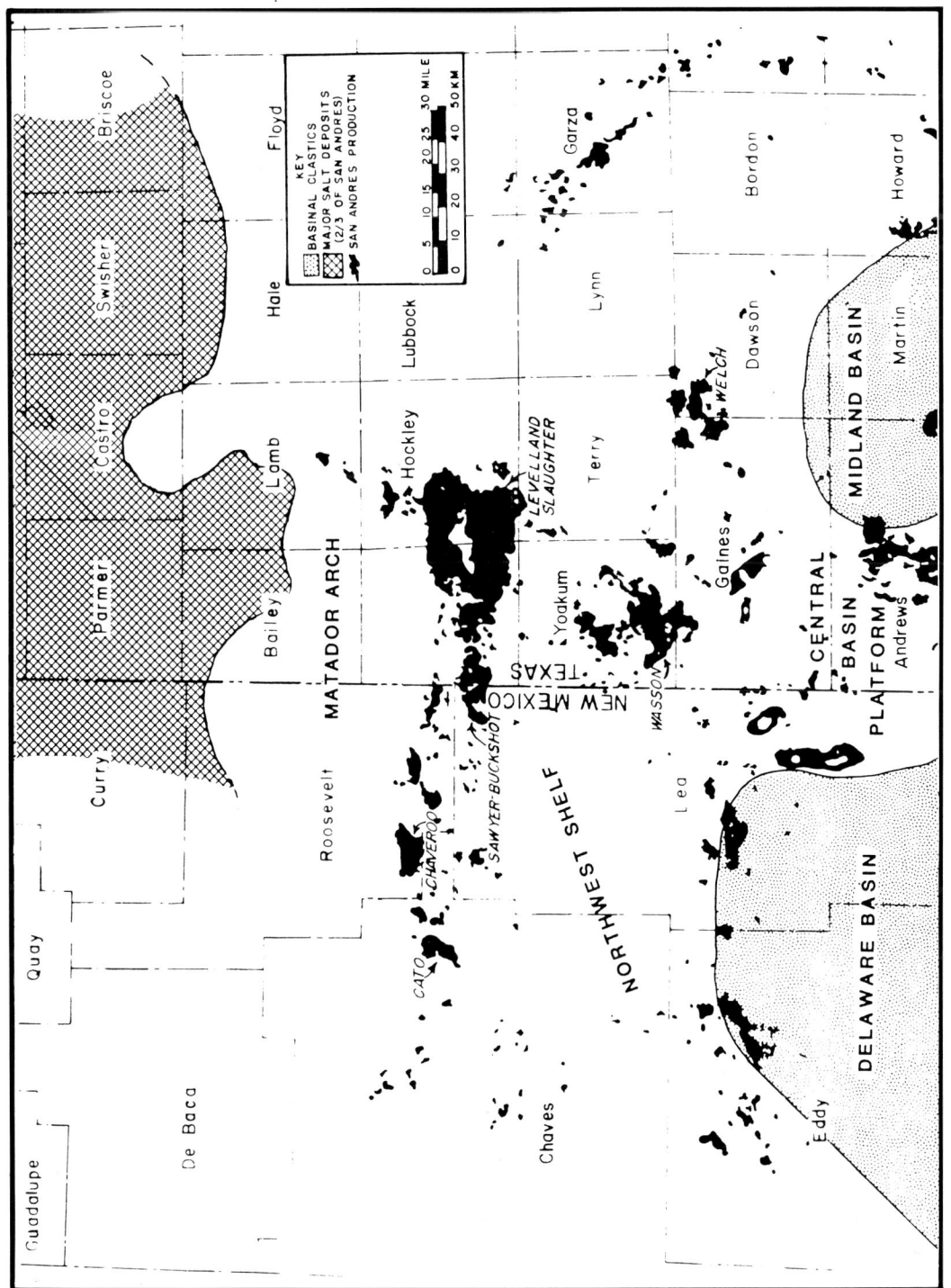

Figure 1 Map showing San Andres paleogeology and production on Northwest Shelf of Texas and New Mexico. Modified after Cowan and Harris, 1986. Shelf dolomites of the San Andres Formation formed on the Northwest Shelf and Central Basin Platform. Clastics were deposited in the Midland and Delaware Basins; evaporites formed over the Matador Arch.

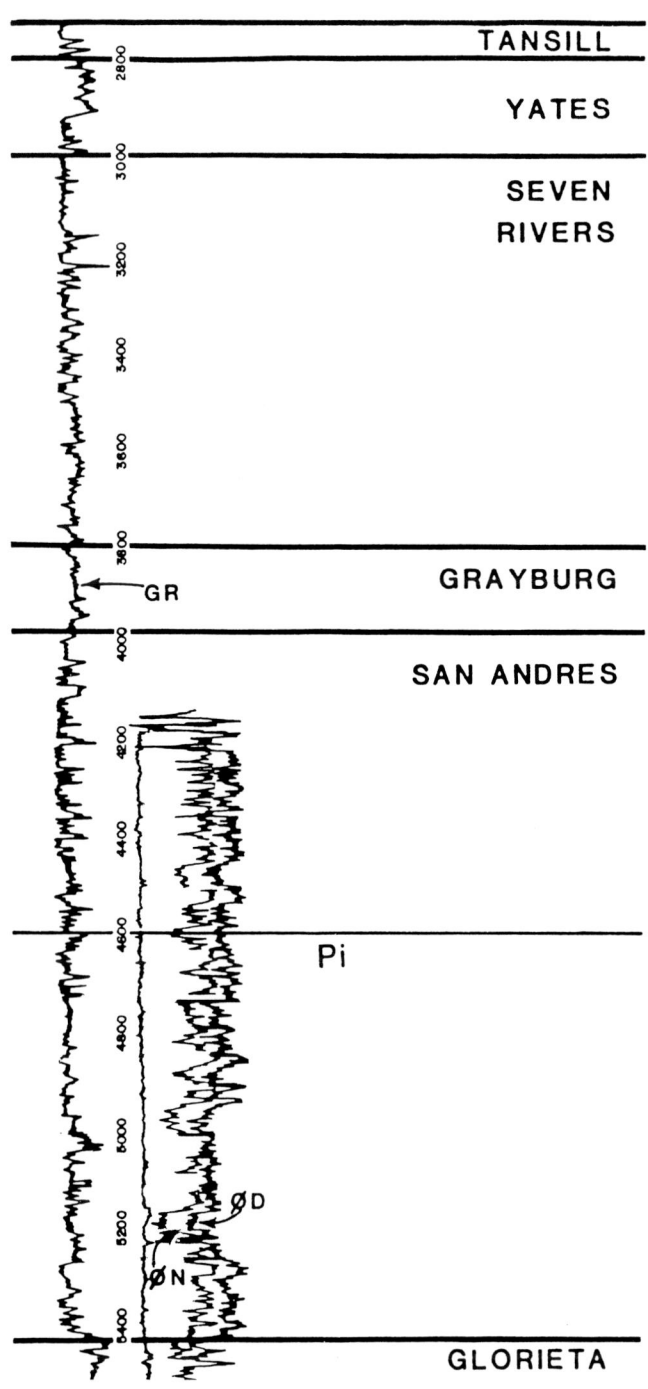

Figure 2 Interpreted borehole logs from a well in Cochran County, Texas. Modified after Cowan and Harris, 1986. The San Andres Formation is part of a thick Upper Permian section; porosity zones located in the lower San Andres, below the Pi marker, are examined in more detail in this paper. GR = gamma-ray log, ϕN = neutron log, ϕD = density log.

depositional cycle that is nearly 800-feet thick (Fig. 3). The depositional facies comprising this cycle have been described in detail by Gratton and LeMay (1969), Silver and Todd (1969), Chuber and Pusey (1972), Meissner (1972), Todd (1976), Dutton et al. (1979), Hafner (1979), and Ramondetta (1982).

Permian shelf deposits have previously been recognized as being highly cyclic (Silver and Todd, 1969; Chuber and Pusey, 1972). The cyclic nature of the restricted shelf and tidal-flat deposits of the lower San Andres has formed stratigraphic traps, resulting in the occurrence of impermeable anhydrite and tight dolostones between vertically stacked porosity zones commonly about 50-feet thick (Gratton and LeMay, 1969; Pitt and Scott, 1981; Ramondetta, 1982; Cowan and Harris, 1986). Anhydrite is thicker to the north (updip), providing an impermeable regional porosity barrier (Fig. 1; Cowan and Harris, 1986). Each of the porous carbonate zones represents a marine, onlapping pulse in the generally regressive San Andres, with the older (deeper) porous zones extending farther to the north (Fig. 4).

Bein and Land (1982) argued that the facies variations seen in the San Andres on the Northwest Shelf can be explained by merely changes in the salinity of the water without changing the water depth. However, core studies in numerous San Andres reservoirs have documented changes in depositional texture and sedimentary structures that argue for cyclical changes being the result of upward-shoaling related to eustatic sea-level change contemporaneous with subsidence as suggested by Silver and Todd (1969). The cyclicity combined with a low-relief depositional profile across the shelf results in great lateral continuity of facies at the same time there is considerable variability vertically.

C. S. DEAN "A," XIT, AND SW LEVELLAND UNITS

Two important structural features have a marked influence on San Andres production on the Northwest Shelf: the Matador Arch and the shallow expression of the Wolfcamp-Pennsylvanian shelf-edge of the Midland Basin (Ramondetta, 1982; Cowan and Harris, 1986). The Matador Arch marks the southern extent of major San Andres halite deposition and the northern extent of reservoir-facies deposition. The arch was a positive feature throughout much of the Mississippian and Pennsylvanian; undoubtedly, it remained a regional high even into the

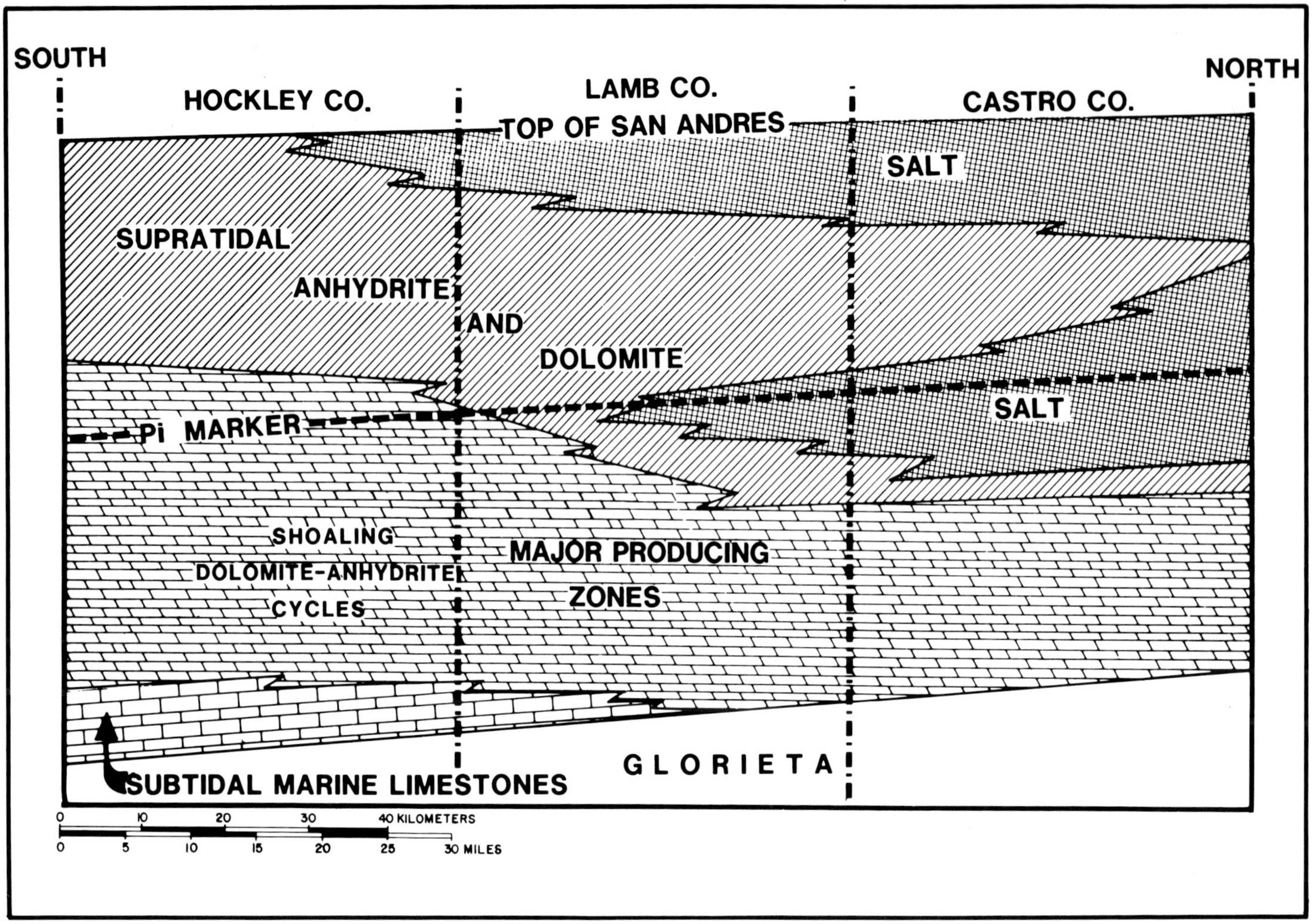

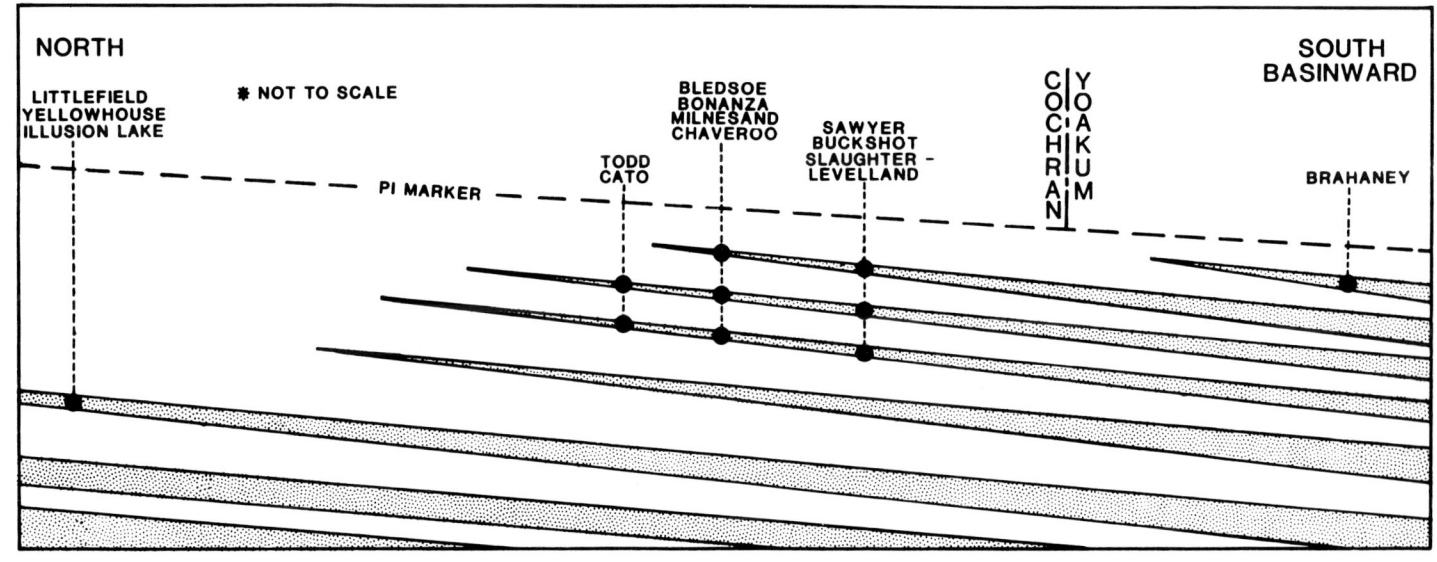

Figure 3 Schematic north-south cross-section indicating lithologies and depositional facies of the San Andres Formation. Modified after Cowan and Harris, 1986. Progradation of shelf during San Andres deposition formed a large-scale shoaling sequence.

Figure 4 Schematic cross-section summarizing production from reservoir zones of San Andres Formation through Cochran County. Modified after Cowan and Harris, 1986. Production occurs in shallower zones in a basinward direction; fields produce only from porosity zones identified with a black circle.

Late Permian. Porosity zones recognized in the Levelland-Slaughter Field pinch out regionally updip onto the Matador Arch, a pattern that is typical for other fields on the Northwest Shelf (Figs. 1 and 4), resulting in an east-west trend of fields which parallels the arch. The eastern and southern limits of Levelland-Slaughter Field are defined by the steepening of dip eastward and southward over the shallow structural expression of the Upper Pennsylvanian-lower Wolfcamp shelf margin of the Midland Basin (Fig. 5). Apparently this change in dip localized San Andres hydrocarbons to a large structural nose or plateau which extends from Hockley County over 100 miles west into New Mexico. Ramondetta (1982) argued, based on structure and isopach mapping and cited core studies, that compaction of sediments above the buried Wolfcamp-Pennsylvanian shelf-edge produced a bathymetrically high belt during San Andres time. This belt, effectively a shelf margin for the San Andres, may have been occupied by a series of shoals or islands during times of moderately low sea level. The San Andres thins over the belt of shoals relative to the restricted shelf to the west, and correlation of thin siliciclastic beds across the shoal is difficult.

The stratigraphic framework of the C. S. Dean "A," XIT, and SW Levelland Units was developed during a core and log study. Since discovery, 21,904,000 BO have been produced from the area of the C. S. Dean "A" Unit from 88 wells (producers and shut-in producers only). Similarly, the area of the XIT Unit has produced over 23,862,000 BO since discovery from 104 wells (producers and shut-in producers only). The area of the SW Levelland Unit has produced over 5,345,000 BO from 45 wells (producers and shut-in producers only) since discovery.

Cores from 19 wells in Cochran County were examined (Fig. 6): 13 cores from the C. S. Dean "A" Unit, five cores from the XIT Unit, and one core from the SW Levelland Unit. The core data are summarized on two cross-sections: one cross-section is an east-west section traversing the XIT and C. S. Dean "A" Units (Fig. 7), whereas the other is an approximately north-south section through the SW Levelland and C. S. Dean "A" Units (Fig. 8). Zones 5, 6, and 7 shown on the cross-sections are reservoir zones, whereas Zones 3 and 4 are nonporous intervals. Thinner nonporous intervals separate both Zones 5 and 6 and Zones 6 and 7. There appears to be a good correlation between the porous

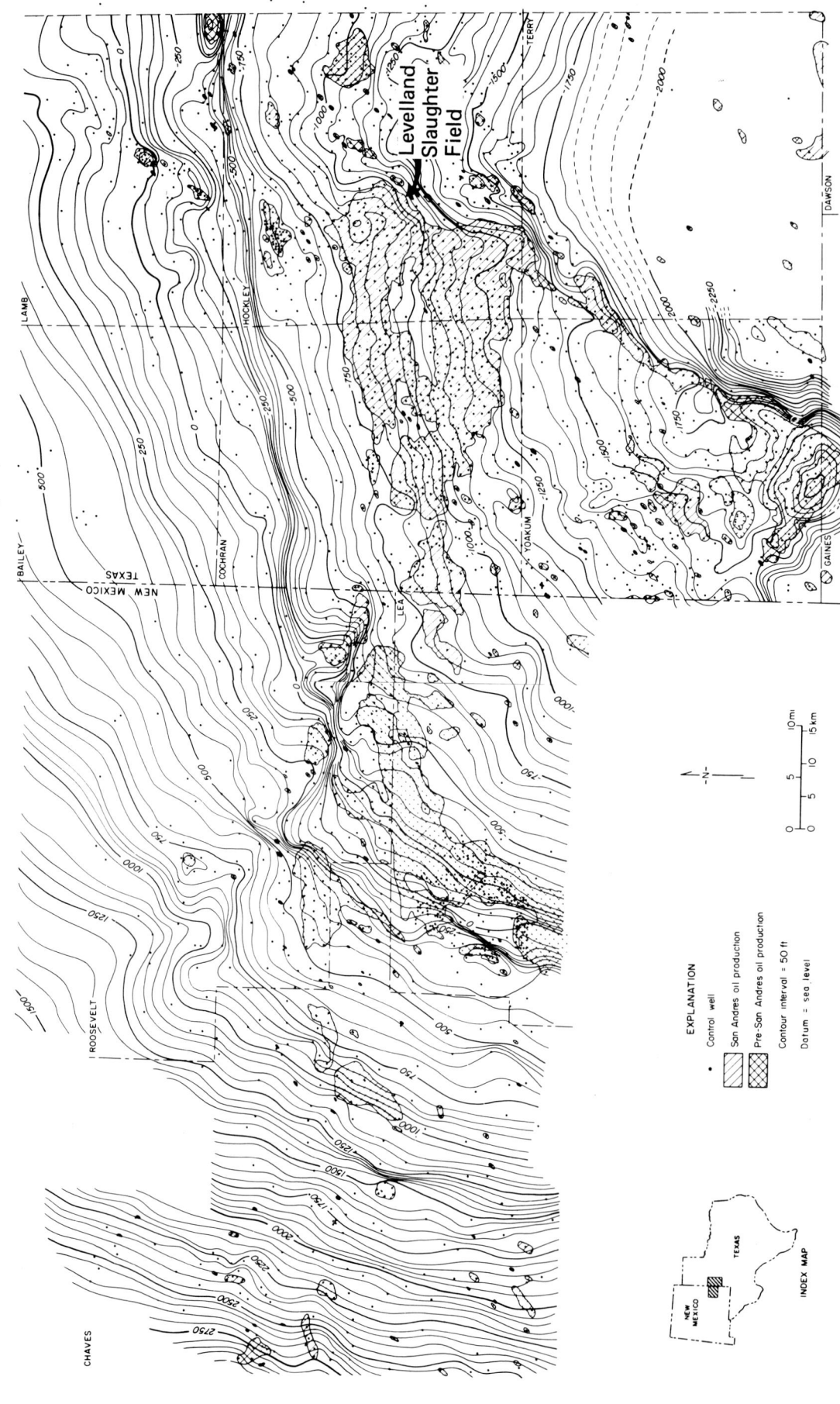

Figure 5 Structure map of Pi marker for Northwest Shelf of Texas and New Mexico. Modified after Ramondetta (1982).

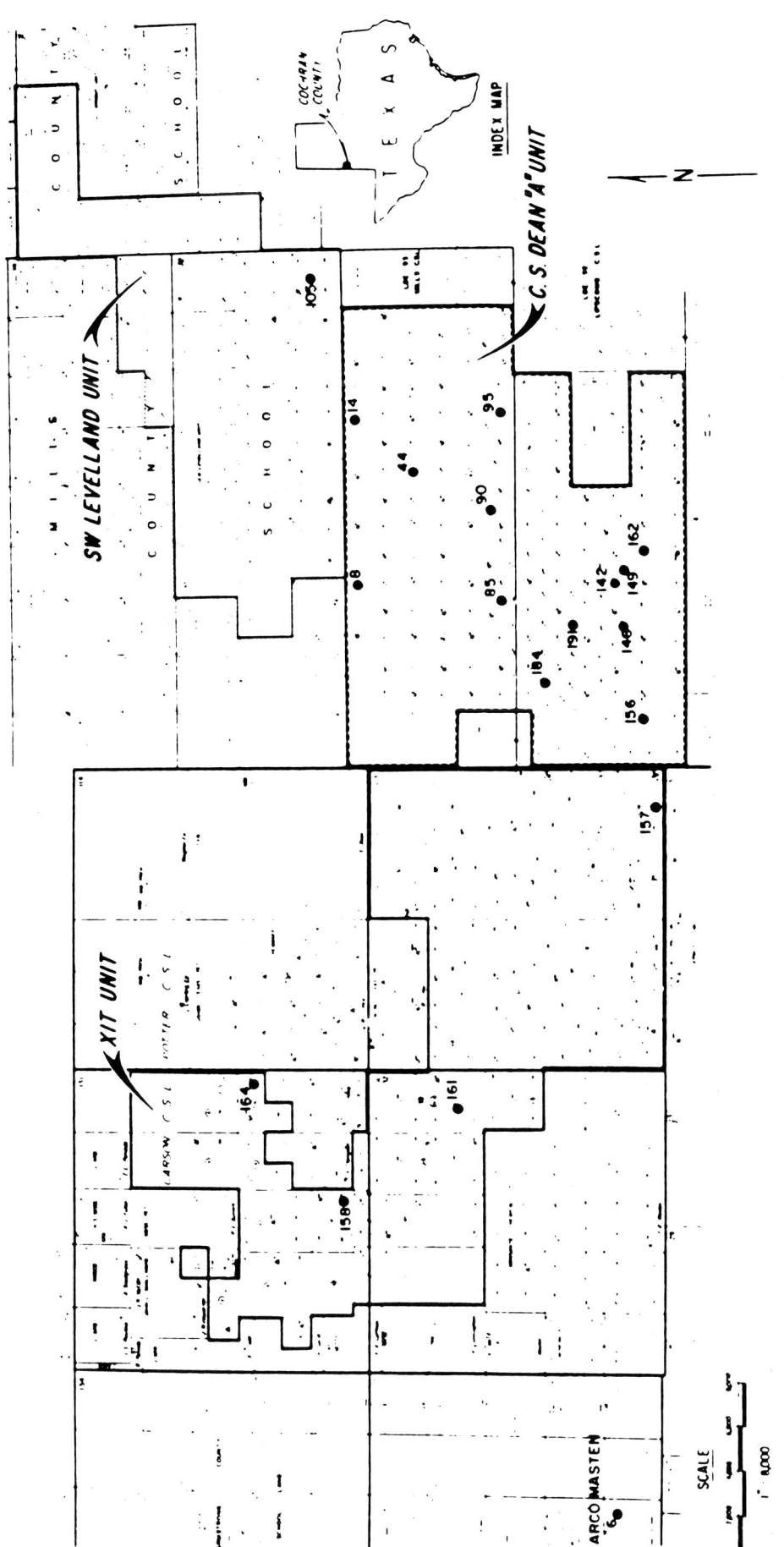

Figure 6 Map showing location of cores described from the C. S. Dean "A," XIT, and SW Levelland Units of Levelland-Slaughter Field.

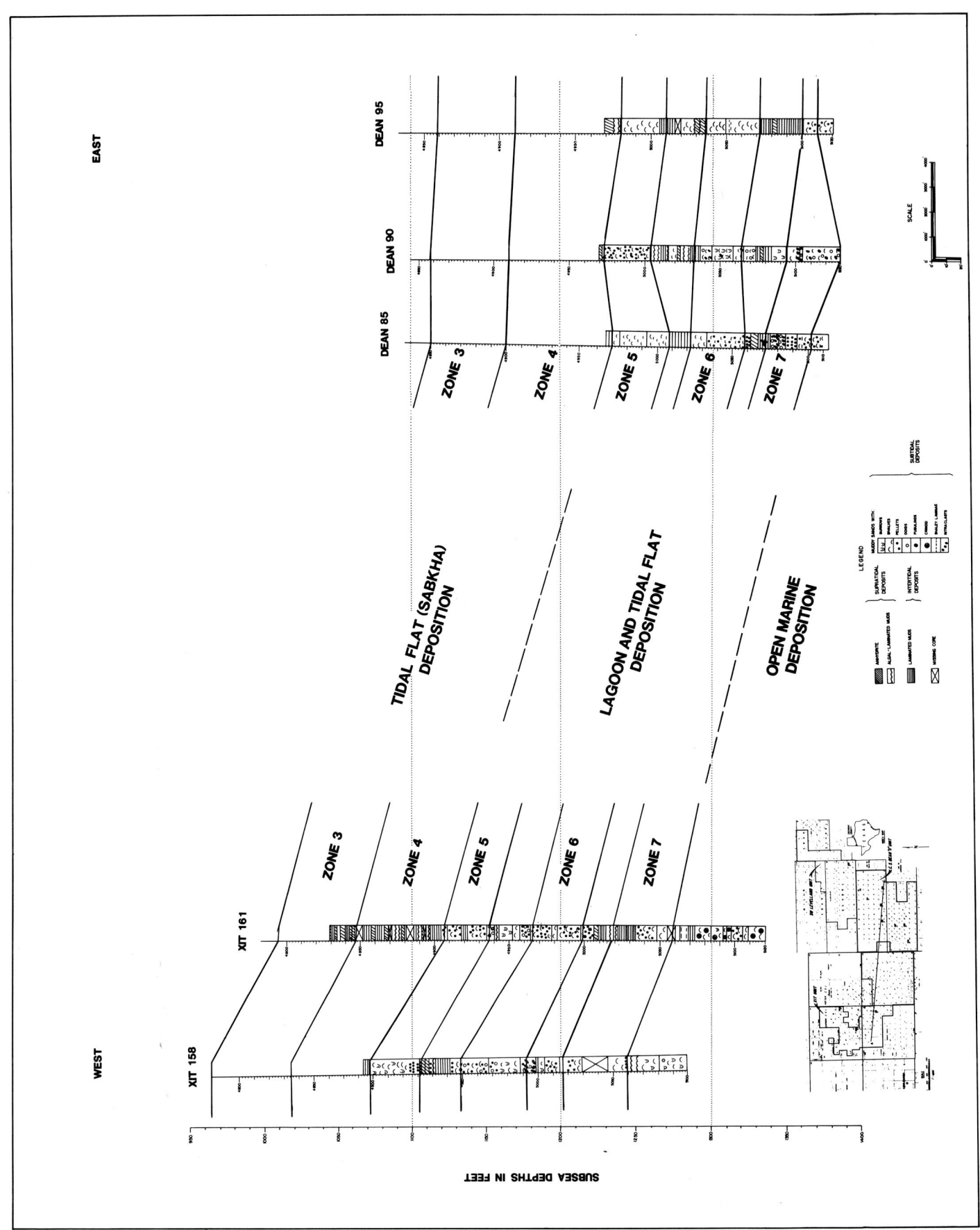

Figure 7 East-west structural cross-section through XIT and C. S. Dean "A" Units of Levelland-Slaughter Field. Lithological data and an environmental interpretation are summarized. Datum is mean sea level. Detailed descriptions for the XIT 161 and Dean 85 wells are shown on Figures 17 and 18 respectively.

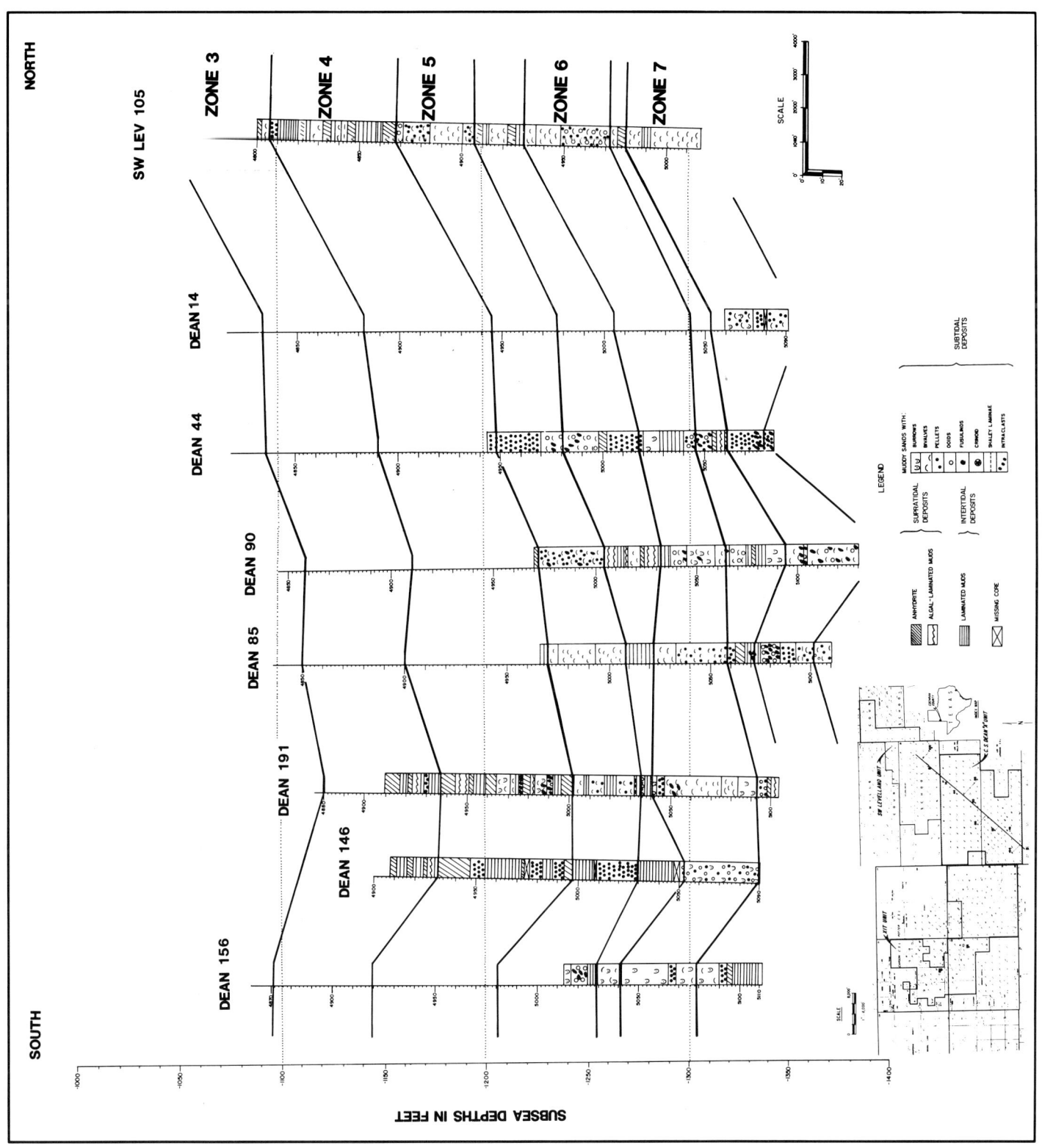

Figure 8 North-south structural cross-section through C. S. Dean "A" and SW Levelland Units of Levelland-Slaughter Field. Lithological data and an environmental interpretation are shown. The Dean 85 well also appears on the cross-section of Figure 7. Datum is mean sea level.

zones and specific facies. The porous zones are shallow, subtidal, restricted shelf deposits; nonporous zones are tidal-flat deposits.

The stratigraphic zones depicted on the cross-sections were picked on electric logs and adjusted by available core data. The designations used for the various zones in this paper conform to that used in field operations. The zones contain both porous and nonporous intervals and represent, from bottom to top, cyclic deposition of carbonates and evaporites which become increasingly evaporitic upward through the section. As previously discussed, these zones are regionally extensive across the Northwest Shelf, and the porous portions of the zones pinch out to the north where they correlate to anhydrite or to anhydrite-plugged dolomites on top of the Matador Arch. Pitt and Scott (1981), Ramondetta (1982), and Cowan and Harris (1986) discussed the regional correlation of these zones across the Northwest Shelf. Despite their efforts to present a consistent nomenclature for the zones, the terminology used by various operators varies between fields and units.

Structure and isopach maps for Zones 4 through 7 through the C. S. Dean "A," XIT, and SW Levelland Units were constructed using depth or thickness data from only the cored wells; thus, the maps provide only a starting point into which data from intervening wells must be incorporated. The maps do indicate general trends in structure and thickness, and they add to the more regional stratigraphic framework developed for the area by Ramondetta (1982) as shown on Figure 5.

Each of the stratigraphic zones dips from the northwest to the southeast as shown on Figures 9 through 12. This apparent dip, measured below present-day mean sea level, is from the shelf toward the Midland Basin. The strata dip approximately 200 feet across the width of the XIT and C. S. Dean "A" Units, or about 30 feet per mile. The amount of dip is essentially unchanged from Zone 7 to Zone 4, suggesting the strata conform to the regional dip and there are no anomalous, widespread, sediment build-ups.

The thicknesses of the stratigraphic zones are irregular and difficult to compare. Zone 7 appears to thicken by a factor of four toward the northwest, but the number of data points is minimal (Fig. 13). Zone 6 also thickens to the northwest toward the XIT Unit and appears to thicken to the north toward the SW Levelland Unit (Fig. 14). The zone also doubles in thickness and thins

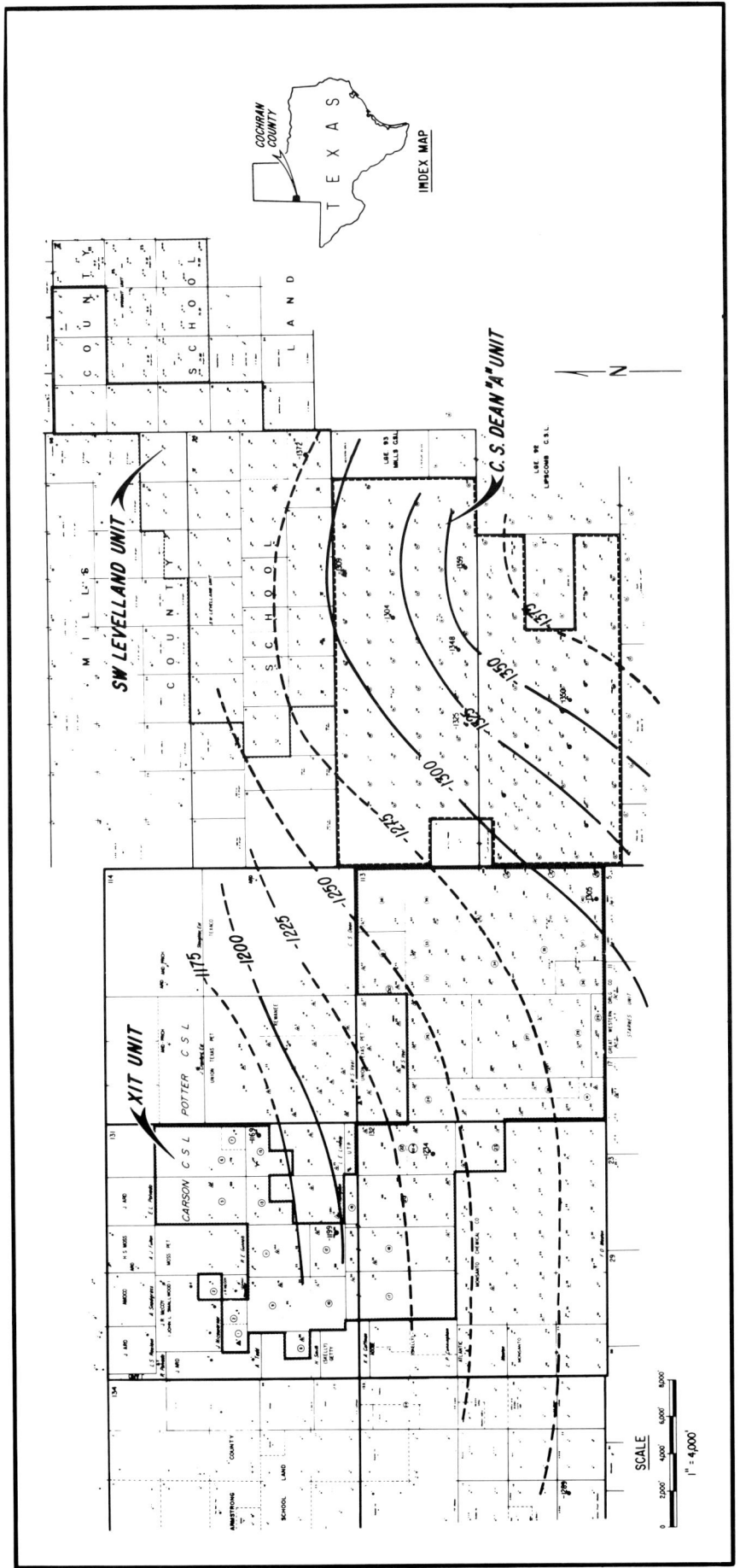

Figure 9 Structure map for the top of Zone 7 in the C. S. Dean "A," XIT, and SW Levelland Units of Levelland-Slaughter Field. Contour interval is 25 ft.

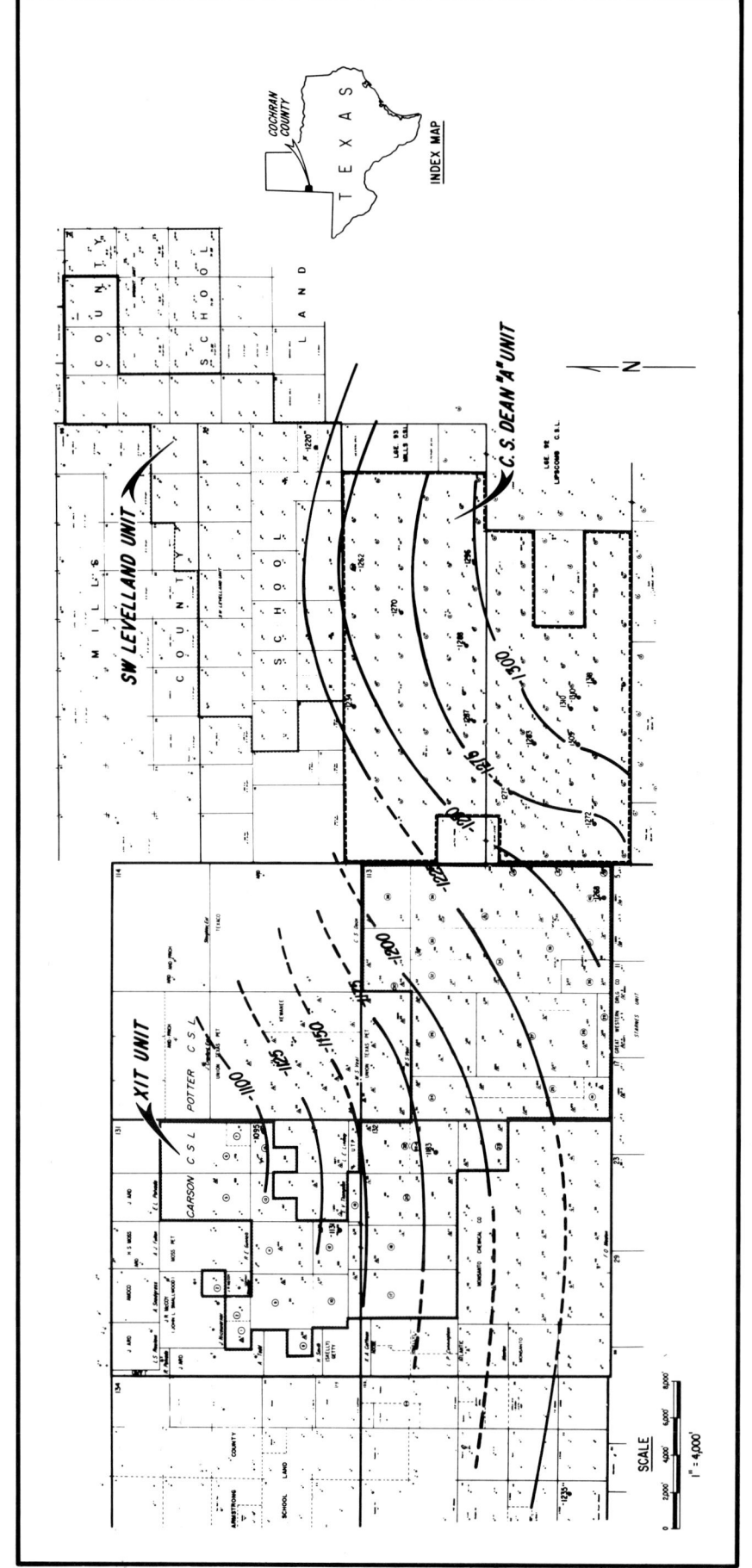

Figure 10 Structure map for the top of Zone 6 in the C. S. Dean "A," XIT, and SW Levelland Units of Levelland-Slaughter Field. Contour interval is 25 ft.

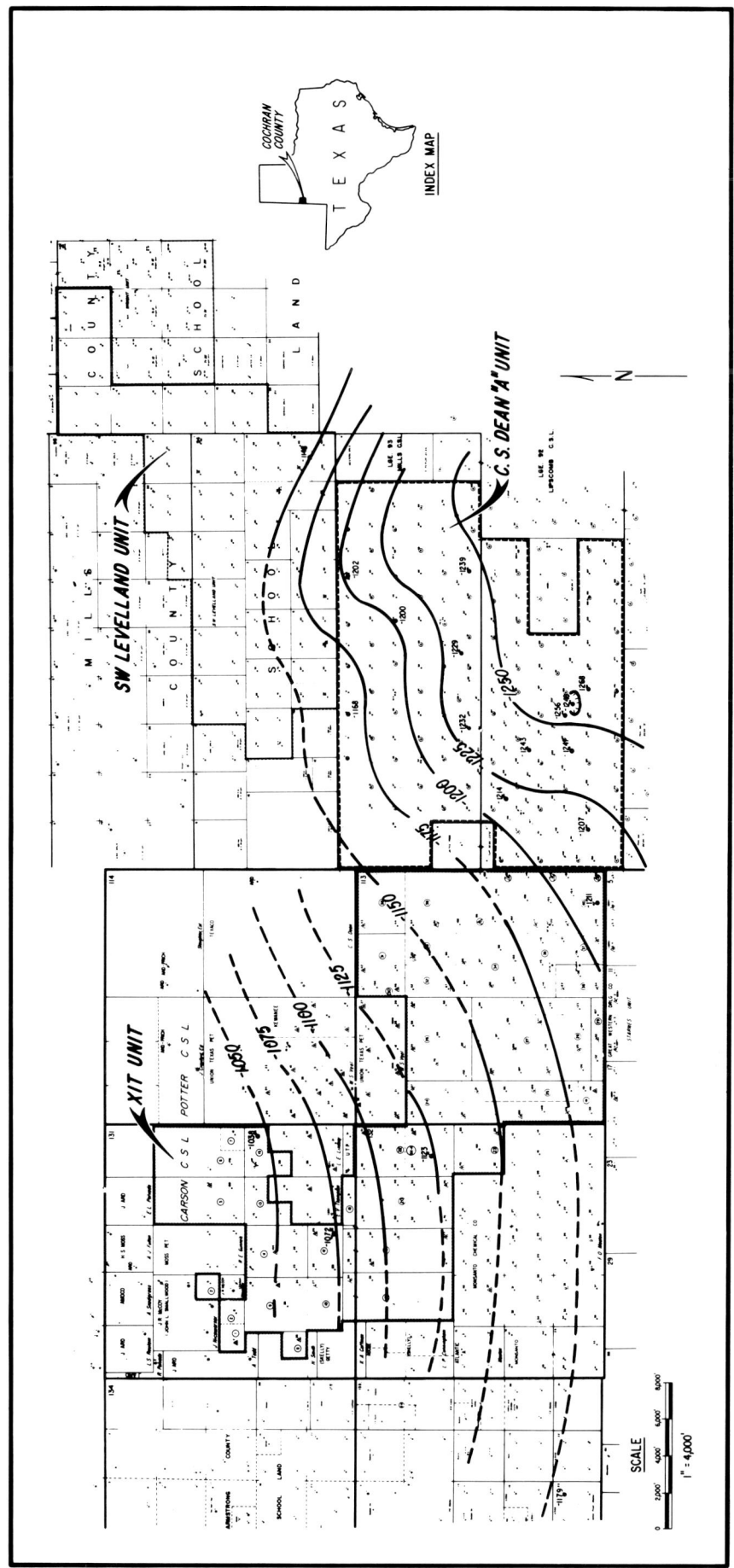

Figure 11 Structure map for the top of Zone 5 in the C. S. Dean "A," XIT, and SW Levelland Units of Levelland-Slaughter Field. Contour interval is 25 ft.

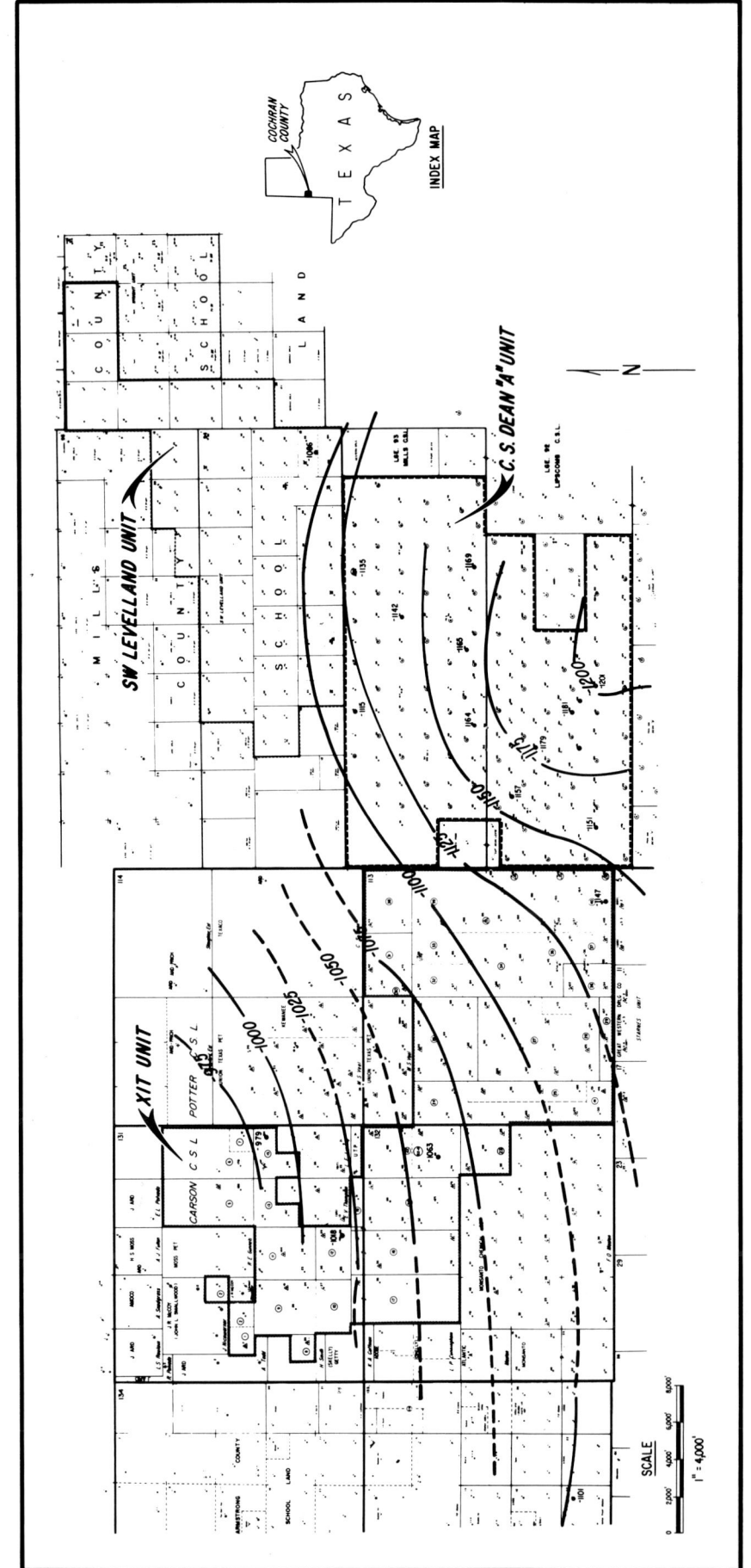

Figure 12 Structure map for the top of Zone 4 in the C. S. Dean "A," XIT, and SW Levelland Units of Levelland-Slaughter Field. Contour interval is 25 ft.

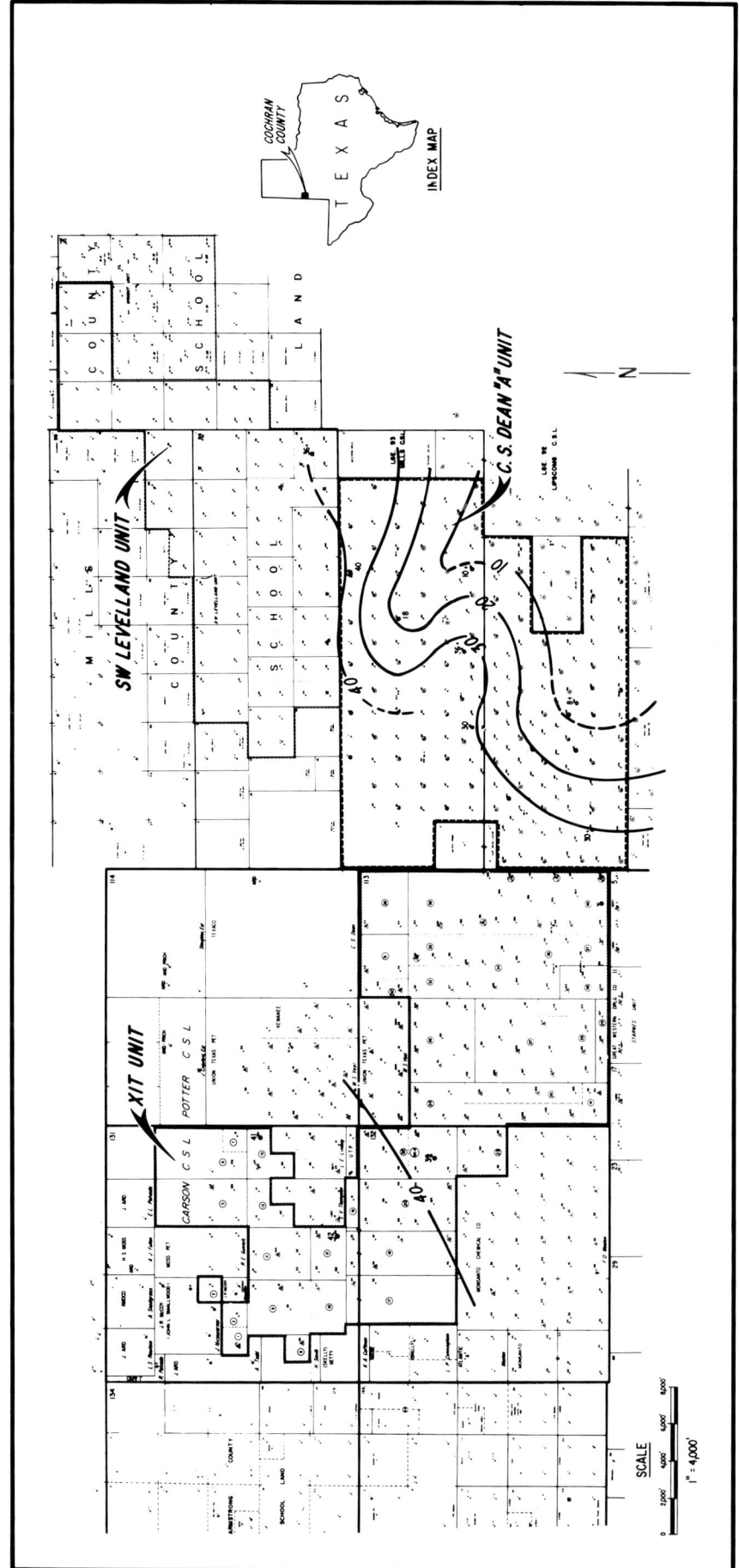

Figure 13 Isopach map of Zone 7 in the C. S. Dean "A," XIT, and SW Levelland Units of Levelland-Slaughter Field. Contour interval is 10 ft.

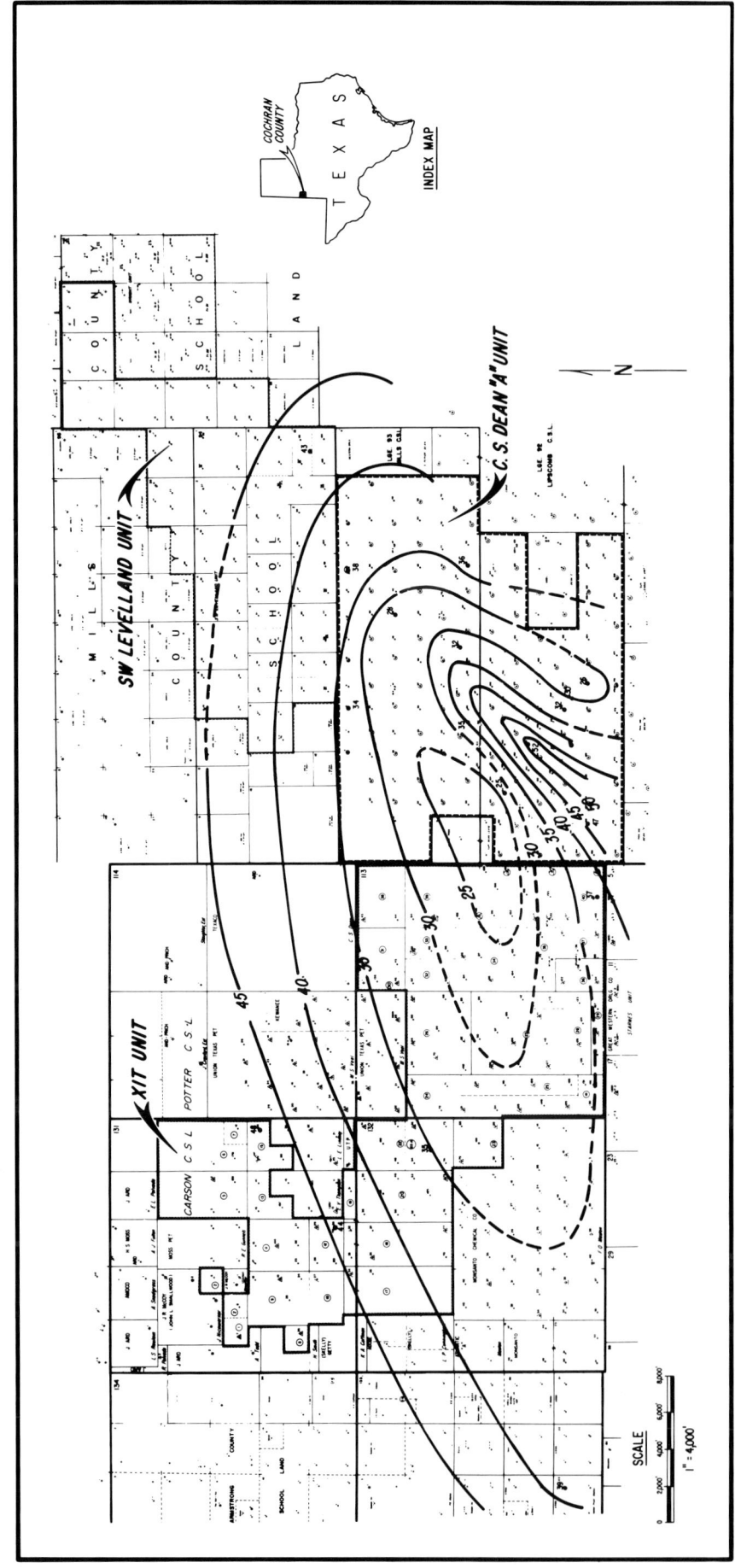

Figure 14 Isopach map of Zone 6 in the C. S. Dean "A," XIT, and SW Levelland Units of Levelland-Slaughter Field. Contour interval is 5 ft.

again across the southern half of the C. S. Dean "A" Unit. The thickness of the overlying Zone 5 seems to have a complementing trend; Zone 5 thins to the northwest and west across the XIT Unit (Fig. 15). The zone is of relatively uniform thickness across the C. S. Dean "A" Unit, although a slightly thicker area in the central portion of the unit corresponds with a slightly thinner area in the underlying zone. Zone 4 thins from south to north across all the units, but the difference in thickness across the area is less than the difference in the other zones (Fig. 16). Only for Zones 5 and 6 are there enough data points to suggest a possible relationship in thickness: basically the overlying Zone 5 tends to thicken in areas where the underlying Zone 6 is thinner.

LITHOFACIES

The overall regressive sequence encountered in the C. S. Dean "A," XIT, and SW Levelland Units contains, from bottom to top: (1) open marine shelf dolowackestones-packstones; (2) intervals of restricted shelf dolowackestones-packstones separated by tidal-flat dolomudstones and anhydrite, and (3) supratidal-flat dolomudstones and anhydrite. Detailed core descriptions from two representative wells illustrate the rock types encountered in this regressive sequence: the Getty XIT 161 is shown on Figure 17A through F and Getty Dean 85 is shown on Figure 18A through C.

The open marine deposits, lying below Zone 7 in Levelland-Slaughter Field, were penetrated in XIT 161. They are occasionally undolomitized and commonly contain crinoids and fusulinids, a fauna more luxurious than that of the overlying strata and indicative of a subtidal sea bottom with relatively open circulation. These dolowackestones and packstones may be vaguely laminated or wispy, but are commonly burrowed with variable amounts of mud (Fig. 19). They occur below Zone 7 and are most often separated from it by intertidal, algal-laminated dolomudstone-wackestone and anhydrite.

Zones 7, 6, and 5 are characterized by skeletal and peloidal dolowackestones-packstones deposited in a shallow restricted shelf (Figs. 20, 21, and 22). The skeletal component is less varied than that of the open shelf deposits, being composed principally of mollusc shell fragments. Beds of abundant mud intraclasts suggest local intertidal conditions where mud

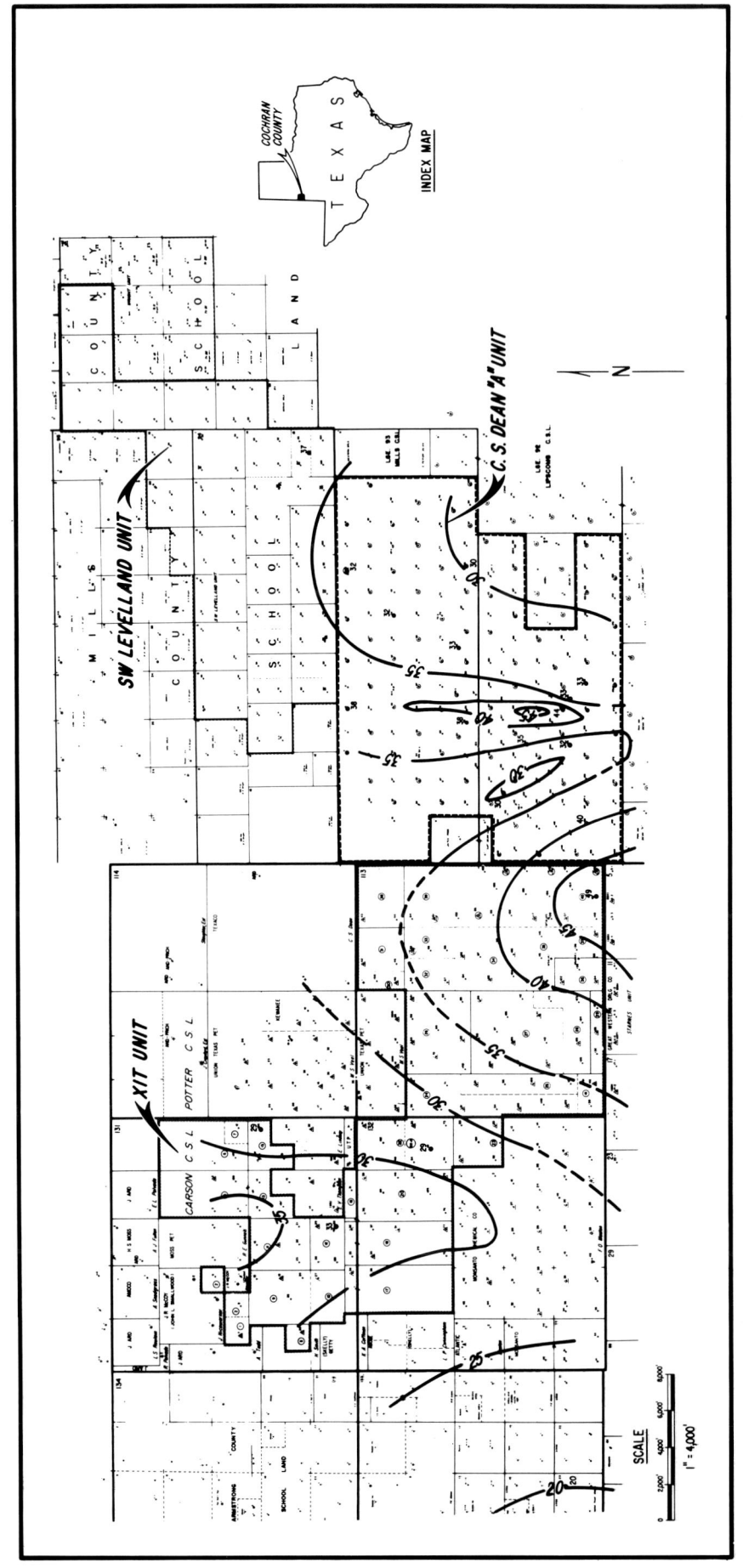

Figure 15 Isopach map of Zone 5 in the C. S. Dean "A," XIT, and SW Levelland Units of Levelland-Slaughter Field. Contour interval is 5 ft.

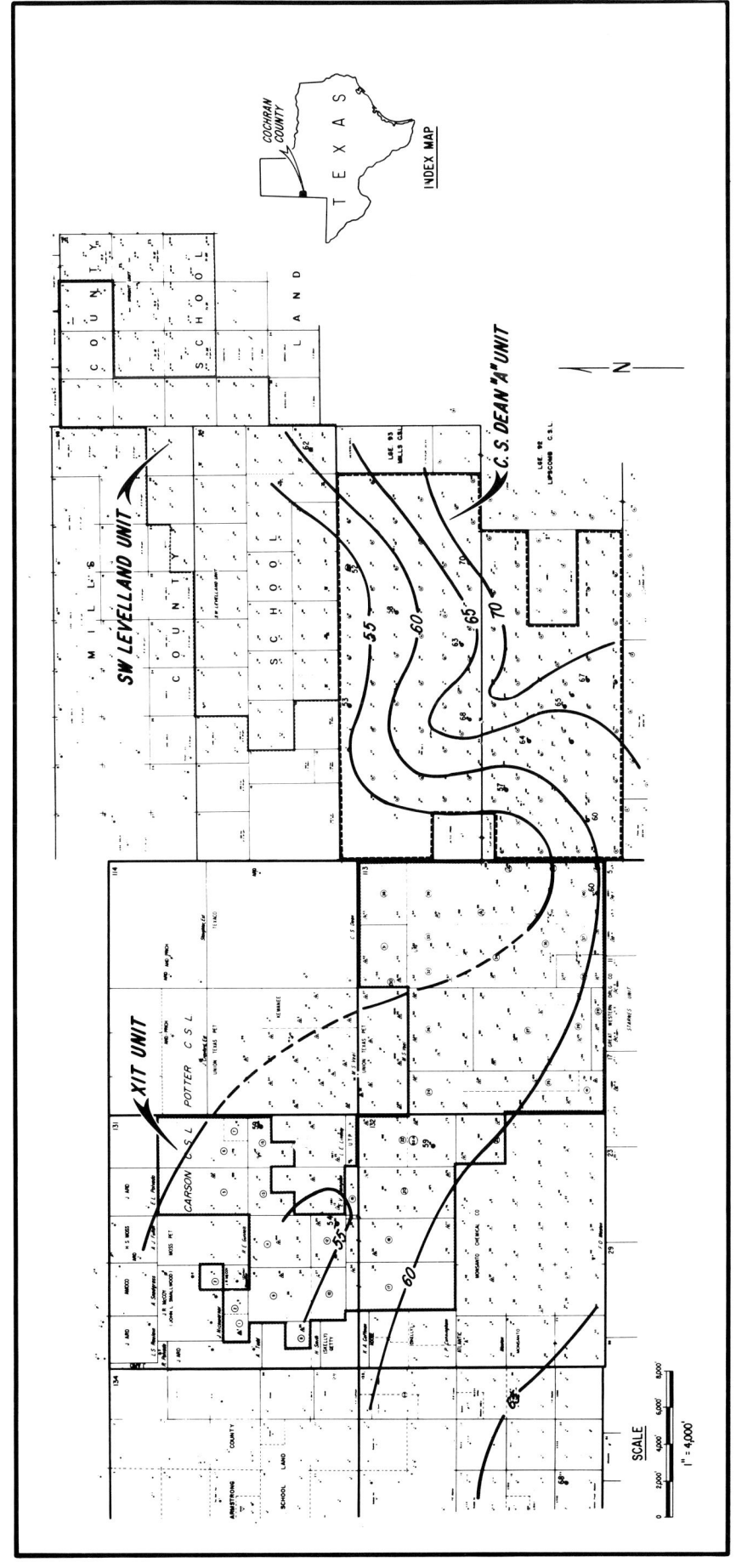

Figure 16 Isopach map of Zone 4 in the C. S. Dean "A," XIT, and SW Levelland Units of Levelland-Slaughter Field. Contour interval is 5 ft.

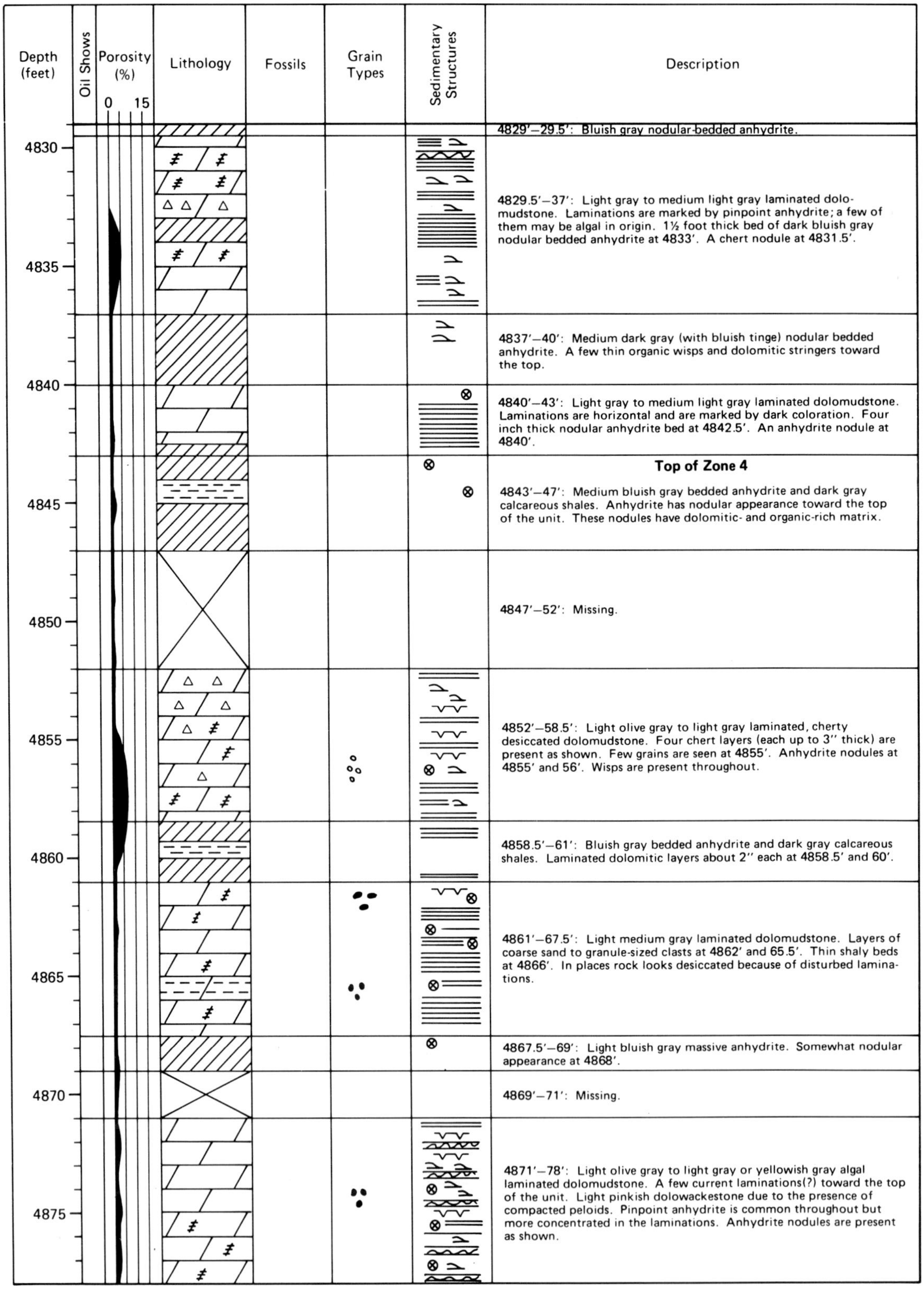

Depth (feet)	Oil Shows	Porosity (%) 0 15	Lithology	Fossils	Grain Types	Sedimentary Structures	Description
4830							4829'–29.5': Bluish gray nodular-bedded anhydrite.
							4829.5'–37': Light gray to medium light gray laminated dolomudstone. Laminations are marked by pinpoint anhydrite; a few of them may be algal in origin. 1½ foot thick bed of dark bluish gray nodular bedded anhydrite at 4833'. A chert nodule at 4831.5'.
4835							
							4837'–40': Medium dark gray (with bluish tinge) nodular bedded anhydrite. A few thin organic wisps and dolomitic stringers toward the top.
4840							4840'–43': Light gray to medium light gray laminated dolomudstone. Laminations are horizontal and are marked by dark coloration. Four inch thick nodular anhydrite bed at 4842.5'. An anhydrite nodule at 4840'.
4845							**Top of Zone 4** 4843'–47': Medium bluish gray bedded anhydrite and dark gray calcareous shales. Anhydrite has nodular appearance toward the top of the unit. These nodules have dolomitic- and organic-rich matrix.
4850							4847'–52': Missing.
4855							4852'–58.5': Light olive gray to light gray laminated, cherty desiccated dolomudstone. Four chert layers (each up to 3" thick) are present as shown. Few grains are seen at 4855'. Anhydrite nodules at 4855' and 56'. Wisps are present throughout.
4860							4858.5'–61': Bluish gray bedded anhydrite and dark gray calcareous shales. Laminated dolomitic layers about 2" each at 4858.5' and 60'.
4865							4861'–67.5': Light medium gray laminated dolomudstone. Layers of coarse sand to granule-sized clasts at 4862' and 65.5'. Thin shaly beds at 4866'. In places rock looks desiccated because of disturbed laminations.
							4867.5'–69': Light bluish gray massive anhydrite. Somewhat nodular appearance at 4868'.
4870							4869'–71': Missing.
4875							4871'–78': Light olive gray to light gray or yellowish gray algal laminated dolomudstone. A few current laminations(?) toward the top of the unit. Light pinkish dolowackestone due to the presence of compacted peloids. Pinpoint anhydrite is common throughout but more concentrated in the laminations. Anhydrite nodules are present as shown.

Figure 17A Detailed core description of the XIT 161 well.

Depth (feet)	Oil Shows	Porosity (%) 0–15	Lithology	Fossils	Grain Types	Sedimentary Structures	Description
4880							4878'–81': Black shale and light blue nodular-bedded mosaic anhydrite. Shale is fissile and noncalcareous.
							4881'–85': Missing.
4885–4890							4885'–91.5': Light gray to medium light gray algal laminated, cherty dolomudstone to dolowackestone. Chert nodules are present throughout. Pinpoint anhydrite is common, concentrated more in the laminations. An anhydrite bed or a big nodule (about 5" thick) at 4886'.
							4891.5'–93': Light bluish gray to pale blue massive to slightly nodular anhydrite.
4895							4893'–95': Missing.
							4895'–98': Light olive gray to medium light gray algal laminated dolomudstone. Anhydrite is present in the form of pinpoints, nodules and layers throughout the unit.
							4898'–99': Missing.
							4899'–99.5': Bluish gray massive to nodular anhydrite.
4900							4899.5'–4901.5': Alternating medium gray and dark gray laminated dolomudstone. Dark laminations are marked by soft muddy material. Pinpoint anhydrite is common throughout.
							4901.5'–02': Dark bluish gray massive anyydrite with 1" thick chert layer.
4905							4902'–05.5': Light olive gray to medium light gray dolopackstone to wackestone. A few mollusc fragments at 4905' are filled with anhydrite. Unit appears to be burrowed and desiccated. A few scattered wisps.
							Top of Zone 5 4905.5'–09': Medium gray to medium dark gray dolomudstone to wackestone. A few scattered fossils and grains. Vaguely laminated in places. Scattered wisps and pinpoint anhydrite.
4910–4915							4909'–18': Medium light gray to olive light gray dolopackstone to wackestone. Mollusc fragments and pellets are scattered throughout. Most of the fossils and grains are leached. Burrows are present throughout. Anhydrite nodules are present at 4909' and 16'. Rock has mottled appearance at 4915'.
4920							4918'–23': Light olive gray laminated dolomudstone to wackestone. A few mollusc fragments toward the top of the unit. Possible burrows at 4919'. A chert nodule is present at 4921'. An anhydrite nodule is present as shown. Desiccated appearance at 4921'.

Figure 17B   XIT 161 core description (contd.).

Depth (feet)	Oil Shows	Porosity (%) 0 15	Lithology	Fossils	Grain Types	Sedimentary Structures	Description
4925–4935							4923'–35': Light olive gray dolowackestone to packstone. Grains are mostly mollusc fragments and pellets; most of them have been leached away. Slightly burrowed at 4934'. Two anhydrite nodules at 4932.5'; pinpoint anhydrite is scattered throughout. A clay nodule(?) 3" across is present at 4931'; its rim is darker and harder than the yellow gray middle material.
4935–4940							4935'–40': Light olive gray to light gray dolomudstone to wackestone. A few coarse sand-sized grains toward the bottom; these are either fusulinids or some clasts. Unit is vaguely laminated in places. Pinpoint anhydrite, thin wisps and stylolites are scattered throughout.
4940–4943							4940'–43': Light olive gray to light gray algal laminated dolomudstone with large anhydrite nodules and pinpoint anhydrite concentrated in laminations. Possible desiccation at 4941.
4943–4952							4943'–52': Light olive gray to medium light gray dolomudstone. Grainy in places, may be due to leached pellets. Vaguely laminated at 4950'. Possible burrowing and desiccation toward the top of the unit. Anhydrite nodules and pinpoint anhydrite are common throughout, more concentrated toward the top of the unit. A few scattered stylolites and wisps. 1" thick shaly bed at 4950.5'.
4952–4957.5							4952'–57.5': Medium light gray to medium gray dolomudstone to wackestone. A few mollusc fragments are seen. Some scattered laminations, probably algal in origin. Nodular mosaic anhydrite toward the top. Pinpoint anhydrite is scattered throughout.
4957.5–4964							4957.5'–64': Medium gray to medium light gray dolopackstone to wackestone. Abundant fine silt-sized leached pellets, ooids and bryozoans. A few mollusc fragments are present; these are replaced by anhydrite. There are a few desiccated/fractured zones; fractures are filled with anhydrite. An anhydrite nodule is present toward the bottom. Two stylolites at 4962' show extensive pressure solution.

Figure 17C XIT 161 core description (contd.).

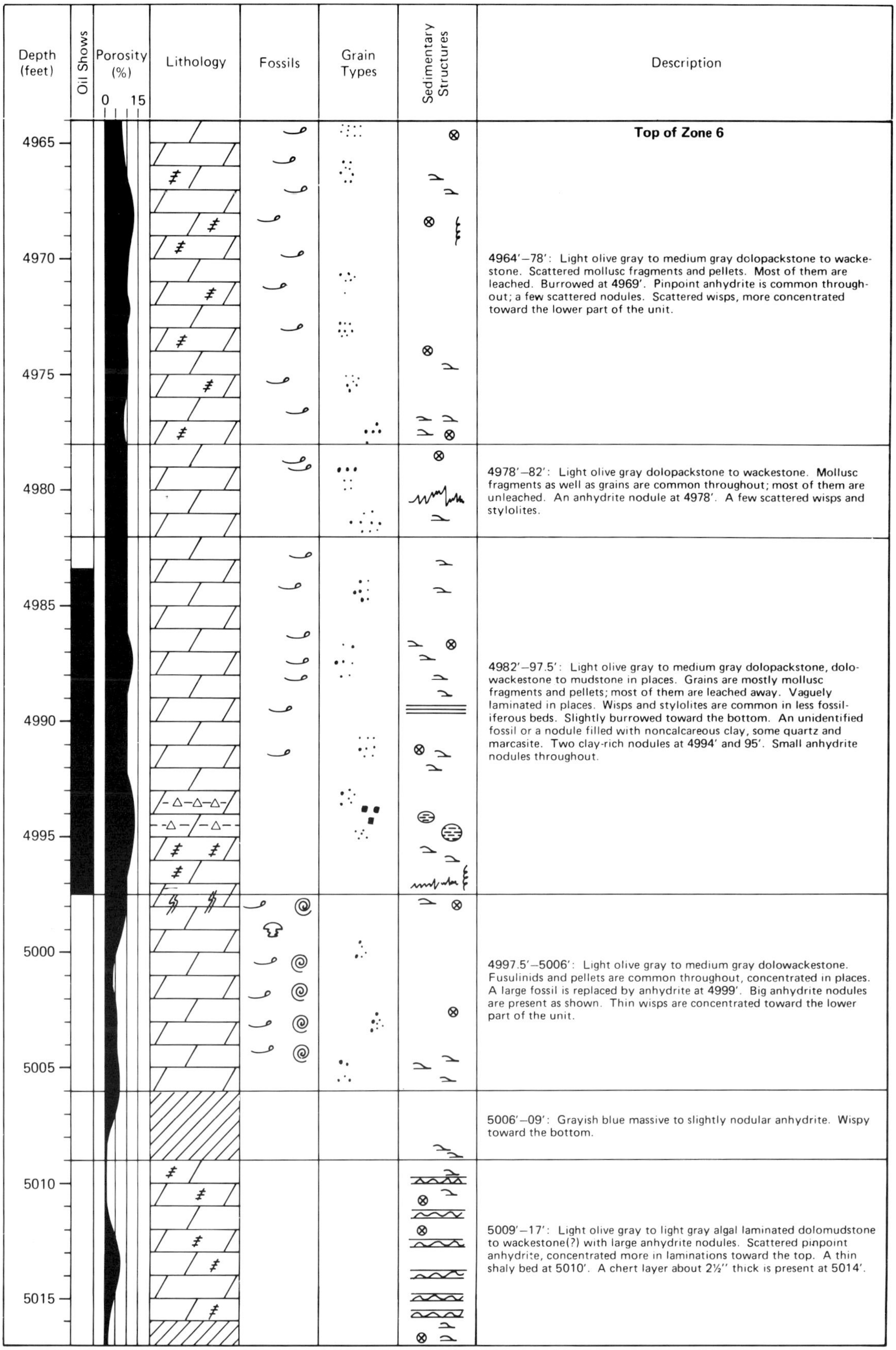

Depth (feet)	Oil Shows	Porosity (%) 0 15	Lithology	Fossils	Grain Types	Sedimentary Structures	Description
4965–4978							**Top of Zone 6** 4964'–78': Light olive gray to medium gray dolopackstone to wackestone. Scattered mollusc fragments and pellets. Most of them are leached. Burrowed at 4969'. Pinpoint anhydrite is common throughout; a few scattered nodules. Scattered wisps, more concentrated toward the lower part of the unit.
4978–4982							4978'–82': Light olive gray dolopackstone to wackestone. Mollusc fragments as well as grains are common throughout; most of them are unleached. An anhydrite nodule at 4978'. A few scattered wisps and stylolites.
4982–4997.5							4982'–97.5': Light olive gray to medium gray dolopackstone, dolowackestone to mudstone in places. Grains are mostly mollusc fragments and pellets; most of them are leached away. Vaguely laminated in places. Wisps and stylolites are common in less fossiliferous beds. Slightly burrowed toward the bottom. An unidentified fossil or a nodule filled with noncalcareous clay, some quartz and marcasite. Two clay-rich nodules at 4994' and 95'. Small anhydrite nodules throughout.
4997.5–5006							4997.5'–5006': Light olive gray to medium gray dolowackestone. Fusulinids and pellets are common throughout, concentrated in places. A large fossil is replaced by anhydrite at 4999'. Big anhydrite nodules are present as shown. Thin wisps are concentrated toward the lower part of the unit.
5006–5009							5006'–09': Grayish blue massive to slightly nodular anhydrite. Wispy toward the bottom.
5009–5017							5009'–17': Light olive gray to light gray algal laminated dolomudstone to wackestone(?) with large anhydrite nodules. Scattered pinpoint anhydrite, concentrated more in laminations toward the top. A thin shaly bed at 5010'. A chert layer about 2½" thick is present at 5014'.

Figure 17D XIT 161 core description (contd.).

Depth (feet)	Oil Shows	Porosity (%) 0–15	Lithology	Fossils	Grain Types	Sedimentary Structures	Description
							Top of Zone 7
5020							5017'–29': Light olive gray to medium gray alternating laminated and burrowed dolomudstone to wackestone. Some laminations may be algal in origin toward the top. These are marked by pinpoint anhydrite. Desiccation in places. A few scattered mollusc fragments. A few annydrite nodules and wisps are scattered throughout.
5025							
5030							5029'–34': Light olive gray to medium light gray vaguely laminated dolopackstone to wackestone. Grains are mostly mollusc fragments and pellets. They are both leached as well as unleached. Wisps and pinpoint anhydrite are common throughout. A 3" thick clay-rich bed at 5029'.
5035							
5040							5034'–48': Medium gray to dark gray dolopackstone to wackestone. Leached mollusc fragments and pellets are scattered throughout. Dolomudstone at 5037' and 38'. Unit appears to be desiccated and fractured, as shown. Large anhydrite nodules are seen.
5045							
5050							5048'–55': Medium gray to medium dark gray leached shaly dolopackstone to wackestone. Fossils are mostly mollusc and brachiopod fragments. A few echinoderms, bryozoans, forams and pellets. Most of the allochems are leached away. Breaking patterns of the rock indicate some fractures along bedding planes. This unit can be differentiated from the unit at 5063'–69' by having larger fossils and more shaly material.
5055							5055'–60': Missing (chewed up by drill bit).
5060							5060'–63': Medium gray dolowackestone with abundant wisps and stylolites. Few scattered white anhydrite nodules.
5065							5063'–69': Medium dark gray shaly dolopackstone. Dominant fauna probably forams(?). A few echinoderm fragments. Fossils are mostly unleached. Shaly and organic-rich zones are present as shown. Abundant wisps and a few small anhydrite nodules.

Figure 17E XIT 161 core description (contd.).

Depth (feet)	Oil Shows	Porosity (%) 0-15	Lithology	Fossils	Grain Types	Sedimentary Structures	Description
5070							5069'–73.5': Yellowish gray to pinkish gray laminated dolomudstone to wackestone. Scattered mollusc fragments which are mostly leached. Some laminations are disrupted by burrows. Large anhydrite nodules at 5072'.
5075–5080							5073.5'–84': Light olive gray to medium light gray dolopackstone. Fossils are mostly crinoids, molluscs and brachiopods. Burrowed in places. Lower part of the unit is vaguely wispy, laminated. Geopetal structures at 5075'. Shaly organic-rich zones at 5078'–83'. An anhydrite nodule at 5071'. Scattered oil patches.
5085–5090							5084'–92': Light gray to medium light gray burrow-reworked packstone. Crinoids and molluscs are common throughout; a few of them are leached toward the top. Some burrows show swirling patterns toward the top. Wisps are abundant in places, giving the rock a dark gray appearance. Big anhydrite nodule at 5085'. Some fossils show iron-staining in places.
							5092'–94': Light olive gray dolowackestone. Fossils are mostly crinoids and mollusc fragments. Unit is less fossiliferous and less burrowed than the unit below. An anhydrite nodule at 5093'.
5095							5094'–97': Light olive gray to medium dark gray burrow-reworked packstone. Dominantly crinoids, molluscs, brachiopods, and a few fusulinids. Dark gray organic-rich bed at 5096'. Wisps are common throughout.
5100–5105							5097'–5105': Light olive gray to light gray dolopackstone. Fossils are mostly echinoderms and a few brachiopods and pellets. Some of them are leached. Highly burrowed, both horizontally and vertically. Large vertical burrows at 5103'. Rich in wisps throughout the unit; more concentrated toward the lower part of the unit.
							5105'–09': Light olive gray packstone to grainstone. Abundant sand-sized molluscs, brachiopods, echinoderms and bryozoans, both leached and unleached. A few burrows toward the top of the unit. A few stylolites at 5108'. Some fossils appear to be iron-stained.
5110–5115							5109'–20': Dusky yellowish brown to olive gray dolowackestone to packstone. Fossils are mostly crinoids, molluscs and brachiopods. They are both leached and unleached. Some fossils are filled with anhydrite. Highly bioturbated in most of the unit. Lime-rich highly fossiliferous zone at 5111'. Dark organic-rich with abundant stylolites and wisps at 5111.5'. A few anhydrite nodules are present. Some good geopetal structures noted, in which anhydrite is filling the cavities.

Figure 17F XIT 161 core description (contd.).

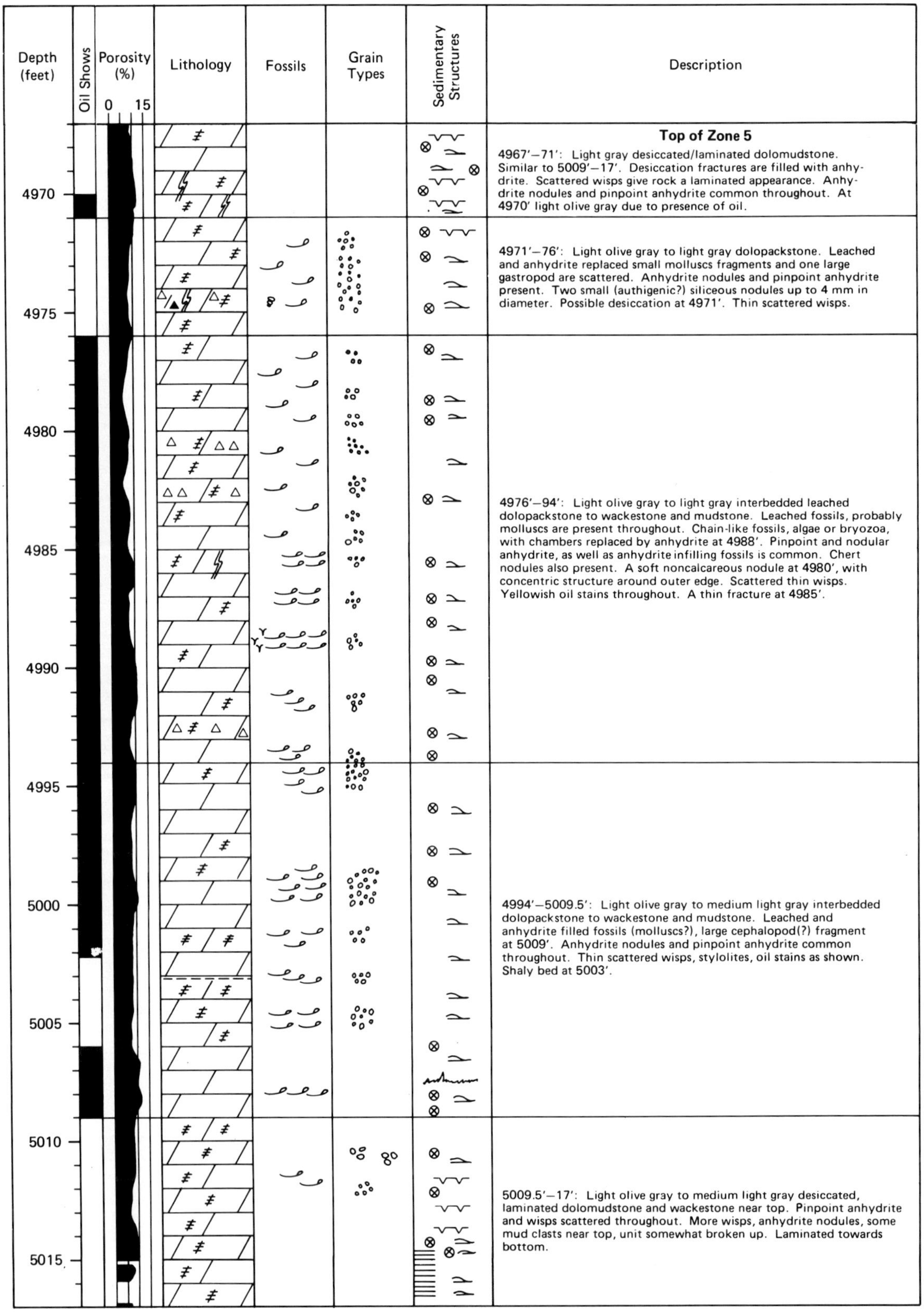

Depth (feet)	Oil Shows	Porosity (%) 0 15	Lithology	Fossils	Grain Types	Sedimentary Structures	Description
4970							**Top of Zone 5** 4967'–71': Light gray desiccated/laminated dolomudstone. Similar to 5009'–17'. Desiccation fractures are filled with anhydrite. Scattered wisps give rock a laminated appearance. Anhydrite nodules and pinpoint anhydrite common throughout. At 4970' light olive gray due to presence of oil.
4975							4971'–76': Light olive gray to light gray dolopackstone. Leached and anhydrite replaced small molluscs fragments and one large gastropod are scattered. Anhydrite nodules and pinpoint anhydrite present. Two small (authigenic?) siliceous nodules up to 4 mm in diameter. Possible desiccation at 4971'. Thin scattered wisps.
4980 4985 4990							4976'–94': Light olive gray to light gray interbedded leached dolopackstone to wackestone and mudstone. Leached fossils, probably molluscs are present throughout. Chain-like fossils, algae or bryozoa, with chambers replaced by anhydrite at 4988'. Pinpoint and nodular anhydrite, as well as anhydrite infilling fossils is common. Chert nodules also present. A soft noncalcareous nodule at 4980', with concentric structure around outer edge. Scattered thin wisps. Yellowish oil stains throughout. A thin fracture at 4985'.
4995 5000 5005							4994'–5009.5': Light olive gray to medium light gray interbedded dolopackstone to wackestone and mudstone. Leached and anhydrite filled fossils (molluscs?), large cephalopod(?) fragment at 5009'. Anhydrite nodules and pinpoint anhydrite common throughout. Thin scattered wisps, stylolites, oil stains as shown. Shaly bed at 5003'.
5010 5015							5009.5'–17': Light olive gray to medium light gray desiccated, laminated dolomudstone and wackestone near top. Pinpoint anhydrite and wisps scattered throughout. More wisps, anhydrite nodules, some mud clasts near top, unit somewhat broken up. Laminated towards bottom.

Figure 18A Detailed core description of the C. S. Dean "A" 85 well.

Depth (feet)	Oil Shows	Porosity (%) 0 15	Lithology	Fossils	Grain Types	Sedimentary Structures	Description
5020							**Top of Zone 6** 5017'–23': Light pinkish gray to light gray desiccated mudstone. No fossils seen. Anhydrite filled fractures. Pinpoint anhydrite throughout, concentrated in laminations, a few nodules separated by organic wisps. Nodular mosaic anhydrite at 5020'.
5025 – 5035							5023'–34.5': Light olive gray to medium light gray interbedded dolopackstone to wackestone and mudstone. Leached fossils and grains, some anhydrite filled. Wisps, scattered anhydrite nodules, laths, and pinpoints throughout. Oil staining, stylolites and burrowing as shown.
5035 – 5059							5034.5'–59': Light olive gray to medium light gray interbedded dolopackstone/wackestone and mudstone. Leached fossils, probably molluscs, and other grains (peloids?), some are anhydrite infilled. Rock breaks more easily at leached fossil zone. Unleached light gray sand-sized grains at 5057'. Burrowed appearance as shown, may be due to differential compaction in wispy zones. Thin wisps present throughout, more towards the top. Scattered anhydrite laths present. A chert layer (nodules?) about 1" thick at 5052'. Oil shows as marked.
5060							**Top of Zone 7** 5059'–63': Yellowish gray dolomudstone to dolowackestone. Desiccated, grainy appearance at 5061'–62'. Big anhydrite nodules present. Some desiccation or disturbed bedding associated with stylolites at 5059'.
5065							5063'–68': Light bluish gray to medium light gray massive anhydrite. Nodules and dense anhydrite at 5064' and 5068', separated by dolomite-rich matrix. Some organic wisps are present.

Figure 18B Dean 85 core description (contd.).

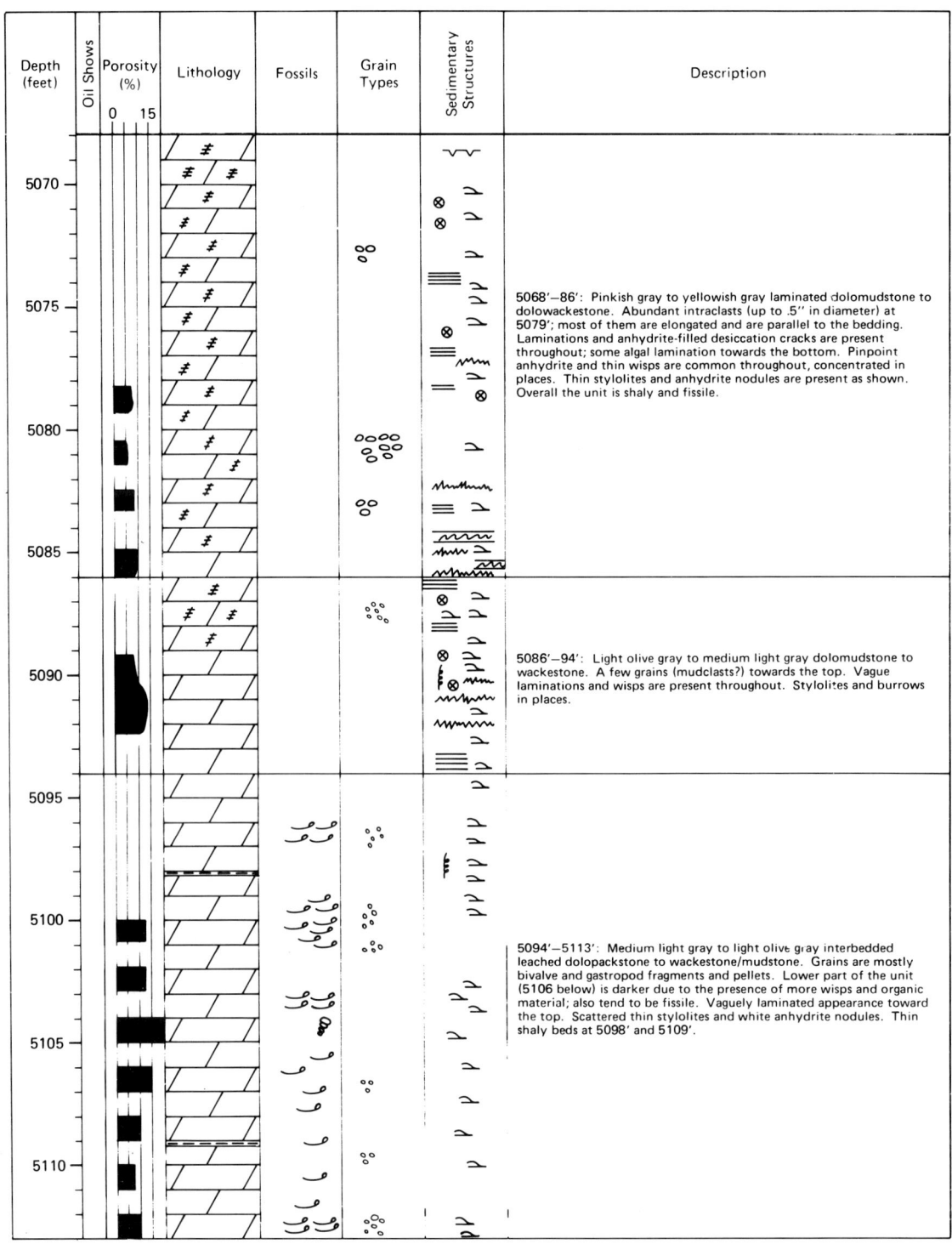

Figure 18C Dean 85 core description (contd.).

Figure 19. Core photos of open marine shelf deposits below Zone 7.

    A. Core slab of light olive gray to light gray dolopackstone. Fossils are mostly echinoderms, molluscs, brachiopods, and few fusulinids. Horizontal and vertical burrows. Fossil layer in middle disrupted by burrows. ϕ = 12.6%; k = 0.4 md. Sample from XIT 161 well, 5071 ft.

    B. Photomicrograph of burrowed, crinoidal, brachiopod dolopackstone. Elongate, whitish shells are mostly brachiopods, with scattered molluscs. Burrowing imparts a color-mottled texture to the matrix. Plain light, sample from XIT 161 well, 5073 ft.

    C. Core slab of light olive gray to medium light gray dolopackstone. Fossils are mostly crinoids, molluscs, and brachiopods. Burrowed in places; vaguely wispy. Patchiness due to oil and organic distribution. ϕ = 13.2%; k <0.1 md. Sample from XIT 161 well, 5099 ft.

    D. Photomicrograph of moldic and vuggy porosity in fossiliferous dolopackstone. Black areas are open pores, porosity averages 13% to 15%. Whitish object in lower left is a partially leached crinoid fragment. Crossed polarized light, sample from XIT 161 well, 5098 ft.

Figure 20. Core photos representative of shallow restricted shelf deposits of Zone 7.

    A. Core slab of light olive gray to medium gray laminated dolomudstone to wackestone. Grains are mostly mollusc fragments and pellets, leached and unleached. Numerous vertical burrows disrupt the laminations. Intertidal to subtidal deposit. ϕ = 10.0%; k = <0.1 md. Sample from XIT 161 well, 5028 ft.

    B. Photomicrograph of anhydritic, laminated, pelletal dolowackestone. The pellets appear as dark gray, subrounded grains, and the whitish areas are anhydrite crystals. Black areas are either anhydrite crystals at extinction and/or open porosity. Crossed polarized light; sample from XIT 161 well, 5030 ft.

    C. Core slab of light olive gray to medium light gray wispy dolomudstone. Burrowed with patchy oil staining. Subtidal deposit. ϕ = 12.5%; k <0.1 md. Sample from SW Levelland 105 well, 4993 ft.

    D. Burrowed, wispy, slightly anhydritic dolomudstone to dolowackestone. Irregular, mottled texture reflects extensive burrow reworking of the original sediment. Whitish elongate objects are open and/or anhydrite-filled fossil molds. Plain light, sample from Dean 95 well, 5100 ft.

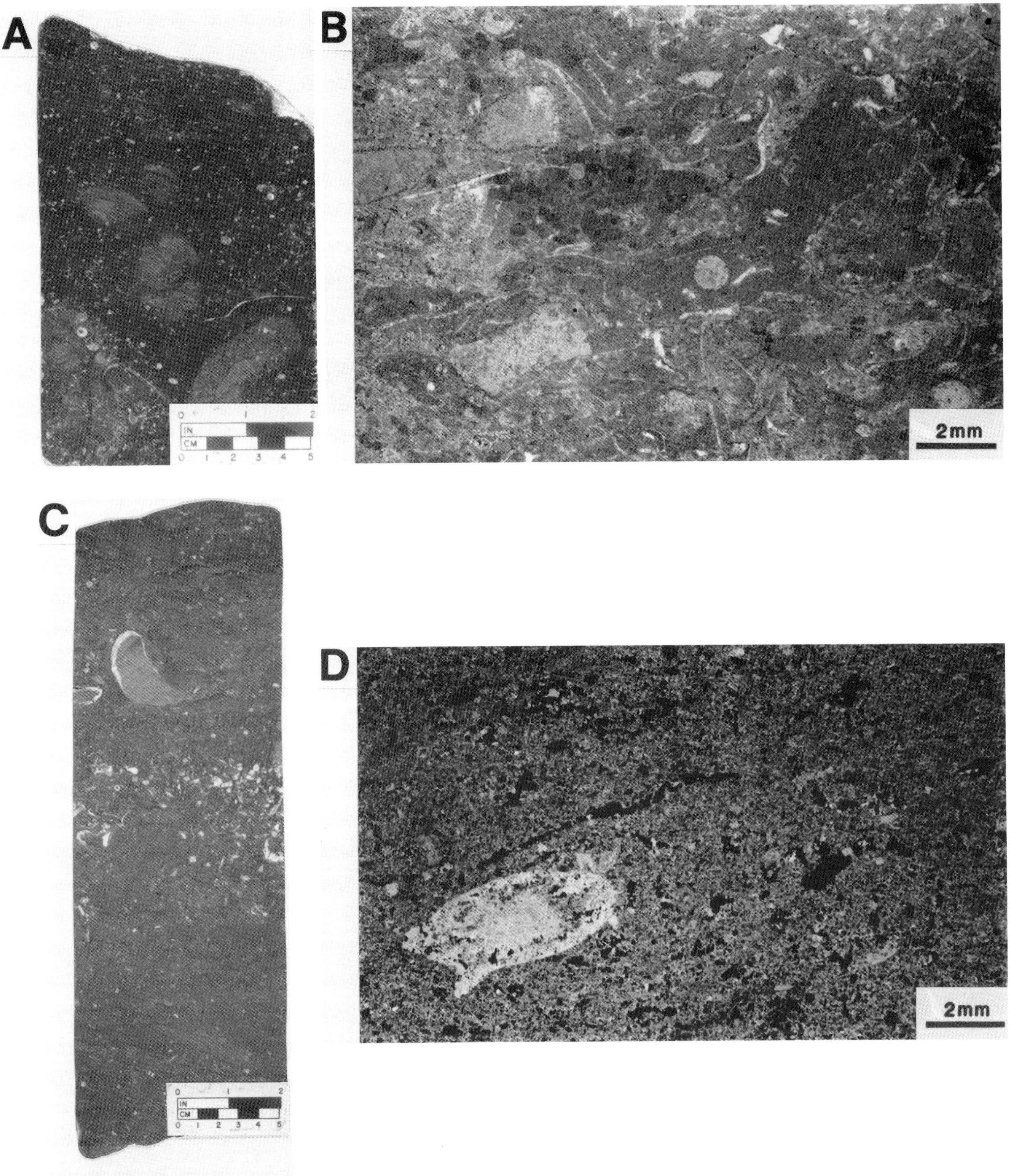

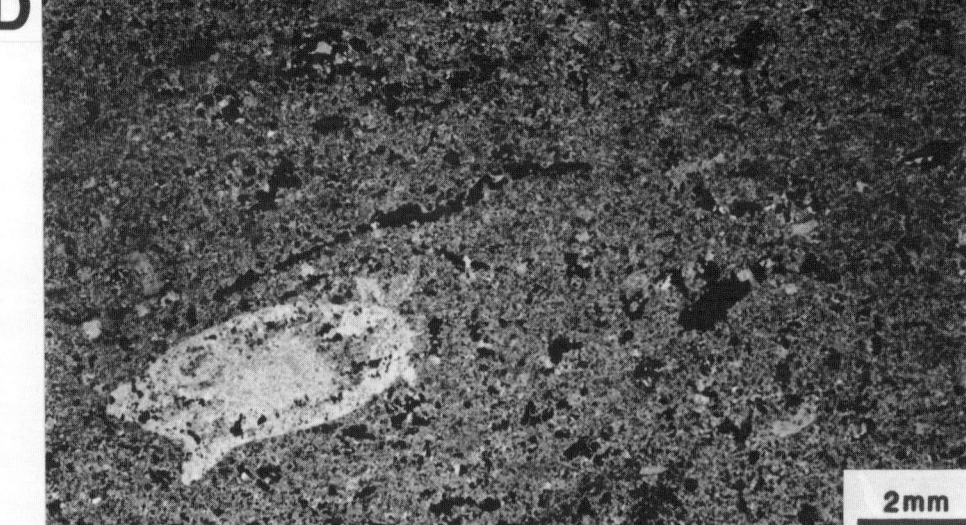

Figure 19

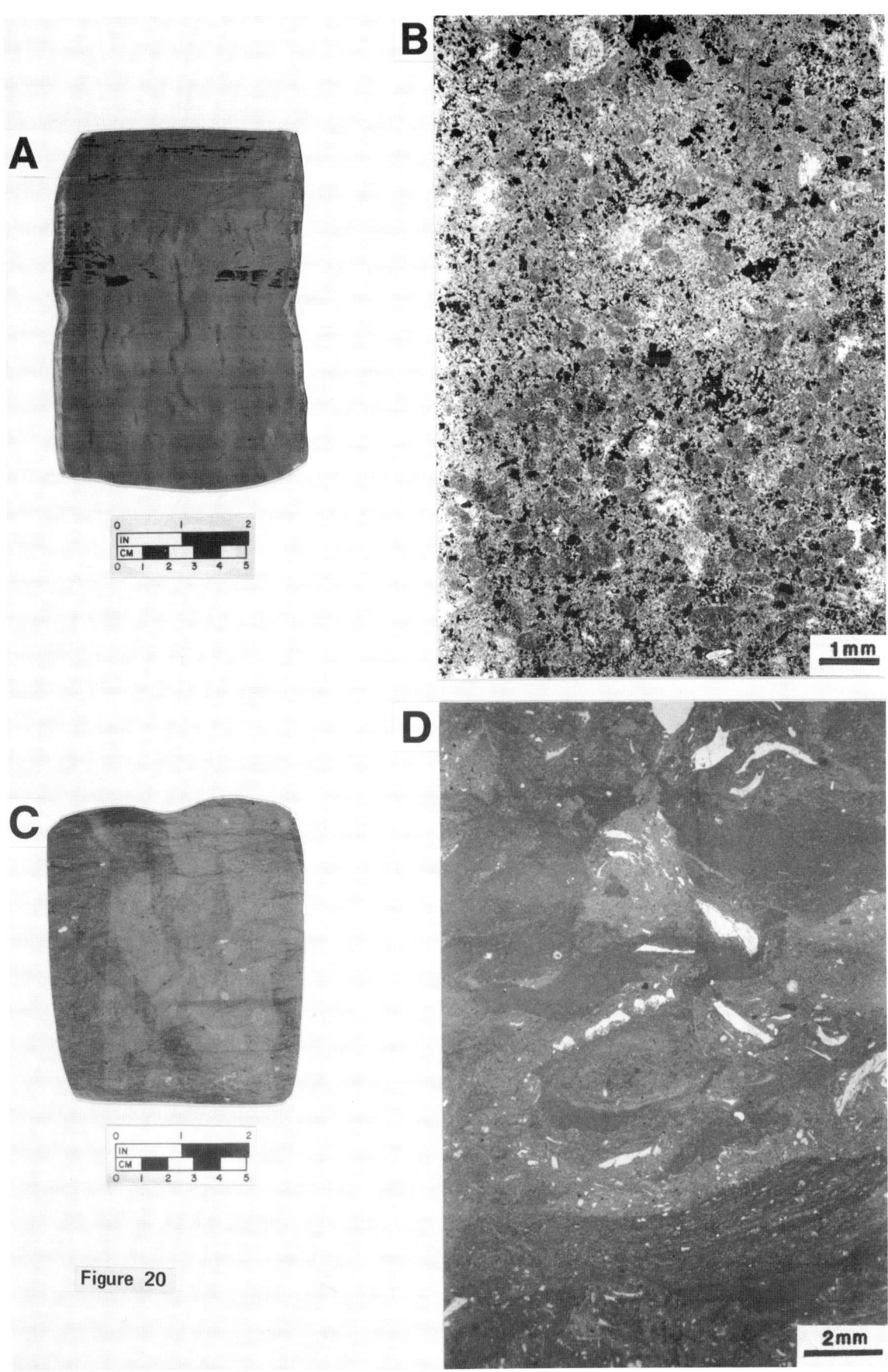

Figure 20

Figure 21. Core photos representative of shallow restricted shelf deposits of Zone 6.

    A. Core slab of light olive gray to medium gray, dense burrowed dolomudstone to wackestone. Scattered leached mollusc fragments and other allochems; some are anhydrite-filled. Scattered wisps and stylolites. Thin vertical fractures. ϕ = 7.1%; k <0.1 md. Sample from Dean 90 well, 5050 ft.

    B. Photomicrograph of burrowed, sparsely fossiliferous dolowackestone. Fossil fragments, mostly molluscs and crinoids, are concentrated in burrow fills. Plain light, sample from Dean 85 well, 5025 ft.

    C. Core slab of light olive gray to medium gray dolopackstone of leached mollusc fragments and pellets. Vaguely laminated in places; burrowed. ϕ = 10.3%; k = 5.6 md. Sample from XIT 161 well, 4996 ft.

    D. Photomicrograph of burrowed, fossiliferous, pelletal dolopackstone. Anhydrite-filled molds are black and white, open porosity is black. Cross-polarized light. Sample from XIT 161 well, 4996 ft.

Figure 22. Core photos representative of shallow restricted shelf deposits of Zone 5.

    A. Core slab of light gray to light olive gray dolopackstone to wackestone. Pellet, mollusc fragments, and intraclasts are common. Burrow mottling reflected in differential oil staining. ϕ = 6.0%; k <0.1 md. Sample from Dean 148 well, 5026 ft.

    B. Photomicrograph of pelletal, fossiliferous, intraclastic dolopackstone. Intraclasts are mostly micritic or micritized, with no recognizable internal grains. Plain light, sample from Dean 148 well, 5027 ft.

    C. Core slab of light olive gray to light gray dolomudstone and packstone. Grains are pellets, with some ooids. Pinpoint anhydrite is common, as are anhydrite-filled fractures. Filled microfractures commonly radiate from anhydrite nodules. ϕ = 6.9%; k = 0.1 md. Sample from XIT 157 well, 5018 ft.

    D. Photomicrograph of anhydritic, fractured pelletal dolowackestone. Whitish areas are anhydrite, developed as nodules, fracture fillings, and pinpoint crystals. Dark objects in matrix are pellets. Plain light, sample from XIT 158 well, 4920 ft.

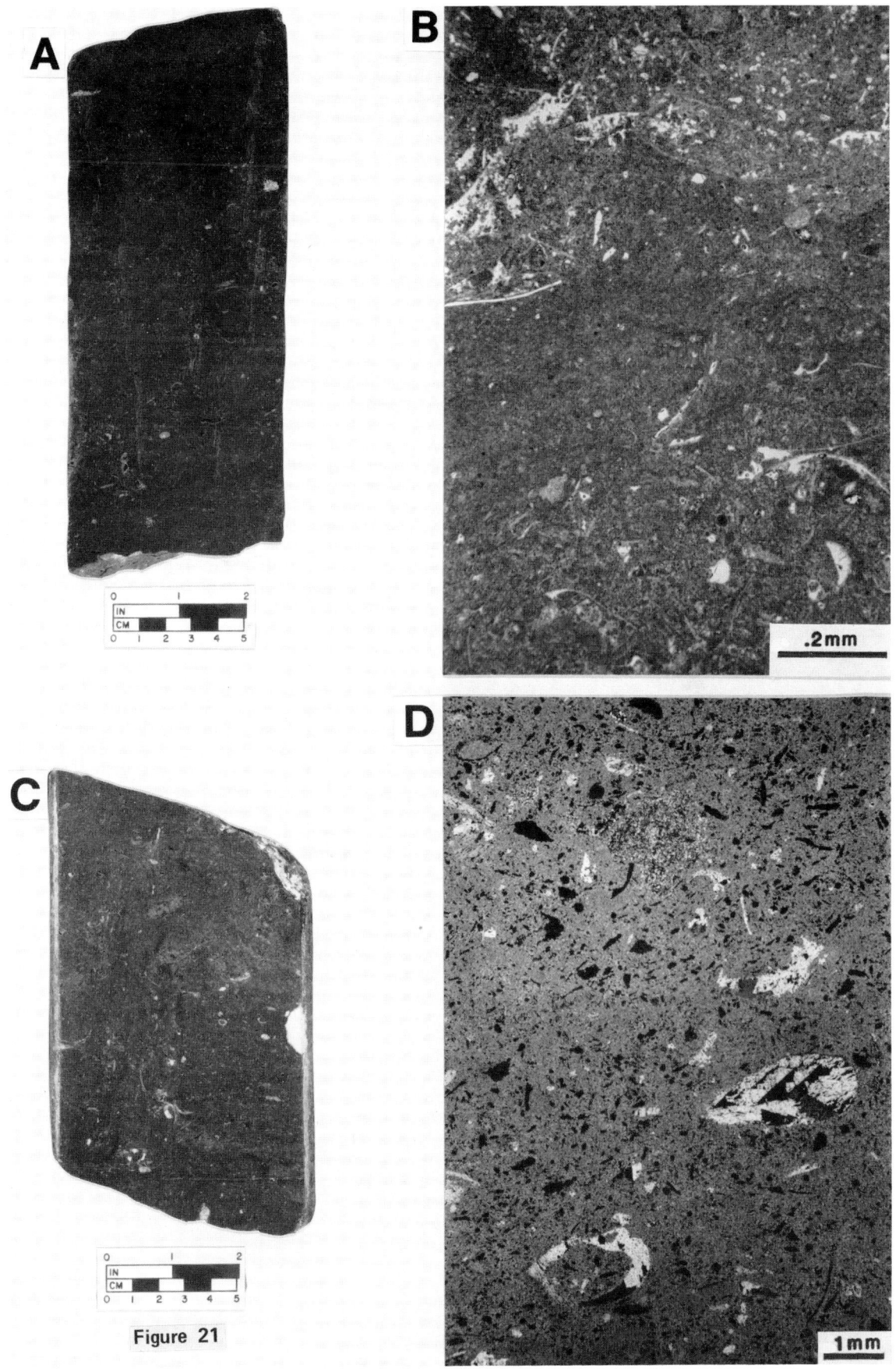

Figure 21

683

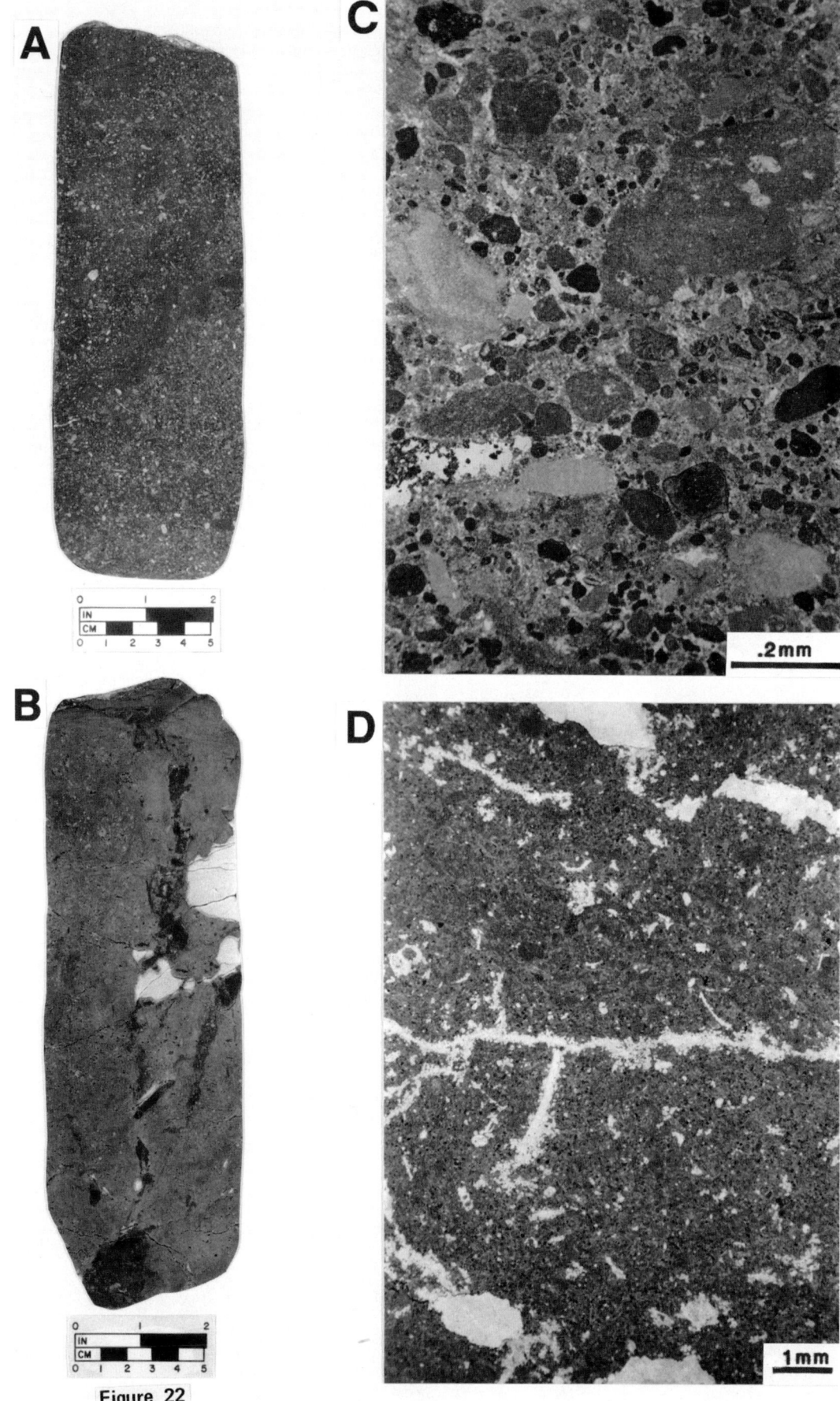

Figure 22

layers were dried and eroded. Occasional ooid beds are unstratified, suggesting they were transported from their site of formation. Most often the subtidal sediments are extensively burrowed, appearing as a churned mixture of mud-free and mud-rich carbonate sands. This churning of the sediments is often reflected by an irregular pattern of oil staining.

These three zones are reservoir zones. Mollusc and peloid dolopackstone-wackestone appears to be the predominant lithology that forms porous intervals, but these intervals are sporadically replaced vertically and horizontally by other subtidal carbonate sands or by intertidal muds and anhydrite (Figs. 7 and 8). Thus we visualize the deposition of these sands as occurring in a widespread subenvironment of the restricted shelf intermixed among a mosaic of lesser subenvironments.

Nonporous intervals separating Zones 7 and 6 and Zones 6 and 5 are characterized by intertidal and supratidal deposits (Fig. 23), similar to the sediments of the thick, overlying, nonporous Zones 4 and 3 (Fig. 24). Intertidal deposits are laminated dolomudstones and dolowackestones with wavy to horizontal planar laminae reflecting either color, size, or sorting variations within the mud, or the presence of pinpoint anhydrite. Supratidal deposits include algal-laminated dolomudstones and dolowackestones and thick intervals of anhydrite. The algal-laminated muds are easily differentiated from laminated muds, and their recognition is a good clue to supratidal deposition. Algal laminae have a crenulated structure, commonly are of irregular thickness, and contain pinpoint or lath anhydrite that reflects a vertical as well as horizontal pattern suggestive of thick mats of blue-green algae. Either type of laminae is indicative of an environment in which bedding is protected from a burrowing community and thus preserved. Fenestral vugs, anhydrite nodules, and rip-up clasts are commonly associated with the algal laminae. The thick intervals of anhydrite are massive or have a nodular mosaic pattern (chicken-wire anhydrite).

DIAGENESIS

Good correlation exists between porosity and a specific lithofacies (and thus depositional environment), despite a considerable amount of diagenesis that has affected the San Andres. Diagenesis has enhanced porosity by creating intercrystalline porosity during dolomitization and moldic porosity by leaching

Figure 23. Core photos representative of intertidal and supratidal deposits between porous zones 5 and 6.

    A. Core slab of yellowish gray to medium light gray desiccated dolomudstone to wackestone with abundant anhydrite nodules, nodular mosaic anhydrite, and pinpoint anhydrite. Algal laminations near bottom; numerous small cracks associated with nodules. No ϕ and k measurements. Sample from Dean 95 well, 5033 ft.

    B. Photomicrograph of anhydritic, siliceous dolomudstone to dolowackestone. Nodular area in upper half of slide is mostly composed of lath anhydrite (finely textured black and white areas). The larger, clearer patches of white, gray, and black are quartz crystals replacing anhydrite. Cross-polarized light, sample from Dean 95 well, 5033 ft.

    C. Core slab of light gray to light olive gray dolomudstone. Algal-laminated, with pinpoint anhydrite mimicking the algal structure. Intraclasts and a stylolite toward the bottom. ϕ = 2.7%; k = 0.1 md. Sample from XIT 157 well, 5043 ft.

    D. Photomicrograph of anhydritic, algal-laminated dolomudstone. Whitish areas are anhydrite crystals growing replacively within the matrix. Note cloudy nature of crystals, due to incorporation of carbonate mud. Plain light, sample from XIT 157 well, 5043 ft.

Figure 24. Core photos representative of supratidal deposits of Zones 4 and 3.

    A. Core slab of light olive gray to light gray or yellowish gray laminated and wispy dolomudstone. Burrowed. Pinpoint anhydrite common. No ϕ and k measurements. Sample from XIT 161 well, 4872 ft.

    B. Photomicrograph of anhydritic, laminated, burrowed, wispy dolomudstone. Abundant replacive anhydrite crystals developed along laminations. Cross-polarized light, sample from XIT 161 well, 4873 ft.

    C. Core slab of light gray to medium light gray laminated dolomudstone to wackestone. Chert nodules and layers are common at bottom. Pinpoint anhydrite is common at top, concentrated in laminae. No ϕ and k measurements. Sample from XIT 161 well, 4885 ft.

    D. Core slab of "chicken wire" anhydrite with remnants of original carbonate mud trapped between the nodules. Sample from XIT 161 well, 4834 ft.

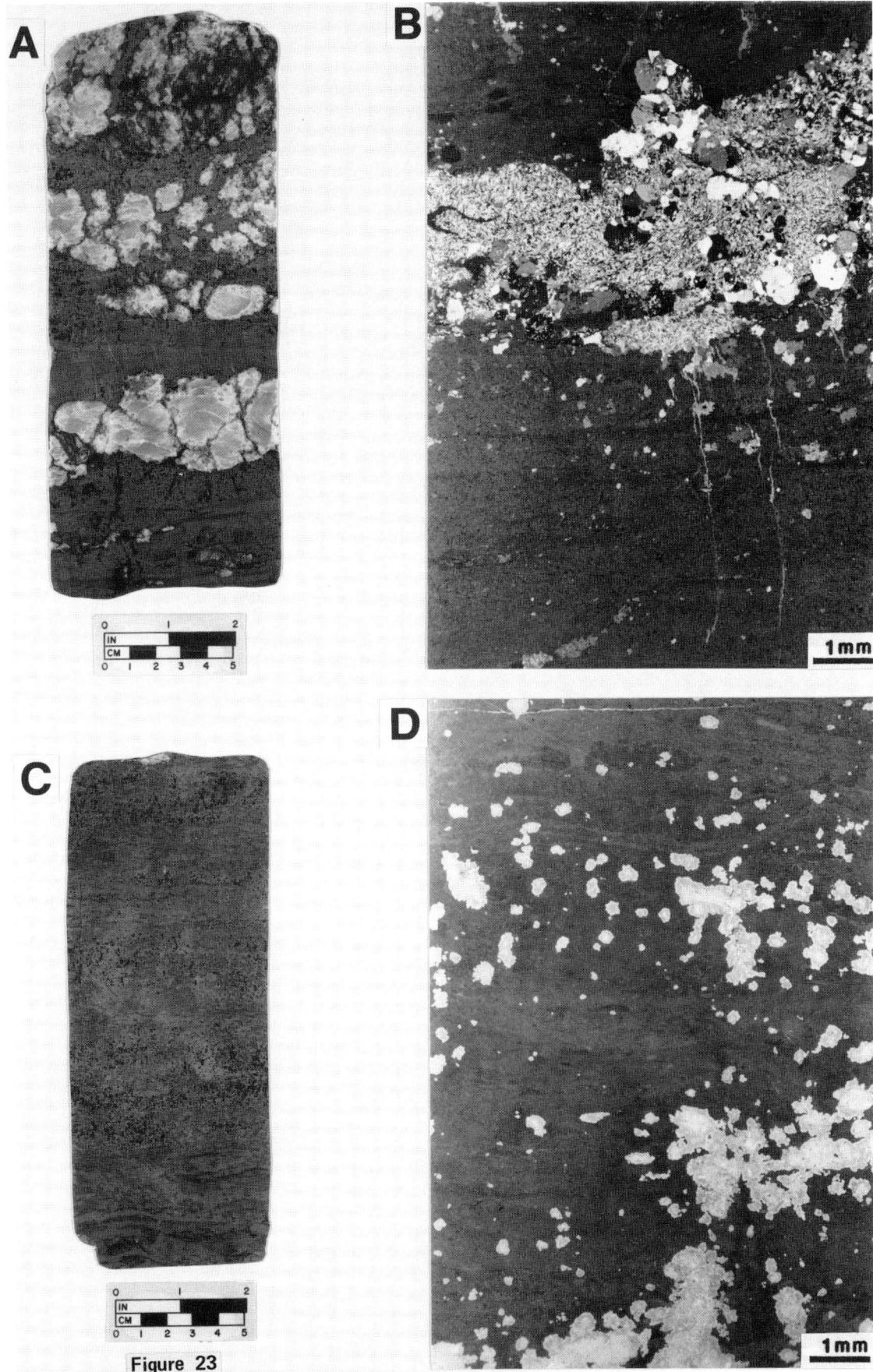

Figure 23

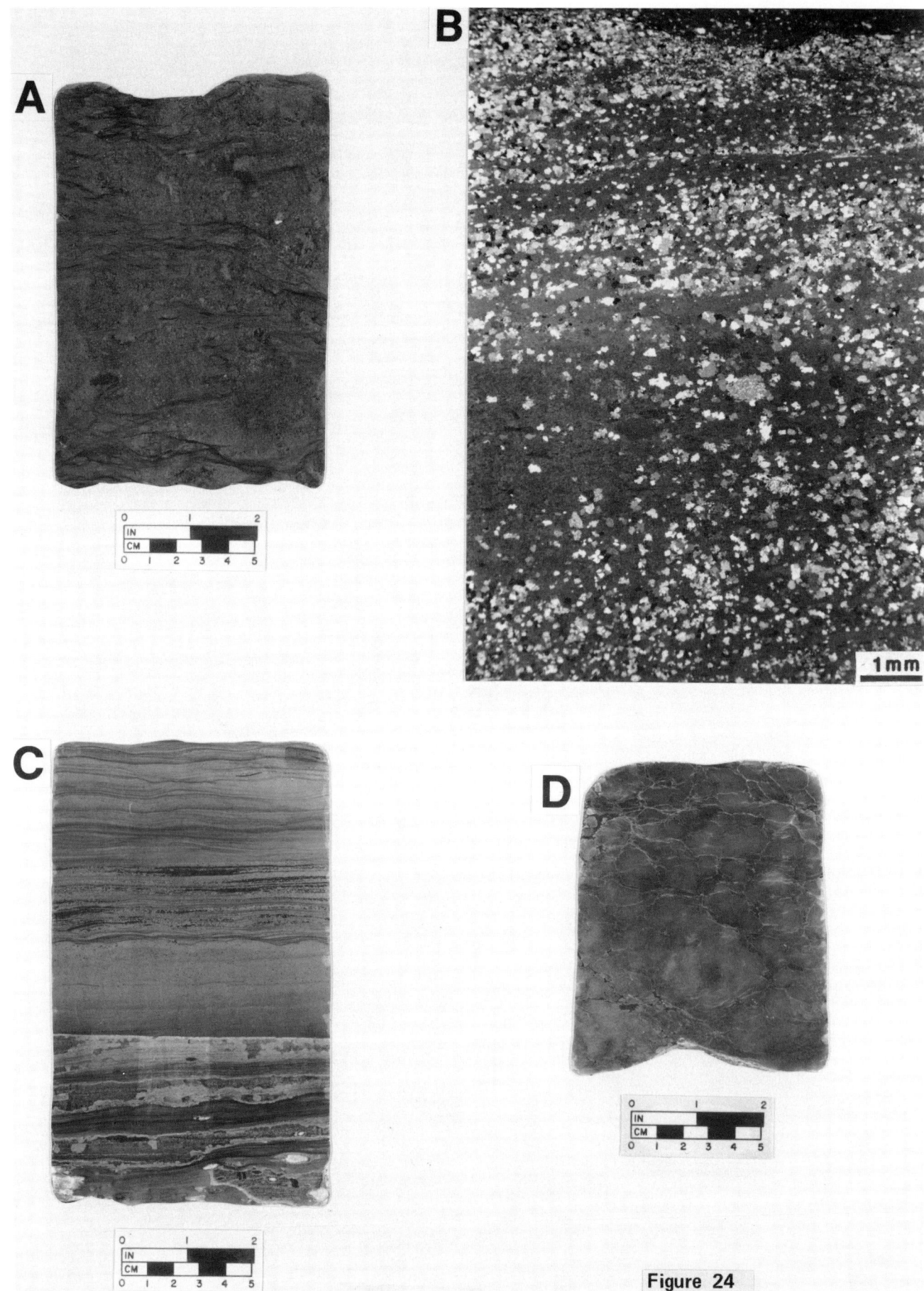

Figure 24

of mollusc shells. The diagenetic history can be briefly summarized in four stages:

1. DEPOSITION - By analogy with Holocene deposits, the subtidal carbonate sands were a mixture of aragonite and magnesian-calcite grains and mud with some calcite skeletal fragments; the supratidal algal-laminated muds were aragonite and magnesian-calcite with laminae or layers of gypsum and anhydrite. Both the subtidal and intertidal-supratidal deposits were heterogeneous mixtures of sediments prior to subsequent diagenesis: subtidal heterogeneity was due to burrowing and minor facies changes, whereas intertidal-supratidal heterogeneity was due to laminations and alternations of carbonates and evaporites.

2. DOLOMITIZATION - The carbonate sands and muds were pervasively dolomitized. Leary and Vogt (1986) presented preliminary carbon, oxygen, and strontium isotope data on San Andres dolomites and anhydrites, suggesting that reflux of Guadalupian brines was the dominant dolomitization mechanism. Dolomite is predominantly finely crystalline and subhedral to euhedral in crystal shape, but differences in original grain size and depositional texture did cause variation as medium-crystalline dolomite coincides with grainier rocks. Intercrystalline porosity is best developed in mollusc and peloid dolopackstones-wackestones.

3. DISSOLUTION - Dissolution of mollusc shells, as well as other originally aragonitic grains, occurred concurrently with or shortly after dolomitization. That the dissolution was fabric-selective is evidence it occurred early, at least prior to mineralogic stabilization. Moldic porosity is also best developed in mollusc and peloid dolopackstones-wackestones.

4. ANHYDRITE FORMATION - Anhydrite occurs in a variety of forms. The distribution of anhydrite beds commonly parallels sedimentary structures (laminae or algal microstructure). Anhydrite also displaces and replaces dolomite matrix as nodules and laths, and occurs as a cement in some of the moldic porosity and microfractures.

Other diagenetic features include wisps and stylolites, microfractures, chert nodules and layers, and clay nodules. Indeed, the rocks have undergone a complex diagenetic history, but the fact that a specific facies with environmental significance was most favorably affected during the diagenesis simplifies the need to understand the diagenesis in order to understand the reservoir potential of the zones.

Waterflooding of a reservoir like Levelland-Slaughter Field can cause added complications to the diagenetic overprint and alter porosity and permeability patterns. Hager and Heathcote (1986) examined pre- and post-waterflood cores from the Mobil H. O. Mahoney Lease of nearby Wasson Field and concluded that a waterflood of 20 years duration had significant effects on the reservoir. Anhydrite was dissolved adjacent to an injection well, gypsum was precipitated as scale in a producing well, and dolomite dissolution and calcitization of anhydrite likely occurred in the reservoir. These types of changes have obvious implications for waterflood efficiency (creation of thief zones), reservoir simulation, and calculation of original oil in place.

SIGNIFICANCE TO ENHANCED RECOVERY OPERATIONS

Most of the high-porosity zones in the reservoir, where porosity ranges from 11 to 14%, are contained in subtidal facies and particularly skeletal and peloidal dolopackstones-wackestones with moldic porosity. Intercrystalline porosities are variable in themselves: 7 to 8% in coarser dolomite, 5 to 7% in burrowed mudstones with medium crystalline dolomite, and 2 to 3% in finely crystalline dolomite. Subtidal facies typically exhibit the most extensive dissolution, yet have the least amount of porosity destruction by evaporite cementation. Porous zones are separated by supratidal anhydrite and muds that are capable of isolating one zone from another. There is little lateral convergence among the porous zones; indeed, the zones may be separate reservoirs. In particular, Zone 7 appears to be separated from Zone 6: the water transition zone in Zone 7 of the XIT wells 161 and 157 occurs structurally above the oil-saturated intervals of Zone 6 in the Dean wells 85, 90, and 95 (Fig. 7). Anhydrite between Zones 6 and 5 is of variable thickness and is sometimes missing. Where absent, or if the supratidal deposits are extensively fractured, there may be a vertical communication between the porous zones.

Lithologic changes occur both laterally and vertically within the porous zones (Figs. 7 and 8). Thus porosity and permeability pinchouts may be expected. Overall, each of the porous zones has attractive reservoir potential, but small-scale porosity variations within a zone may cause localized variations during waterflooding. The source of the variation will most likely be a "permeability barrier" established by a lens of supratidal deposits within the predominantly subtidal zone. If anhydrite is missing and the barrier is formed only of less-porous laminated muds, the barrier may be difficult to detect without a core study.

Anhydrite cement is also a major control over porosity in the reservoir zones. Anhydrite preferentially fills larger pores within the upper portions of the subtidal portion of each depositional cycle. The distribution of anhydrite in the rock causes fluid flow to be dominated by the intercrystalline pore system of the dolomite. The more permeable portions of the reservoir have yielded most of their oil during primary and secondary recovery; therefore, as discussed by Ebanks (1986), enhanced processing must displace oil from parts of the reservoir that water did not reach.

Waterflooding of the reservoir in the C. S. Dean "A," XIT, and SW Levelland Units has been less successful than anticipated, in part because individual porous zones were not completed separately in many wells, but also due to geological considerations. Due to very low permeabilities and porosities (averaging 1 md and 7.8%), fluid movement through certain units is limited, oil is bypassed or trapped in low permeability rock, and there is probable channeling along high-porosity pathways. The sweep efficiency of the waterflood is thus seriously affected by the spatial arrangement of lithofacies in the reservoir, and recovery from this type of reservoir is lower than that from one with more uniform permeability distribution. Previous studies of other San Andres reservoirs (for example, Wasson Field by Mathis and Sears, 1984; Reeves Field by Chuber and Pusey, 1972; Garza Field by Hild, 1986; Suniland Field by Beaver, 1986; and the Mallet Lease of Slaughter Field by Ebanks, 1986) have shown the permeability distribution to be in a patchy pattern often with laterally discontinuous permeability zones. Cross flow between reservoir zones is unlikely, unless fractures, natural or induced, are present. Therefore, it necessary to perforate all of the porous zones to insure that they are drained or flooded.

ACKNOWLEDGMENTS

Chevron Oil Field Research Company and Texaco, Inc. permitted publication of the paper. R. O. Hafner and M. Bukhari helped immensely in the core studies.

REFERENCES

BEAVER, J. L., 1986, Deposition, Porosity Occurrence, and Reservoir Properties of the San Andres Formation-Suniland Field, Lynn County, Texas: in Bebout, D. G. and Harris, P. M. (eds.), Hydrocarbon Reservoir Studies - San Andres/Grayburg Formations, Permian Basin, Permian Basin Section SEPM Publ. No. 86-26, p. 9-16.

BEIN, A. and LAND, L. S., 1982, The San Andres Carbonates in the Texas Panhandle: Sedimentation and Diagenesis Associated with Mg-Ca Chloride Brines: Univ. of Texas, Bureau of Economic Geology Report of Investigations No. 121, 48 p.

CHUBER, S. and PUSEY, W. C., 1972, Cyclic San Andres Facies and Their Relationship to Diagenesis, Porosity, and Permeability in the Reeves Field, Yoakum County, Texas: in Elam, J. G. and Chuber, S. (eds.), Cyclic Sedimentation in the Permian Basin, West Texas Geological Society Special Publication No. 72-60, p. 135-150.

COWAN, P. E. and HARRIS, P. M., 1986, Porosity Distribution in San Andres Formation (Permian), Cochran and Hockley Counties, Texas: AAPG Bull., v. 70, p. 888-897.

DUTTON, S. P., FINLEY, R. J., GALLOWAY, W. E., GUSTAVSON, T. C., HANDFORD, C. R., and PRESLEY, M. W., 1979, Geology and Geohydrology of the Palo Duro Basin, Texas Panhandle: Univ. of Texas, Bureau of Economic Geology Geological Circular 79-1, 99 p.

EBANKS, W. J., 1986, Geologic Description of the San Andres Reservoir, Mallet Lease, Slaughter Field, Hockley County, Texas: Implications for Reservoir Engineering Projects: in Bebout, D. G. and Harris, P. M. (eds.), Hydrocarbon Reservoir Studies - San Andres/Grayburg Formations, Permian Basin, Permian Basin Section SEPM Publ. No. 86-26, p. 1-3.

GRATTON, P. J. F. and LEMAY, W. J., 1969, San Andres Oil East of the Pecos: in Summers, W. K. and Kottlowski, F. E. (eds.), The San Andres Limestone, A Reservoir for Oil and Water in New Mexico, New Mexico Geological Society Special Publ. 3, p. 37-43.

HAFNER, R. O., 1979, Lithofacies and Depositional Setting of the San Andres Formation, Cochran County, West Texas: Unpub. M.S. Thesis, Univ. of New Orleans, 72 p.

HAGER, R. C. and HEATHCOTE, R. C., 1986, Petrography of Recent Waterflood Effects on the San Andres Formation-Mobil H. O. Mahoney Lease, Wasson Field, Yoakum County, Texas: in Bebout, D. G. and Harris P. M. (eds.), Hydrocarbon Reservoir Studies - San Andres/Grayburg Formations, Permian Basin, Permian Basin Section SEPM Publ. No. 86-26, p. 39-41.

HILD, G. P., 1986, The Relationship of San Andres Facies to the Distribution of Porosity and Permeability - Garza Field, Garza County, Texas: in Bebout, D. G. and Harris, P. M. (eds.), Hydrocarbon Reservoir Studies - San Andres/Grayburg Formations, Permian Basin, Permian Basin Section SEPM Publ. No. 86-26, p. 17-20.

LEARY, D. A. and VOGT, J. N., 1986, Diagenesis of the Permian (Guadalupian) San Andres Formation, Central Basin Platform, Texas: in Bebout, D. G. and Harris, P. M. (eds.), Hydrocarbon Reservoir Studies - San Andres/ Grayburg Formations, Permian Basin, Permian Basin Section SEPM Publ. No. 86-26, p. 67-68.

MATHIS, R. L. and SEARS, S. O., 1984, Effect of CO_2 Flooding on Dolomite Reservoir Rock, Denver Unit, Wasson (San Andres) Field, Texas: Soc. Pet. Eng. Annual Tech. Conf., Paper SPE 13132.

MEISSNER, F. F., 1972, Cyclic Sedimentation in Middle Permian Strata of the Permian Basin, West Texas and New Mexico: in Elam, J. G. and Chuber, S. (eds.), Cyclic Sedimentation in the Permian Basin, West Texas Geological Society Special Publication No. 72-60, p. 203-232.

PITT, W. D. and SCOTT, G. L., 1981, Porosity Zones of Lower Part of San Andres Formation, East-Central New Mexico: New Mexico Bureau of Mines and Mineral Resources Circular 179, 20 p.

RAMONDETTA, P. J., 1982, Facies and Stratigraphy of the San Andres Formation, Northern and Northwestern Shelves of the Midland Basin, Texas and New Mexico: Univ. of Texas, Bureau of Economic Geology Report of Investigations No. 128, 56 p.

SILVER, B. A. and TODD, R. G., 1969, Permian Cyclic Strata, Northern Midland and Delaware Basins, West Texas and Southeastern New Mexico: AAPG Bull., v. 53, p. 2223-2251.

TODD, R. G., 1976, Oolite-Bar Progradation, San Andres Formation, Midland Basin, Texas: AAPG Bull., v. 60, p. 907-925.

MISSION CANYON (MISSISSIPPIAN) RESERVOIR STUDY, WHITNEY CANYON-CARTER CREEK FIELD, SOUTHWESTERN WYOMING

P. M. HARRIS
Chevron Oil Field Research Company,
La Habra, California 90631

P. E. FLYNN and J. L. SIEVERDING
Chevron U.S.A., Inc., Northern Region,
Denver, Colorado 80111

ABSTRACT

Geologic data from conventional cores and wireline logs have been used to better understand the controls on reservoir-quality porosity within the Mission Canyon Formation of Whitney Canyon-Carter Creek Field.

The Mission Canyon Formation (Madison Limestone Group) is a thick, regressive sequence that has been subdivided by previous workers into three major stratigraphic units: (1) a lower section of very fine-grained, black to dark gray limestone and dolomite deposited in a starved, anaerobic basin; (2) a middle "main" porosity zone of dolomite and limestone that accumulated on an oxygenated shelf; and (3) an upper fractured and brecciated section of sabkha dolomite and anhydrite. The upper unit was extensively altered during subsequent solution brecciation. All three stratigraphic units were observed in cores from the field, but the middle section was studied in most detail.

Most Mission Canyon production in Whitney Canyon-Carter Creek Field occurs from porous dolomites of the middle "main" producing zone. This section is 300 to 350 feet thick and consists of porous dolomites interbedded with tight, finely crystalline calcitic dolomites or tight lime grainstones. The reservoir dolomites contain intercrystalline, moldic, fracture, and vug porosities. The best reservoir intervals have skeletal-moldic and intercrystalline porosities of greater than 9% and permeabilities exceeding 0.7 md. Although the earliest stages of diagenesis seem to have created the porosity patterns observed in the reservoir, the carbonates were intensely stylolitized and fractured during burial and subsequent thrusting. Porosity was partially filled with later-stage calcite and anhydrite cements that are possibly related to fracturing.

A similar vertical succession of facies was recognized in all of the cores. The wells are oriented roughly parallel to depositional strike of the facies and little variation occurred in that direction; therefore, the facies and porous dolomite zones are correlated easily where faults or other structural problems do not limit communication. Good stratigraphic horizontal communication apparently exists throughout the reservoir as only minor rock variability and differential cementation are present along layers. Vertical communication may be restricted by thin, tight limestone layers.

INTRODUCTION

Whitney Canyon-Carter Creek Field, Lincoln and Uinta Counties, Wyoming, was discovered in 1976 and currently produces from 36 sour natural gas and condensate wells within an area 13 miles long and 2 miles wide (Fig. 1). The field is volumetrically the largest natural gas field in the U.S. Rocky Mountains. Total field-wide in-place reserves have been estimated at 2 TCF, 60 MMBC, and 12 MMLTS (see Hoffman and Kelly, 1981; Bishop, 1982; and Hoffman and Balcells-Baldwin, 1982, for details). Lying in the Wyoming Thrust Belt province, hydrocarbons in the field are trapped in a series of anticlinal folds which are bounded on the east by reverse faults. The field is divided into two large anticlinal structures: Carter Creek to the north and Whitney Canyon to the south. These folds formed to the west of where the nearly flat-lying Absaroka fault cuts up-section through Paleozoic sedimentary rocks and planes out again in Triassic shales. Along the fault ramp, Paleozoic rocks were placed in fault contact with underlying Cretaceous black shales which served as the source for the hydrocarbons produced in the field (Hoffman and Kelly, 1981). Although porous dolomites of the Mission Canyon Formation contain over 70% of the total recoverable reserves for the field, reservoir zones are also found in the Ordovician Bighorn dolomite, the Mississippian Lodgepole Formation, the Pennsylvanian Weber Sandstone, and the Triassic Thaynes Formation.

The main objectives of this study were to provide a geologic description of the Mission Canyon Formation within the field by (1) describing cored intervals in detail, dividing the sequence into lithofacies, and constructing a depositional model for the formation; (2) determining the types of porosity and showing relations to depositional and diagenetic facies; and (3) determining the continuity of facies and porosity between wells. Dolomitization is largely responsible for early porosity development; therefore, it is also important to have some understanding of the controls over dolomite distribution.

STRATIGRAPHY

Correlation of the Mission Canyon Formation (Madison Limestone Group) regionally through Wyoming and elsewhere in the Overthrust Belt is complicated by structure and the coincidence of the shelf margin trend and its parallel facies tracts with that of the thrust belts. Some of the problems with

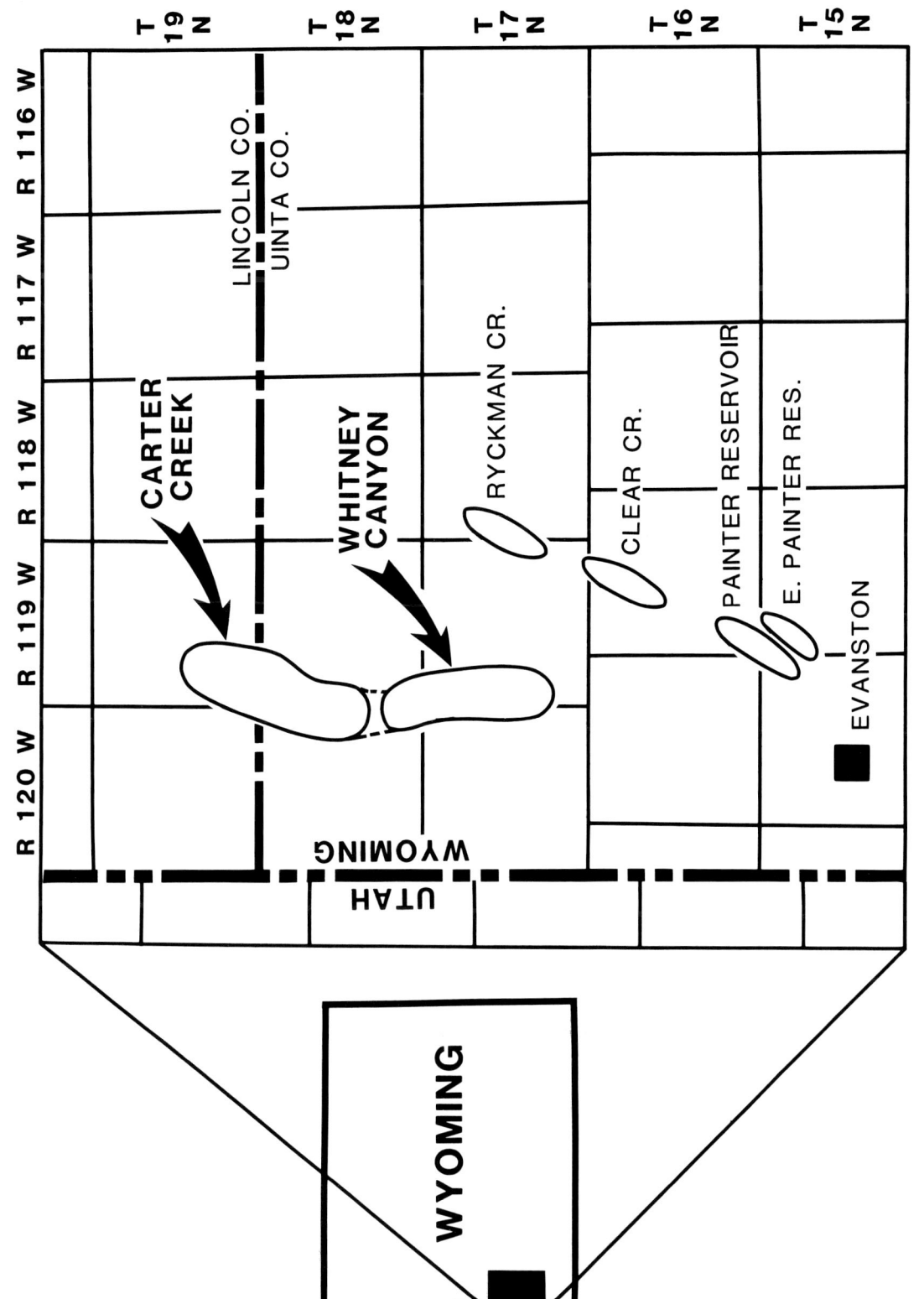

Figure 1 Location map for Whitney Canyon-Carter Creek Field, Southwestern Wyoming.

interpretations of the Mississippian strata are reviewed by Rose (1977). Regional stratigraphic and paleogeographic work by Sandberg and Gutschick (1980), Sandberg et al. (1982), and Gutschick and Sandberg (1983) have shown that the Mission Canyon Formation is a thick, regressive sequence. During Kinderhookian to middle Meramecian time, deposition in Wyoming and elsewhere in the Rocky Mountain region represents transgressions and regressions across the Cordilleran Platform, a broad carbonate ramp grading to the west into a foreland basin. Southwestern Wyoming was situated in an equatorial position with a crudely north-south oriented platform "margin" lying just to the west of the state boundary. Deeper-water sedimentation of the Cordilleran Miogeosyncline occurred to the west of the platform and, toward the east, the platform was bordered by the Transcontinental Arch, a linear positive feature extending southwestward from the Canadian Shield. Uplift of the craton and lowering of sea level in middle Meramecian time formed a karst plain in most of the area previously occupied by the carbonate platform (Sando, 1976; Gutschick et al., 1980; Gargallo-Quinones, 1985).

The Mission Canyon has been subdivided by previous workers (see Rose, 1977, for example) into three generalized stratigraphic units, in ascending order: (1) dark gray to black, fine-grained limestone and dolomite that were deposited in a starved, anaerobic basin or deeper shelf environment; (2) variable limestones and dolomites of shallower-water shelf origin that collectively form the "main" porosity zone; and (3) anhydritic tidal-flat deposits that were subsequently brecciated and karstified. It is the middle unit of the Mission Canyon Formation that is of most interest since it contains most of the reservoir facies of Whitney Canyon-Carter Creek Fields. Portions of all three generalized stratigraphic units have been examined from various cores in the field, although no single well contains core from all three intervals.

Rock types, recognized in cores from 13 wells in the field (see Fig. 2 for well locations), can be subdivided by depositional texture and sedimentary structures into seven facies that relate to depositional setting and subsequent diagenesis. Figure 3 is an idealized vertical section of these facies, and Figure 4 is a chart summarizing the data base for the core study; that is, the facies recovered in core are shown for each well. Conceptually, the seven facies can best be understood in the context of the overall regressive sequence

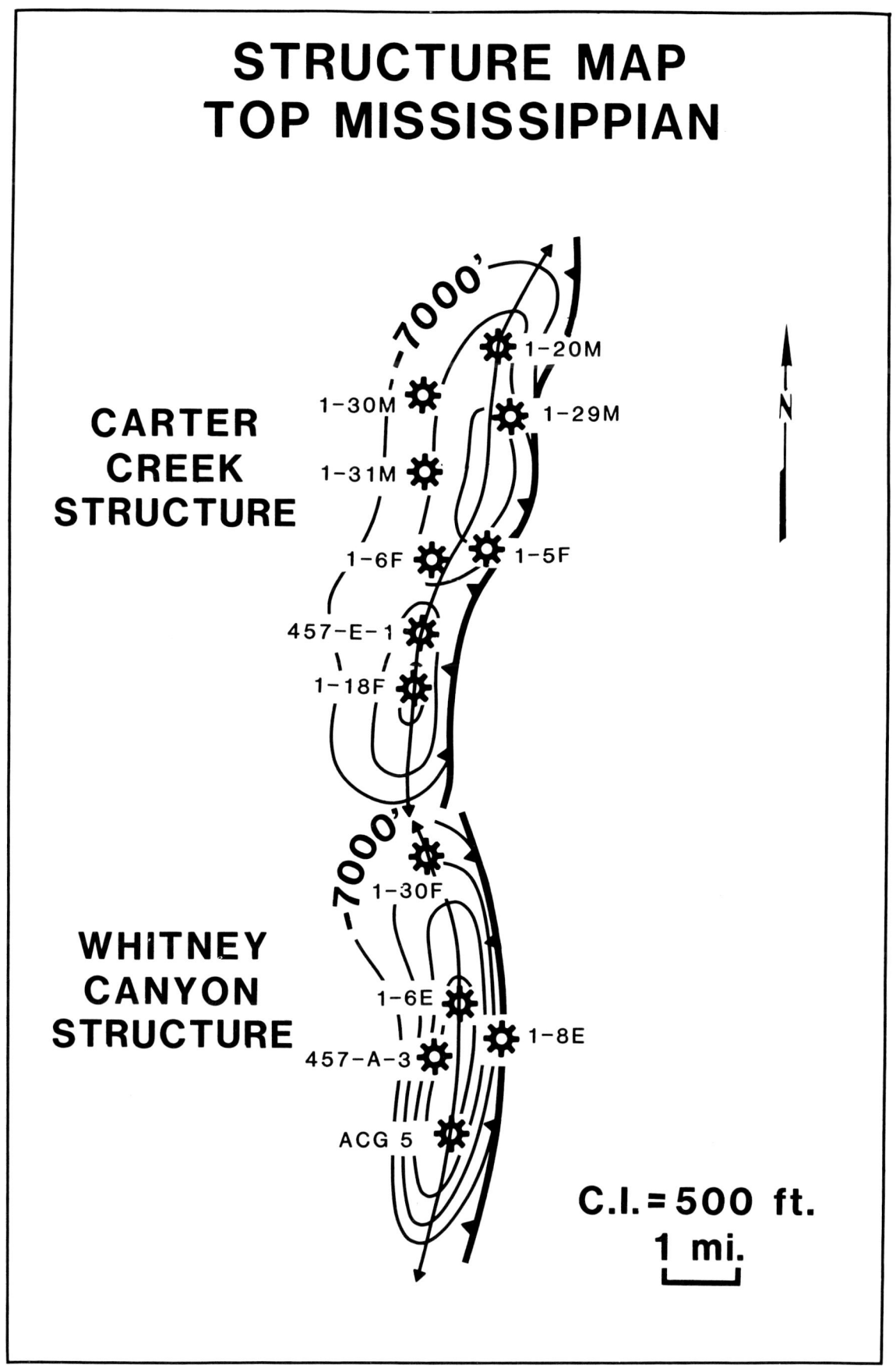

Figure 2 Locations of cored wells in Whitney Canyon-Carter Creek Field.

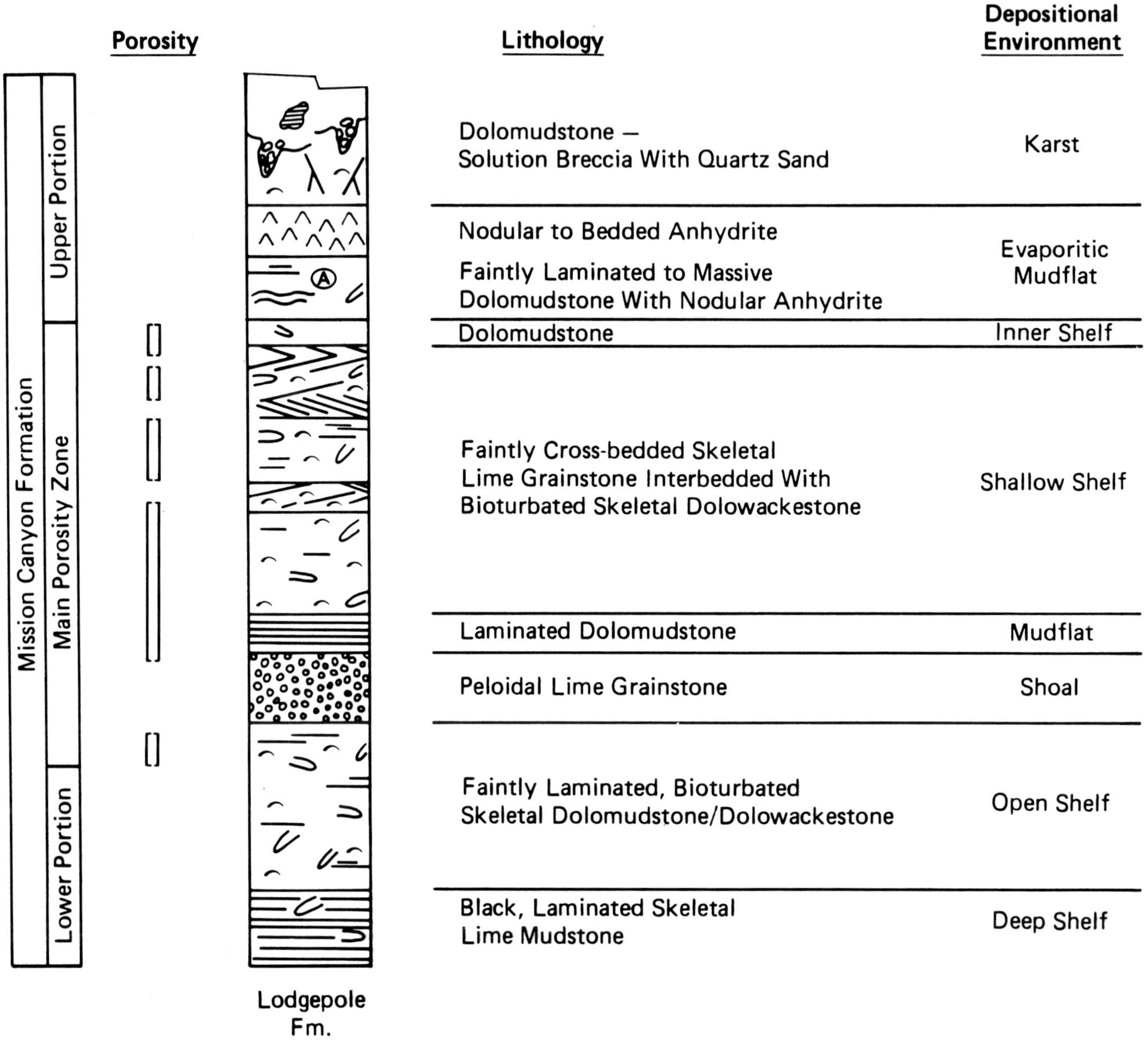

Figure 3 Idealized composite section of Mission Canyon Formation in Whitney Canyon-Carter Creek Field.

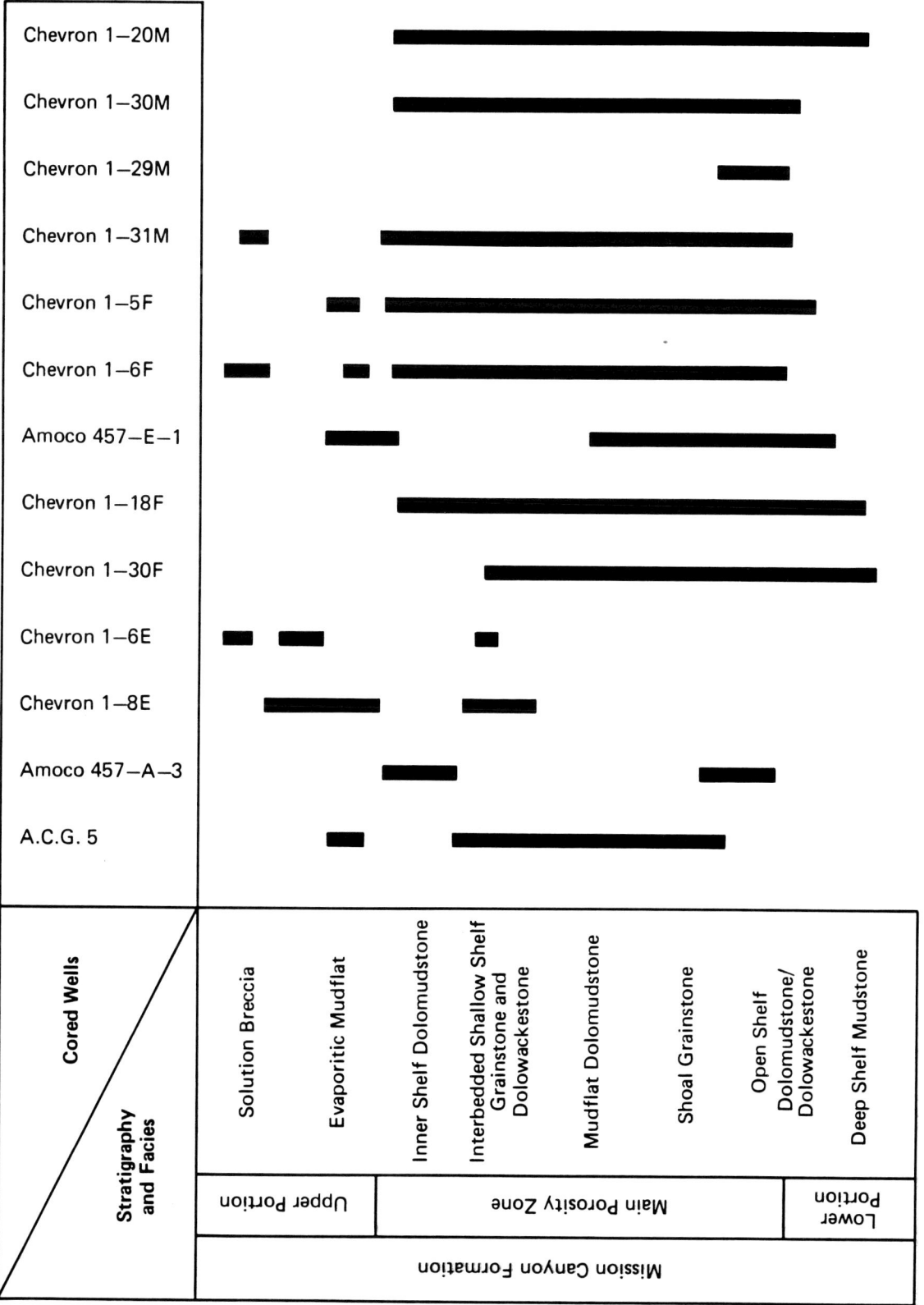

Figure 4 Chart showing stratigraphic interval found in each cored well from Whitney Canyon-Carter Creek Field.

that typifies the Mission Canyon and its three generalized stratigraphic units briefly described above:

(1) Deep shelf mudstone recognized during the core study is found in the lower portion of the generalized stratigraphic sequence, whereas the open shelf dolomudstone/dolowackestone contains the gradational contact between the lower and middle portions of the stratigraphic sequence;

(2) Shoal grainstone, mudflat dolomudstone, interbedded shallow shelf grainstone and dolowackestone, and inner shelf dolomudstone found in cores from the field typify the middle portion of the generalized stratigraphic sequence; and, lying above another gradational contact,

(3) Evaporitic mudflat anhydrite and dolomudstone, commonly altered toward the top of the section to a karst-related solution breccia, seen in the cores are representative of the upper portion of the generalized stratigraphic sequence.

The schematic block diagram of Figure 5 portrays the facies in simplified form as they may have occurred across the shelf and gentle slope setting. The progradation of the entire shelf facies tract toward the west in a basinward direction produced the regressive vertical succession of facies that typifies the Mission Canyon Formation in cores from Whitney Canyon-Carter Creek Field. In detail, the progradational history may have been complicated by minor sea-level variations and related shifts of facies belts, as suggested by Rau (1982), who interpreted two stacked shallowing sequences within the same stratigraphic interval during her core studies.

The main producing interval of the Mission Canyon Formation in the field encompasses most of the facies recognized during the core study. The base of the producing interval occurs within the open shelf dolomudstone/dolowackestone unit, and production occurs from throughout the overlying interval extending upward into the inner shelf dolomudstone unit (Fig. 3).

The stratigraphy within the main porosity zone of the field varies little between cores. The field is oriented parallel to depositional strike (Fig. 2), thus facies variation that likely would be seen in a dip direction is not encountered with so little well control in that direction. Conversely, the trend of the field and well control in a strike direction has encountered

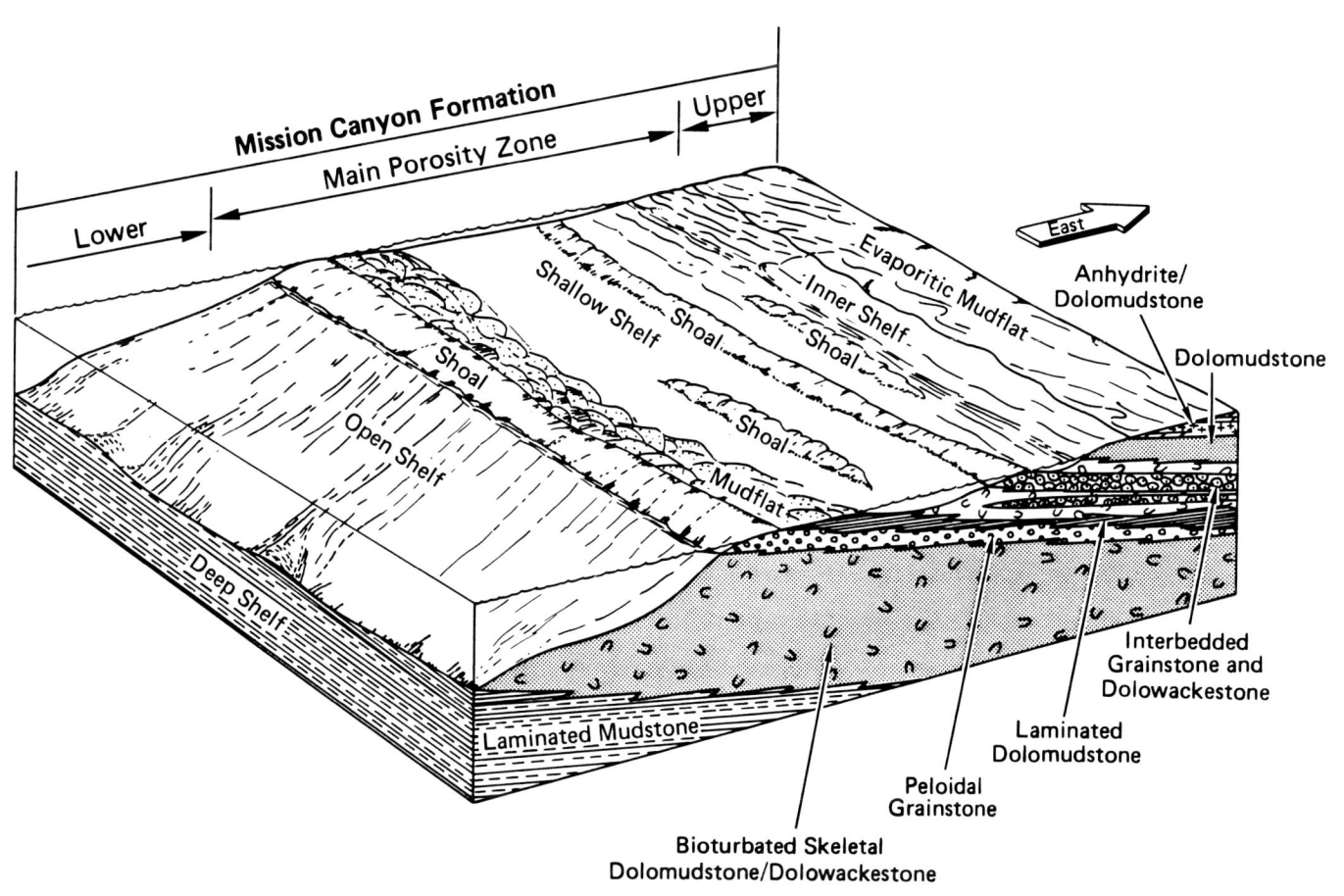

Figure 5   Schematic block diagram showing depositional facies interpreted for Whitney Canyon-Carter Creek Field.

little facies variation due to uniformity of depositional environments along strike.

The following brief sections summarize important descriptive and interpretive points about each of the facies, with the discussions presented in the same relative order that the facies occur from bottom to top in the cores. Additional details of most of the facies are presented in a detailed core description of the Chevron 1-18F well (Fig. 6).

Lower Portion of Mission Canyon Stratigraphic Sequence

Facies 1 - Deep Shelf Mudstone

Cores of deep shelf mudstone contain dark gray to black, finely crystalline mudstone and dolomudstone. Commonly these muds are well laminated, but occasionally they become contorted and wavy or even structureless. Skeletal debris are rare where present, consisting of thin, discontinuous streaks of fine ostracode, crinoid, and coral fragments. Fractures are common and usually are healed by dolomite or anhydrite. Anhydrite also occurs as nodules scattered throughout the interval. The contact with overlying sediments is marked by a distinct increase in the quantity of fossil debris and a color change from nearly black to a medium gray. Also, laminations are replaced by burrowed, structureless, or faintly laminated beds. With the exception of minor open fracture porosity, the deep shelf mudstone interval is tight.

The well-laminated, mud-rich texture of deep shelf mudstones suggests deposition occurred in a relatively deeper-water and lower-energy environment, probably below both daily and storm wave base. In addition, the lack of many burrows and much skeletal debris indicate the environment was relatively hostile to organisms. The rare, discontinuous skeletal lenses may represent exceptional storm lags; several of the lenses have microscours as basal contacts. The wavy, contorted laminae that were sometimes observed were probably produced by dewatering of the sediments or other soft-sediment deformation. The dark color hints of organic richness, but little organic carbon was detected during Rock-Eval analysis, and thin sections show abundant pyrobitumen and other opaque material.

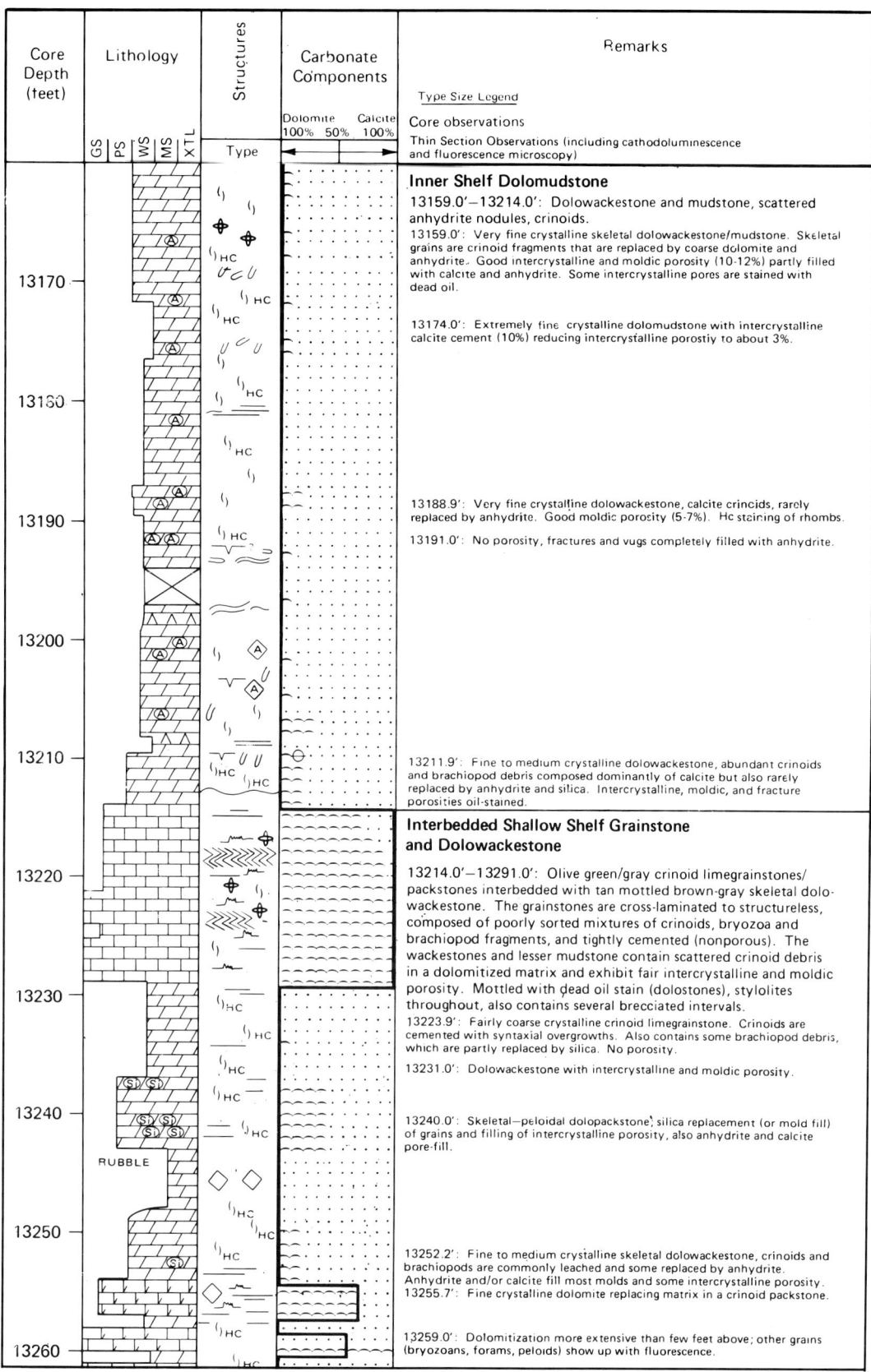

Figure 6A Detailed core description of Chevron 1-18F well.

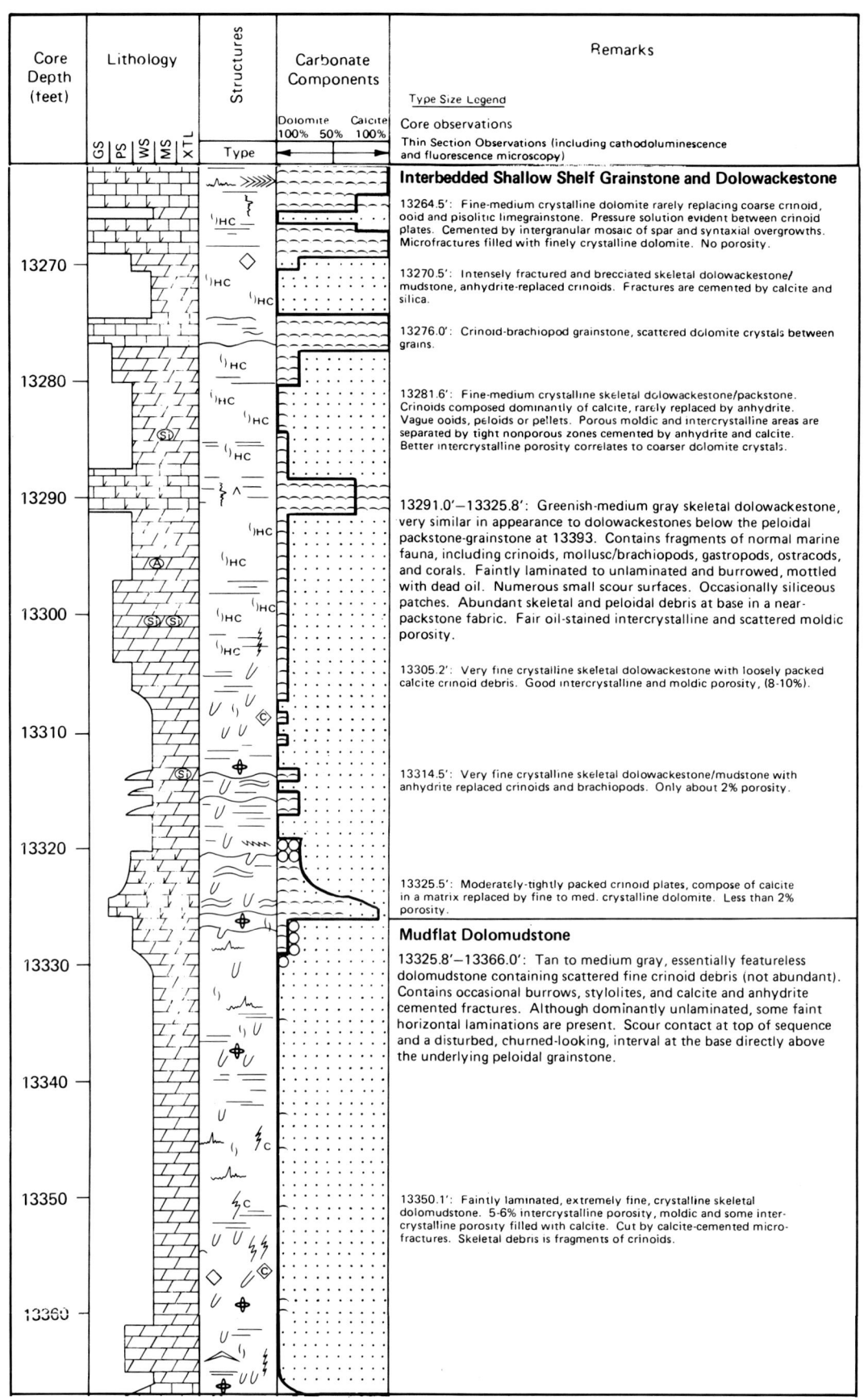

Figure 6B Detailed core description of Chevron 1-18F well (contd.).

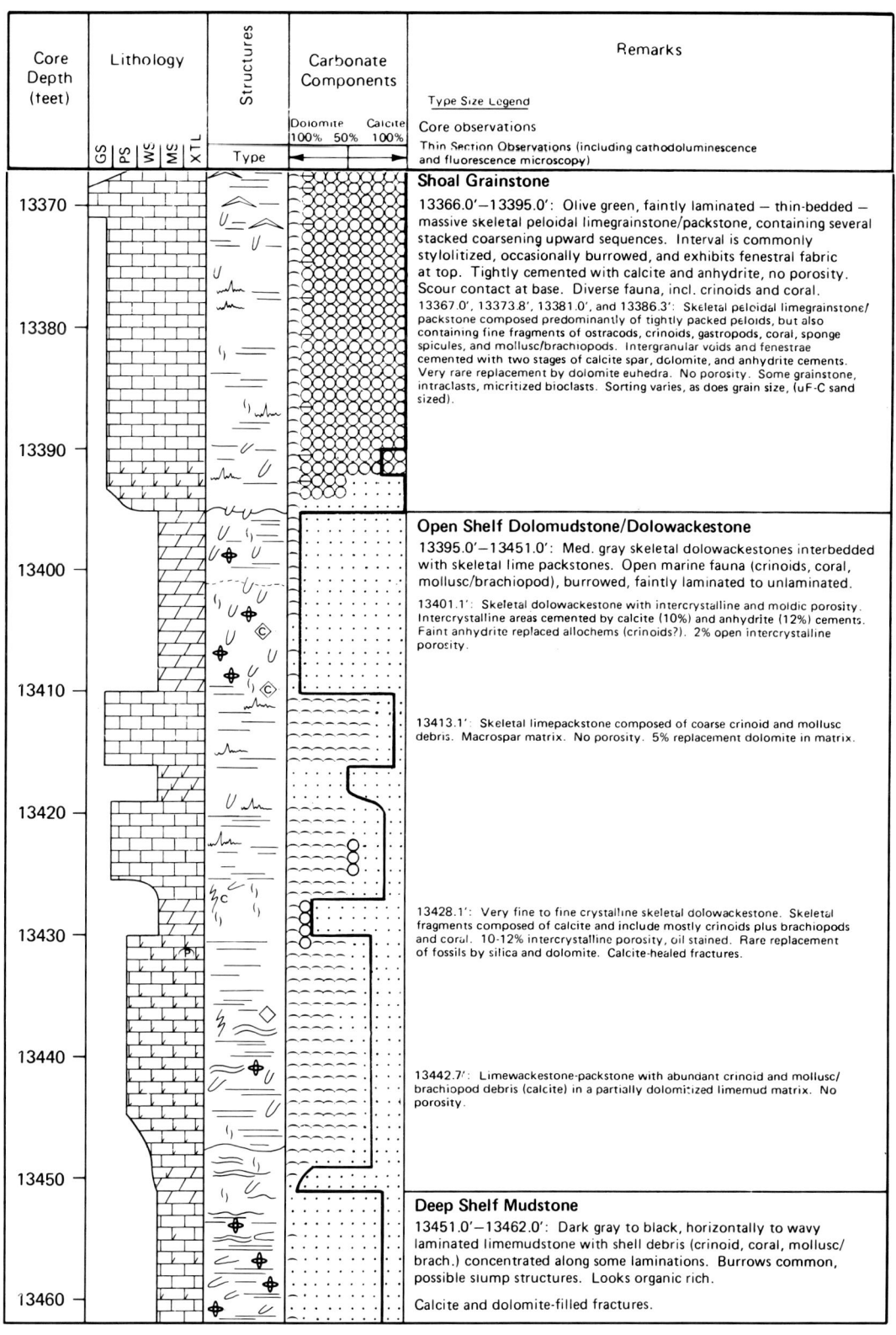

Figure 6C Detailed core description of Chevron 1-18F well (contd.).

Facies 2 - Open Shelf Dolomudstone/Dolowackestone

The gradational change from lower to middle portions of the regressive stratigraphic sequence of the Mission Canyon Formation occurs within the open shelf dolomudstone/dolowackestone recognized in cores. Intercrystalline and moldic porosity of up to 12% are frequently developed in the upper dolomudstone and dolowackestone intervals and, in these cases, the section forms the lower part of the producing interval in the field. Porosity within the open shelf dolomudstone/dolowackestone facies varies considerably in thickness between wells, and porosity also appears to be discontinuous both vertically and horizontally. Nevertheless, production data indicate the unit contributes significantly to flow in certain parts of the field, and it must be considered one of the better portions of the reservoir.

Sediments in cores are characteristically brownish gray to olive gray, seldom medium gray in color, and formed of skeletal dolomudstone, skeletal dolowackestone, and less commonly of dolomitic skeletal packstone (Fig. 7). The unit varies from faintly laminated to bioturbated and massive; well-preserved outlines of burrows are rare. Skeletal debris are common locally, consisting of crinoids and fewer corals including Syringopora, brachiopods, gastropods, and ostracodes. Fractures cemented with calcite, anhydrite, or rarely dolomite are ubiquitous. Nodules of anhydrite are scattered, and horizontal stylolites are common.

Mineralogy as inferred from thin-section examination and acid etching of core slabs varies from dolomite in the mudstone and skeletal wackestone intervals to dolomitic limestone in the skeletal packstones. The matrix in the mudstones and wackestones is formed of finely crystalline subhedral dolomite rhombs; rarely are the crystals more coarsely crystalline. Skeletal grains remain as calcite or have been removed by solution to form moldic porosity; less often the grains are replaced by anhydrite, dolomite, or even silica. Porosity variation is due in part to plugging by calcite, anhydrite, or minor dolomite cements. Where present, the pores are often stained with pyrobitumen. Typically the packstones are tight and formed of tightly packed calcitic crinoid fragments within a dolomitized, fine-grained matrix.

The mud-supported depositional texture, relatively diverse biota, and bioturbation indicate that the dolomudstone/dolowackestone interval was

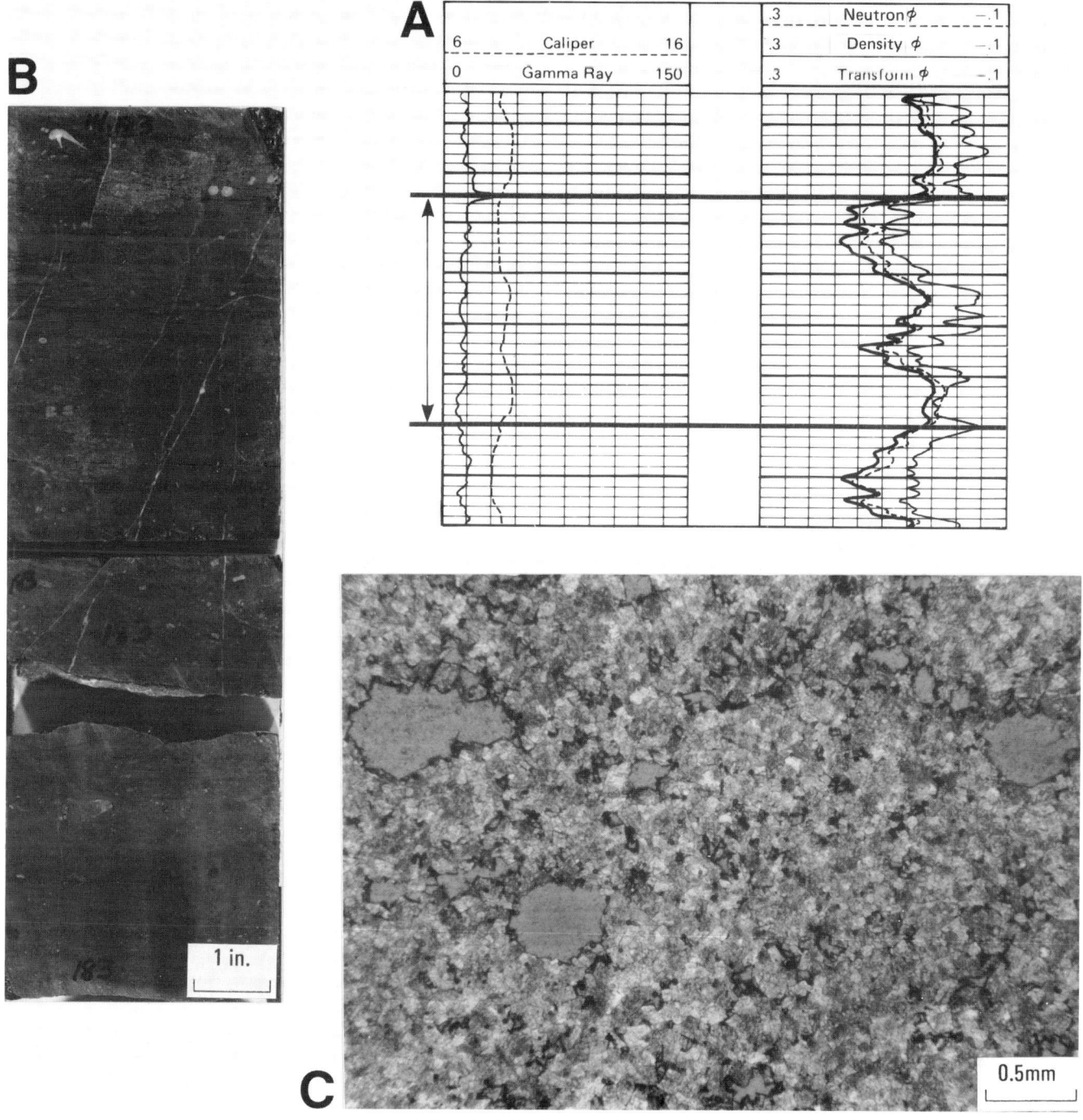

Figure 7 Log response (A), core photo (B), and photomicrograph (C) of open shelf dolomudstone/dolowackestone.

deposited in a deep, low-energy, unrestricted marine shelf. Skeletal packstones probably formed during storm-intensified current winnowing on the sea bottom, and their more frequent occurrence in comparison to the underlying unit is suggestive of relative shallowing. The subtle changes of depositional texture, abundance of biota, and sedimentary structures are hints that deposition was gradually changing from the deeper shelf and even perhaps basin environment of the underlying unit, through an open and shallower, but still relatively deep shelf for this unit, to a progressively shallower shelf where an even more varied suite of rock types formed. These important varied lithologies form the middle portion of the Mission Canyon stratigraphic sequence and are discussed in the following sections.

Middle Portion of Mission Canyon Stratigraphic Sequence

Facies 3 - Shoal Grainstone

Immediately overlying the open shelf dolomudstone/dolowackestone portion of the reservoir, the shoal grainstone is an easily recognizable unit in cores, being a prominent nonporous limestone of relatively uniform thickness (18-24 feet). These medium olive green peloidal grainstones and packstones, faintly laminated to massive but rarely burrowed, contain numerous fractures and stylolites of varying orientations (Fig. 8). Peloids are very fine to fine sand-sized grains that occur in variably sorted mixtures with lesser amounts of other grains. Less abundant allochems include intraclasts, grapestones, skeletal grains (corals, crinoids, brachiopods, gastropods), micritized bioclasts, and ooids. The abundant peloids have been compacted in certain intervals to the point where they coalesce to form a pseudomudstone texture.

Although occurring within the lower portion of the main porosity zone in the field, the shoal grainstone unit is tight and contributes little to production as indicated by production logs. The grainstones and associated packstones contain little dolomite; where present, the minor dolomite occurs as a replacement of matrix or cement. The interparticle porosity between grains is cemented with two stages of blocky calcite: an isopachous rim of "dogtooth" spar surrounds the grains and is in turn overlain by a later stage, coarser crystalline spar mosaic. Less often the interparticle porosity is filled by anhydrite, a cement that also fills minor fenestral porosity in the grainstones. Open, randomly spaced microfractures were observed in the cores,

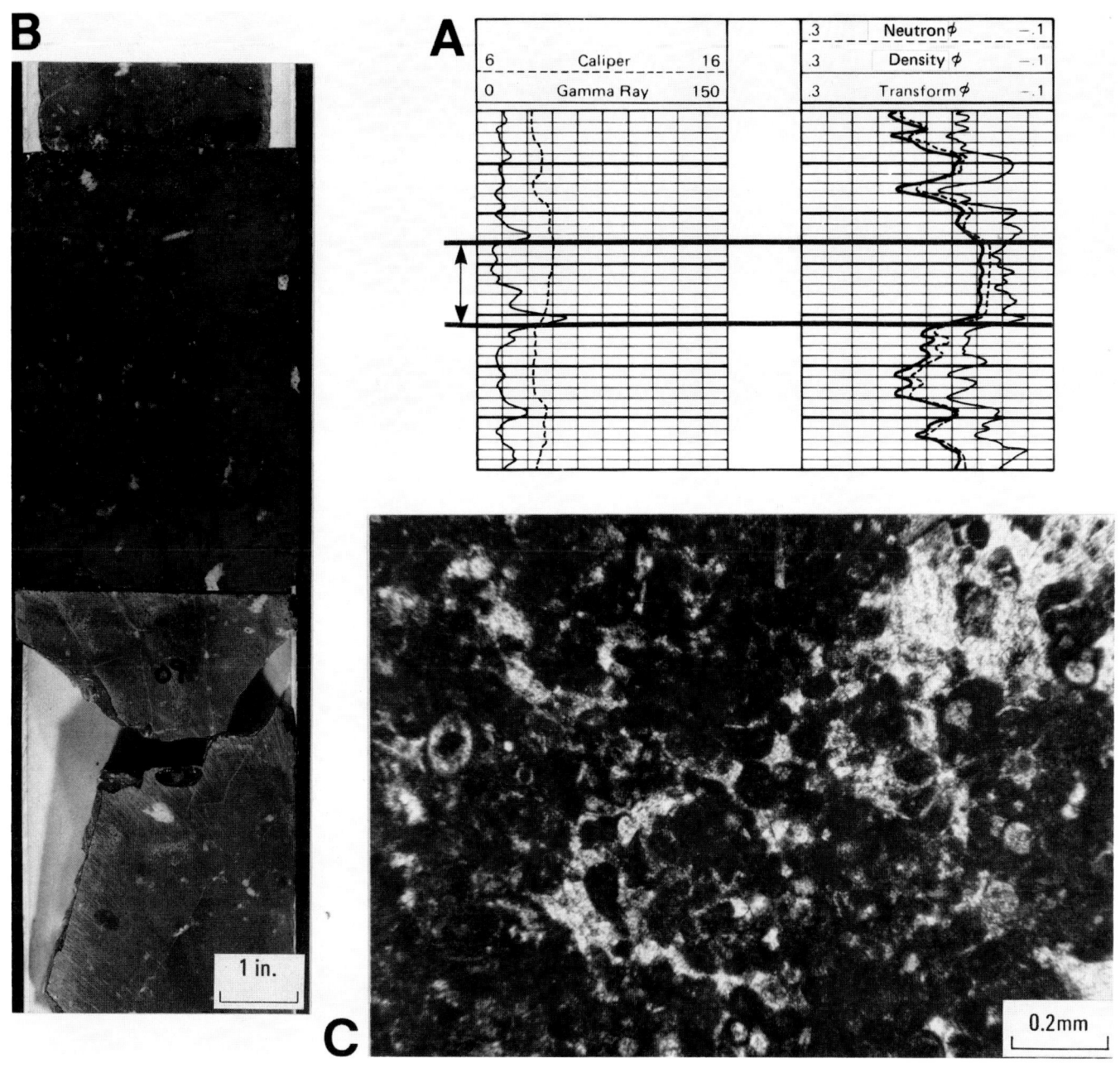

Figure 8 Log response (A), core photo (B), and photomicrograph (C) of shoal grainstone.

but many appear to be induced fractures, and the natural fractures are not abundant or large enough to be significant.

The shoal grainstones formed on an agitated sea bottom, and the thickness and widespread occurrence of the unit argue against an episodic storm-generated origin in slightly deeper water as proposed for the relatively minor carbonate sands in the underlying depositional units. The grainstones are interpreted to have formed within a major shallow subtidal to intertidal facies belt that fronted the diverse shallow shelf environments lying further to the east.

In the cores, the shoal grainstone is actually a complex aggregate of several stacked, coarsening-upward, graded sequences in which the size and sorting of the grains improves from bottom to top. Cement-filled porosity that is strikingly similar to fenestral porosity occurs near the top of many of these smaller sequences, and the contact with the overlying mudflat mudstone unit in some cores is a disrupted zone containing mud clasts. These small-scale sequences within the shoal grainstone unit are interpreted as shoaling cycles having formed during a relative shallowing of the sediment surface to perhaps within the intertidal range. Further support for a very shallow, and even intertidal, origin for the sands is the early-generation cementation. The cement likely represents partial lithification in a meteoric phreatic or beachrock diagenetic environment, therefore suggestive of intermittent exposure and rapid localized cementation along bar crests or ephemeral islands. The smaller shoaling cycles are likely locally developed, and therefore would not be individually correlatable over large lateral distances, but collectively they form a widespread carbonate sand layer.

Facies 4 - Mudflat Dolomudstone

Lying directly above the shoal grainstone unit in the cores is an interval of medium gray to tan, variably laminated but rarely burrowed, hydrocarbon mottled mudstone (Fig. 9). Porosity within the mudflat dolomudstone unit varies greatly between the cores studied; porosity may exceed 12% where the intercrystalline porosity is enhanced by fine moldic porosity or may occur in only trace amounts where plugging by calcite and anhydrite cements occurs. This interval is a part of the productive section in the field, but suffers from irregular porosity development.

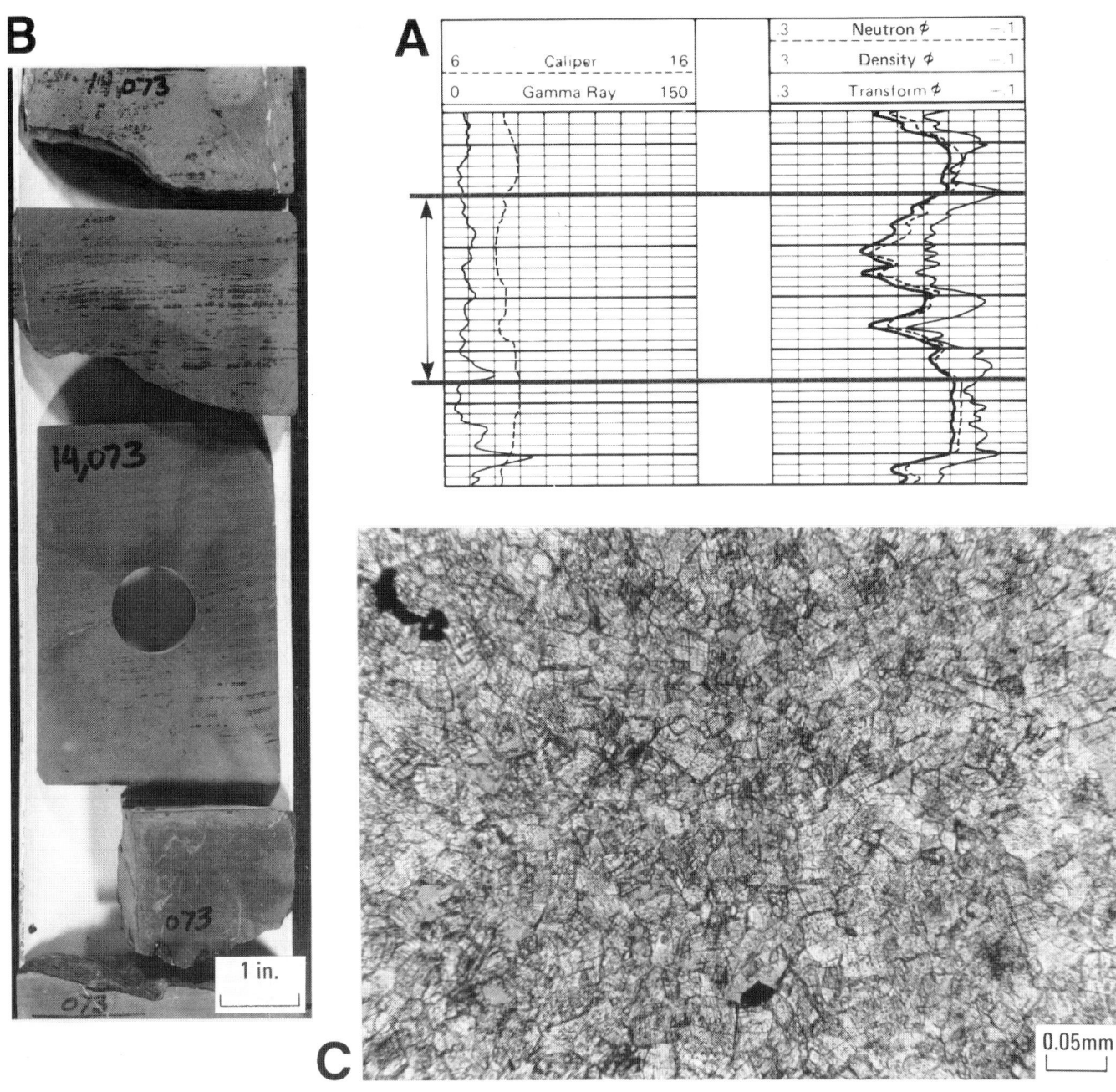

Figure 9 Log response (A), core photo (B), and photomicrograph (C) of mudflat dolomudstone.

Where preserved, the basal contact of the mudflat mudstone is very distinct, consisting of a 10 inch-thick sequence of wavy, contorted, chaotic-looking laminations that have the appearance of a combination of soft-sediment deformation and rip-up clasts. The upper contact of the interval is variable, composed either of an abrupt scour contact overlain by a peloidal, skeletal grainstone or as a more gradational contact marked by an increase in amount of skeletal debris.

Skeletal grains are rare, but, where present, are usually fine-grained crinoid debris. Fractures and horizontal stylolites are common, the former commonly cemented with anhydrite and calcite. The dolomudstones are formed of very fine, cloudy dolomite crystals. Textures of the dolomite are variable: interlocking crystals forming a xenotopic fabric are found along with more open sucrosic textures containing well-developed intercrystalline porosity.

The mud-rich depositional texture, paucity of fossils, presence of laminations, and absence of burrows found in the mudstone interval collectively suggest a protected depositional setting where organisms were virtually excluded. A shallow subtidal to intertidal mudflat closely associated with the underlying shoal grainstones is a probable depositional environment. Conceptually the muds would accumulate in the lee of subtle topographic irregularities formed by the carbonate sands, and organisms would be excluded by frequent exposure above sea level. Exposure is also supported by the light color of the mudstones; shallow marine muds are typically darker in color, more fossiliferous, and burrowed.

Facies 5 - Interbedded Shallow Shelf Grainstone and Dolowackestone

This unit differs from the underlying section by being more fossiliferous and containing skeletal grainstones that are interbedded with dolowackestones/ packstones and dolomudstones. Well-developed intercrystalline and moldic porosities give the lower portion of this unit perhaps the best reservoir quality in the field within the Mission Canyon pay zone. Porosities, where best developed, reach 16%, but anhydrite and calcite cementation cause some variation. Upwards in the section, more numerous, discontinuous, tight grainstone beds add significant heterogeneity to that part of the reservoir.

Unlike the carbonate sands found in lower portions of the cores, the interbedded grainstones of this unit are composed almost exclusively of well-sorted crinoid fragments (Fig. 10). Peloids and brachiopods are minor components. The grainstones are olive green, thin-bedded, occasionally cross-bedded, and exhibit a flaggy, blocky appearance in core. Bed thickness ranges from 1 to 10 feet, with the thicker beds more likely to be continuous between wells. Healed fractures and horizontal stylolites are numerous; scour contacts are associated with many of the grainstone beds.

The crinoidal grainstones are well cemented with syntaxial overgrowths of calcite and are essentially nonporous. Rarely, the grainstones are partially dolomitized, but no porosity is developed. In these cases, fine to medium crystalline euhedral dolomite crystals have incompletely replaced the calcite cement to occasionally form rings of dolomite rhombs around the crinoid fragments. Minor partial replacement of skeletal grains by silica also does not effect porosity.

The interbedded dolowackestone/packstone intervals are medium gray to greenish gray, mottled with dark hydrocarbon stain, faintly laminated to burrowed and massive, and occasionally wavy-laminated (Fig. 11). Composed of both skeletal dolowackestones and dolomudstones, the interval also contains stringers of poorly sorted, fine skeletal dolopackstones. The lower portions of this unit are thus similar to the open shelf dolomudstone/dolowackestone of facies 2 seen lower in the cores. The dolowackestones are frequently compacted to the point that many of the skeletal grains are in contact, thus locally appearing in core or thin section as a dolopackstone. Skeletal grains include crinoids, brachiopods, corals, and probable bryozoans. Nodular anhydrite, silica replacement of skeletal fragments, and irregular patches of silica are present throughout. The replacement silica, a characteristic feature of this facies, preserves original depositional textures in the skeletal dolowackestone.

The fine-grained matrix of this unit has been completely dolomitized and consists of very fine to finely crystalline, subhedral, cloudy dolomite rhombs. Where not dissolved to form moldic pores, the skeletal grains are commonly calcite, but may be replaced by anhydrite, silica, or dolomite. Intercrystalline porosity is well developed in the dolomite and combines with

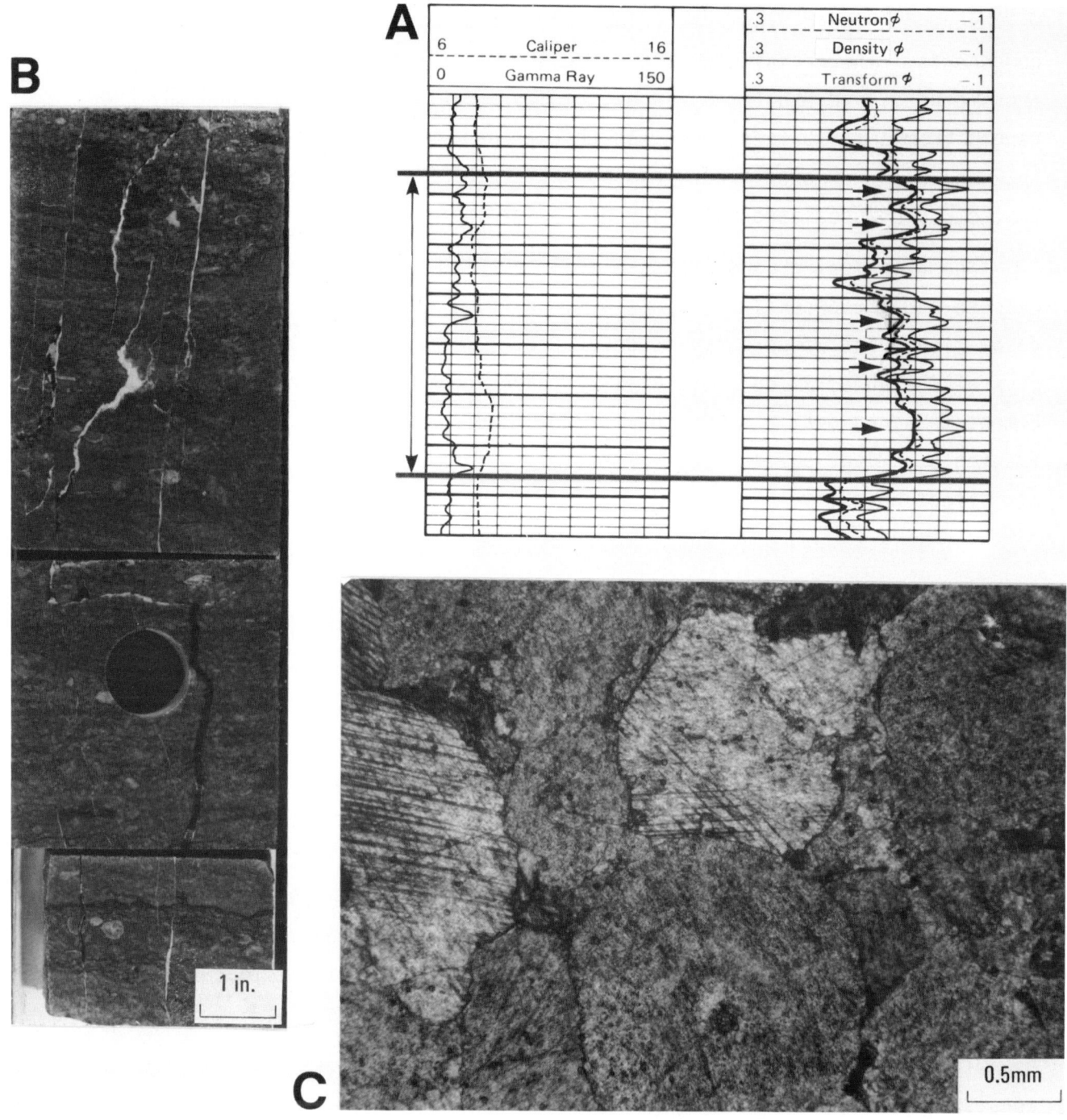

Figure 10 Log response (A), core photo (B), and photomicrograph (C) of shallow shelf grainstone.

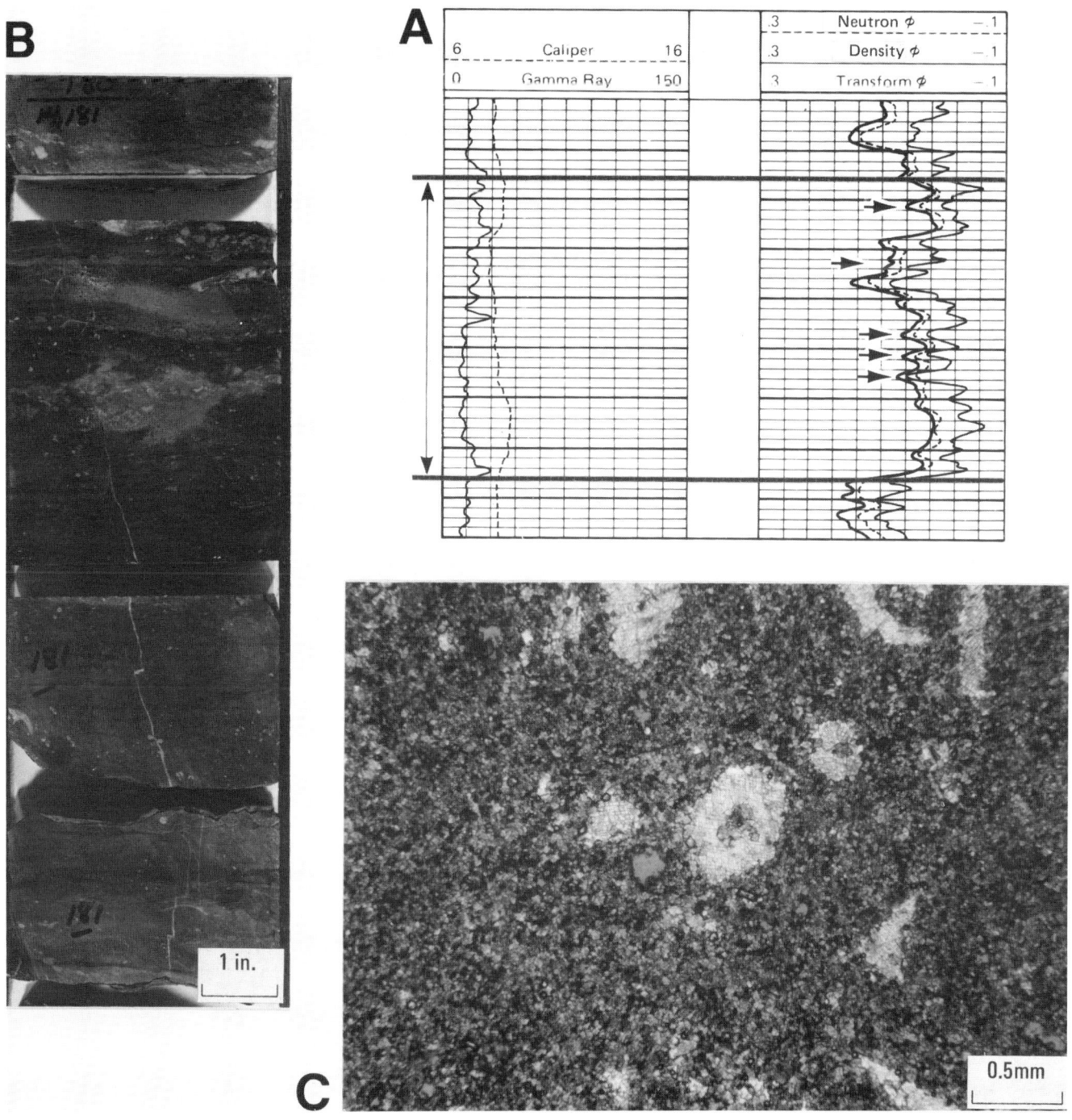

Figure 11 Log response (A), core photo (B), and photomicrograph (C) of shallow shelf dolowackestone.

the skel-moldic porosity to form what are commonly the most porous intervals observed in the cores.

The frequent alternation in this unit between skeletal grainstone and dolowackestone infers a dynamic depositional setting in which wave and current energies are variable. These sediments are interpreted to have formed on a shallow marine shelf consisting of several topographic highs and intervening lows. This wide, complex shelf was positioned behind or updip of the shoal grainstone deposits observed lower in the cores, but was not restricted by that shoal to the point of exclusion of organisms or widespread precipitation of evaporites across the shelf. The interbedded dolowackestones formed within the less agitated topographic lows between the bars found on the shelf.

The grainstones found across the shelf represent mobile bar or shoal deposits as suggested by the cross-bedding and lack of fines, but many of these bars were not as long-lived or well developed as the seawardmost shoal grainstone. Several of the thicker (6 to 10 feet) and better-developed bars within the shelf extend parallel to depositional strike for a considerable distance, whereas many of the thinner (2 to 4 feet) bars were more localized and pinch out between wells. Correlations suggest that many skeletal grainstones grade laterally into skeletal dolopackstones often with only 2% to 4% porosity due to calcite and anhydrite plugging.

Facies 6 - Inner Shelf Dolomudstone

The top of the main porosity zone in the field is positioned within the inner shelf dolomudstone unit. Intercrystalline and, within the lower part of the unit, minor moldic porosity combine to give maximum porosities of 10% to 15%. Porous zones can be correlated laterally for several miles. Porosity, however, is contained in finer crystalline dolomite than occurs lower in the cores, and is also variable vertically on all scales due to cementation by both calcite and anhydrite. As a result, producibility for this unit as indicated by flow meter data is low.

Inner shelf dolomudstone/dolowackestone differs from the dolowackestones of the immediately underlying unit by being less fossiliferous. Scattered skeletal debris rapidly decrease in abundance upward in the unit so it typically is a burrowed dolomudstone, usually tan, but mottled to gray by hydrocarbon stain (Fig. 12). Crinoid fragments are the most common fossil, but the coral <u>Syringopora</u> is abundant. The dolomite is composed of extremely fine crystalline, cloudy crystals. Nodules and thin beds of anhydrite, open and healed fractures, and stylolites are common. Mudcracks and algal laminations in a few wells occur in the lower portion of the unit.

The dolomudstones and dolowackestones of this unit are interpreted to have been deposited at the landward end of the marine shelf, a depositional environment containing a mixture along strike of an irregular coastline of shallow, relatively open, subtidal and intertidal subenvironments. Support for this interpretation comes from the occurrence of <u>Syringopora</u>, suggesting adequate bottom circulation during deposition of the subtidal muds along with algal laminations, mudcracks, beds of anhydrite, and the paucity of skeletal debris in the intertidal sections.

Upper Portion of Mission Canyon Stratigraphic Sequence

Facies 7 - Evaporitic Mudflat

An evaporitic mudflat section forms the upper portion of the Mission Canyon Formation and lies above the main reservoir zone of the field. Bedded anhydrite and dolomudstone typify the unit (Fig. 13). The anhydrite beds are characteristically white and have a massive to chicken-wire nodular texture. Deposition is interpreted to have taken place on an intertidal-supratidal flat.

As observed in the cores, dissolution has altered the upper 150 to 200 feet of the Mission Canyon Formation and also less commonly lower parts of the section. The evaporitic mudflat interval often appears as a chaotic mixture of angular dolomite blocks and quartz sand that is cemented with calcite, dolomite, or anhydrite (Fig. 14). Reddish coloration is common and, where the disruption is most severe, the cores are fractured and rubbly.

The chaotic appearance in cores has a twofold interpretation, as dissolution and resultant brecciation may accompany subaerial exposure or tectonism. That the upper Mission Canyon of Whitney Canyon-Carter Creek Field was exposed subaerially is not in question; the interval represents the culmination of the overall shallowing-upward cycle or regressive sequence discussed in the previous sections. Sando (1976) and numerous other authors have discussed the

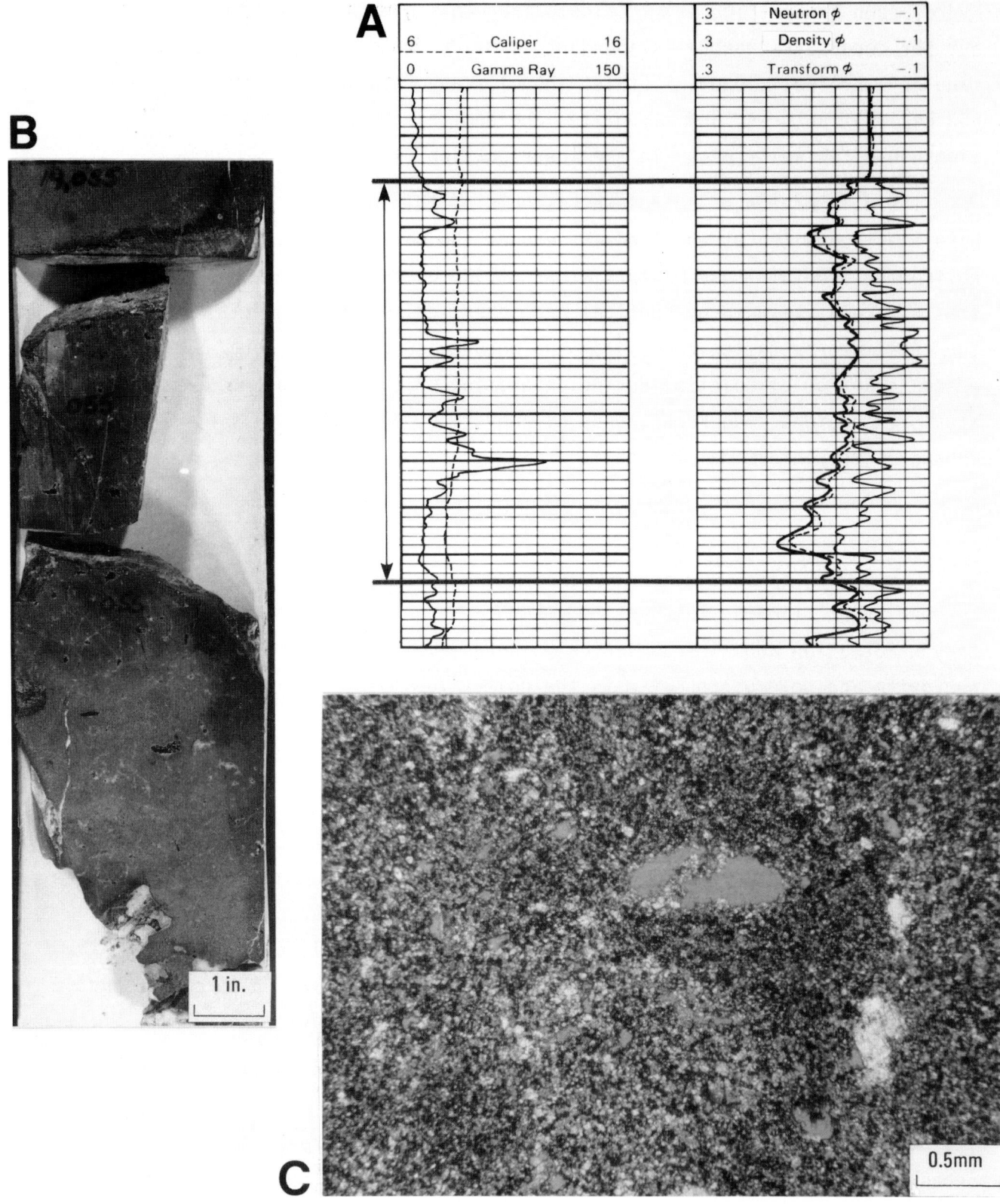

Figure 12 Log response (A), core photo (B), and photomicrograph (C) of inner shelf dolomudstone.

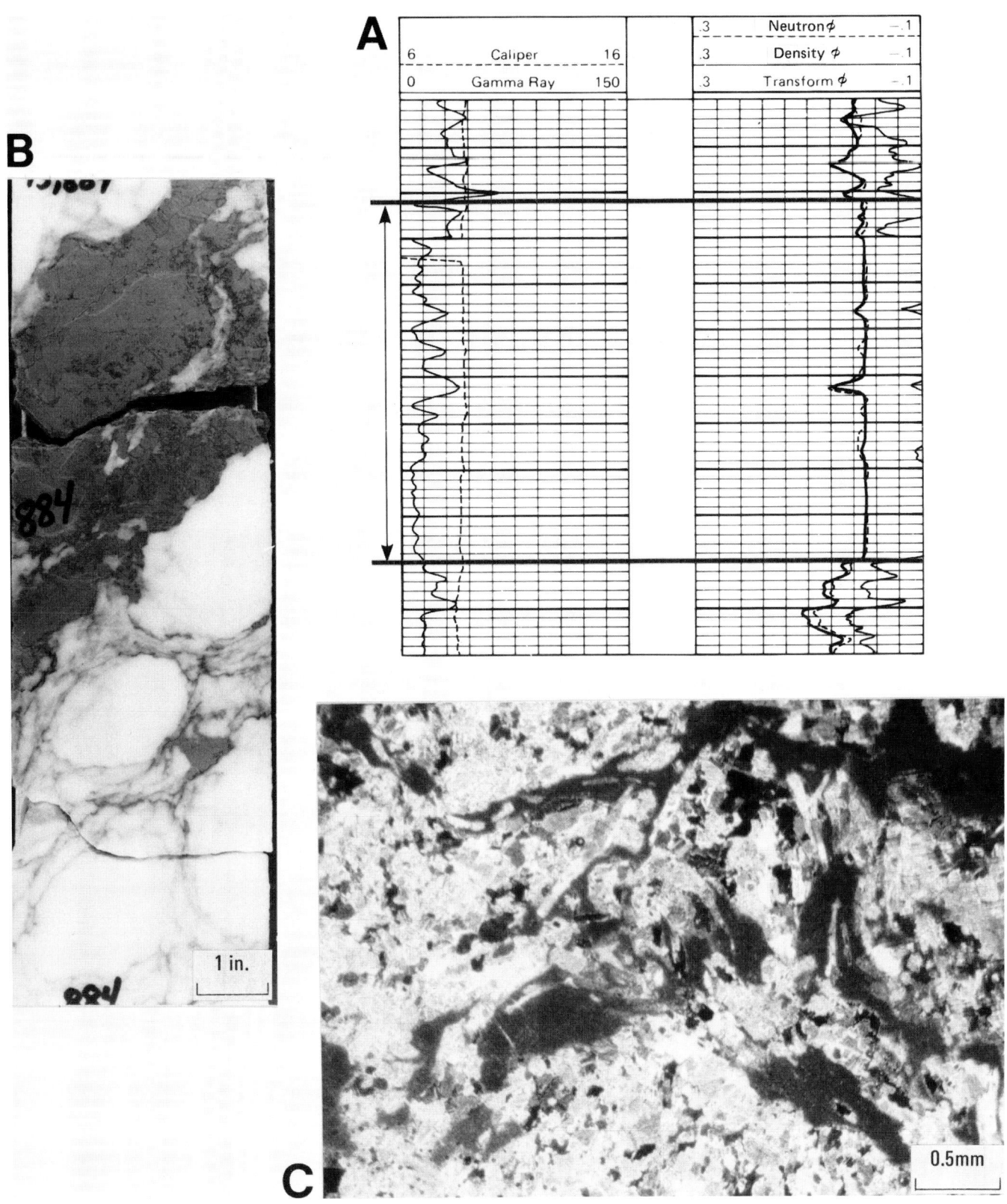

Figure 13 Log response (A), core photo (B), and photomicrograph (C) of evaporitic mudflat.

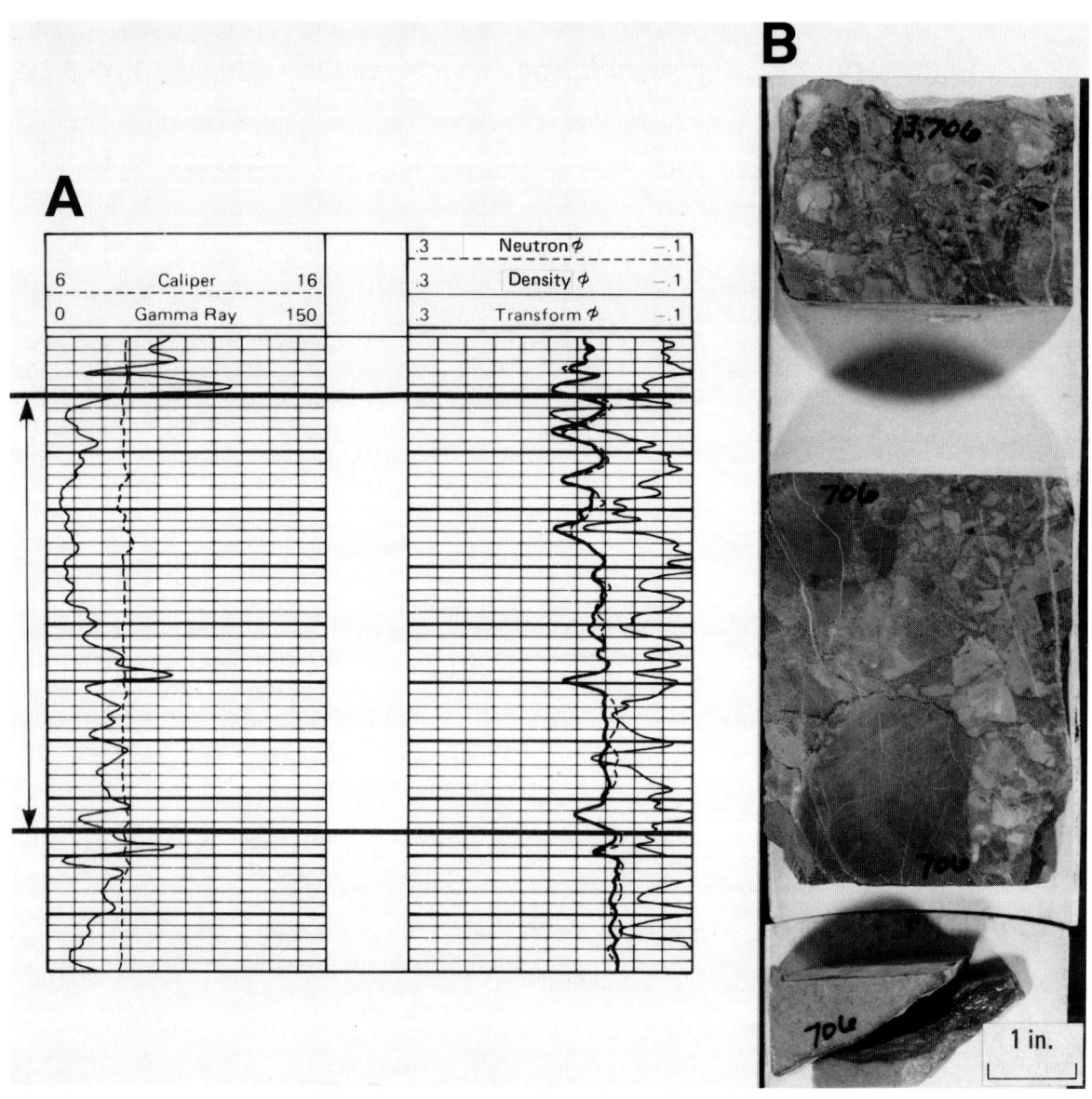

Figure 14 Log response (A) and core photo (B) of karstified, evaporitic mudflat.

late Mississippian karstification of the Cordilleran Platform, an event caused by regional uplift of the platform perhaps in some combination with a lowering of sea level. The quartz sands seen in cores were probably reworked into the upper Mission Canyon from overlying deposits during subaerial exposure and karstification.

Rau (1982) believed, based on four lines of evidence, that the breccias observed during her core study were the result of solution-collapse: (1) correlation between wells of an evaporite zone to a breccia zone; (2) remnant evaporites within the breccia itself; (3) solution features; and (4) the existence of an overlying exposure surface of post-Mission Canyon, but pre-Darwin Sandstone age. Budai (1985) also commented on the collapse beds in the upper portion of the Mission Canyon. She believed evaporite solution may have occurred during the late Mississippian unconformity, but the presence of clasts of Amsden sandstone within some breccia beds indicated that some amount of collapse followed deposition and lithification of the overlying Amsden Formation.

The origin and timing of brecciation in the upper portions of the Mission Canyon Formation have not been completely resolved. It is likely that the breccias are formed by the solution of evaporite beds toward the top of the Mission Canyon shallowing-upward depositional sequence and the subsequent collapse of overlying beds. Late Mississippian and Pennsylvanian karsting could have caused the brecciation, but this simple scenario is complicated by the additional likelihood that fluids may have moved along faults and these same breccia beds during thrusting, causing further solution collapse. Gargallo-Quinones (1985) investigated solution features including breccias on outcrops in Wyoming, and suggests that those with sand matrix are related to karstification occurring after initial deposition of the Darwin Sandstone. But she goes on to interpret breccias and collapse zones lacking the sands as being related to evaporite removal, the timing of which cannot be constrained. Similar breccia zones, possibly related to fracturing, are present in both the main porosity zone and upper Mission Canyon in cores from the field. The timing of brecciation may be as old as Late Mississippian or as young as the Cretaceous-Tertiary thrusting.

CYCLIC DEPOSITION OF THE MISSION CANYON FORMATION

The seven facies recognized during the core study have been discussed and illustrated using a generalized block diagram, in the context of the overall large-scale regressive sequence of the Mission Canyon Formation recognized by numerous previous workers. This thick, shallowing-upward cycle, probably a third-order cycle of Vail et al. (1977), contains numerous smaller-scale cycles that are especially well developed in the main porosity zone of the field, and give rise to the alternations between carbonate sands and muds. Similar smaller-scale cycles have previously been recognized by Wilson (1975) and others in the Mission Canyon Formation of the Williston Basin. More recently, Dorobek (pers. comm.) has recognized similar cycles in central and southwestern Montana, and Bartberger (1985) has discussed their occurrence in Whitney Canyon-Carter Creek Field. In general, the smaller-scale cycles are regarded by these authors as having coarse-grained carbonate sand facies in their lower parts, grading upwards into progressively finer-grained facies that may show evidence of exposure. The smaller-scale cycles form during short duration sea-level fluctuations that are more apparent upwards in the larger-scale regressive sequence, after the platform has had time to aggrade to near the initial maximum sea-level rise that triggered the larger cycle (Read and others, 1986).

The interbedded shallow shelf grainstone and dolowackestone interval recognized during our core study is formed of small-scale alternations like those described for the smaller cycles above, with some important exceptions. There is no consistent pattern of carbonate sands grading upward into muds for the alternations viewed in the cores, and the muds in core show no evidence of exposure above sea level. In fact, we suggest that the muds likely form the lower portions of the cycles and the sands the upper parts. This pattern in the mud-to-sand relation is more consistently observed in shallow, sand-dominated carbonate shelves (see Halley et al., 1983, for a discussion of modern examples) and has been interpreted previously for the Mission Canyon Formation in the Williston Basin by Lindsay and Kendall (1985) and numerous others.

The thinking, in our interpretation of a shallow marine shelf setting with grainstones forming as mobile sand bars and muds forming in the intervening lows, does not discount in any way the likelihood that low-amplitude sea-level variations were occurring. Water-depth-dependent deposition and sea-level oscillation are primary driving forces to the accumulation of carbonate cyclic sequences as shown by the computer modeling of Read and others (1986). We prefer to emphasize, however, the fact that complex variations found across a marine shelf could in themselves have caused many of these changes during the progradational history of the shelf facies tract. The precise reason for the small-scale cyclicity is probably not as important as the observation that porosity development is controlled by the distribution of mud-rich facies and these facies have relatively good continuity along depositional strike.

DIAGENESIS AND POROSITY DEVELOPMENT

The Mission Canyon Formation in Whitney Canyon-Carter Creek Field has undergone three major stages of diagenesis that have formed the final porosity and permeability patterns in the main porosity zone of the reservoir. The diagenetic stages can be related to the depositional, burial, and structural history of the area: (1) the depositional stage of diagenesis occurs during accumulation of the sediments in the marine setting and during shallowest burial when the strata are still within easy reach of surface-related waters; (2) the burial stage of diagenesis follows the sediments as they are buried to their maximum burial depth; and (3) the structural stage of diagenesis occurs during and after thrusting. A burial history plot (Fig. 15) suggests the first diagenetic stage lasted through approximately the end of the Permian, the second stage where burial depths may have exceeded 20,000 feet subsea extended into the Late Cretaceous, and the last stage of thrusting and uplift to present reservoir depths was concentrated in the latest Cretaceous and Tertiary. This paragenetic sequence of early diagenesis through burial and deformation of the Mission Canyon Formation is in good agreement with that developed during a more regional study through the Wyoming and Utah thrust belt by Budai et al. (1987). Estimated burial depths exceeding 20,000 feet subsea may be considered a maximum amount. Budai et al. (1984) suggested, by stratigraphic reconstruction of the Paleozoic and early Mesozoic sections, that pre-thrusting burial depths of the Mission Canyon were similar to present burial depths.

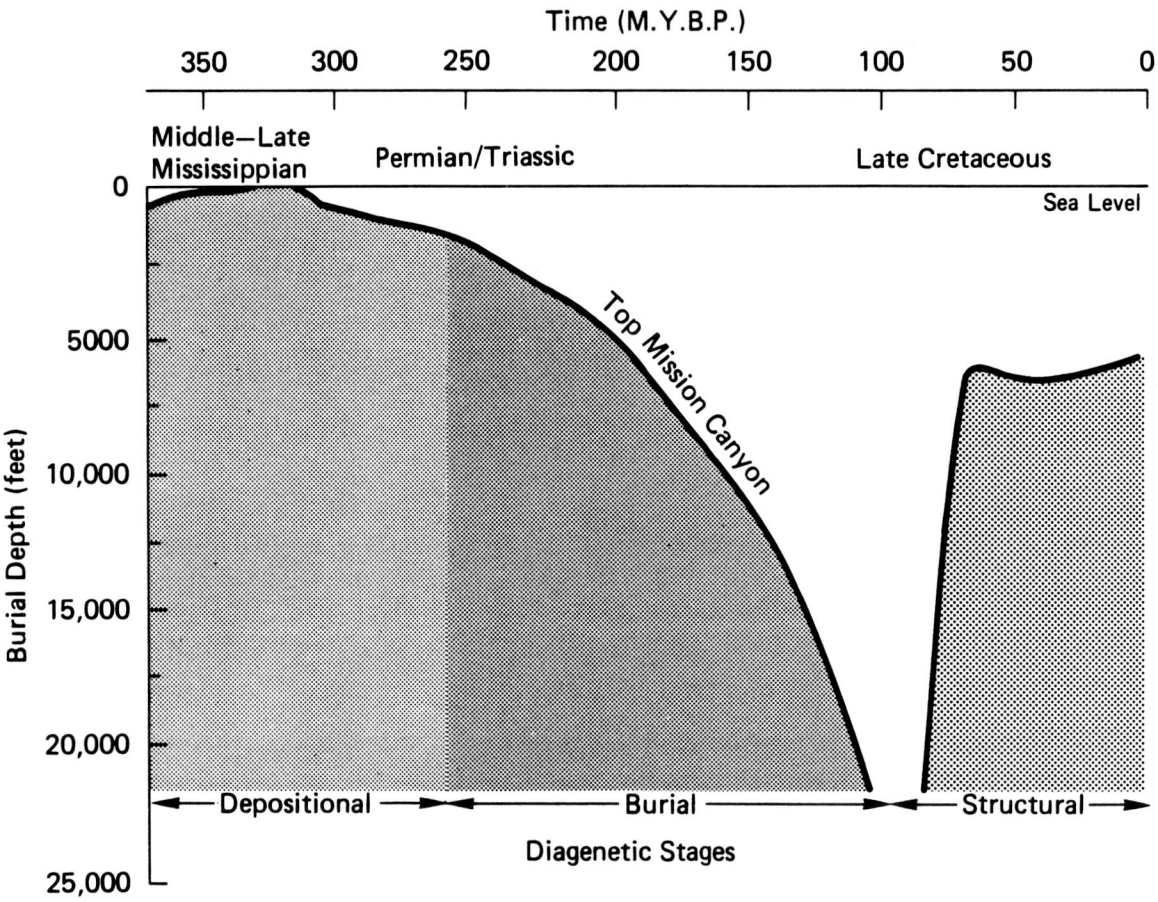

Figure 15 Diagenetic stages of Mission Canyon Formation superimposed on burial history curve.

Depositional Stage of Diagenesis

The depositional stage of diagenesis was most important in development of porosity and permeability within the field, as suggested by the facies control over porosity development and the types of common porosity. Intercrystalline and moldic porosity in dolomite are the main types of early-formed pores found in the reservoir. Included in this diagenetic stage are cementation by calcite of the interparticle porosity in peloidal and skeletal grainstones, dolomitization of mud-rich sediments, and solution of skeletal fragments.

The grainstones and perhaps mud-poor packstones contained the highest primary porosities and were therefore preferentially cemented both syndepositionally and shortly after deposition. Isopachous rims of "dogtooth" spar rimming the grains in the peloidal grainstones are clearly the first-generation cement in those sediments, as are the syntaxial overgrowth cements in the crinoidal grainstones. We infer, based on analogy with modern examples discussed by Halley and Harris (1979), that the rimming cements were calcite formed from meteoric waters probably in the phreatic zone and are thus indicative of short-term exposure of bar crests above sea level and development of a freshwater lens. An alternative interpretation would place the cementation in the marine environment, perhaps beachrock formation in the intertidal zone, and invoke a magnesian calcite mineralogy for the crystals. In either case, the cementation is syndepositional and does not significantly reduce the overall porosity. The remaining interparticle porosity in the peloidal grainstones is filled by an equant calcite spar, also interpreted to be a product of meteoric phreatic cementation. The overgrowth cements formed on crinoid fragments, also generally considered to be an early product of meteoric cementation, completely occlude the porosity in crinoidal grainstones.

Dolomitization occurred before significant compaction and, preferentially in the mud-rich lithologies, the grainstones and mud-poor packstones were already well cemented as described above. The facies control over dolomite occurrence, along with petrography and the isotopic composition of the dolomite, indicate that the pervasive dolomitization and with it intercrystalline porosity are relatively early diagenetic products. That the upper portions of the Mission Canyon Formation become increasingly evaporitic strongly suggests that magnesium-rich hypersaline brines may have been generated and refluxed through the underlying marine carbonate interval.

The pervasive dolomite crystals are commonly uniformly finely crystalline, preserve original sedimentary structures and grain textures, and pre-date all fracturing and stylolitization. The dolomite is nonferroan, and thin sections observed under cathodoluminescence show the dolomite crystals to have uniformly dull luminescence; rarely zoned crystals having dull cores and brighter rims were observed. These results are consistent with those of Budai et al. (1984) who analyzed the Chevron 1-5F well.

Some of the dolomite observed during our study was fabric-destructive, but petrography, including fluorescence microscopy, gives us confidence in our interpretation of original depositional textures. In only two cases were relict grain textures highlighted in thin sections of completely dolomitized intervals using fluorescence microscopy, and these grains were vaguely recognizable due to inclusions or slight staining using only normal, plane-polarized light. We interpret from these observations that grainstones and packstones can be recognized even in the rare case where they are completely dolomitized and our interpretation of dolomudstones, dolowackestones, and, less often, dolopackstones is reasonably correct.

The precise timing for dolomitization cannot be determined from our data, but the overall regressive sequence in which the dolomitized intervals reside, the regional geology, and petrographic-geochemical observations strongly suggest that the dolomitization was likely completed before any significant burial depth and possibly by the end of Mission Canyon deposition. The isotopic composition of dolomite from dolomudstones, dolowackestones, and from minor dolomitic packstones and grainstones is tightly clustered on a carbon-oxygen plot (Fig. 16). δC^{13} values from the 1-18F well vary from +2 to +4, and corresponding δO^{18} values range between -1 and -3 (14 samples analyzed); data very suggestive of early surface-related dolomitization. These values are in close agreement with stable isotope data presented for matrix dolomites from the Chevron 1-5F well by Budai et al. (1984 and 1987).

Budai et al. (1987) recognized two stages of pre-tectonic dolomitization regionally in the Mission Canyon of the Wyoming and Utah thrust belt. A first stage, with average δC^{13} of +2.5 to 4.5 and δO^{18} of +1.0 to 3.0, was

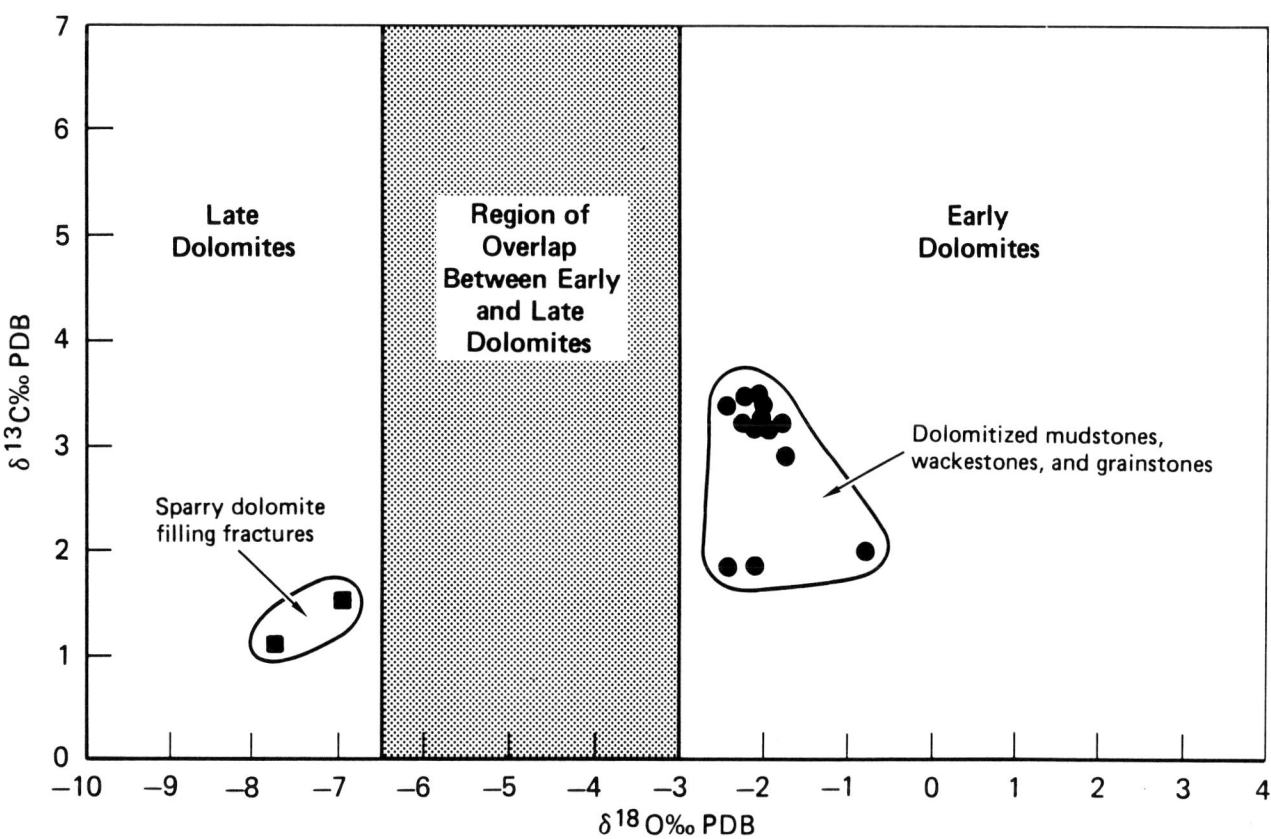

Figure 16 Plot of carbon and oxygen isotopic composition of dolomite samples from Chevron 1-18F well.

interpreted as marine to hypersaline dolomitization occurring syndepositionally with the Mission Canyon in peritidal and sabkha settings. Regionally, the volumetrically more important second stage of dolomite was suggested to be the product of a marine-meteoric mixing zone formed during exposure of the Mission Canyon shelf and continuing during deposition of the overlying Amsden Formation. Average stable carbon and oxygen isotopic composition for the second stage dolomite is δC^{13} of +0.5 to 7.0 and δO^{18} of -2.0.

Skeletal fragments including crinoids, brachiopods, corals, gastropods, and other minor grains are commonly dissolved in dolowackestone and dolopackstone intervals to form moldic porosity. The relative timing of moldic porosity formation is during or after dolomitization; the outlines of the molds are preserved, the molds not having been filled with dolomite. From our observations, we cannot determine the precise timing of development of the moldic porosity. The dissolution may have occurred much later than dolomitization; the only constraint on timing being the presence of oil stain in both moldic and intercrystalline pores.

If the dissolution is temporally related to dolomitization, then perhaps the dolomitizing fluids generated in updip and subsequently overlying evaporitic facies or mixing zone environments were both magnesium-rich and calcite-poor. Such fluids may have dolomitized matrix and preferentially dissolved grains simultaneously. Localized freshwater lenses associated with any of the higher-standing carbonate sand shoals would have provided opportunities for dissolution, but this would require the dolomitization to be nearly syndepositional. The major period of exposure that marks the top of the Mission Canyon Formation would be an opportunity to flush fresh waters into the underlying sediments. These waters, however, would be required to deeply invade the underlying strata using the intercrystalline porosity in the dolomites as pathways, flowing around or through the intervening tight limestone beds to selectively dissolve the remaining calcite skeletal grains.

As a result of the depositional stage of diagenesis, the limestones within the reservoir zone of Whitney Canyon-Carter Creek Field are nonporous, whereas the dolomites have variable reservoir quality. Subsequent stages of diagenesis did little to modify this general pattern of porosity distribution. In general, finely crystalline dolomites have porosities that range from 1% to

4% and permeabilities from 0.1 to 5 md. Dolomites with better porosities are somewhat coarser crystalline and contain intercrystalline and additional moldic porosity. Porosities range from approximately 4% to 16% and permeabilities from 0.3 to 1 md for dolomites having only intercrystalline porosity. Those dolomites having both intercrystalline and well-formed moldic pores have porosity values from 9% to 16% and permeabilities from 1 to 10 md.

Burial Stage of Diagenesis

Stylolites, minor fracturing, retention of early-formed porosities, minor cementation, and emplacement of oil all happened during the burial stage of diagenesis. Budai (1985) recognized a complex paragenetic sequence with multiple episodes of stylolitization, fracturing, and fracture-filling in her studies of the Mission Canyon in Wyoming and Utah. Our observations on the complicated sequence of fracturing and stylolitization, superimposed on the earlier diagenetic overprint of the Mission Canyon in Whitney Canyon-Carter Creek Field, are less complicated, but consistent with those of Budai (1985).

Horizontal stylolites and small, vertically or obliquely oriented microfractures observed in our cores may have formed as a result of overburden stress at any time during burial. The horizontal stylolites are small amplitude and generally offset the vertical microfractures. Physical compaction or the chemical compaction associated with stylolitization caused little porosity loss, and the microfractures created no significant effective porosity. The intercrystalline and moldic porosities within dolomite remained open and interconnected during burial; in thin section, these pore types, as well as some of the vertically oriented microfractures, contain pyrobitumen. The observation of residual hydrocarbon emplaced into these early-formed pore types, but rarely in the later-formed fracture and stylolite-related porosity to be discussed in the following section, suggest an episode of oil migration and emplacement occurred during the burial stage of diagenesis prior to thrusting. Burial-history reconstruction, assuming a thick Mesozoic section based on regional geology, indicates that oil sourced from a lower Paleozoic (pre-Mission Canyon) source interval could have been generated in the Jurassic and Cretaceous.

The composition of the pore fluids changed during burial and with the migration of hydrocarbons. Further change likely occurred as the oil-saturated section was buried deeply enough to form pyrobitumen. Minor diagenetic alteration perhaps associated with these fluid changes are dissolution of anhydrite, precipitation of calcite adjacent to stylolites, silica replacement of skeletal fragments, and cementation of intercrystalline and moldic pores by anhydrite, silica, calcite, or dolomite. These later-stage cements also fill microfractures, discussed in the next section, and therefore may in part be post-structuring in origin.

Stable isotope analysis of calcite cements filling various types of porosity has potentially differentiated earlier-formed from later-formed, perhaps post-thrusting, diagenesis (Fig. 17). A plot of the carbon versus oxygen isotopic composition of calcite, measured on skeletal grains and cements that fill molds, as well as microfractures and minor vugs interpreted to be early, shows a tight cluster of the data. δC^{13} ranges from +0.5 to +2.5, and δO^{18} varies between -3.5 and -5.5 (eight samples analyzed), a tight grouping of values that differs from calcites filling fractures and vugs interpreted to be later and discussed below.

Structural Stage of Diagenesis

During and after thrusting in the region, the Mission Canyon Formation of Whitney Canyon-Carter Creek Field was extensively fractured and stylolitized, and gas was later emplaced into the reservoir. The timing of stylolitization relative to fracturing is often ambiguous, suggesting that both processes occurred complexly on an intermittent basis. Microfractures, sometimes gash-like in outline adjacent to the stylolites, are oriented vertically, obliquely, or horizontally. Microfractures that cross-cut all other fracture sets and stylolites in a sample, therefore the last event to have occurred, are commonly the thickest fracture. We cannot suggest from our geological description of the cores how much effective fracture porosity was created in the field during this diagenetic stage. In cores, the microfractures are characteristically small and discontinuous, but some thin rubble zones are present. The stylolites, offsetting some sets of microfractures as well as earlier horizontal stylolites, are oriented to reflect stress associated with the fracturing and chemical compaction from several directions. The stylolites

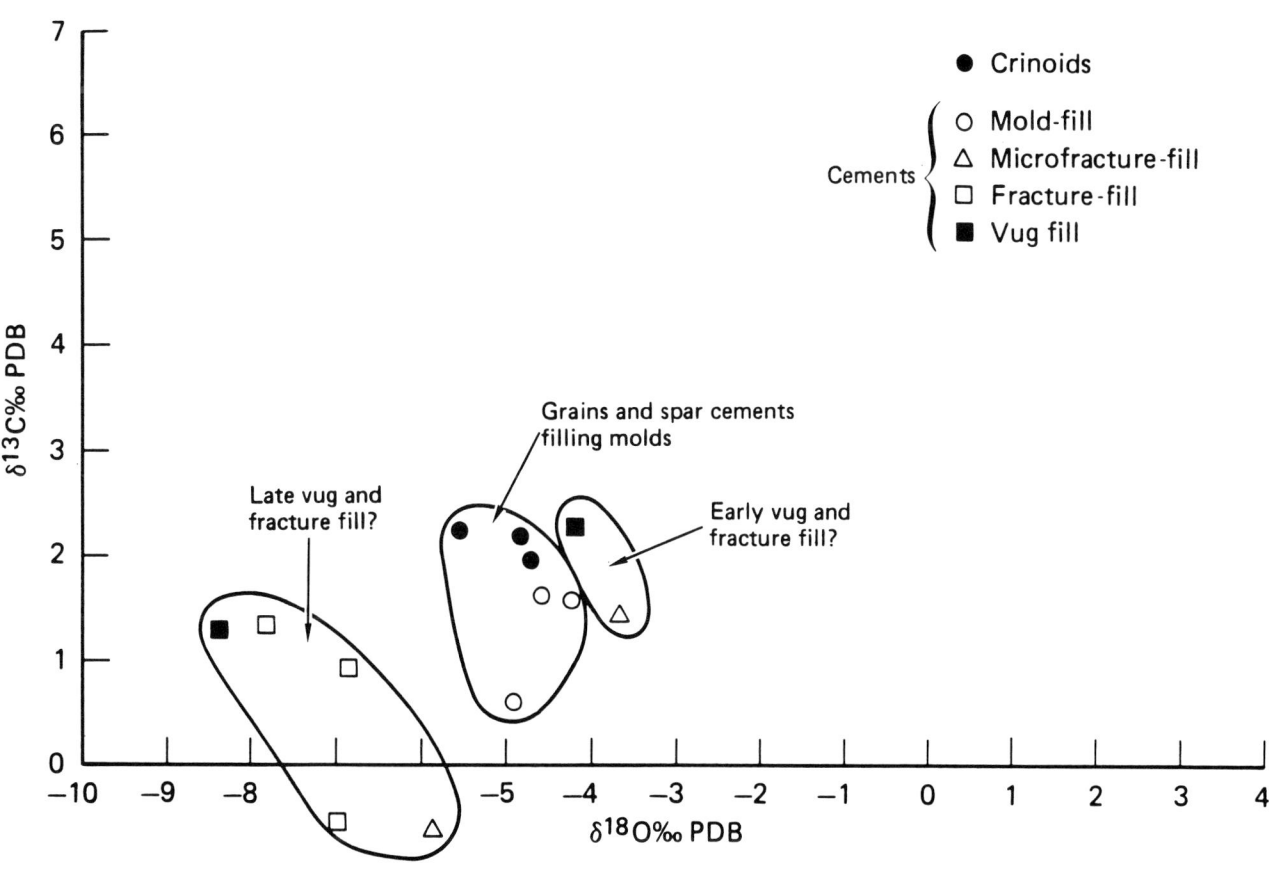

Figure 17 Plot of carbon and oxygen isotopic composition of limestone samples from Chevron 1-18F well.

oriented vertically or at oblique angles are a larger amplitude than their horizontal counterparts.

Our observations of cores suggest a pre-thrusting event of hydrocarbon migration, but cannot confirm a second event associated with or following thrusting. Pyrobitumen along some of the later tectonic-related stylolites may have previously been trapped in the porosity of the host dolomite and later collected as an insoluble residue along the stylolite seam during its formation. Budai (1985), however, observed bitumen, calcite, and anhydrite complexly filling enlarged seams of tectonic stylolites in the Chevron 1-5F well and reasoned then that the fluids associated with deformation also contained migrating hydrocarbons.

The fractures in cores are often cemented by calcite and anhydrite and less commonly by finely crystalline dolomite; in fact, the fractures may have acted as conduits for the pore fluids from which these cements were formed. Budai (1985), in her more regional study of fracture-filling cements of the Mission Canyon Formation in the Overthrust Belt, noted the earlier fractures are filled by dolomite or calcite, whereas subsequent veins and fractures were filled only with calcite. These results agree with our observations; fractures in cores from Whitney Canyon-Carter Creek Field that were shown by cross-cutting relationships to be the latest event were always filled by calcite. Our studies and those of Budai (1985) agree that the fracture-filling calcite and dolomite cements exhibit no luminescence or distinctive fluorescent zoning.

The carbon and oxygen isotopic composition of both dolomite and calcite cements in most fractures clearly show that these cements differ from their counterparts in earlier-formed fractures and other porosity types. Dolomite cements filling fractures have an oxygen isotopic composition that is depleted relative to the earlier-formed pervasive dolomite (Fig. 16). Sparry dolomite filling fractures has δC^{13} values of +1.0 to +1.5 and δO^{18} values of -7.0 to -8.0 (two samples analyzed), the later being distinctly different from pervasive matrix dolomites with δO^{18} values of -1.0 to -3.0. Similarly, calcite cements that fill later-fractures and minor vuggy porosity have δC^{13} values of +1.5 to -0.5 and δO^{18} values of -6.0 to -8.5 (five samples analyzed). The oxygen isotopic composition, especially, discriminates these cements from the

δO^{18} range of -3.5 to -5.5 typical for calcites filling earlier-formed fractures, vugs, and other pore types (Fig. 17). We did not observe the broad range of depleted carbon compositions reported by Budai (1985) for calcite cements associated with bitumen and filling fractures and stylolite-related porosity. She inferred the depleted carbon records the effects of hydrocarbon degradation and release of light carbon dioxide or bicarbonate during the calcite cementation.

Calcite and anhydrite are the most abundant of the later-stage cements and fill porosity of all types. The loss of porosity due to later-stage cementation is quite irregular in distribution, apparently not related to a particular facies or depositional texture, but locally plugging porosity wherever it occurred. The degree of remaining porosity as a result of this continuing pore-filling cementation is varied both between the same facies in different wells and within the same facies in a well. In some cases, samples taken several feet apart in the same well have very different porosities due to the later cements, although the depositional texture and earlier diagenetic overprint is the same.

Calcite not only fills fractures, but commonly fills the intercrystalline and moldic porosity in dolomite adjacent to the fracture. Upon closer examination in thin section, one can see not only the calcite-filling porosity, but also calcite intergrown with dolomite rhombs that exhibit corrosion or partial dissolution. This replacement calcite, or dedolomite, has not created additional porosity in the reservoir; conversely, together the calcite cement and dedolomite have locally reduced porosity.

Budai et al. (1984) described dedolomite from both outcrop and subsurface sections of the Mission Canyon in the Wyoming and Utah Overthrust Belt, and argued that it occurred in four stages during burial at higher temperatures. The dedolomite is associated with fractures, stylolites, and anhydrite nodules. Petrographic evidence for the dedolomite, as summarized by Budai et al. (1984), includes (1) corroded edges on dolomite crystals; (2) partial dolomite rhombs floating in calcite cement; (3) calcite patches within dolomite rhombs, infrequently filling a rhombohedral pore; and (4) rhombohedral calcite after dolomite. Their petrographic and geochemical study shows the fracture-related dedolomite formed adjacent to two sets of fractures that are temporally separated by their timing relative to stylolites. They suggest the earlier-fracture dedolomite is shallow-burial in origin, and the dedolomite and calcite cements associated with the later-fractures are deep burial and tectonic-vein mineralization related to late Tertiary tectonic fracturing and uplift. Budai (1985) suggests the fluids associated with thrusting contained hydrocarbons and brines high in calcium, but low enough in magnesium and sulfate to dissolve dolomite and anhydrite.

CORRELATION OF POROUS ZONES

Previous sections have examined the lithofacies and sequence in which they occur in cores from the field, a depositional model has been proposed, and the porosity development and alteration during diagenesis have been discussed. As previously mentioned, we feel the porosity development seen in the cores is largely the product of relatively early dolomitization and dissolution controlled in distribution by the occurrence of mud-rich lithologies. The stratigraphic cross-section of Figure 18 suggests both porous and tight zones as recognized on logs that have been integrated with cores are continuous through the field. Several observations from our core study are reemphasized here as they pertain to porosity continuity in the field.

The orientation of the field and, therefore, distribution of cores parallel to depositional strike and lack of any significant facies variation along strike are reasons for the similar vertical sequence and generalized occurrence of porous zones in all the cores. This apparent similarity is reinforced by the correlations shown on Figure 18. Porous dolomites within the main porosity zone of the Mission Canyon Formation appear to be well developed and continuous; the intervening tight limestones are thinner and most, but not all, are also continuous between wells. This large-scale pattern of laterally continuous, stacked porous and tight layers should result in good horizontal communication within layers that are relatively isolated in a vertical sense. Vertical communication between porous dolomite layers would occur where the thin limestones pinch out or are extensively fractured, the latter not being obvious from the core studies.

On the scale of correlatable stratigraphic units, the intercrystalline and moldic porosities within dolomites are directly related. Porosity and permeability development in the dolomite is heterogeneous, however, on a

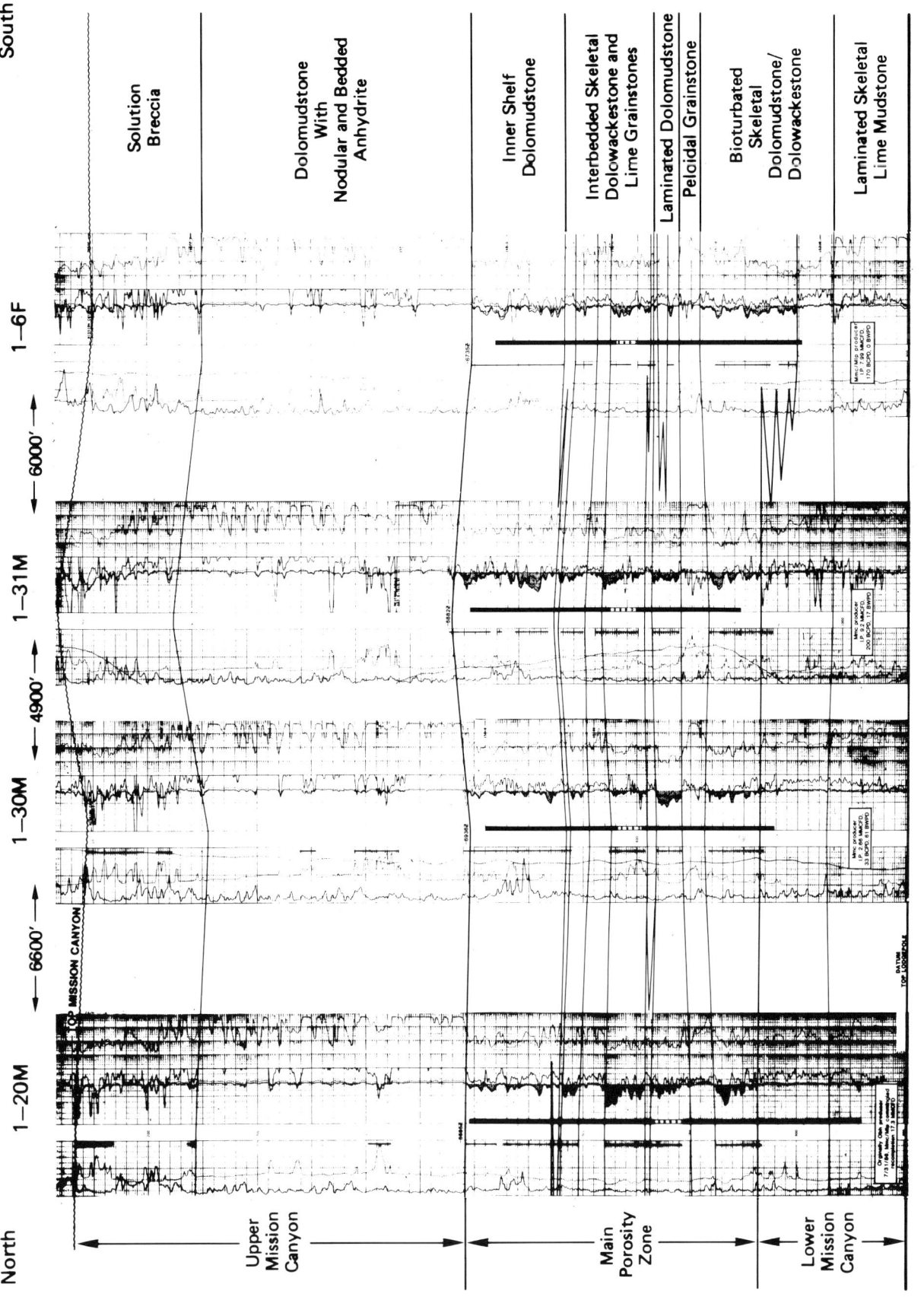

Figure 18 Cross-section highlighting continuity of porous and tight beds in main porosity zone of Mission Canyon Formation, Whitney Canyon-Carter Creek Field.

smaller-scale due to minor rock variability, differential cementation, and the observation in thin section that intercrystalline and moldic porosities behave independently. Because dolomudstones-dolowackestones and dolowackestones-dolopackstones are so commonly intermixed in the cores, moldic porosity can be quite variable within a core sample or within a porous layer. To the extent that original depositional texture controls the crystal size of the replacement dolomite, the permeability which is closely related to crystal size can be equally varied. Thus dolomitization and dissolution are strictly facies controlled in Whitney Canyon-Carter Creek Field, but this early diagenetic overprint, variable itself on a smaller scale, further accentuates heterogeneity related to depositional texture.

Later cementation, possibly related to stylolites and microfractures, further complicates porosity patterns. Anhydrite and calcite cements, similar to those filling fractures, were seen plugging intercrystalline and moldic pores in the cores. These cements are not equally well developed in corresponding units between wells; therefore, they are not strictly speaking facies-related cements, but to what extent they may be localized surrounding fractured zones is not presently known.

REFERENCES

BARTBERGER, C. E., 1985, Facies Control of Mississippian Porosity, Whitney Canyon-Carter Creek Field, Wyoming Overthrust Belt: (abs.) AAPG Rocky Mountain Section, AAPG Bull., v. 69, p. 842-843.

BISHOP, R. A., 1982, Whitney Canyon-Carter Creek Gas Field, Southwest Wyoming: in Powers, R. B. (ed.), Geologic Studies of the Cordilleran Thrust Belt: Rocky Mountain Assoc. Geologists, Field Conference Guidebook, v. 2, p. 591-599.

BUDAI, J. M., 1985, Evidence for Rapid Fluid Migration During Deformation, Madison Group, Wyoming and Utah Overthrust Belt: in Longman, M. W., Shanley, K. W., Lindsay, R. F., and Eby, D. E. (eds.), Rocky Mountain Carbonate Reservoirs: SEPM Core Workshop No. 7, p. 377-407.

BUDAI, J. M., LOHMANN, K. C., and OWEN, R. M., 1984, Burial Dedolomite in the Mississippian Madison Limestone, Wyoming and Utah Thrust Belt: Jour. Sed. Petrol., v. 54, p. 276-288.

BUDAI, J. M., LOHMANN, K. C., and WILSON, J. L., 1987, Dolomitization of the Madison Group, Wyoming and Utah Overthrust Belt: AAPG Bull., v. 71, p. 909-924.

GARGALLO-QUINONES, M., 1985, Study of the Ancient Karstification and Brecciation in the Madison Limestone (Mississippian) in Wyoming: Unpub. M.S. Thesis, SUNY Stony Brook, 208 p.

GUTSCHICK, R. C. and SANDBERG, C. A., 1983, Mississippian Continental Margins of the Conterminus United States: in Stanley, D. J. and Moore, G. T. (eds.), The Shelfbreak - Critical Interface on Continental Margins: SEPM Spec. Publ. No. 33, p. 79-96.

GUTSCHICK, R. C., SANDBERG, C. A., and SANDO, W. J., 1980, Mississippian Shelf Margin and Carbonate Platform from Montana to Nevada: in Fouch, T. D. and Magathan, E. R. (eds.), Paleozoic Paleogeography of the West-Central United States, SEPM, Rocky Mountain Section: Rocky Mountain Paleogeography Symposium 1, p. 111-128.

HALLEY, R. B. and HARRIS, P. M., 1979, Fresh-Water Cementation of a 1,000-Year Old Oolite: Jour. Sed. Petrol., v. 49, p. 969-988.

HALLEY, R. B., HARRIS, P. M., and HINE, A. C., 1983, Bank Margin Environment: in Scholle, P. A., Bebout, D. G., and Moore, C. H. (eds.), Carbonate Depositional Environments: AAPG Memoir 33, p. 463-506.

HOFFMAN, M. E. and KELLEY, J. M., 1981, Whitney Canyon-Carter Creek Field, Uinta and Lincoln Counties, Wyoming: in Reid, S. G. and Miller, D. D. (eds.), Energy Resources of Wyoming: Wyoming Geological Assoc., 32nd Annual Field Conference Guidebook, p. 99-107.

HOFFMAN, M. E. and BALCELLS-BALDWIN, R. N., 1982, Gas Giant of the Wyoming Thrust Belt: Whitney Canyon-Carter Creek Field: in Powers, R. B. (ed.), Geologic Studies of the Cordilleran Thrust Belt: Rocky Mountain Assoc. Geologists, Field Conference Guidebook, v. 2, p. 613-618.

LINDSAY, R. F. and KENDALL, C. G., 1985, Depositional Facies, Diagenesis, and Reservoir Characteristics of Mississippian Cyclic Carbonates in the Mission Canyon Formation, Little Knife Field, Williston Basin, North Dakota: in Roehl, P. O. and Choquette, P. W. (eds.), Carbonate Petroleum Reservoirs: Springer-Verlag, New York, p. 177-190.

RAU, R., 1982, Stratigraphy of the Mississippian Madison Group of the Whitney Canyon-Carter Creek Field of Southwest Wyoming: Unpub. M.S. Thesis, Kent State University, 156 p.

READ, J. F., GROTZINGER, J. P., BOVA, J. A., and KOERSCHNER, W. F., 1986, Models for Generation of Carbonate Cycles: Geology, v. 14, p. 107-110.

ROSE, P. R., 1977, Mississippian Carbonate Shelf Margins, Western United States: in Heisey, E. L., Lawson, D. E., Norwood, E. R., Wach, P. H., and Hale, L. A. (eds.), Rocky Mountain Thrust Belt Geology and Resources: Wyoming Geological Assoc., 29th Annual Field Conference Guidebook, p. 155-172.

SANDBERG, C. A. and GUTSCHICK, R. C., 1980, Sedimentation and Biostratigraphy of Osagean and Meramecian Starved Basin and Foreslope, Western United States: in Fouch, T. D. and Magathan, E. R. (eds.), Paleozoic Paleogeography of the West-Central United States, SEPM, Rocky Mountain Section: Rocky Mountain Paleogeography Symposium 1, p. 129-147.

SANDBERG, C. A., GUTSCHICK, R. C., JOHNSON, J. G., POOLE, F. G., and SANDO, W. J., 1982, Middle Devonian to Late Mississippian Geologic History of the Overthrust Belt Region, Western United States: in Powers, R. B. (ed.), Geologic Studies of the Cordilleran Thrust Belt: Rocky Mountain Assoc. Geologists, Field Conference Guidebook, v. 2, p. 691-719.

SANDO, W. J., 1976, Mississippian History of the Northern Rocky Mountains Region: U.S. Geological Survey, Jour. of Research, v. 4, p. 317-338.

VAIL, P. R., MITCHUM, R. M., JR., and THOMPSON, S., III, 1977, Seismic Stratigraphy and Global Changes in Sea Level: in Payton, C. E. (ed.), Seismic Stratigraphy - Application to Hydrocarbon Exploration: AAPG Memoir 26, p. 51-62.

WILSON, J. L., 1985, Carbonate Facies in Geologic History: Springer-Verlag, New York, 471 p.

STRUCTURAL HISTORY AND RESERVOIR CHARACTERISTICS (MISSISSIPPIAN) OF NESSON ANTICLINE, NORTH DAKOTA

Robert F. Lindsay
Chevron U.S.A. Inc.
P.O. Box 599
Denver, Colorado 80201

Sidney B. Anderson and Julie A. LeFever
North Dakota Geological Survey
University Station
Grand Forks, North Dakota 58202

Lee C. Gerhard
Kansas Geological Survey
1930 Constant Ave.
Lawrence, Kansas 66046

Richard D. LeFever
Department of Geology
University of North Dakota
Grand Forks, North Dakota 58202

ABSTRACT

Nesson anticline is the largest hydrocarbon productive structure in North Dakota portions of the Williston basin. It was discovered in 1951 just a few months after discovery of oil in the Williston basin. Fifty-four fields, producing from fourteen lower to middle Paleozoic formations are scattered along the north-south length of the anticline. Nesson has produced a total of 377 MMBO, with the Madison Group accounting for two-thirds of the total production. Central and southern portions of the anticline were subdivided into nine areas, which revealed episodic and independent structural movement since the Late Precambrian.

All Phanerozoic periods are present within the stratigraphic section. Unconformity bound, major tectonic-eustatic sequences were mapped along the length of Nesson, with sedimentary tectonics documented for the entire Phanerozoic. Greatest amounts of tectonic development of the anticline was during the Devonian into early Mississippian. Post-Greenhorn, Laramide tectonism was responsible for the last major structural deformation of the anticline.

Selected oil fields, productive from the Madison Group, were studied where they are productive from Mission Canyon Formation and the Rival ("Nesson") subinterval. These intervals record sediment infill of a slowly shrinking epeiric sea, as a series of shorelines separated by brief transgressions, prograded toward the center of the basin. Mission Canyon Formation can be characterized as a major shallowing upward sequence, which upsection is: 1) shallow open marine, 2) transitional open to restricted marine, 3) restricted marine, and 4) fringing marginal marine. Barrier island/shoreline buildup complexes developed along the shoreline, with bedded evaporites located at a distance away from the anticline toward the east and south. Production is from high energy open marine and barrier island/shoreline buildup limestone facies to the north, and is from interbedded limestones and dolostones deposited in transitional open to restricted and restricted marine facies to the south. After a brief transgressive event deposited the State A marker the Rival ("Nesson") subinterval progradation covered the southern half of Nesson. The Rival ("Nesson") subinterval is subdivided into lower and upper halves. This is due to a subtle transgressive event with the Rival. The lower half was deposited as barrier island/shoreline buildup complexes to the north and as bedded evaporites to the south. The upper half was deposited in an offshore shallow marine setting to the north and center and in a restricted to marginal marine setting to the south. Both shoreline and offshore limestone beds are productive.

INTRODUCTION

Nesson anticline is the largest producing structural feature in North Dakota portions of the Williston basin (Fig. 1). It is approximately 110 miles (175 km) long, with nearly continuous production along a north-south line from just south of the Canadian border (T163N) to the Killdeer Mountains (T146N) south of the Missouri River (Fig. 2). All Paleozoic formations in the basin are present along at least part of the anticline (Fig. 3). Fifty-four oil fields have been discovered along the length of the anticline, producing from fourteen formations ranging in age from Cambrian through Mississippian. Nesson has produced 377 MMBO (as of December 31, 1986) since its discovery, with Madison Group production accounting for 248 MMBO (Rygh, 1987).

Oil was first discovered in North Dakota at Nesson anticline on April 4, 1951, by the Amerada Clarence Iverson No. 1. Drilling along the crest of Nesson revealed a 75-mile long (120 km) trend before drilling the first dry hole. From 1951 to early 1955 more than 70 oil fields were discovered in the Williston basin.

Nesson is similar in size to northwest trending Cedar Creek anticline which is located in southeastern Montana, southwestern North Dakota and northwestern South Dakota (Fig. 1). Detailed mapping of Nesson has revealed fault zones and linear elements which appear to have controlled its structural and stratigraphic development (Fig. 4). Different portions of the anticline appear to have had separate histories of movement.

DISCOVERY OF NESSON ANTICLINE

Nesson anticline was originally identified in 1917 by a U.S. Geological Survey field party which was mapping lignite beds. In 1918, A.J. Collier of the USGS continued mapping the anticline with an alidade and plane table, formally naming the anticline "Nesson" (Collier, 1919). Collier thought that chances of finding oil in Nesson anticline were small (p. 216). Dove (1922) continued work on Nesson north of the Missouri River.

In 1937, large amounts of acreage were leased for oil and gas exploration in western North Dakota, with several reconnaissance surveys and one seismic

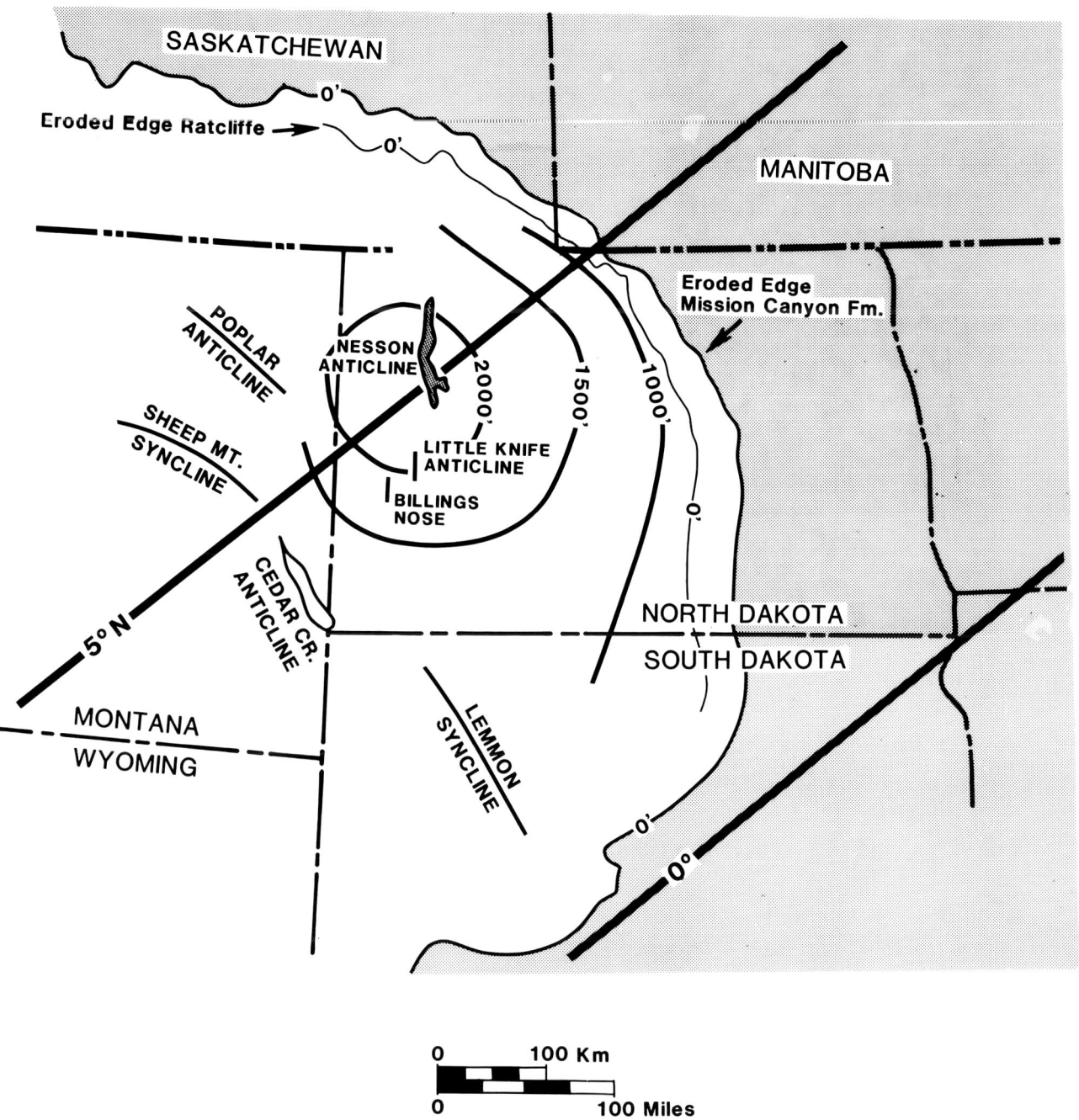

Figure 1 Index map of the Williston basin, delineating the position of Nesson anticline with respect to: 1) major surface and subsurface structures, 2) isopach thickness of the Mississippian Madison Group (Carlson and Anderson, 1965), 3) subcrop eroded edge of Mission Canyon Formation and Ratcliffe interval (Proctor and Macauley, 1968), and Carboniferous paleolatitudes (Habicht, 1979).

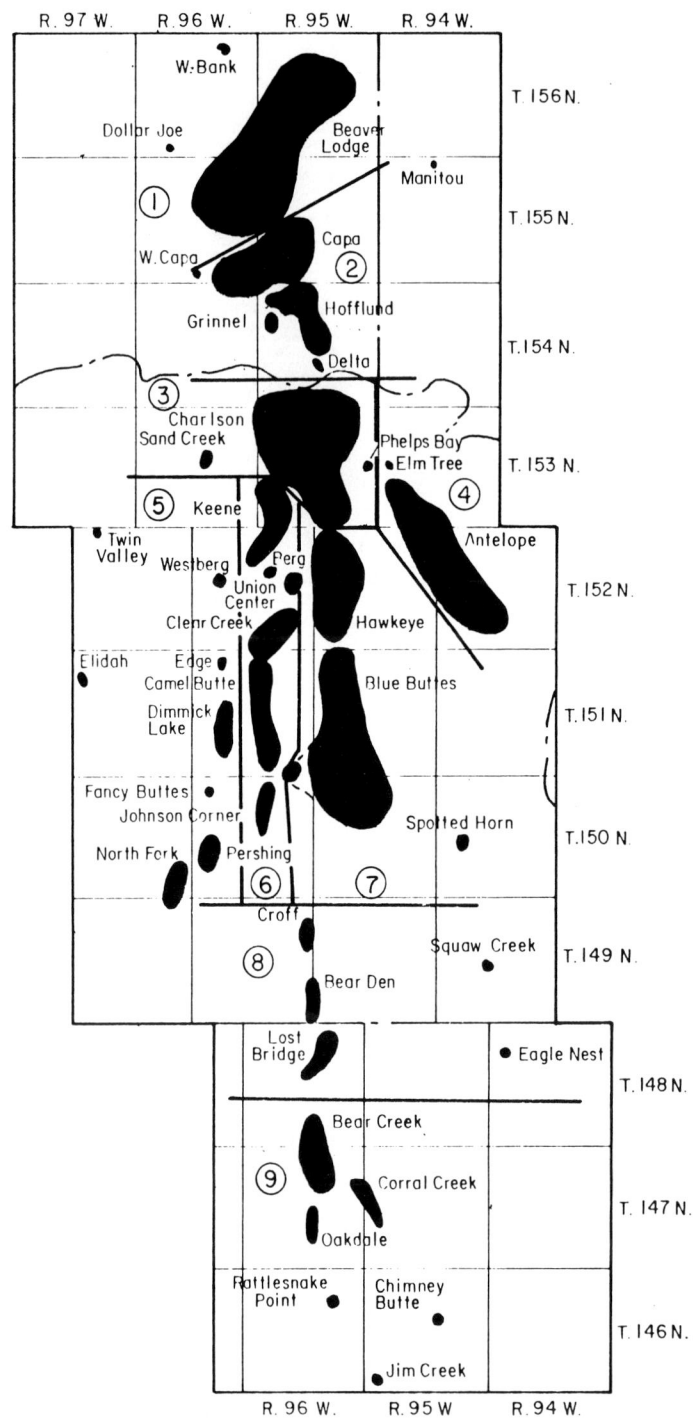

Figure 2 Principle oil fields located along central to southern portions of Nesson anticline. Outlined and numbered areas are individual areas Nesson was subdivided into to study relative amounts of tectonic uplift or subsidence.

Figure 3 Paleozoic and Mesozoic stratigraphic column at Nesson anticline. Shading areas are formations that are hydrocarbon productive.

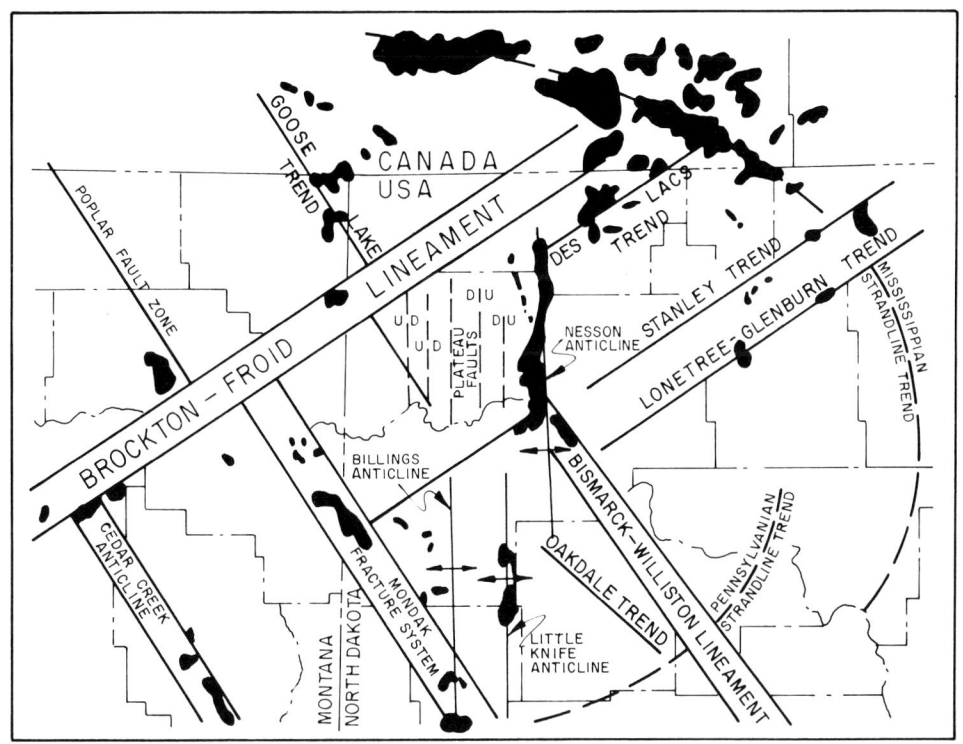

Figure 4 Major structural trends throughout the Williston basin and their relationship to Nesson anticline. Prepared from subsurface and proprietary seismic data.

survey (Nevin, 1946). Seismic work by Amerada in 1946 resulted in the discovery of Beaver Lodge Field in 1951 (Laird et al, 1955; also see Laird, 1946).

First drilling on Nesson was in 1935 by Big Viking Oil Company (SW, NE Sec. 3, T145N, R96W). The well drilled to a depth of 4642 feet, yielding non-commercial quantities of gas. In 1938, the California Company drilled the No. 1 Kamp well (NW, NE Sec. 3, T154N, R96W) to a total depth of 10,287 feet, but abandoned the well due to stuck drill pipe (Laird et al, 1956). This location was just immediately west of Capa Field.

Discovery of oil in Nesson anticline by Amerada in 1951 opened Beaver Lodge Field. The discovery well was originally completed in the Silurian Interlake Formation at 11,630-11,660 feet with initial production flowing 307 BOPD and 25 BWPD. The well was plugged back to the Devonian Duperow Formation at 10,490-10,530 with initial production of 290 BOPD. Two successful drillstem tests in the Madison Group at Nesson (H.O. Bakken No. 1 and Palmer Dilland No. 1) caused Amerada to re-enter the discovery well and recomplete it in the Madison (8520-8528 feet) with an initial production of 667 BOPD in December, 1951. This was the first Madison Group well completion in North Dakota. The H.O. Bakken No. 1 was the discovery well for Tioga field, located just north of Beaver Lodge (Fig. 2), and was completed in the Madison on April 14, 1952. By 1956 both Beaver Lodge and Tioga fields were almost completely developed.

STRUCTURE

Nesson anticline is predominantly a subtle south-plunging fold (Figs. 1 and 2). North of the Missouri River the anticline exhibits more closure, with the hinge surface inclined slightly to the west. South of the Missouri River the fold becomes more open and maintains an upright position. Although the fold follows an overall north-south direction the hinge line bends along the trace of the fold. A notable change in direction of Nesson occurs south of Beaver Lodge Field, where the hinge line changes from a northeasterly orientation to a northwesterly orientation. Further south, at Charlson Field, the single fold splits into three folds (Fig. 2). One fold is oriented to the southeast and is named Antelope anticline. The second, or main fold, continues south through Hawkeye and Blue Buttes Fields. While the third is a secondary fold, that is not well developed, running through Clear Creek and Camel Butte Fields.

On the west side of the anticline the western Nesson fault bounds the structure from Beaver Lodge Field to just south of T150N (Fig. 5). This fault has been active since the Precambrian, moving intermittently through time, with its west side downthrown (Gerhard et al, 1982; also see Gerhard et al, 1987). A second fault occurs along the northeast side of Antelope anticline, with its northeast side downthrown. It appears to have only affected Devonian and younger strata.

Central and southern portions of Nesson anticline have been divided into nine areas, from north to south, with large fields, such as Beaver Lodge located in area one, to the north and small fields, such as Rattlesnake Point, Chimney Butte and Jim Creek located in area nine, to the south (Fig. 2). Structural contour maps were constructed on 23 different formations or marker beds. Isopach maps of principal units in the stratigraphic column were constructed, along with structural relief plots (LeFever et al, 1987). Where control permitted the structural relief was determined for each area (Fig. 6). To ensure that only structural movement was being recorded only conformable formation tops were used. For comparative purposes, only wells which penetrated all conformable units down to the lower Devonian Ashern Formation (19 units total) were used. Due to paucity of data below the Ashern, no formation or marker bed tops were used for relief determination. Relief on a given unit is a function of net tectonic activity since deposition of the unit, so continued uplift in an area should produce an increase in relief with increasing age, while continued subsidence should produce the opposite trend.

It is evident from Figure 6 that amounts and timing of uplift or subsidence occurring in each area along Nesson anticline were independent of events in other areas. No two areas appear to have had similar histories of tectonic activity. Because of this, it is most likely that each of these

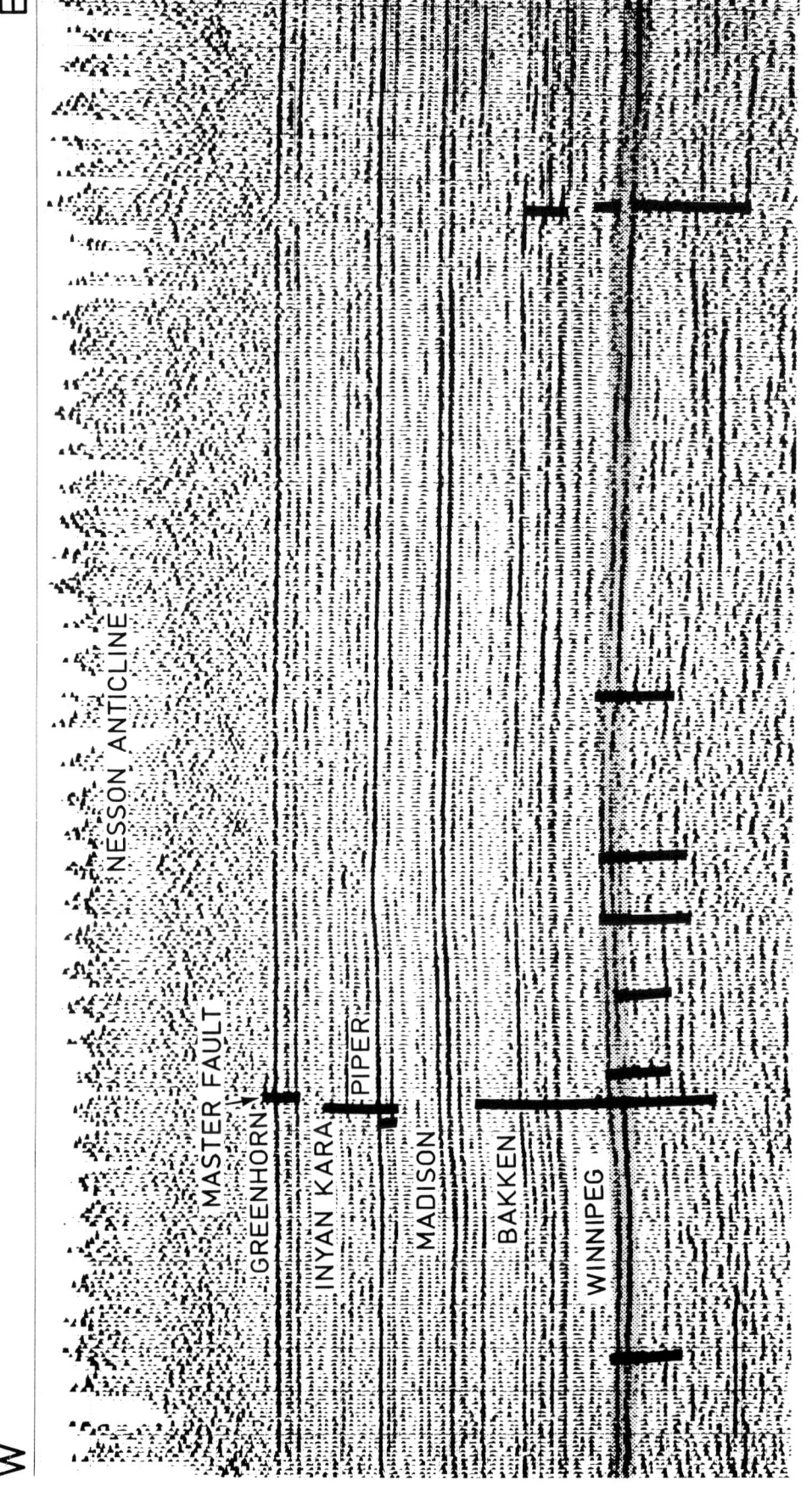

Figure 5 West to east seismic cross section across Nesson anticline, showing major basement-controlled faults.

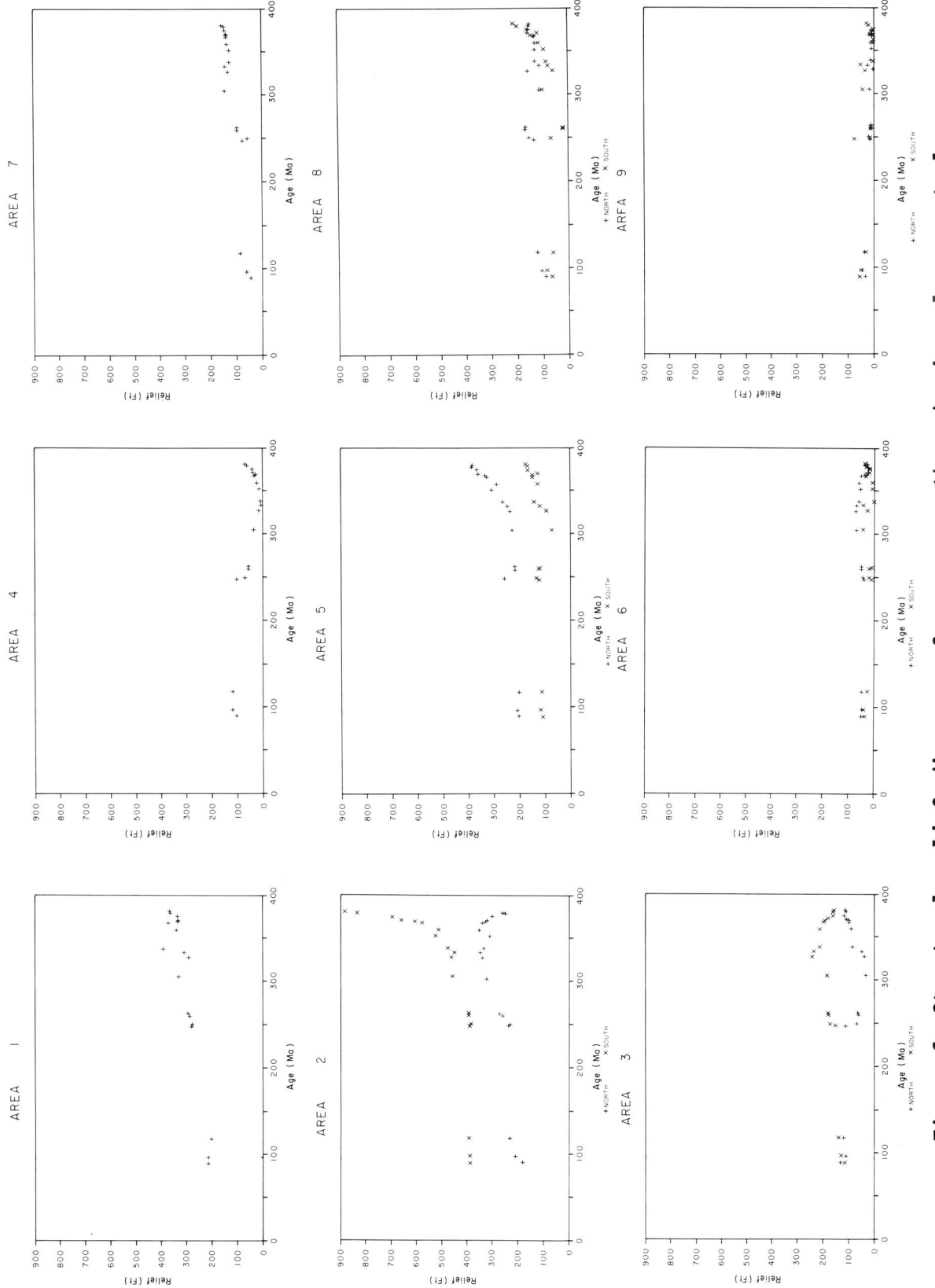

Figure 6 Structural relief diagrams of areas one through nine along central and southern portions of Nesson anticline.

areas behaved as independent blocks. If more detailed data were available, it would probably reveal even smaller scale areas which acted independently throughout time.

In general, greatest amounts of uplift along Nesson anticline occurred during the Devonian or, in some cases, in the Early Mississippian. Due to a lack of data below the Devonian it is not possible to ascertain when Nesson anticline first became a positive feature. In most areas movement, whether as uplift or subsidence, was episodic with few instances of prolonged movement. Generally, uplift alternated with quiescence and subsidence.

STRATIGRAPHY

All Phanerozoic periods are present at Nesson anticline (Fig. 3). Petroleum production is from the Paleozoic section with major petroleum production from Ordovician, Silurian, Devonian and Mississippian age rocks. Carbonate rocks dominate the lower and middle Paleozoic, while detrital silisiclastic rocks are predominant in the upper Paleozoic.

Major unconformities subdivide the stratigraphic section into convenient tectonic-eustatic cycles (Sloss, 1963). Sloss named six stratal sequences which are: 1) Sauk (Cambrian-Early Ordovician), 2) Tippecanoe (middle Ordovician-Silurian), 3) Kaskaskia (Devonian-Mississippian), 4) Absaroka (Pennsylvanian-Triassic), 5) Zuni (Jurassic-Paleocene), and 6) Tejas (Eocene-Quaternary) (Fig. 7). These sequences are used in this paper to compare rates of sedimentation with rates and timing of tectonic movement.

SEDIMENTARY TECTONICS

SAUK SEQUENCE

Despite limited control for Sauk sequence (Cambrian-Early Ordovician) rocks, several structural implications are apparent from both isopach and structural mapping (Figs. 8 and 9). Nesson anticline appears on Sauk maps as

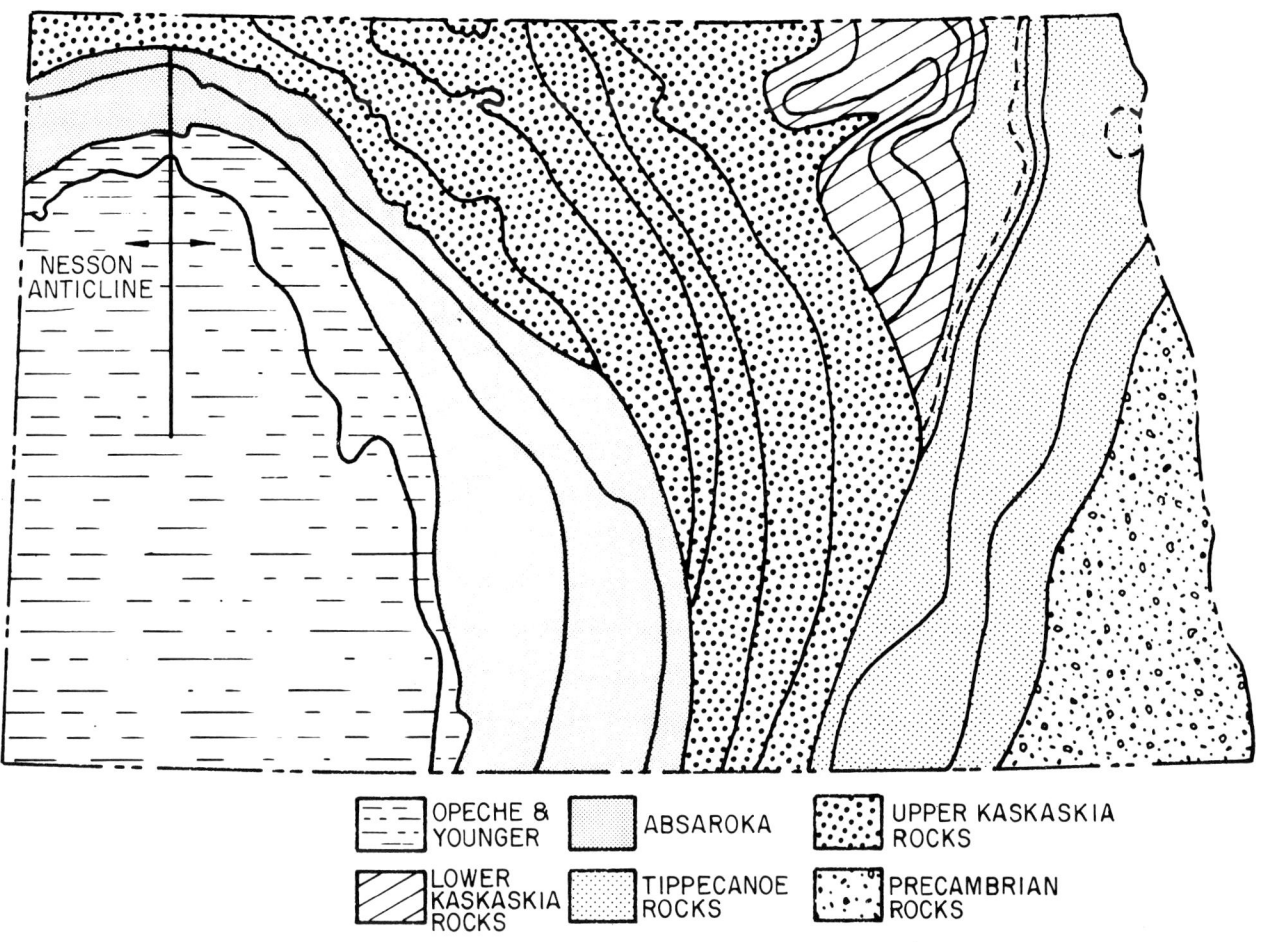

Figure 7 Pre-Mesozoic subcrop map of North Dakota portions of the Williston basin and the position of Nesson anticline.

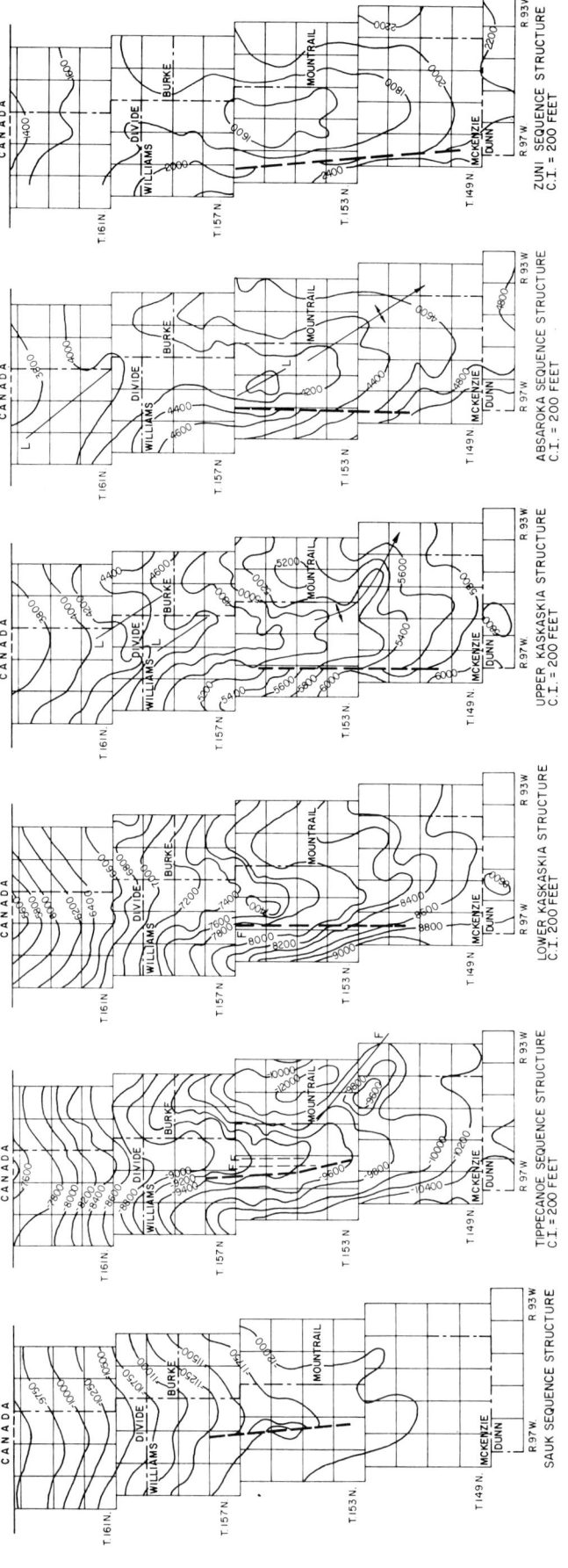

Figure 8 Structural maps drawn on top of the Sauk, Tippecanoe, lower and upper Kaskaskia (Three Forks Formation and Otter Formatiosn, respectively), Absaroka (Spearfish Formation), and Zuni (Greenhorn Formation) sequences along the length of Nesson anticline.

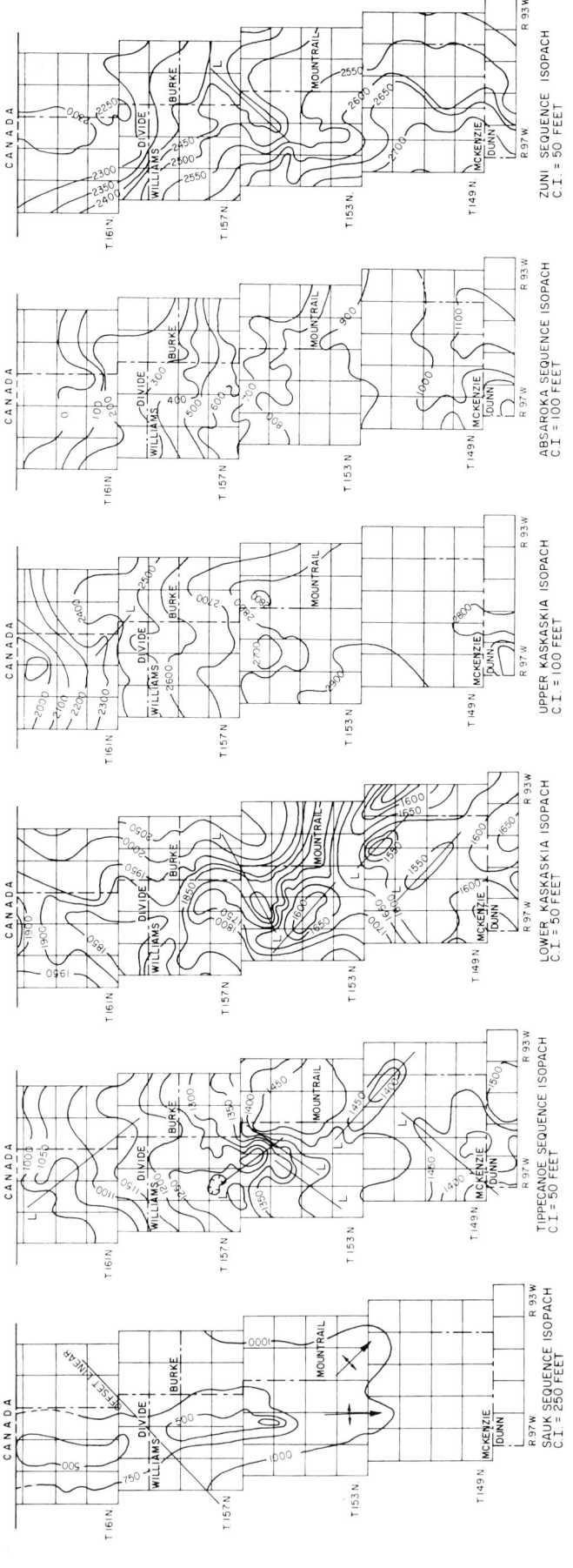

Figure 9 Isopach maps of the Sauk, Tippecanoe, lower and upper Kaskaskia, Absaroka, and Zuni sequences along the length of Nesson anticline.

a roughly north-south trending structure, with both a master fault along the western boundary of the anticline and smaller normal faults with complex tilting of individual blocks at the crest of the fold. Nesson apears to have had a Precambrian ancestry. Carlson (1960) showed that the lower part of Deadwood Formation thins over a Precambrian high, representing syndepositional relief (Fig. 9).

Rapid thickening of the Deadwood west of the crest of the structure also suggests, but does not establish, that the master fault on the western side of the anticline had its origin in Sauk or earlier time. Of greatest significance in the anticline's history is the offset of two thin areas in the Deadwood, in Divide and Williams counties, respectively. These two thins, probably separate blocks of Nesson anticline, appear to be offset by left-lateral movement, perhaps along a lineament that has been previously described in that location. In addition to offset of the Nesson trend, Antelope anticline appears as a subdued feature, suggesting that the fault mapped on its northeastern flank did not affect sedimentation. The very thin area (T155 and 156N, R96W), is considered to be the result of syndepositional or pre-depositional movement along complex faults along crestal portions of Nesson.

These Sauk structures are congruent with a left-lateral origin for the Williston basin. The master fault on the west side of Nesson, coupled with additional north-south faults nearer the crest of the anticline, represents extensional strain of the left-lateral Brockton-Froid and Colorado-Wyoming couple (Fig. 4). Erosion of Deadwood Formation during post-Sauk and pre-Tippecanoe time, represents plate-wide emergence during a major low sea stand of the Paleozoic.

TIPPECANOE SEQUENCE

Silisiclastic deposition is characteristic of the initial Tippecanoe transgression, which was followed by general basin subsidence and carbonate sedimentation (Figs. 8-10). Isopach mapping of Red River Formation demonstrates the existence of Nesson during Red River time (Carroll, 1979). The basin was well differentiated during deposition of the Tippecanoe sequence

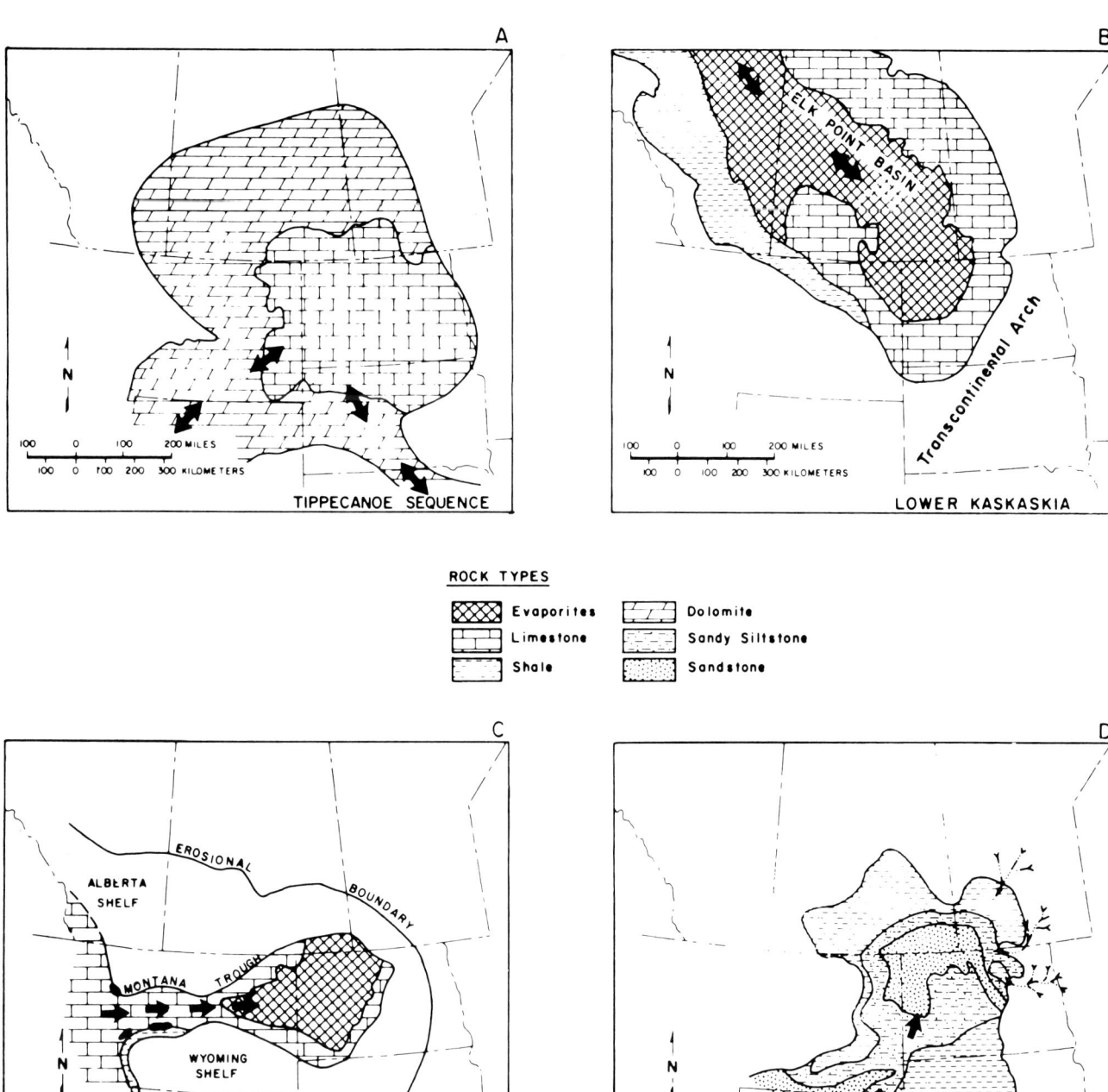

Figure 10 Patterns of sedimentation throughout time in the Williston basin. Sequence maps show various directions of communication or lack of communication with open marine conditions: A) Tippecanoe sequence was connected with the Cordilleran miogeosyncline from the southwest, B) Lower Kaskaskia sequence was connected to the Elk Point basin in Canada, C) Upper Kaskaskia was connected with the Cordilleran miogeosyncline through the Montana Trough, and D) Absaroka sequence was connected with silisiclastics entering the basin from the southwest and with drainage from the northeast and east.

(middle Ordovician-Silurian) (Fig. 10). Deepest parts of the basin lie to both the immediate southwest and east of southern portions of Nesson anticline. Shallowing (thinning) of the basin occurs toward the south. Some thinning occurred due to the presence of the Little Knife anticline southwest of Nesson.

Several subsidiary elements of Nesson were active during Tippecanoe sedimentation. Quite prominent is the Beaver Lodge Field area fault complex, previously seen in the Sauk (T155 and 156N, R96W), which is thin and contains a strong northeasterly linear trend. A second northwesterly trend is in the northern part of the mapped area. Several north-south faults also occur in this small area.

Other thickness anomalies are Antelope anticline and a southeast-trending nose in the northern part of the area. Crestal portions of Nesson actually lie west of isopached crestal thins from about T158N to T160N, just south of a southeastern lineament. This may have resulted from westward tilting of the basin in post-Tippecanoe time or, more likely, as a result of more recent movement of the Nesson master fault system on the west side of the anticline. These two crests of the anticline are separated by about 4 miles (6.4 km). Significant movement of the master fault appears to be post-Tippecanoe, since only small thickness changes indicate pre-Tippecanoe movement.

LOWER KASKASKIA SEQUENCE

Thickness of the lower Kaskaskia sequence (Devonian) demonstrates a significant change from earlier structural patterns (Figs. 8-10). Both northeast- and northwest-trending linear features show as thins in this isopach interval, with elongation of the northwest-southeast Antelope trend and a duplicate parallel trend to the south of the Antelope feature. Beaver Lodge Field still persists as a thin.

A radical change of overall thickness patterns is present. Whereas the thickest stratigraphic sections of pre-early Kaskaskia rocks was in the southern part of the mapped area, where the basin was deepest, during Early

Kaskaskia time the stratigraphic section thickened to the north. There was approximately 350 feet (106.7 m) of thickening in the lower Kaskaskia sequence from south to north, making a total minimum structural reversal of over 800 feet (243.8 m) when compared to the Tippecanoe. This was caused by regional uplift along the Transcontinental Arch during the Early Devonian, which tilted the Williston basin northward and established normal marine communication with the Elk Point basin in central and western Canada (Fig. 10).

Termination of Early Kaskaskia sedimentation by a regional unconformity took place near the end of the Devonian. The next overlying unit is the Bakken Formation, which straddles the Devonian-Mississippian boundary. Significance of the pre-upper Kaskaskia unconformity is not totally clear. Near margins of the Williston basin this unconformity is clearly present (Bjorlie, 1979; Bjorlie and Anderson, 1976), but in deeper parts of the basin there remains some argument as to the actual presence of an erosional surface. Clement (1987) has clearly demonstrated that there is extensive erosion of Cedar Creek anticline in southeastern Montana at this time, which breached Devonian Three Forks Formation and lower units down into Silurian Interlake Formation.

UPPER KASKASKIA SEQUENCE

Mostly carbonates and associated evaporites form the upper Kaskaskia sequence (Mississippian) (Figs. 8-10). Basal units of the sequence are the Bakken Formation, the major source rock for Mississippian carbonate reservoirs, and Lodgepole Formation, a transgressive carbonate sequence. Overlying rocks are progradational carbonates and evaporites of the Mission Canyon and Charles Formations. A few thin sands apparently derived from the shield and bypassed by drainage into the marine environment are present. An extensive and significant unconformity overlies the upper Kaskaskia sequence.

Reversal of northward tilting of the basin, which occurred in Early Kaskaskia time, took place during the Late Kaskaskia as a seaway opened toward the west along the old Belt basin. Progradational carbonate deposits in the Late Kaskaskia are the most important hydrocarbon-producing units in the basin. Evaporites associated with these carbonates form excellent reservoir seals.

Thicknesses of the upper Kaskaskia (Fig. 9) show Nesson anticline to be present as a low relief feature. Sharp thinning in the northern portion of the study area is due to post-Late Kaskaskia erosion. Thinning seen at the approximate location of Beaver Lodge Field (T156N, R96W) is partly due to erosion and partly due to thinning over that persistently active feature. A northeast-trending linear feature extends from that field area to the eastern map boundary; a structural discontinuity (fault?) also occurs at that site (Figure 8).

In the southern part of the study area thickness changes are small and the overall aspect is that of a broad and shallow depression with sharp thinning to the southwest. Antelope anticline is still present. The center of the basin is located just southwest of the study area. Over 2950 feet (899 m) of upper Kaskaskia rocks are present in deep portions of the basin.

Structural throw on the Nesson master fault is about 800 feet (243.8 m) at this horizon (Fig. 11). Most of this throw can be accounted for by post-Absaroka motion, but a reversal of motion during Absaroka time damped the apparent throw.

Truncation (Ancestral Rocky Mountain deformation) of upper Kaskaskia rocks in the northern portion of the Nesson anticline averages 42 feet/mile (7.95 m/km) of thinning. Average apparent depositional thinning along the crest of Nesson in the upper Kaskaskia is about 12.5 feet/mile (2.4 m/km).

ABSAROKA SEQUENCE

Deposition of the Absaroka sequence (Pennsylvanian) was greatest in southern portions of the basin (Figs. 8-10). While depositing lower parts of the sequence the basin contained normal marine water with southern drainage entering through an estuary and offshore bar system (Tyler Formation). Later

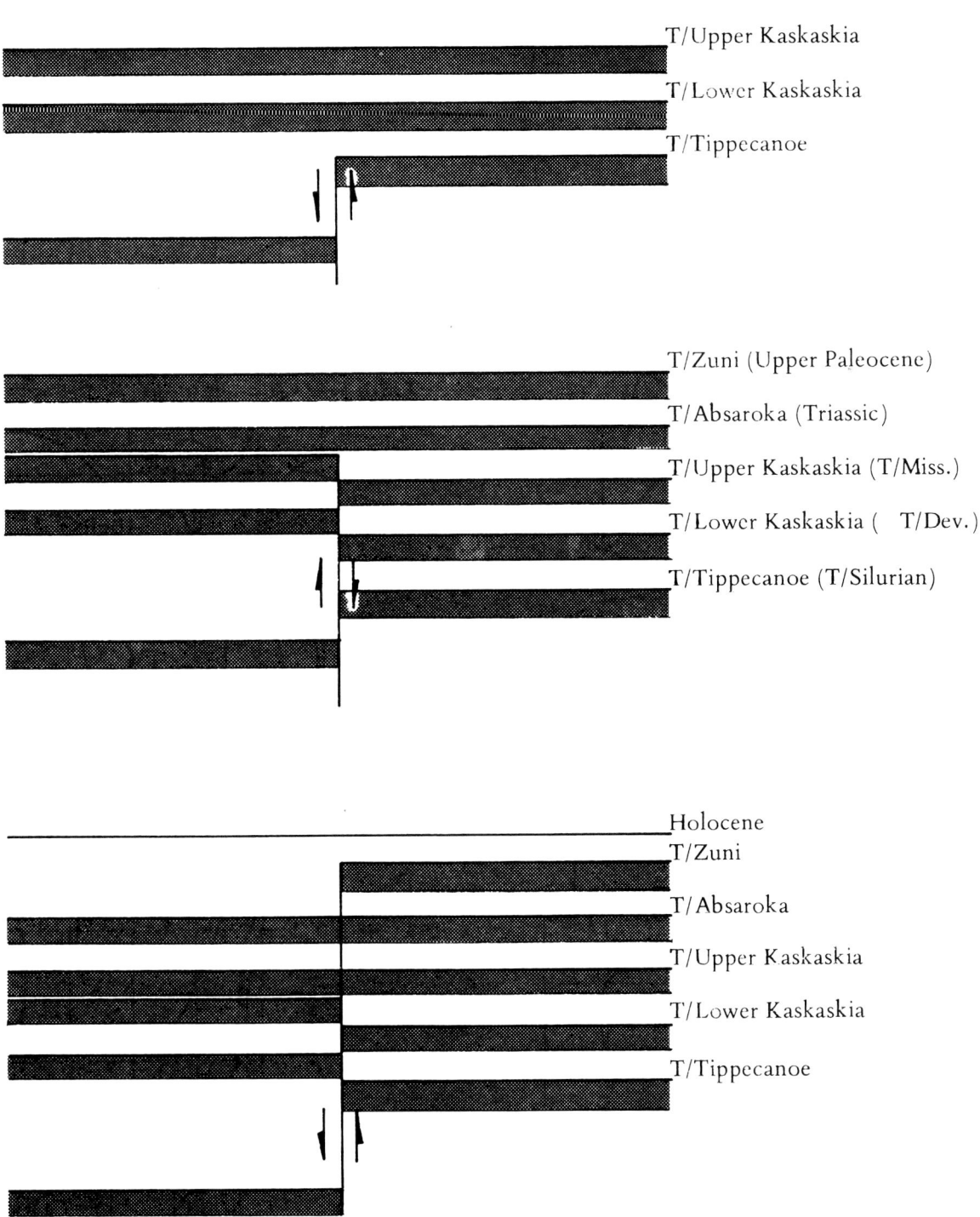

Figure 11 Episodic fault block motions through time along the west side of Nesson anticline.

sedimentation was progressively more saline as the basin filled with detritus and circulation of marine water from the Cordilleran seaway was less effective.

During the Permian an unconformity developed, which resulted in a regional angular discordance between underlying Minnelusa and overlying Opeche beds. Pre-Opeche erosion thinned the Pennsylvanian and Lower Permian section. In northern portions of the study area thinning is the result of pre-Zuni erosion, which stripped all Absaroka rocks from northern Nesson anticline.

Thickness of the Absaroka sequence on Nesson anticline ranges from zero to over 1100 feet (335.3 m) (Fig. 9). Filling of the basin by silisiclastics obscured much of the structure. Some influence of the Nesson master fault may be seen in the north-striking contours in west-central portions of the anticline. However, during Absaroka time there was at least one episode of structural reversal on the master fault system.

Structure on top of the Absaroka sequence is relatively simple (Figure 8). A south-plunging anticline is present only from about T149N to T160N. There is about 500 feet (152.4 m) of displacement on the western master fault. About 150 feet (45.7 m) of displacement is present at the Antelope anticline border fault. Beaver Lodge area is structurally high and is the high point of Nesson anticline during upper Absaroka time. Silisiclastic deposition subdued the paleostructural expression, in contrast with earlier carbonate sedimentation which frequently enhanced paleostructural expression.

Erosion and a time-stratigraphic hiatus mark an end of the Absaroka sequence at Nesson anticline. Structural distortion is minimal when compared to past events affecting the Williston basin.

ZUNI SEQUENCE

The last major cycle of sedimentation upon this part of the North American craton occurred as two sub-cycles: 1) a Jurassic cycle, which ended with a disconformity on the Morrison Formation in Montana and Swift Formation in North Dakota; and 2) a Cretaceous through Paleocene and Eocene cycle.

Sufficient control is available to map the Greenhorn and since it is a common seismic maping horizon, this study has used the Greenhorn as the horizon for structural mapping (Fig. 8). Isopach mapping of the Zuni sequence (Fig. 9) used the Greenhorn as the top of the sequence.

Thickness of the Zuni sequence (Jurassic-Paleocene) indicates that Nesson anticline had ceased to be a major influence on sedimentation, but instead acted as a "sill" to sedimentation (Figs. 8-10). East of the anticline thickness changes are relatively regular, approximately 5-10 feet/mile (0.9-1.8 m/km), with gradual thickening to the west and south. At Nesson this pattern changes abruptly, with thicknesses increasing to about 25 feet/mile (4.7 m/km) along its western margin. This rapid thickening suggests that the basin was filled with sediments during Zuni time. The basin could only accommodate limited sedimentation east of Nesson, with continued but slight subsidence of the basin to the west.

Only central portions of Nesson had definition as an anticlinal structure during Zuni deposition. Overall, it appears to have been a gentle, broad, open anticline at the Greenhorn level, with closure only over the Beaver Lodge feature and south to T153N.

Deformation of the Greenhorn and of the Zuni sequence in general took place as a direct consequence of Laramide orogenesis. Isopach mapping of the Zuni does not show a clearly defined Nesson feature, whereas structural mapping of the Greenhorn within the Zuni does. This illustrates that Nesson anticline was reactivated in post-Greenhorn time.

The master fault on the west side of Nesson has controlled much of the anticline's history. There are many faults associated with Nesson, but the western fault system seems to have dominated. Motion along this fault system can be summarized as major normal (west side down) movement in Late Precambrian-Sauk time, normal motion at the end of Tippecanoe time, reverse motion (west side up) during the Absaroka sequence, and normal motion since, including major down-dropping during Laramide deformation.

MADISON GROUP RESERVOIR CHARACTERISTICS IN SELECTED FIELDS

Several large unitized oil fields, which produce from rocks in the Madison Group of the upper Kaskaskia sequence, that are scattered along crestal portions of Nesson anticline were studied (Fig. 1). From north to south these fields are: North Tioga, Tioga, Beaver Lodge, Capa, Hofflund, Hawkeye, Blue Buttes, Antelope and Clear Creek (Figs. 2, 12 and 13). In many cases, these fields along Nesson are separated by just a few plugged and abandoned wells. These individual fields can be depicted as being part of a larger segregated reservoir. Besides producing hydrocarbons from the Madison Group, which is composed upsection of Lodgepole, Mission Canyon and Charles Formations, other formations deeper in the section are also productive, which are upsection: Cambrian Deadwood; Ordovician Winnepeg Group, Red River, Stony Mountain; Ordovician-Silurian Stonewall; Silurian, Interlake; Devonian Winnipegosis, Dawson Bay, Duperow, Birdbear, Three Forks; and Devonian-Mississippian Bakken.

Within the Madison Group oil production is from upper portions of the Mission Canyon Formation and the Rival ("Nesson") subinterval, which is equivalent to the Frobisher-Alida interval, and Midale subinterval, which is the lower part of the Ratcliffe interval (Fig. 14). Reservoirs within the Mission Canyon and the Rival ("Nesson") subinterval are reported on in this paper. Rival ("Nesson") beds define the top of the Frobisher-Alida interval and is equivalent to the Frobisher evaporite. State A marker separates overlying Rival ("Nesson") subinterval from underlying Mission Canyon carbonates. Midale subinterval, of the Ratcliffe interval, overlies the Rival ("Nesson") subinterval. These two subintervals, Rival ("Nesson") and Midale, are considered part of Mission Canyon Formation if deposited as carbonates in a marine to shoreline setting. They are, however, considered to be a part of the Charles Formation when deposited as evaporites along or behind the paleoshoreline.

OVERVIEW

Madison Group deposition in the Williston basin can be characterized as sediments infilling a subsiding cratonic basin, with carbonates followed by evaporites (Figs. 10, 15 and 16). This is regarded by McCabe (1959, p. 58) as a large cycle of basin development. Lodgepole Formation records deeper water-starved basin facies. With lower Lodgepole deposition recording maximum transgression. Mission Canyon Formation records a change of water depth to a shallow marine setting. Shoreline regression began with evaporitic basin margin deposits of the Charles Formation. These grade into basinal evaporite deposition interspersed with carbonates and argillaceous muds. These three formations intertongue. Uppermost beds of Lodgepole Formation grade laterally shelfward into Mission Canyon Formation, and uppermost beds of Mission Canyon Formation grade laterally into Charles Formation. Because of basin-wide intertonguing of facies, marker beds, such as the State A marker and other markers, cut across formational and lithologic boundaries. Age of the Madison Group is lower Mississippian, specifically the Lodgepole is Kinderhookian, the Mission Canyon is Late Kinderhookian and Osagian, and the Charles is Osagian and Meramecian (Peterson, 1984).

At Nesson Anticline, the Mission Canyon Formation and the Rival ("Nesson") subinterval (Frobisher-Alida and Ratcliffe intervals) were deposited in a slowly shrinking epeiric sea. This was not a simple regression but was composed of a series of shoreline progradations and smaller transgressions with an overall regressive theme, which slowly infilled the basin (Figs. 17 and 18). Various Mission Canyon shoreline progradations, such as the Glenburn, Mohall, Sherwood and Bluell (Harris et al., 1966), did not reach Nesson anticline. While the Rival ("Nesson") subinterval shoreline

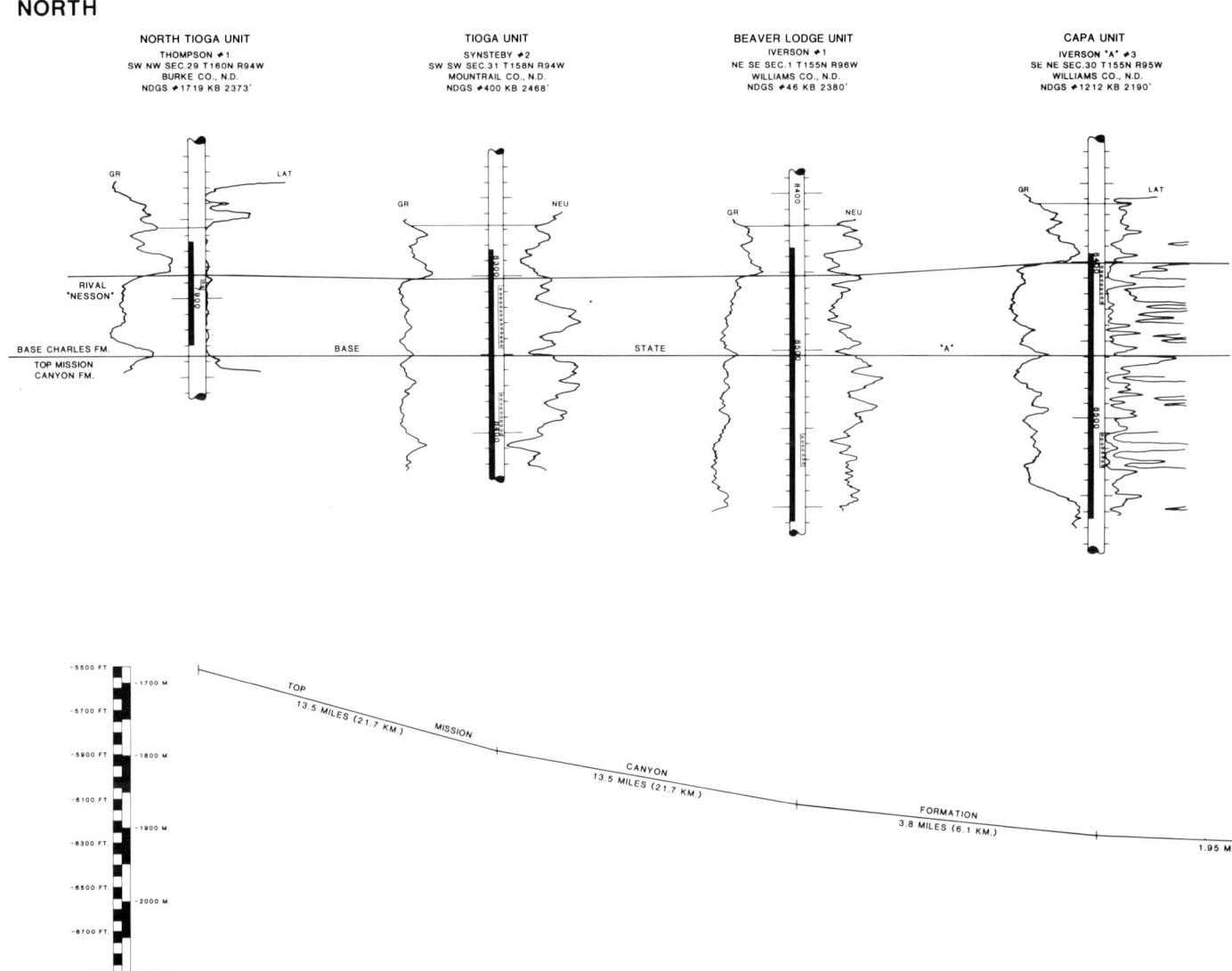

Figure 12 Fifty-four mile north to south stratigraphic log cross section through major oil fields along Nesson anticline. Intervals shown are upper portions of the Mission Canyon Formation and the Rival ("Nesson") subinterval. Stratigraphic hang line is the base of the State A marker. Each wellbore shows perforated intervals as a box with circles and cored intervals as a black bar. Beneath each log is the present-day structural position of the State A marker.

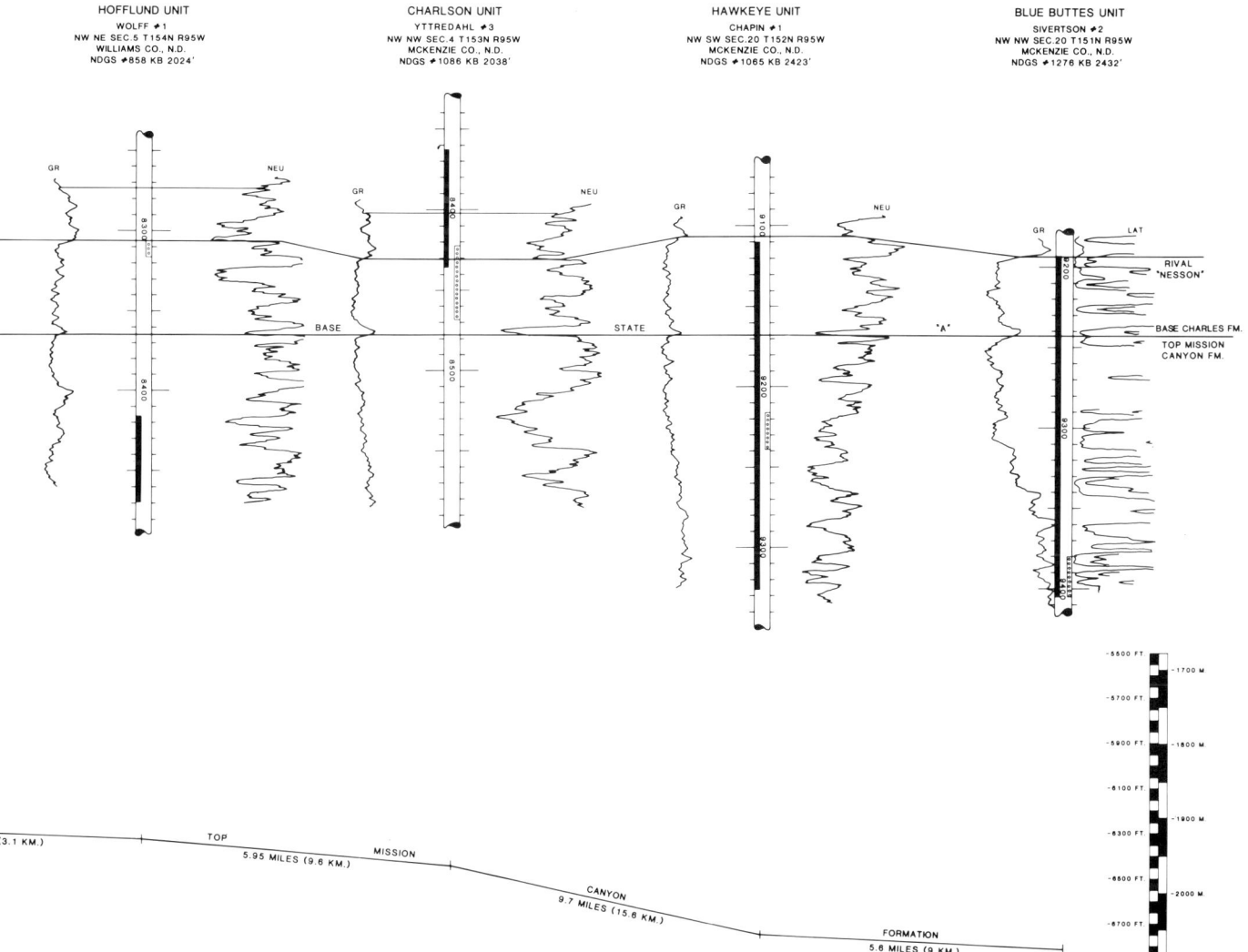

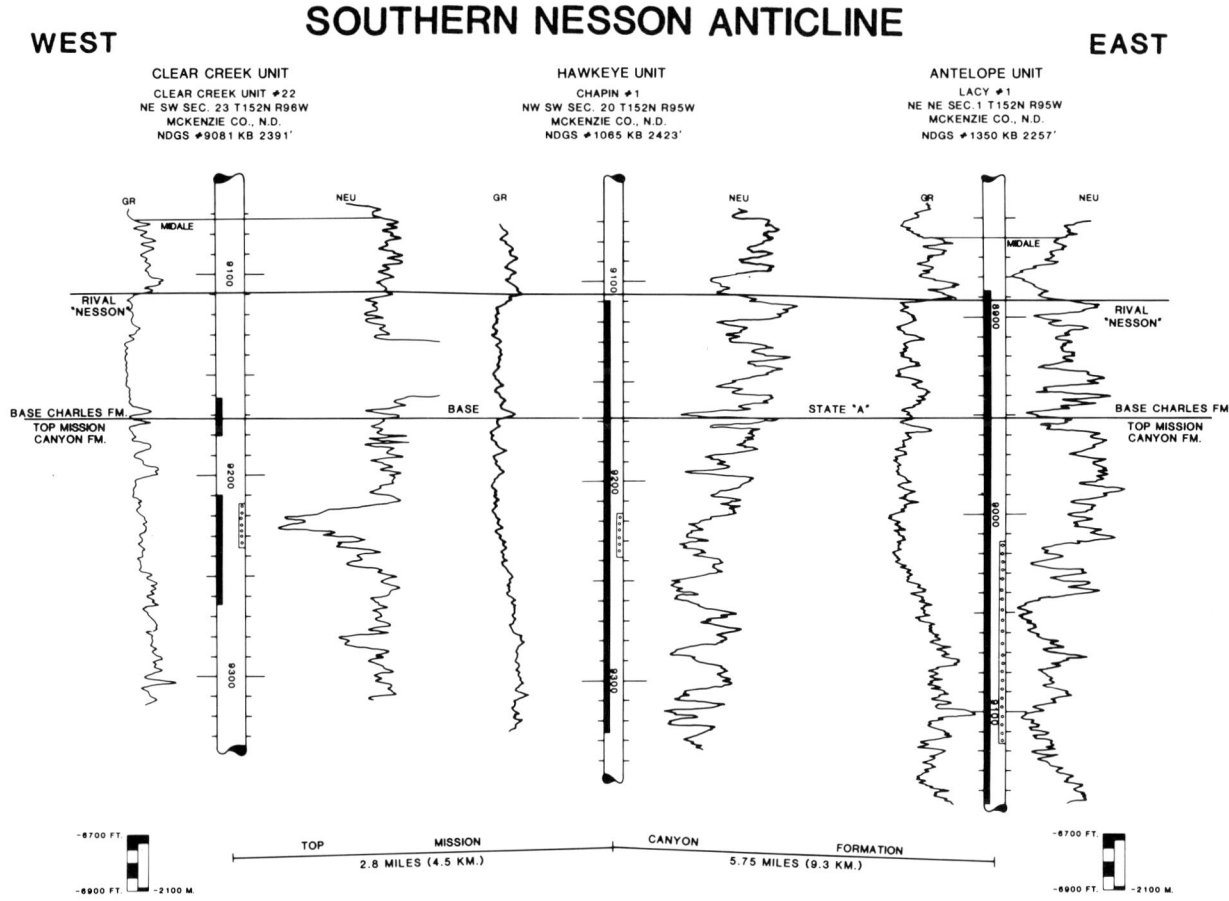

Figure 13 Eight and one-half mile west to east stratigraphic log cross section through three oil fields along southern portions of Nesson anticline. Intervals shown are upper portions of the Mission Canyon Formation and the Rival ("Nesson") subinterval. Stratigraphic hang line is the base of the State A marker. Each wellbore shows perforated intervals as a box with circles and cored intervals as a black bar. Beneath each log is the present-day structural position of the State A marker.

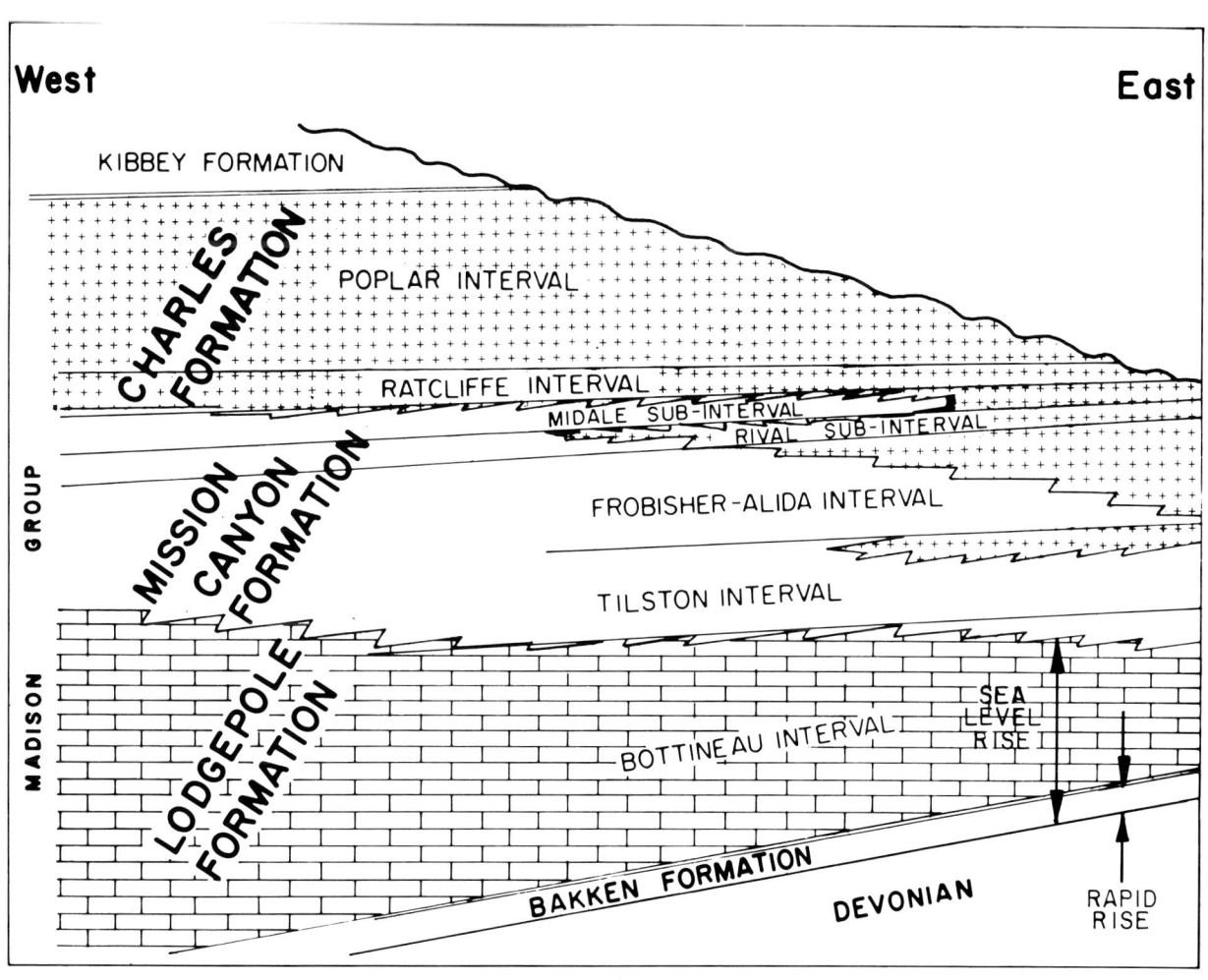

Figure 14 Cross section of the Madison Group in the Williston basin. The group is subdivided into Lodgepole, Mission Canyon and Charles Formations. These Formations are referred to as Bottineau, Tilston, Frobisher-Alida, Ratcliffe and Poplar intervals and Rival and Midale subintervals in Saskatchewan. The section progressively subcrops to the east, which was caused by erosion in the Permian-Triassic. Lodgepole, Mission Canyon and Charles facies climb progressively higher in the section toward the west at the center of the basin. Modified from Carlson and Anderson (1966).

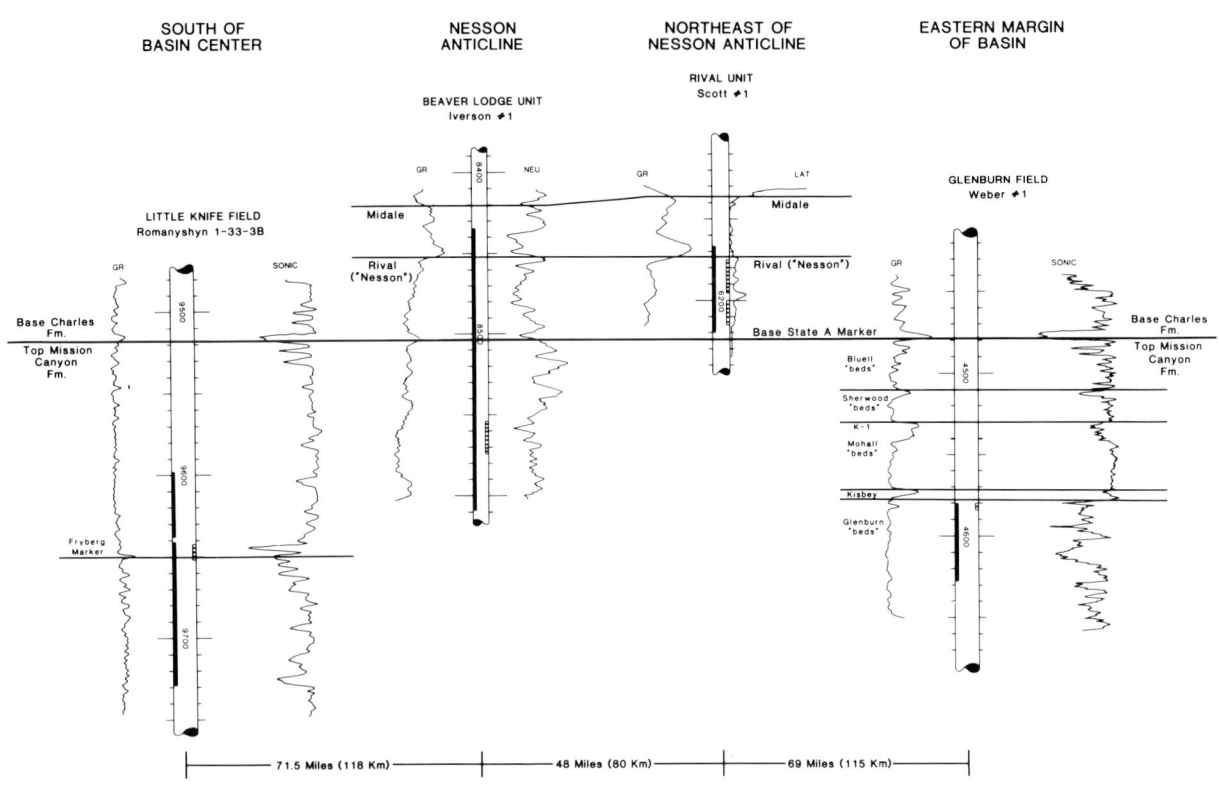

Figure 15 One hundred eighty-eight mile stratigraphic log cross section of upper Mission Canyon Formation, Rival ("Nesson") and Midale subintervals from south of Nesson anticline, through the anticline, and extending northeast and east of Nesson. Stratigraphic datum is the base of the State A marker. Wellbores show perforated intervals as boxes with circles and cored intervals are black bars.

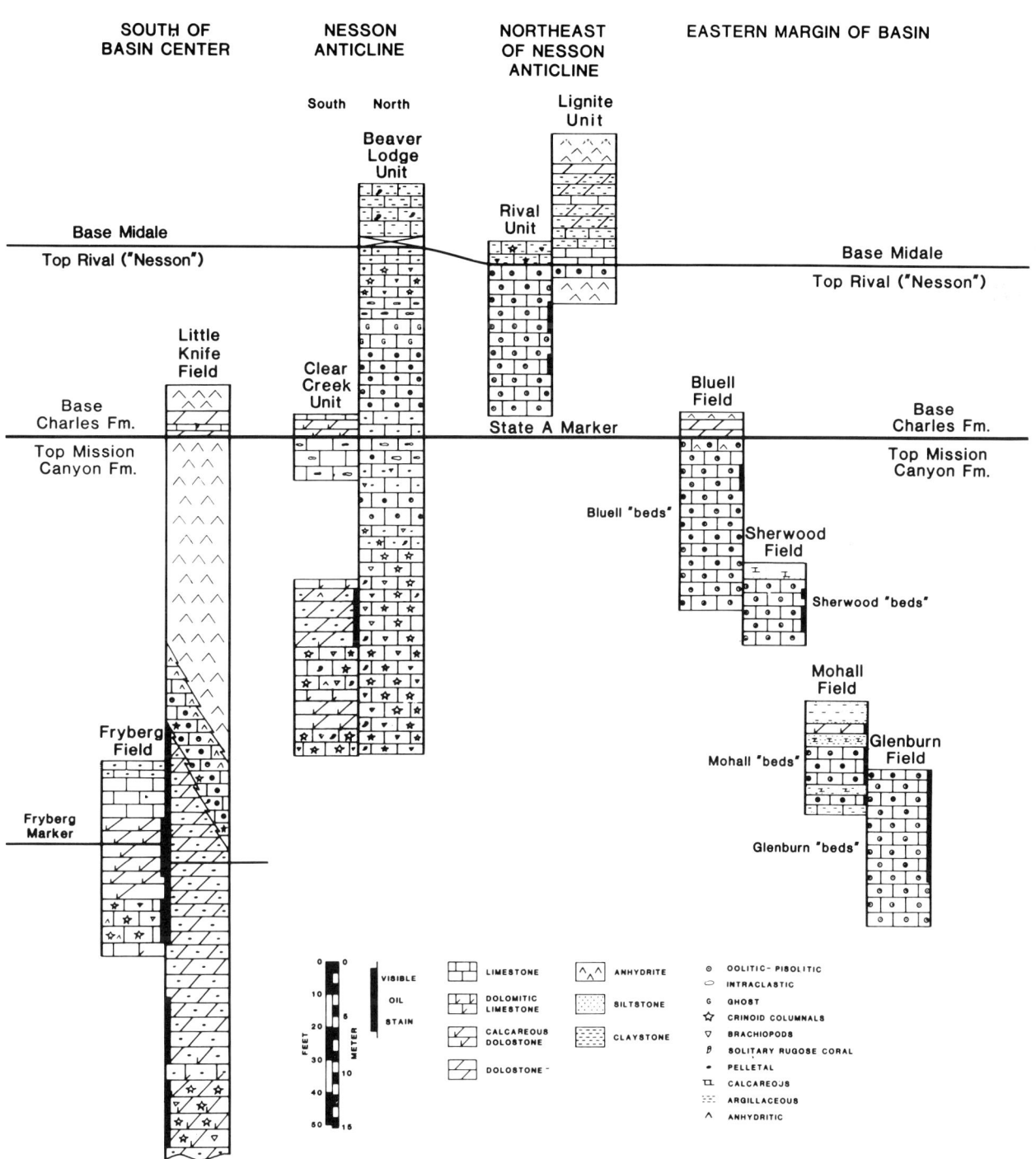

Figure 16 Lithologic cross section through the same area of Figure 15. Stratigraphic datum is the base of the State A marker. Additional nearby fields are incorporated to illustrate similarities of lithologies. Vertical black bars within lithologies are porous productive intervals.

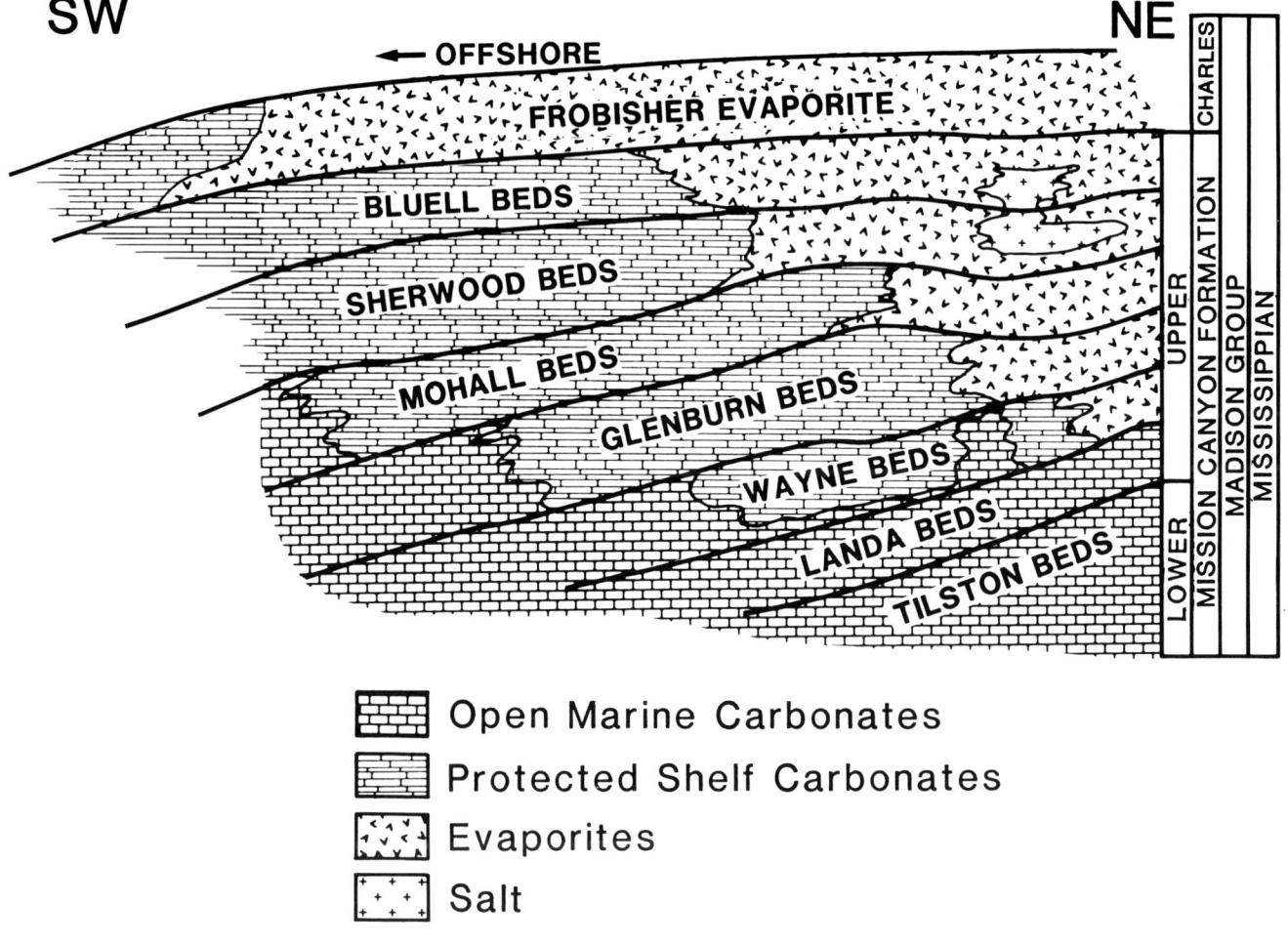

Figure 17 Southwest to northeast cross section of Mission Canyon and Charles formations. Shoreline areas are where carbonates transition into evaporites. Each interval of "beds" have prograded basinward and are segregated by abrupt short-lived transgressions. These transgressions reworked fine silisiclastics into blanket deposits that subdivide the Mission Canyon and Rival ("Nesson") subinterval. Rival ("Nesson") subinterval is equivalent to the Frobisher evaporite. Note that many individual shorelines prograded and then retrograded back some distance before coming to equilibrium with sea level rise and fall and basin subsidence. Modified from Harris et al (1966).

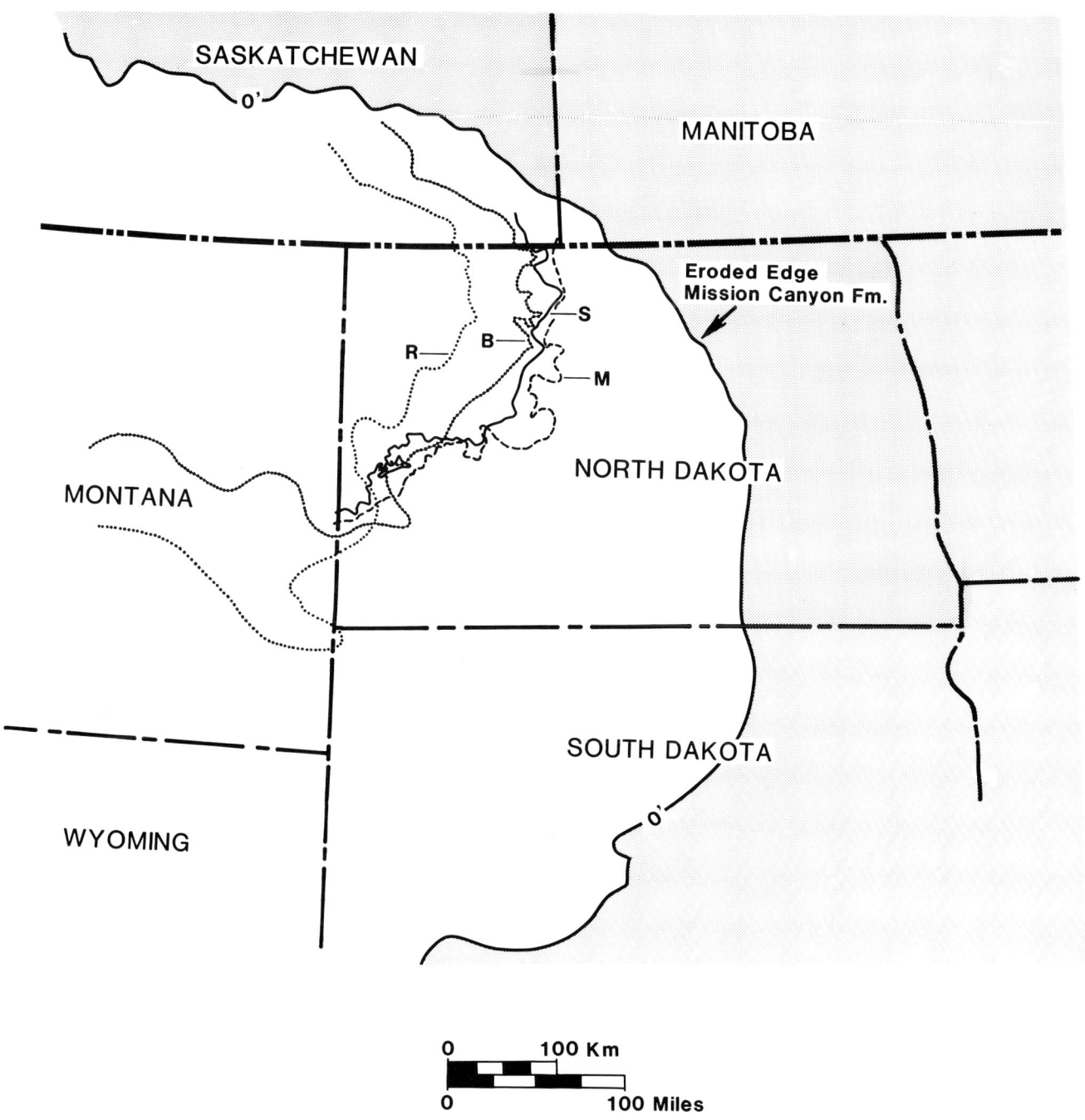

Figure 18 Map view of various Mission Canyon shorelines and the Rival ("Nesson") subinterval shoreline. Each shoreline is the position where basinward carbonates transition laterally into shoreline evaporites. M = Mohall beds, S = Sherwood beds, B = Bluell beds, and R = Rival ("Nesson") subinterval.

progradation covered the southern half of Nesson. These mappable paleoshorelines are picked where beds of shoreline carbonate rocks transition laterally into lagoonal to supratidal evaporite rocks.

Deposition of the Mission Canyon responded to these repeated shoreline progradations, which did not reach Nesson anticline, by the development of an overall shallowing upward carbonate sequence which was punctuated by open marine cyclic sedimentation. At the top of the Mission Canyon subtidal deposition gave way to deposition upon exposed barrier island/shoreline buildup complexes, with evaporites deposited behind the buildups at a distance away from Nesson.

A relatively quick transgression reworked sediments and deposited the State A marker, which was followed by progradation of the Rival ("Nesson") subinterval shoreline. This prograding shoreline built out away from earlier Glenburn through Bluell shorelines and covered the southern half of Nesson anticline. As progradation ceased the shoreline position was out of equilibrium with respect to sea level rise and fall and rates of basin subsidence. This caused an approximate 20 mile adjustment by shoreline retrogradation to the east. This re-equiliberation created a small transgressive event within the Rival ("Nesson") subinterval, which divides it into lower and upper halves. At end of Rival ("Nesson") deposition the entire area was covered by the Midale transgression. The small transgressive event within the Rival ("Nesson") subinterval and the Midale transgression are two separate events.

Depositional Model

The Mission Canyon depositional facies at Nesson anticline start with open marine skeletal wackestone/packstones (Figs. 19-26). This facies gives way upsection to skeletal wackestone deposition in a transitional open to restricted marine setting, which in turn give way to restricted marine pelletal wackestone/packstones. The thicknesses of each facies upsection varies, but the overall theme remains the same. No deep cores, revealing lowermost Mission Canyon facies were available. In most cases the transition

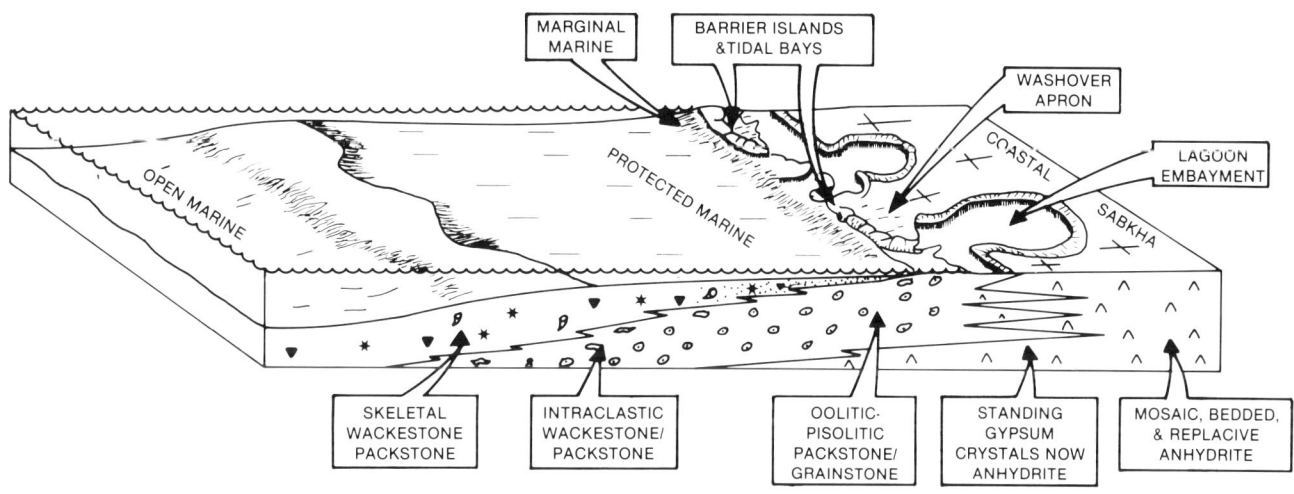

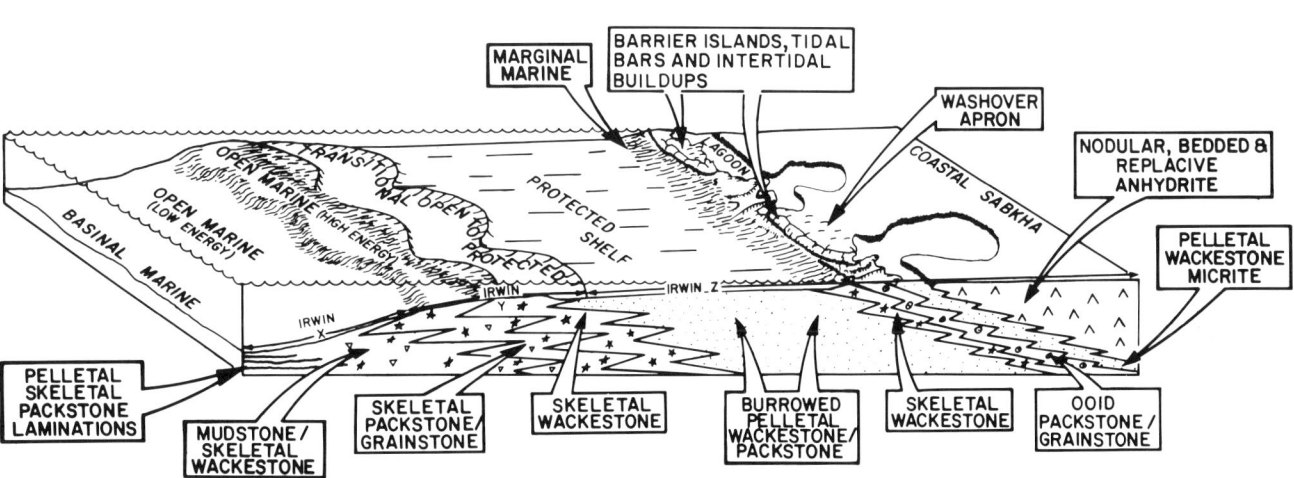

Figure 19 Top is idealized depositional setting of the Rival ("Nesson") subinterval at Nesson anticline. The offshore setting is most typical of northernmost portions of Nesson, while the barrier island/shoreline buildup complexes are typical of central portions of Nesson with southern portions of Nesson in an evaporite setting as the paleoshoreline crosses the anticline. Bottom is idealized depositional setting of the Mission Canyon Formation at Nesson anticline. This setting was at least several ten's of kilometers to 100 kilometers across. The protected shelf located behind shallowest open marine areas was very narrow to the north and much broader to the south. Barrier island/shoreline buildup complexes cap the Mission Canyon at Nesson anticlne with lagoonal, tidal flat and supratidal evaporites located at a distance south and east of the anticline. Top modified from Lindsay (1985) and bottom modified from Lindsay and Kendall (1985).

Figure 20 Mission Canyon productive facies at central protions of Nesson Anticlne. A. Amerada Synsteby No. 2 (C, SW, SW Sec. 31, T158N, R94W), Tioga Field, NDGS No. 400, 8392 feet, rock is porous oolitic-pisolitic-oncolotic, intraclastic wackestone/grainstone. B. Amerada Iverson No. 1 (C, NE, SE Sec. 1, T155N, R96W), Beaver Lodge Field, NDGS No. 46, 8563 feet, rock is partially porous coated grain, skeletal packstoe/grainstone. C. Amerada Iverson No. 3A (C, SE, NE Sec. 30, T155N, R95W), Capa Field, NDGS No. 1212, 9510 feet, rock is skeletal wackestone/packstone. D. Amerada Wolff No. 1, (C, NW, NE Sec. 5, T154N, R95W), Hofflund Field, NDGS No. 858, 8415 feet, rock is sparsely porous skeletal packstone. Scales are 1 centimeter.

Figure 21 Mission Canyon productive facies at central-southern portions of Nesson Anticline. A. Amerada Chapin No. 1 (NW, SW Sec. 20, T152N, R95W), Hawkeye Field, NDGS No. 1065, 9225 feet, rock is porous, bioturbated, intraclastic, skeletal wackestone/packstone. B. Amerada Sivertson No. 2 (C, NW, NW Sec. 20, T151N, R95W), Blue Buttes Field, NDGS No. 1276, 9382 feet, rock is porous bioturbated skeletal wackestone, with anhydrite cement infilling some burrow traces. C. Amerada Lacey No. 1 (NE, NE Sec. 1, T152N, R95W), Antelope Field, NDGS No. 1350, 9060 feet, rock is porous dolomitized, bioturbated, pelletal wackestone/packstone. D. Amerada Lacey No. 1, 9087 feet, rock is porous, bioturbated, skeletal, pelletal wackestone/packstone. Scales are 1 centimeter.

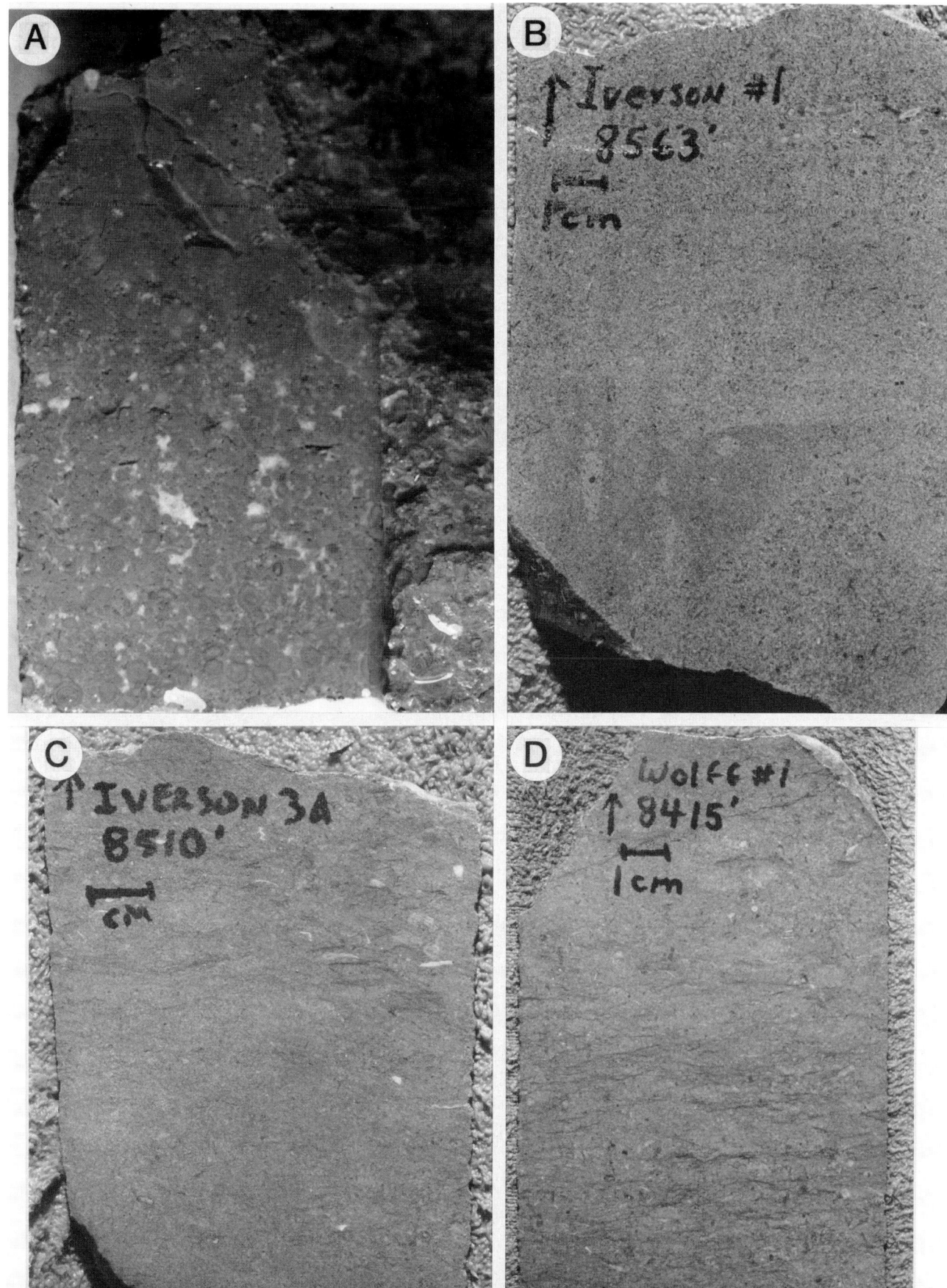

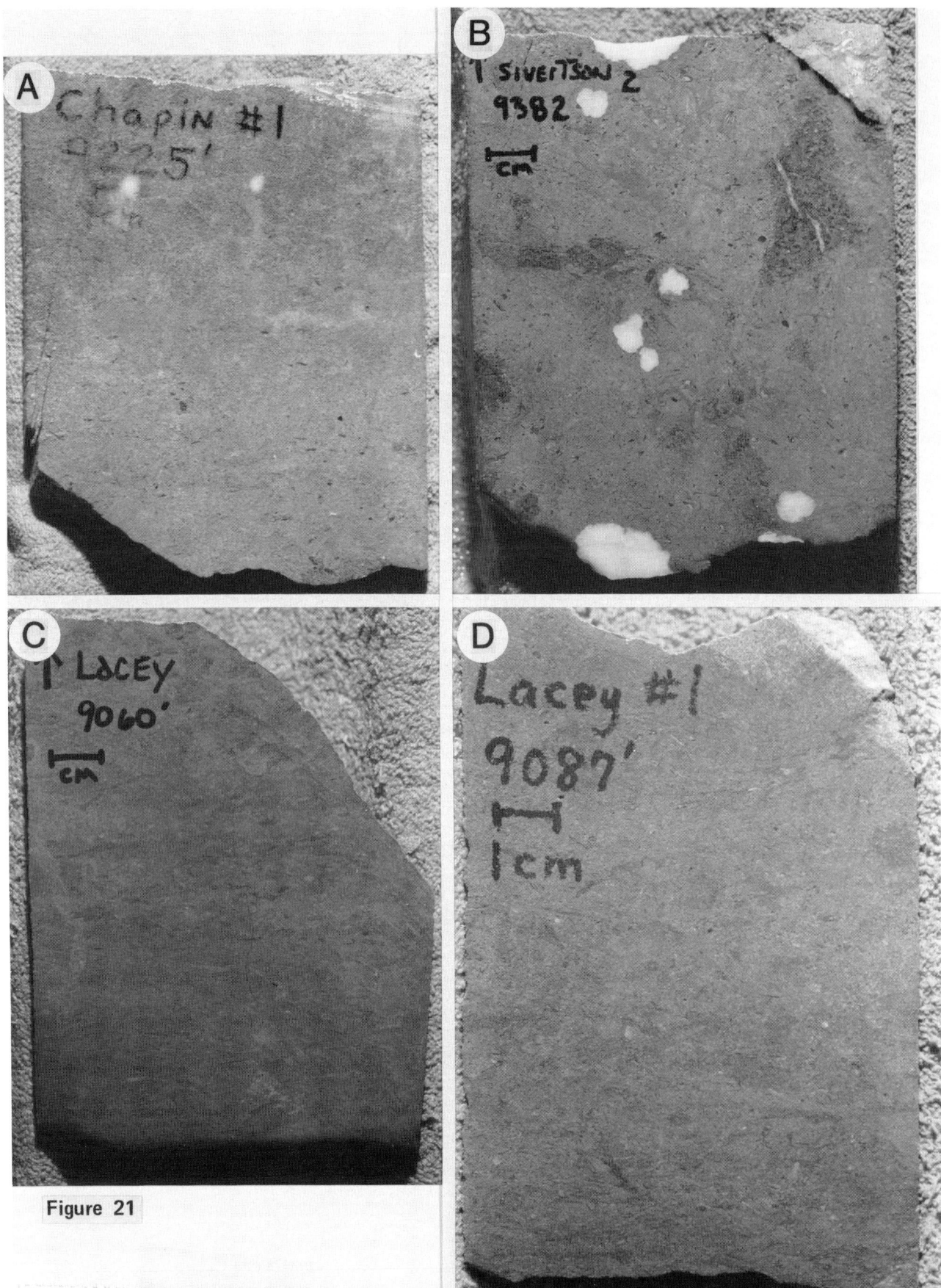

Figure 21

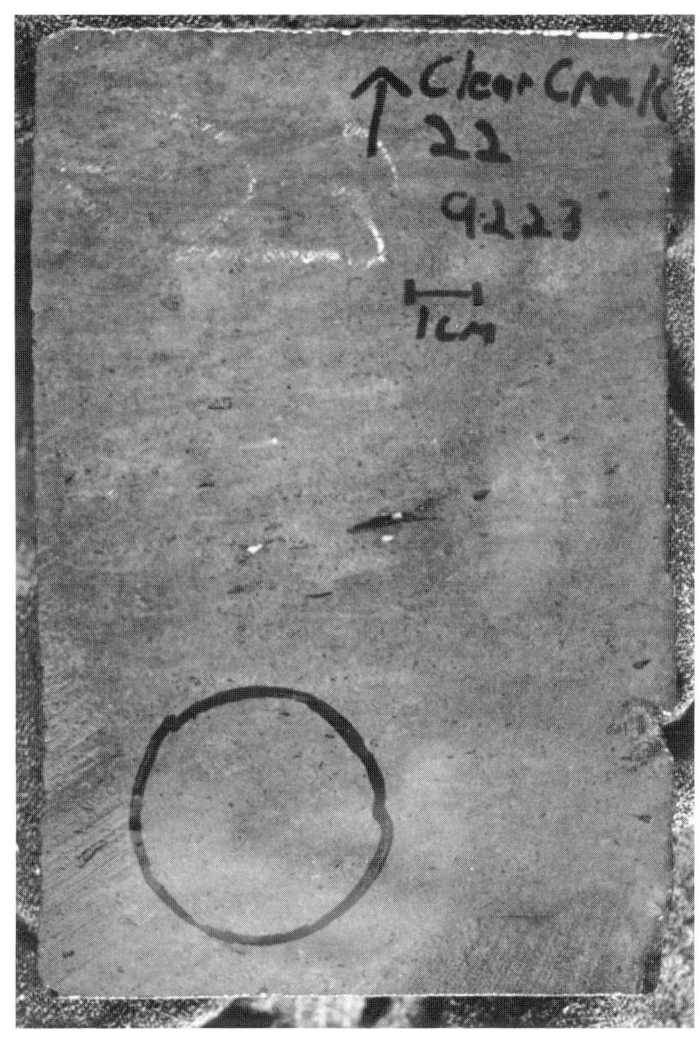

Figure 22 Mission Canyon productive facies at Clear Creek Field, in central-southern portion of Nesson anticline. Getty Clear Creek Unit No. 22 (NE, SW Sec. 23, T152N, R96W), NDGS No. 9801, 9223 feet, rock is porous, dolomitized, bioturbated, pelletal wackestone. Circle is thin section location. Scale is 1 centimeter.

Figure 23 Photomicrographs of porous, productive Mission Canyon Formation at Tioga, Capa and Beaver Lodge Fields.

A. Porous, productive Mission Canyon Formation, 32 feet beneath the State A marker, from the Amerada Synsteby No. 2 in Tioga Field (C, SW, SW Sec. 31, T158N, R94W), NDGS No. 400, 8392 feet, composed of oolitic-pisolitic-oncolitic, intraclastic packstone/grainstone. Note the large microstalactite at top. Porosity is interparticle and moldic. Oversized pores were fenestral pores that were filled by anhydrite and replaced part of the matrix. The anhydrite was partially leached with resulting moldic porosity mostly filled late diagenetic saddle dolomite. Completely stained by Alizarin Red S. Plane light. Bar scale is 2.5 mm.

B. Porous, productive Mission Canyon Formation, 53 feet beneath the State A marker, from the Amerada Iverson No. 3A in Capa Field (C, SE, NE Sec. 30, T155N, R95W), NDGS No. 1212, 8510 feet, composed of skeletal packstone. Porosity is intraparticle in skeletal detritus. Compaction has created a network of subtle microstylolites which run around many of the skeletal particles in the field of view. Completely stained by Alizarin Red S. Plane light. Bar scale is 500 microns.

C. Porous, productive Mission Canyon Formation, 53 feet beneath the State A marker, from the Amerada Iverson No. 1 in Beaver Lodge Field (C, NE, SE Sec. 1, T155, R96W), NDGS No. 46, 8563 feet, composed of coated grain, skeletal grainstone. Porosity is interparticle. Low magnification view. Completely stained by Alizarin Red S. Plane light. Bar scale is 500 microns.

D. Close-up view of the center of Figure 23C, illustrating micritized coated grains. Some particles are skeletal detritus that were coated, others are peloidal ghosts. Interparticle porosity is between the particles where cement has most likely been leached away. Plane light. Bar scale is 500 microns.

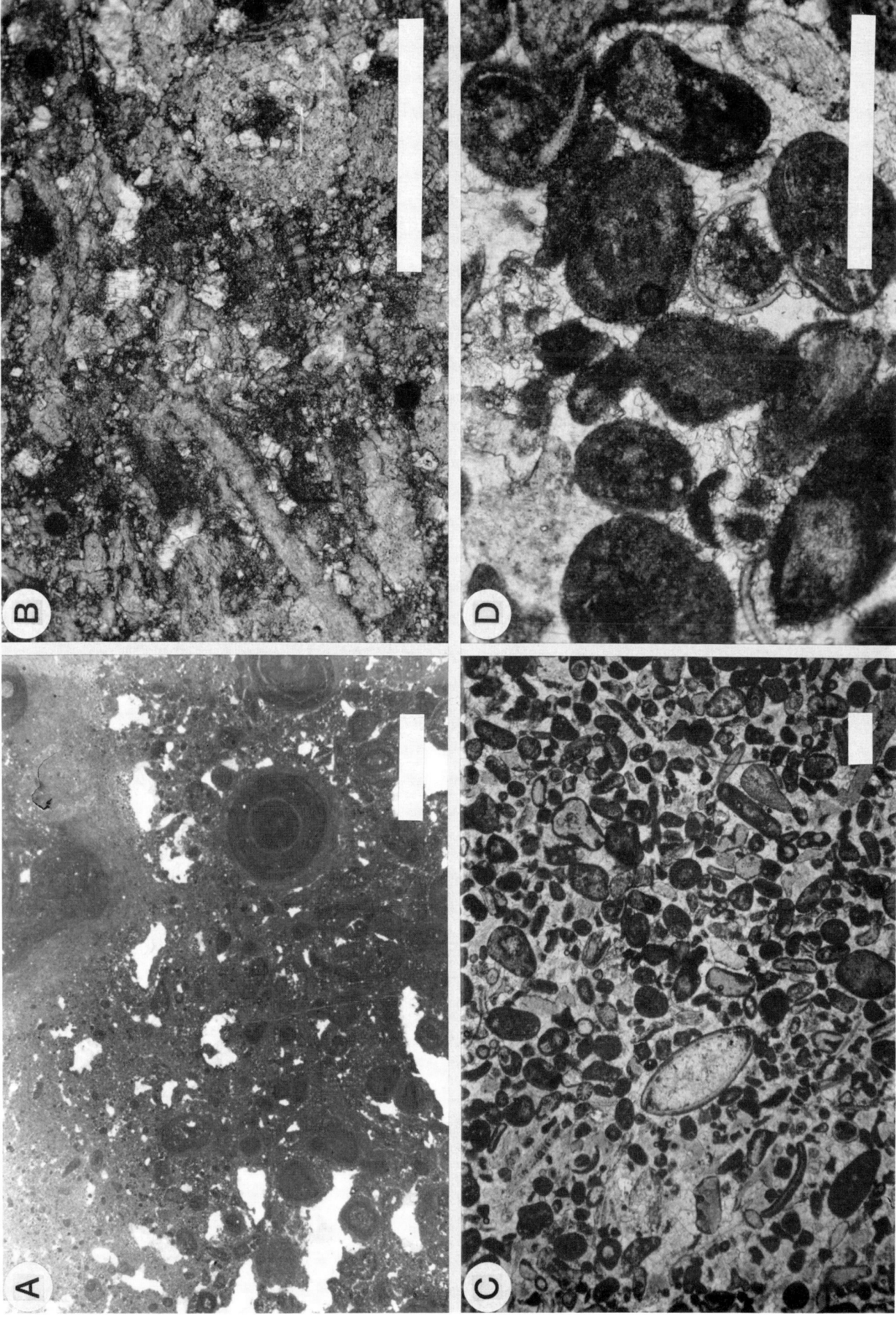

Figure 24. Photomicrographs of porous, productive Mission Canyon Formation at Hofflund, Hawkeye and Blue Buttes Fields.

A. Porous, productive Mission Canyon Formation, 54 feet beneath the State A marker, from the Amerada Wolff No. 1 in Hofflund Field (C, NW, NE Sec. 5, T154N, R95W), NDGS No. 858, 8415 feet, composed of pelletal, skeletal packstone. Porosity is interparticle and intraparticle. Completely stained with Alizarin Red S. Plane light. Bar scale is 500 microns.

B. Porous, productive, slightly dolomitized Mission Canyon Formation, 55 feet beneath the State A marker, from the Amerada Chapin No. 1 in Hawkeye Field (NW, SW Sec. 20, T152N, R95W), NDGS No. 1065, 9225 feet, composed of intraclastic, skeletal packstone. Porosity is interparticle and intraparticle. Completely stained with Alizarin Red S. Plane light. Bar scale is 500 microns.

C. Porous, productive, partially dolomitized Mission Canyon Formation, approximately 130 to 140 feet beneath the State A marker, from the Amerada Sivertson No. 2 in Blue Buttes Field (C, NW, NW Sec. 20, T151N, R95W), NDGS No.1276, 9382 feet, composed of pelletal, skeletal wackestone/packstone. Anhydrite has partially replaced some skeletal particles. Porosity is interparticle, intraparticle and intercrystalline. Completely stained by Alizarin Red S. Plane light. Bar scale is 500 microns.

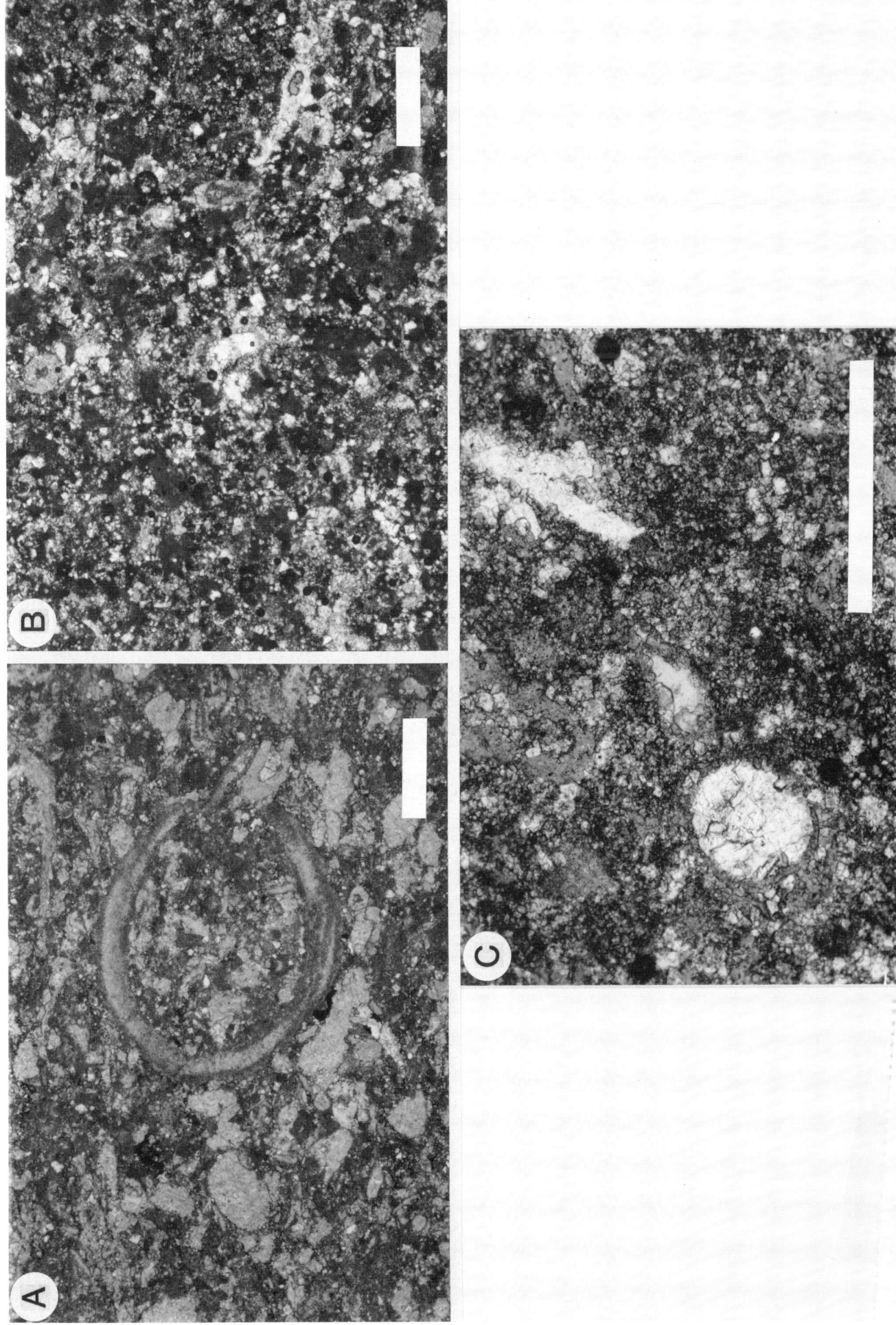

Figure 25 Photomicrographs of porous, productive Mission Canyon Formation at Antelope Field in the Lacy No. 1 (NE, NE Sec. 1, T152N, R95W), NDGS No. 1350.

A. Dolomitized, porous, productive Mission Canyon Formation composed originally of lightly skeletal, pelletal wackestone/packstone. This sample is from 9060 feet. Porosity is intercrystalline between finely crystalline euhedral dolomite crystals. Completely stained by Alizarin Red S. Plane light. Bar scale is 500 microns.

B. Porous, productive, slightly dolomitized Mission Canyon Formation composed of pelletal, skeletal packstone. This sample is from 9087 feet. Porosity is interparticle and intraparticle. Completely stained by Alizarin Red S. Plane light. Bar scale is 500 microns.

C. Closer view of the same sample. Here a few fragments of crinoid columnals have syntaxial rim cements encasing them. Porosity is between skeletal and smaller pelletal particles, as well as a few dolomite crystals (light areas). Completely stained by Alizarin Red S. Plane light. Bar scale is 500 microns.

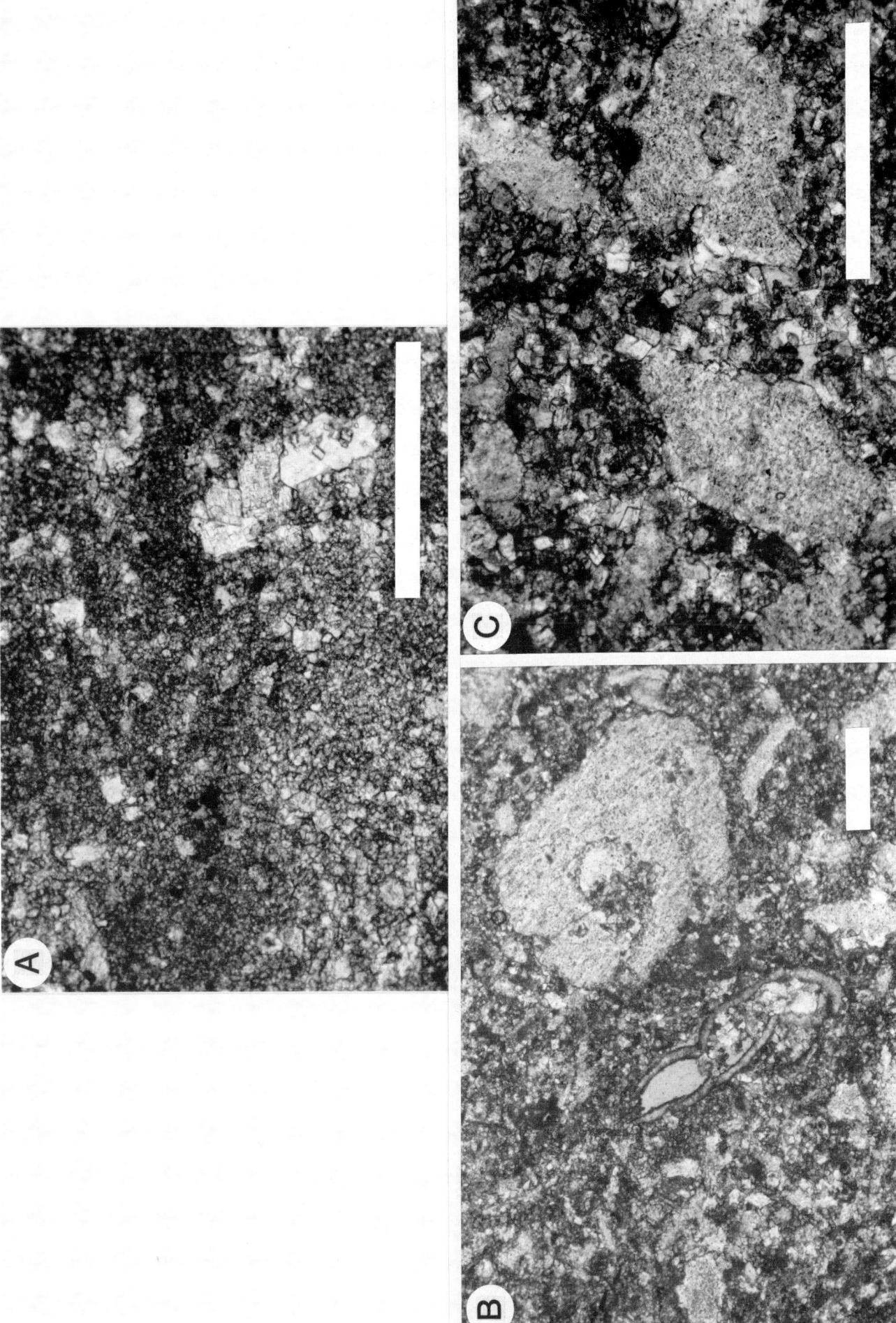

Figure 26 Photomicrographs and Scanning-electron micrographs
(SEM) of porous, productive Mission Canyon Formation
at Clear Creek Field in the Getty Clear Creek
Unit No. 22 (NE, SW Sec. 23, T152N, R96W),
NDGS No. 9081, 8223 feet.

A. Porous, productive Mission Canyon Formation, 55 feet beneath the State A marker, composed of dolomitized sparsely skeletal, pelletal wackstone/packstone. Brachiopod skeletal fragments are partially leached away. Pellets and skeletal matrix have been completely dolomitized. Completely stained with Alizarin Red S. Plane light. Bar scale is 500 microns.

B. Close-up view of Figure 26a, illustrating finely crystalline euhedral dolomite crystals. Porosity is both moldic and intercrystalline. Moldic pores may have been finely broken skeletal detritus that have been leached away. Late dolomite infilling of moldic pores is commonly seen but does not occlude much porosity. Completely stained by Alizarin Red S. Plane light. Bar scale is 250 microns.

C. SEM view of a moldic pore, which is similar in size to smaller moldic pores shown in Figure 26b. Bar scale is 10 microns.

D. SEM view of a relief pore cast. This was made by injecting acid resistant epoxy into the pore system, letting it harden then dissolving the rock in diluted HCl acid. This gives a three-dimensional view of the pore system. Larger, many-sided pores are polyhedral pores, similar in shape to a complex polyhedron. Smaller sized traingular shaped pores are tetrahedral pores. Narrow sheet-like pores are interboundary-sheet pores. Bar scale is 20 microns.

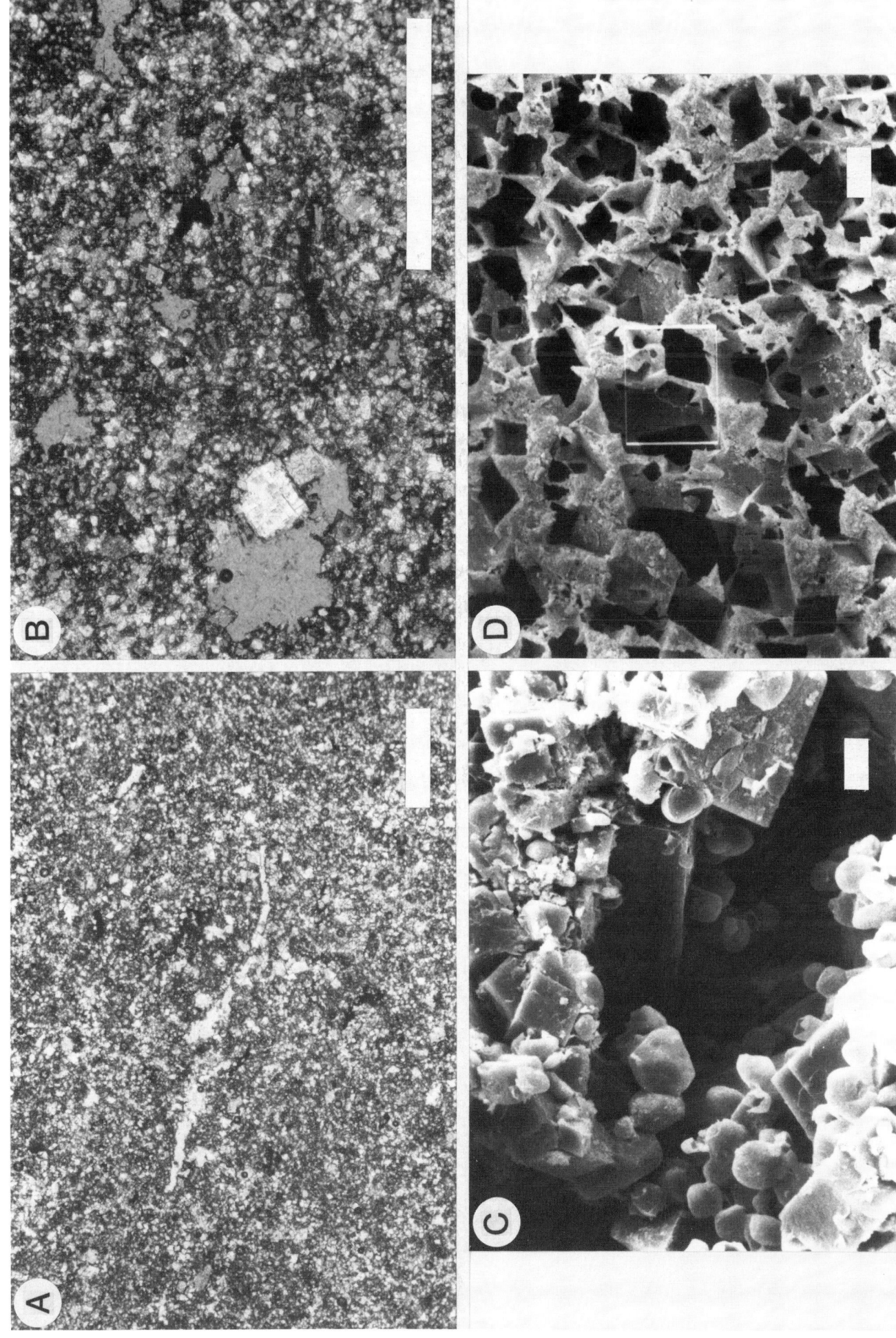

from a more open marine setting to a more restricted marine setting was near the top of the Mission Canyon with restricted marine pelletal wackestone/packstone beds of limited thickness. A bed of skeletal mudstone to skeletal packstone forms the uppermost bed in the subtidal section, which represents a fringing marginal marine setting just seaward of the paleoshoreline. At the top of the Mission Canyon, oolitic to pisolitic, intraclastic packstone/grainstones cap the sequence and represent low-lying barrier island/shoreline buildup complexes, varying in thickness, which emerged above the subtidal setting along the shoreline.

The lower Rival ("Nesson") subinterval is composed of intraclastic, oolitic to pisolitic packstone/grainstones along northern and central portions of Nesson anticline (Figs. 19 and 27-29). Occasionally, some skeletal and pelletal constituents will be intermixed with intraclasts. In southern portions of Nesson anticline lower Rival ("Nesson") subinterval beds change laterally into anhydrite beds (Frobisher Anhydrite), with occasional overlying beds of intraclastic packstone which represent thin barrier island/shoreline buildup complexes and mudstones of lagoonal affinities. This southward change from low-lying buildups of intraclastic, oolitic to pisolitic packstone/grainstone into anhydrite beds of lagoonal, tidal flat and some supratidal origins represents the Rival ("Nesson") subinterval shoreline cutting across the center of Nesson anticline. The upper part of the Rival ("Nesson") subinterval is composed of skeletal wackestone/packstones, which occasionally contain intraclasts, along northern and central portions of Nesson anticline. These represent nearshore marginal marine to offshore open to restricted marine deposits which slowly transgressed over low-lying buildups. To the south, mudstones to skeletal wackestones, sometimes containing intraclasts, were deposited by a marginal to restricted marine couplet as it transgressed over beds of anhydrite and thin buildups.

These fields that were studied are located along a 50 mile length of Nesson anticline. Because of this length it was felt best to discuss each field separately, describing field discovery, reservoir characteristics and specific reservoir rock facies.

Figure 27 Rival ("Nesson") subinterval productive facies at northern to central portions of Nesson anticline. A. Spartan Thompson No. 1 (SW, NW Sec. 29, T160N, R94W), North Tioga Field, NDGS No. 1719, 7902 feet, rock is porous skeletal, intraclastic, pelletal packstone. B. Amerada Synesteby No. 2 (C, SW, SW Sec. 31, T158N, R94W), Tioga Field, NDGS No. 400, 8317 feet, rock is porous slightly skeletal, pelletal, pisolitic-oncolitic, intraclastic wackestone/packstone. C. Amerada Iverson No. 1 (C, NE, SE Sec. 1, T155N, R96W), Beaver Lodge Field, NDGS No. 46, 8488 feet, rock is porous slightly skeletal, intraclastic, oncolitic-oolitic-pisolitic packstone. D. Amerada Iverson No. 3A (C, SE, NE Sec. 30, T155N, R95W), Capa Field, NDGS No. 1212, 8421 feet, rock is porous calcispheric, intraclastic packstone. Scales are 1 centimeter.

Figure 28 Photomicrographs of porous Rival ("Nesson") subinterval at North Tioga, Tioga and Beaver Lodge Fields.

A. Porous, productive uppermost Rival, ("Nesson") subinterval from the Spartan Thompson No. 1 in N. Tioga Field (SW, NW Sec. 29, T160N, R94W), NDGS No. 1719, composed of skeletal, intraclastic, pelletal packstone. Porosity is interparticle and intraparticle. Completely stained by Alizarin Red S. Plane light. Bar scale is 500 microns.

B. Close-up view of Figure 28a, with a micritized endothyroid foraminifera at the upper right, along with a crinoid columnal and calcisphere also at top. Compacted intraclasts and a few pellets form most of the view, with a few crushed calispheres in the center. Plane light. Bar scale is 500 microns.

C. Porous, productive upper Rival ("Nesson") subinterval from the Amerada Synsteby No. 2 in Tioga Field (C, SW, SW Sec. 31, T158N, R94W), NDGS No. 400, composed of slightly skeletal, pelletal, intraclastic grainstone. Light areas are oversized interparticle pores, which may really be moldic pores after leaching a particle or replacive cement. Large pore near the center is filled by one crystal of saddle dolomite and a smaller calcite crystal. To the right is a crinoid columnal with a slight amount of syntaxial calcite. Completely stained by Alizarin Red S. Plane light. Bar scale is 500 microns.

D. Porous, lower middle Rival ("Nesson") subinterval from the Amerada Iverson No. 1 in Beaver Lodge Field (C, NE, SE Sec. 1, T155N, R966W), NDGS No. 46, composed of intraclastic, oncolitic-oolitic-pisolitic packstone/grainstone. Oversized fenestral pores appear solution widened. These pores were most likely formed by anhydrite filling originally, smaller, fenestral pores and replacing nearby grains as well. Later leaching created these oversized moldic pores. Coarse to very coarsely crystalline calcite and saddle dolomite has partially filled the pores. Completely stained by Alizarin Red S. Plane light. Bar scale is 2.5 mm.

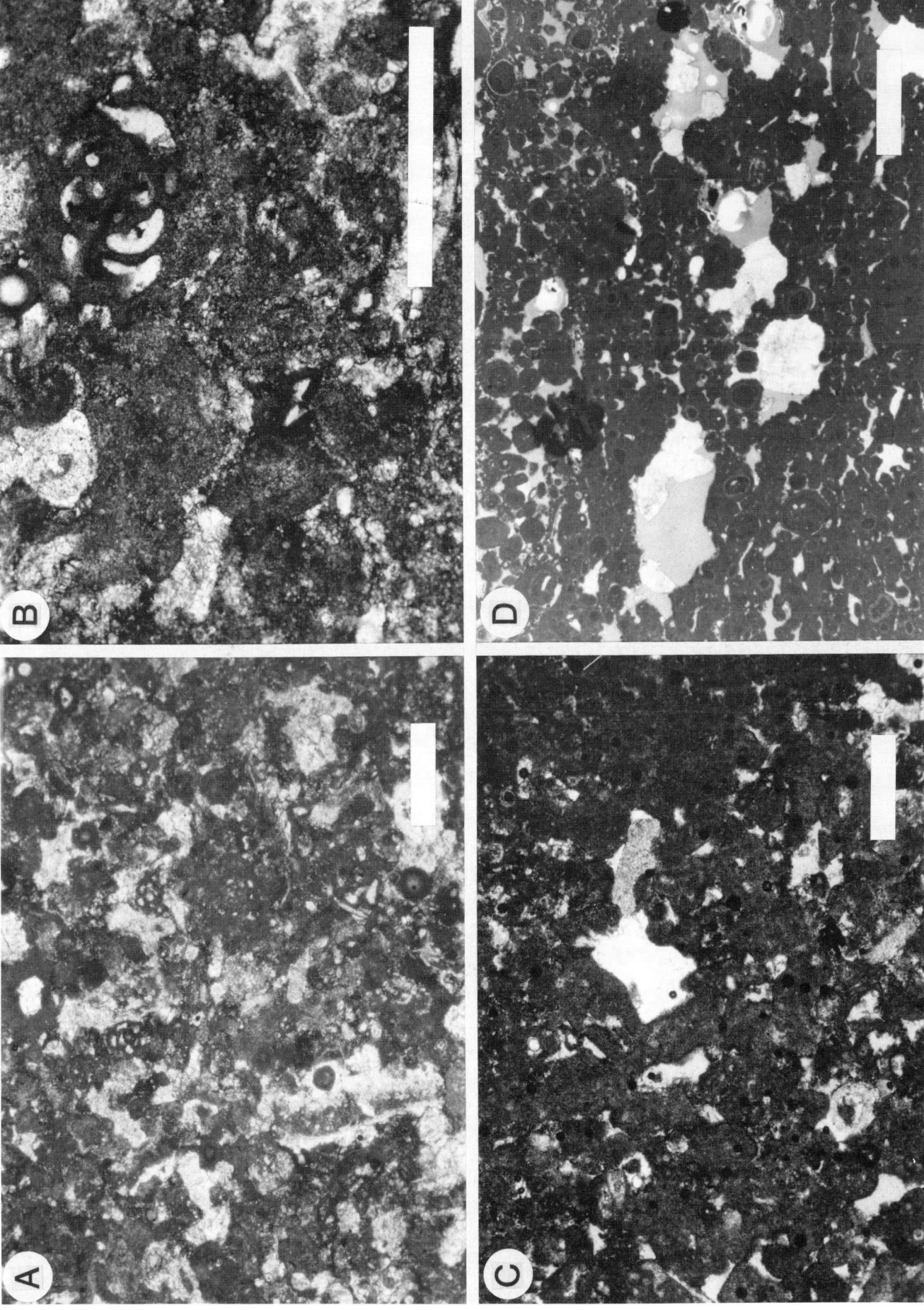

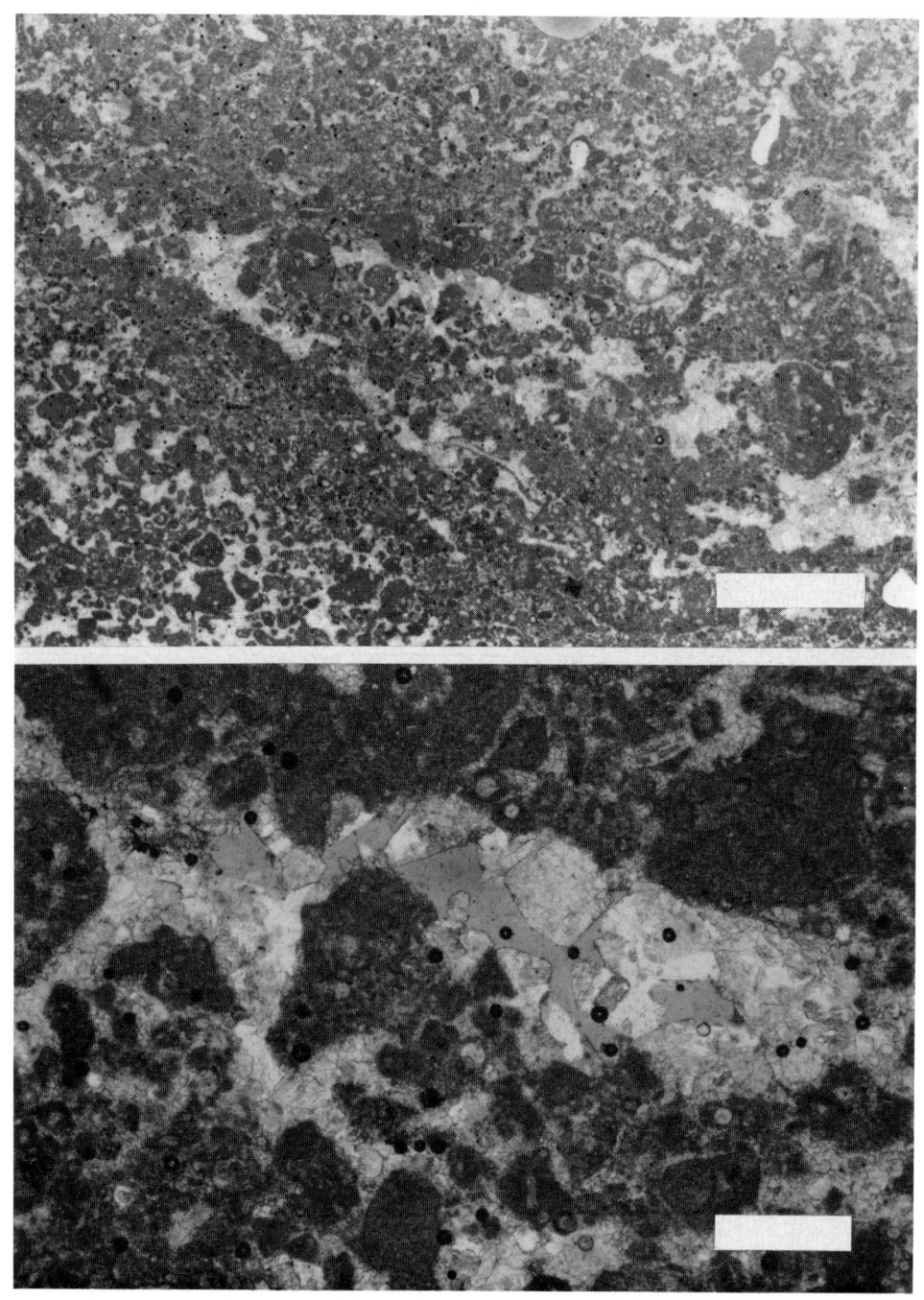

Figure 29 Photomicrograph of porous, productive Rival ("Nesson") subinterval from the Amerada Iverson 3A in Capa Field (S, SE, NE Sec. 30, T155N, R95W), NDGS No. 1212, 8421 feet. top is low magnification view and bottom is close-up view of calcispheric, intraclastic packstone/grainstone. Pore space (fenestral?) is partially filled in by very coarsely crystalline calcite cement. Unstained thin section. Plane light. Upper bar scale is 2.5 mm. Lower bar scale is 500 microns.

NORTH TIOGA FIELD

North Tioga was discovered by the Calvert Tande No. 1 (NE, NE Sec. 29, T159N, R94W), which drilled to a total depth of 8,102 feet (Tyler and George, 1962). The field extends through portions of Burke, Divide, and Williams counties. The well was completed on April 6, 1957, and had an initial potential flowing of 480 BOPD (41.8° API). Average porosity is 15 percent and average permeability is 15 millidarcies. The reservoir had an original gas-oil ratio of 1270:1. Reservoir drive mechanism is by solution gas. The trapping mechanism is a combination structural and stratigraphic trap. Structure is provided by a large anticlinal nose, plunging gently to the south. Down the length of the nose there is approximately 365 feet of drop from the north to south end of the field. North Tioga Field was unitized in April, 1962, with waterflood operations begun in December of the same year. Since its discovery North Tioga has produced 15.6 MMBO from the Madison Group (Rygh, 1987).

North Tioga produces from the Rival "Nesson" subinterval. Reservoir rocks are composed of beds of skeletal, intraclastic, pelletal packstones and skeletal wackestone/packstones (Fig. 27a and 28a,b). Rocks are dark brown to brownish-black and are medium bedded. Some beds are bioturbated and almost homogenized. Beds composed of skeletal, intraclastic and pelletal constituents transition upward into beds of skeletal detritus. These record a slow transgression which covered barrier island/shoreline buildups and lagoonal areas with offshore open-restricted marine sediments during upper Rival ("Nesson") deposition (Fig. 19). Overlying Midale beds record another transgressive event, which resulted in quiet, low energy, open marine deposits.

The Rival ("Nesson") subinterval at North Tioga is significantly different from Rival ("Nesson") subinterval in central portions of Burke County, approximatly 15 miles to the northeast (Lindsay, 1985). Reservoir rock at North Tioga contains more skeletal and pelletal constituents and fewer intraclasts, with few, if any, ooids and pisolites. Whereas the Rival ("Nesson") to the northeast is productive from intraclastic, oolitic-pisolitic grainstones deposited upon barrier island/shoreline buildup complexes. This reflects a more dominant marine setting at North Tioga. Porosity is intraparticle and may be moldic if leaching of cements has occurred in central portions of calcispheres and interstacies of skeletal fragments such as endothyrid foraminifera. Due to a loss of porosity preservation or development there is a transition from porous to tight limestone at the north end of the field.

TIOGA FIELD

Tioga was discovered by the Amerada Bakken No. 1 (SW, NW Sec. 12, T157N, R95W), which drilled to a total depth of 13,709 feet (Tyler and George, 1962). The field is at the intersection of Burke, Mountrail and Williams counties. The well was completed on April 19, 1952, with an initial potential flowing of 25 BOPH (42.4° API). Average porosity is 8 percent. The original gas-oil ratio of the reservoir was 1900:1. Reservoir drive mechanism is a water drive. The trapping mechanism is partially controlled by four-way anticlinal closure in portions of the field, but in part must also be controlled by stratigraphic trapping. North Tioga is directly north of Tioga along the same structural trend with only a few plugged and abandoned wells separating the two fields. South of Tioga is Beaver Lodge Field. These two fields are only one-half mile apart and share the same unit boundary, with one intervening plugged and abandoned well. Tioga was unitized in August, 1958, and a waterflood operation was begun in February, 1959. Since its discovery Tioga has produced 56.1 MMBO from the Madison Group (Rygh, 1987).

Production is from the Mission Canyon Formation and the Rival ("Nesson") subinterval (Frobisher-Alida interval). Mission Canyon reservoir rock is composed of oolitic-pisolitic-oncolitic intraclastic wackestone/grainstones (Figs. 20a and 23a). Reservoir rocks were deposited as part of low-lying barrier island/shoreline buildup complexes (Fig. 19). These rocks are light to medium grayish-brown. Porous beds are leached of anhydrite cements with porous intervals segregated by tight intervals. Some anhydrite cements are not leached and some moldic pores are partially filled by saddle dolomite (Fig. 23a). Moldic pores are visible to the unaided eye.

Rival ("Nesson") subinterval reservoir rock is composed of slightly skeletal, pelletal, intraclastic wackestone/packstones, with packstone textures being dominant (Fig. 27b and 28c). The base of the productive interval was deposited in a transitional open-restricted marine to restricted marine setting which gave way to restricted marine deposition (Fig. 19). Stacked on top of these beds are beds deposited in a marginal marine setting and as barrier island/shoreline buildup complexes. These rocks are light to dark brown. Portions of the section are slightly dolomitized. The pore system is composed of finely moldic and in some cases solution enlarged moldic pores. Muddy intervals were bioturbated.

BEAVER LODGE FIELD

Beaver Lodge was discovered by the Amerada Iverson No. 1 (SE, SE Sec. 16, T155N, R95W, Williams County), which drilled to a total depth of 11,955 feet (Tyler and George, 1962). Initial potential flowing was 290 BOPD from the Devonian Duperow Formation and 307 BO (17 hour test) from the Silurian Interlake Formation. In the following year, 1952, the Dilland No. 1 was the discovery well for the Madison Group. Initial potential flowing of that well was 515 BOPD (42.9° API).

Within the Madison Group, production is from the upper Mission Canyon Formation and Rival ("Nesson") and Midale subintervals of the Ratcliffe interval. The original gas-oil ratio was 1175:1. The reservoir drive mechanism is by water drive. The trap is structural, with complete anticlinal closure. Near the field's edges some well locations may have received some stratigraphic assistance. Beaver Lodge is located just south of Tioga Unit and is just north of Capa Unit, along the same structural trend. Beaver Lodge and Capa share the same unit boundary line, with one plugged and abandoned well between them. Beaver Lodge was unitized in July, 1958, with a waterflood operation started in January, 1959. Since its discovery Beaver Lodge has produced 50.5 MMBO from the Madison Group (Rygh, 1987).

Mission Canyon reservoir rocks are composed of coated grain, skeletal packstone/grainstones (Fig. 20b and 23c,d). Skeletal constituents are crinoid columnals, brachiopods, and sparse corals. Only some coated grains and rare ooids are present. Original sediments were deposited in a stable, open marine, slightly shoaling, high energy setting with fairly continuous deposition (Fig. 19). These rocks are brown and thick to massively bedded, and are only laminated in places. Interparticle porosity is not well developed. Some cement has been dolomitized.

Porous Rival ("Nesson") subinterval beds were not tested in the cored well that was studied. These beds are composed of slightly skeletal, intraclastic, oncolitic-oolitic-pisolitic packstones (Fig. 27c and 28d). Skeletal fragments were washed up onto low-lying barrier island/shoreline buildup complexes and mixed with ooids and pisolites (Fig. 19). A few carbonate crusts, suggestive of subaerial exposure, are present. Some beds contain pisolites which have been altered to structureless ghosts. Porosity is both moldic and interparticle. Cementation of pore space was by calcite, anhydrite and late saddle dolomite. Some cement has been leached, producing moldic pores (Fig. 28d). The uppermost two beds are composed of offshore shallow marine skeletal wackestone/packstones. These two beds indicate a gradual transgression occurred during deposition of the upper Rival ("Nesson") subinterval.

CAPA FIELD

Capa was discovered by the Amerada North Dakota "F" No. 1 (NW, NW Sec. 36, T155N, R96W, Williams County), which drilled to a total depth of 8,467 feet (Tyler and George, 1962). The well was completed on December 16, 1953, with an initial potential of 616 BOPD (43.1° API). The original gas-oil ratio was 1201:1. The reservoir drive mechanism is by water drive. Reservoir trapping mechanism is a combination structural and stratigraphic trap. Structural trapping is confined to a gentle, southwest plunging anticlinal nose. Stratigraphic trapping must also be present updip to the north, since there is no closure in that direction. Capa is directly south of Beaver Lodge Unit and just north of Hofflund Unit. Capa Field shares its northern and southern boundary with the Beaver Lodge and Hofflund Fields, respectively. Capa was unitized in November, 1960, with waterflood operations beginning in September, 1961. Since its discovery Capa has produced 11.5 MMBO from the Madison Group (Rygh, 1987).

Capa produces from the Rival ("Nesson") subinterval, with some production from the Midale subinterval. Some tests of the Mission Canyon were also conducted. The Mission Canyon is composed of skeletal wackestone/packstones (Fig. 20d and 23b). These rocks were deposited in a shallow open marine setting just beneath wave base (Fig. 19). Skeletal constituents are crinoid columnals and small brachiopod fragments. The rocks are brown and are medium to thick-bedded. Bioturbation is common with mostly horizontal burrow traces.

The Rival ("Nesson") subinterval is composed of slightly intraclastic, oolitic to pisolitic packstone/grainstones (Fig. 27d and 29). These rocks were deposited upon a barrier island/shoreline buildup complex (Fig. 19). Sparse skeletal constituents include gastropods and calcispheres. The rock is gray to brown and is thin to medium-bedded. Porosity is fenestral and moldic (Fig. 29). Upper portions of the Rival ("Nesson") subinterval are made up of intraclastic, skeletal wackestone/packstones. Slow transgression during upper Rival ("Nesson") deposition placed offshore to nearshore shallow marine beds over the barrier island/shoreline buildup facies. These uppermost beds are also porous and productive.

HOFFLUND FIELD

Hofflund was discovered by the Amerada North Dakota "E" No. 1 (NW, NW Sec. 16, T154N, R95W, Williams County), which drilled to a total depth of 8928 feet (Tyler and George, 1962). The well was completed on October 24, 1952 with an initial potential of 50 BOPD (42.8° API) and 33 BWPD. Average porosity is 8.5 percent. The reservoir had an original gas-oil ratio of 1100:1. The reservoir drive mechanism is a combination of solution gas and a partial water drive. Reservoir trapping mechanism is a combination of a southward plunging anticlinal nose with no updip closure, which indicates some stratigraphic trapping is assisting. From the top of the nose to the lowest productive limit there is a structural drop of approximately 135 feet. Hofflund was unitized in July, 1967, with a waterflood underway in November, 1967. Since its discovery Hofflund has produced 2.8 MMBO from the Madison Group (Rygh, 1987).

Hofflund is productive from the Mission Canyon and is partially productive from the Rival ("Nesson") subinterval (Frobisher-Alida and Ratcliffe intervals). The well that was studied was completed in the Rival ("Nesson") Subinterval but was cored in the Mission Canyon.

Porous Mission Canyon beds are composed of crinoid and brachiopod bearing skeletal packstones (Figs. 20d and 24a). These beds were deposited as part of a carbonate cycle in a transitional open to restricted marine setting (Fig. 19). Some chert has replaced a few skeletal particles, but is sparse, as is some anhydrite and dolomite. The first porous bed is located 55 feet beneath the State A marker. Porosity is not well developed and is only fair, composed of interparticle porosity and to a lesser extent intraparticle porosity.

HAWKEYE FIELD

Hawkeye was originally the northern half of Blue Buttes Field. For administrative purposes Hawkeye was unitized separately (Anderson, 1983, personal communication) at a structural bottleneck, which divides Hawkeye and Blue Buttes fields. A typical well is the Amerada No. 1 Grimstad (SE, SE Sec. 19, T152N, R95W, McKenzie County), which drilled to a total depth of 9,291 feet. The well was completed on November 23, 1965, with an initial potential flowing of 244 BOPD (40° API) and 78 BWPD. Average porosity is 9 percent and permeability ranges between 1 to 15 millidarcies. Original gas-oil ratio was 1264:1. The reservoir drive mechanism is a water drive. Trapping mechanism is a combination trap within a southward plunging (one-half to two-thirds of one degree) anticlinal nose. Hawkeye was unitized in September, 1964, with a waterflood operation commencing in January, 1965. Since its discovery Hawkeye has produced 13.2 MMBO from the Madison Group (Rygh, 1987).

Hawkeye produces from the Mission Canyon Formation. Porous beds are approximately 50 to 75 feet beneath the State A marker. Reservoir rock is intraclastic, skeletal wackestone/packstones (Fig. 21a and 24b). Sediments were originally deposited in a shallow transitional setting between open and restricted marine environments (Fig. 19). Skeletal constituents are crinoid columnal fragments and a few rugose corals. These subtidal rocks are bioturbated. Compaction laminae are present and form swarms in places. These rocks are brown and are medium to thick-bedded. Porosity is interparticle and is intraparticle where leaching of cement and particles has occurred.

BLUE BUTTES FIELD

Blue Buttes was discovered by the Texaco Helle No. 1 (NE, NE Sec. 8, T151N, R95W, McKenzie County), which drilled to a total depth of 9,418 feet (Tyler and George, 1962). The well was completed on August 22, 1955, with an initial potential of 314 BOPD (40.9° API). Average porosity is 8 percent and average permeability is 3 millidarcies. The gas-oil ratio was originally 1200:1. Reservoir drive mechanism is by solution gas. The reservoir trapping mechanism is a combination trap, within a gently plunging anticlinal nose with porosity and stratigraphic influences. Blue Buttes Field was unitized in August, 1967, and a waterflood project was started in August, 1968. Since its discovery Blue Buttes has produced 29.7 MMBO from the Madison Group (Rygh, 1987).

Blue Buttes produces from the Mission Canyon formation, from porous intervals approximately 135 to 160 feet beneath the State A marker. Reservoir rocks are interbedded skeletal wackestone/packstones and dolomitized mudstones to skeletal mudstones (Figs. 21b and 24c). Sediments were originally deposited in a low energy open marine setting which changed shoreward into a restricted marine setting (Fig. 19). Dolomitized mudstone and skeletal mudstones are more porous and permeable, while skeletal wackestone/packstones are less porous. Skeletal constituents are crinoid detritus, brachiopods and a few rugose corals. Original sediments were bioturbated, in places being completely churned. Compacted burrows form laminae in some intervals. In places fingernail to fist-sized nodules of anhydrite are common. These rocks are generally dark brown and are medium bedded. Porosity is intercrystalline and moldic. Moldic pores formed after leaching of intraparticle areas.

ANTELOPE FIELD

Antelope was discovered by the Amerada Lacey-Norby No. 1 (SE, SE Sec. 1, T152N, R95W, McKenzie County), which drilled to 10,277 feet (Tyler and George, 1962). The well was completed on December 18, 1953, with an initial potential flowing of 520 BOPD (40° API). The original gas-oil ratio of the reservoir was 700:1. The reservoir drive mechanism is solution gas. The trapping mechanism is structural within the Antelope anticline, which trends northwest to southeast. Antelope produces from the Silurian Interlake Formation, Devonian Duperow Formation, Devonian Three Forks Formation ("Sanish Sand"), and the Mississippian Mission Canyon Formation (Frobisher-Alida and Ratcliffe intervals). Since its discovery Antelope has produced 15.3 MMBO from the Madison Group (Rygh, 1987).

Mission Canyon reservoir rocks are composed of interbedded limestones and dolostones which are approximately 65 to 165 feet beneath the base of the State A marker bed (Fig. 21c, d). Original sediments were deposited in a shallow transitional open to restricted marine setting which gave way upsection to a marginal marine setting (Fig. 19). This subtidal section then gave way to a barrier island/shoreline buildup complex of intraclastic, oolitic to pisolitic packstone/grainstones, which cap the top of the Mission Canyon. Limestones are mudstones and skeletal, pelletal wackestone/packstones (Figs. 21d and 25b, c). Skeletal constituents are mostly crinoid detritus with less common rugose corals and sparse syringaporoid corals. These rocks are mostly brown to dark brown and are medium to thick bedded. Dolostones were originally pelletal wackestone/packstones (Fig. 21c and 25a). They are dark brown and are thick bedded. Both limestones and dolostones have been bioturbated and nearly churned. Porosity is very finely intercrystalline and moldic, with some interparticle and intraparticle pores. Moldic porosity formed within completely leached, very small pieces of non-descript skeletal detritus. Intraparticle porosity formed within larger skeletal constituents, such as, endothyrid foraminifera chambers.

CLEAR CREEK FIELD

Clear Creek was discovered by the Skelley Tank No. 1 (NE, NW Sec. 34, T152N, R96W, McKenzie County), which drilled to a total depth of 9,305 feet (Tyler and George, 1962). This well was completed on October 3, 1958, with an initial potential flowing of 218 BOPD (44° API) and 17 BWPD. Average porosity is 15 percent and average permeability is 1 millidarcy. The original gas-oil ratio was 1700:1. Reservoir drive mechanism is a combination of solution gas and water drives. Clear Creek is located just west of Hawkeye Unit and trends from the northeast to the southwest. The trapping mechanism is a combination of an updip permeability restriction across a low relief anticlinal nose, which plunges gently to the south. The permeability restriction is across the north and northwest updip field limit and may represent a lateral facies change. Clear Creek was unitized in July, 1962; a waterflood began in February, 1964. Since its discovery Clear Creek has produced 8.4 MMBO from the Madison Group (Rygh, 1987).

Clear Creek produces from the Mission Canyon and to a lesser extent from the Ratcliffe interval. The Mission Canyon produces from porous intervals 45 to 70 feet beneath the State A marker. Mission Canyon reservoir rocks are composed of dolomitized pelletal wackestone (Fig. 22 and 26). Deposition was in a restricted marine setting which graded up into a marginal marine setting (Fig. 19). Rocks are dark brown to black and are thick bedded. Sediments were bioturbated with many horizontal compactional laminae. Porosity is intercrystalline and moldic. Moldic porosity was generated by leaching of anhydrite, which was disseminated within the dolostone matrix (Fig. 26b). Moldic pores are 10 to over 100 microns in width. Intercrystalline pores are mostly polyhedral pores and interboundary-sheet pores, with tetrahedral pores less common (Fig. 26c, d). Polyhedral pores are 30 to 10 microns in width. Tetrahedral pores are 10 to 3 microns in width. Interboundary-sheet pores are approximately 1 micron in width.

CONCLUSIONS

Fifty-four large to small oil fields are productive along the length of the Nesson anticline. Nesson is the largest structural feature in North Dakota portions of the Williston basin. Various fields are productive from fourteen formations, which range from Cambrian to Mississippian in age. Nesson anticline is a subtle north-south surface feature. The anticline is a single fold in the north splitting into three folds to the south. Nesson appears to have had a very late Precambrian ancestry, with maximum structural growth occurring during the Devonian and Early Mississippian. Latest phases of structural development post-dated Cretaceous Greenhorn deposition and is associated with Laramide tectonics. Individual areas along Nesson anticline experienced episodic uplift or subsidence, with each area acting independently throughout time.

All Phanerozoic periods are present at Nesson, with oil production dominantly from lower and middle Paleozoic rocks, mainly carbonates. The stratigraphic section has been subdivided into six tectonic-eustatic cycles, with tectonic events and associated styles and patterns of deposition documented.

Large oil fields located along the crest of Nesson anticline are productive from Madison Group, upper Kaskaskia sequence. Mission Canyon deposition was in an open shallow marine setting which shallowed upsection into transitional open to restricted marine and restricted marine settings. To the north open marine conditions were slightly higher in the section, with restricted marine conditions subordinate. To the south open marine conditions were slightly deeper in the section, with a protected shelf of transitional open to restricted marine and restricted marine settings being more common upsection. Width of the restricted marine setting appears to have widened further south. Barrier island/shoreline buildup complexes cap subtidal sections with evaporite beds at a distance away from Nesson anticline. To the north Mission Canyon production is from barrier island/shoreline buildup complexes and open marine high energy settings. To the south transitional open to restricted marine and restricted marine settings are productive. Productive limestone beds are more grain rich, while dolostone beds are more mud and pellet rich. Dolomitizing brines may have been sourced downward from overlying Rival ("Nesson") subinterval evaporite beds, since Mission Canyon evaporites are located at a distance, to the east and south, away from Nesson anticline. The State A marker, a small transgressive event, overlies this carbonate sequence.

A second thin progradational event is the Rival ("Nesson") subinterval. This interval is divided into lower and upper halves due to a subtle transgression within the Rival ("Nesson") subinterval. Lower beds were deposited as barrier island/shoreline buildup complexes at the northern end of Nesson anticline. To the south this facies changes laterally into bedded anhydrite of Frobisher evaporite. The upper interval in northern and central portions of Nesson anticline was deposited in an offshore, shallow marine setting. At the southern end of Nesson the upper Rival was deposited in a marginal to restricted marine setting. Both barrier island/shoreline buildup complexes and skeletal rich shallow marine beds are productive.

REFERENCES

ANDERSON, S.B., 1974, Pre-Mesozoic Paleogeographic Map of North Dakota: North Dakota Geol. Survey Misc. Map 17.

ANDERSON, S.B., 1983, Personal Communication: North Dakota Geological Survey.

BJORLIE, P.F., 1979, The Carrington Shale Facies (Mississippian) and its Relationship to the Scallion Subinterval in Central North Dakota: North Dakota Geol. Survey Rept. Inv. 67, 46 p.

BJORLIE, P.F., and ANDERSON, S.B., 1978, Stratigraphy and Depositional Setting of the Carrington Shale Facies (Mississippian) of the Williston Basin: in The Economic Geology of the Williston Basin: Williston Basin Symposium, Montana Geol. Soc., p. 165-177.

COLLIER, A.J., 1919, The Nesson Anticline, Williams County, North Dakota: U.S. Geol. Survey Bull. 691-G, p. 211-217.

CARLSON, C.G., 1960, Stratigraphy of the Winnipeg and Deadwood Formations in North Dakota: North Dakota Geol. Survey Bull. 35, 145 p.

CARLSON, C.G., and ANDERSON, S.B., 1966, Sedimentary and Tectonic History of North Dakota Part of Williston Basin: AAPG Bull., v. 49, p. 1833-1846.

CARROLL, W.K., 1979, Depositoinal Environments and Paragenetic Porosity Controls, Upper Red River Formation, North Dakota: North Dakota Geol. Survey Rept. Inv. 66, 51 p.

CLEMENT, J.H., 1987, Cedar Creek: A Significant Paleotectonic Feature of the Williston Basin: in Peterson, J.A., Kent, D.A., Anderson, S.B., Pilatzke, R.H., and Longman, M.W. (eds.), Williston Basin: anatomy of a Cratonic Oil Province: Rocky Mountain Assoc. Geologists, p. 323-336.

DOVE, L.P., 1922, The Geology and Structure of the East Side of the Nesson Anticline: Quarterly Journal of the University of North Dakota, v. 12, no. 3, p. 238-249.

GERHARD, L.C., ANDERSON, S.B., CARLSON, C.G., and LE FEVER, J.A., 1982, Geological Development, Origin, and Energy Mineral Resources of Williston Basin, North Dakota: AAPG Bull. v. 66, p. 989-1020.

GERHARD, L.C., ANDERSON, S.B., and LE FEVER, J.A., 1987, Structural History of the Nesson Anticline, North Dakota: in Peterson, J.A., Kent, D.A., Anderson, S.B., Pilatzke, R.H., and Longman, M.W. (eds.), Williston Basin: Anatomy of a Cratonic Oil Province: Rocky Mountain Assoc. Geologists, p. 337-353.

HARRIS, S.M., LAND, C.B. Jr., and MCKEEVER, J.H., 1966, Relation of Mission Canyon Stratigraphy to Oil Production in North-Central North Dakota: AAPG Bull., v. 50, n. 10, p. 2269-76.

LAIRD, W.M., 1946, The Subsurface Stratigraphy of the Nesson Anticline: North Dakota Geological Survey Bull. 21, part II, p. 11-25.

LAIRD, W.M., HANSON, M., FOLSOM, C.B. JR., and ANDERSON, S.B., 1955, The Beaver Lodge and Tioga Fields, Mountrail and Williams Counties, North Dakota: AAPG Rocky Mountain Sec. 1955 Geol. Record, p. 37-54.

LAIRD, W.M., and FOLSOM, C.B. JR., 1956, North Dakota's Nesson Anticline: North Dakota Geological Survey Report of Investigation 22, 12 p.

LE FEVER, J.A., LE FEVER, R.D., and ANDERSON, S.B., 1987, Structural Evolution of the Central and Southern Portions of the Nesson Anticline, North Dakota: in Carlson, C.G., and Christopher, J.E. (eds.), Fifth International Williston Basin Symposium: Saskatchewan Geol. Soc., p. 147-156.

LINDSAY, R.F., 1985, Rival, North and South Black Slough, Foothills and Lignite Oil Fields: Their Depositional Facies, Diagenesis and Reservoir Character, Burke County, North Dakota: in Longman, M.W., Shanley, K.W., Lindsay, R.F., and Eby, D.E. (eds.), Rocky Mountain Carbonate Reservoirs - A Core Workshop: SEPM Core Workshop No. 7, p. 217-263.

LINDSAY, R.F., and KENDALL, C.G. St. C., 1985, Depositioal Facies, Diagenesis, and Reservoir Characteristics of Mississippian Cyclic Carbonates in the Mission Canyon Formation, Little Knife Field, Williston Basin, North Dakota: in Roehl, P.O. and Choquette, P.W. (eds.), Carbonate Petroleum Reservoirs: Springer-Verlag, New York, p. 177-190.

MCCABE, H.R., 1959, Mississippian Stratigraphy of Manitoba: Manitoba Dept. Mines and Nat. Resources Mines Br. Pub. 58-1, 99 p.

NEVIN, C., 1946, The Keene Dome Northeast McKenzie County, North Dakota: North Dakota Geological Survey Bull. 21, p. I, p. 1-10.

PETERSON, J.A., 1984, Stratigraphy and Sedimentary Facies of the Madison Limestone and Associated Rocks in Parts of Montana, Nebraska, North Dakota, South Dakota and Wyoming: U.S. Geol. Survey Prof. Paper 1273-A, 34 p.

RYGH, M., 1987, Personal Communication: North Dakota Geological Survey.

SLOSS, L.L., 1963, Sequences in the Cratonic Interior of North America: Geol. Soc. America Bull., v. 74, p. 93-114.

SMITH, G.E., SUMMERS, G.E. JR., WILLINGTON, D., and LEE, J.L., 1958, Mississippian Oil Reservoirs in Williston Basin: in Weeks, L.G. (ed.), Habitat Of Oil - A Symposium: AAPG, p. 149-177.

TYLER, D.C., and GEORGE, R.S., 1962, Oil and Gas Fields of North Dakota - A Symposium: North Dakota Geol. Soc., 227 p.

THE DEVONIAN SWAN HILLS FORMATION AT SWAN HILLS FIELD AND ADJACENT AREAS, CENTRAL ALBERTA, CANADA

CHRISTIAN A. VIAU
Shell Canada Ltd.
Calgary Research Centre
3655 - 36th Street N.W.,
Calgary, Alberta, Canada

ABSTRACT

The Swan Hills Field is located in the west-central Plains of Alberta, 115 miles NW of the city of Edmonton. Light oil (42 °API) and some gas have been produced since the discovery in 1957. The field produces from the Middle and Upper Devonian Swan Hills Formation at depths ranging from 7400 to 8700 feet. The limestones of the Swan Hills Formation consist of stacked stromatoporoid-rich reefal and shoal complexes. A laterally extensive platform forms the base of the Swan Hills Formation and is covered by localized buildups. The passage from the platform to the buildup at the Swan Hills Field was controlled by subtle syn-sedimentary structural disturbance. Most of the porosity in the Swan Hills Formation is primary in origin. The distribution of porosity and permeability is related to original sedimentological fabrics and patterns. The tight carbonaceous shaly limestones of the overlying Waterways Formation form the seal for this stratigraphic trap. Zoned dolomite and anhydrite cementation in the Swan Hills Field was caused by early (pre-burial) recirculation of hot, modified Devonian sea water through the extensional structural framework. The Swan Hills Field is the second largest oil field in Canada. It is the largest oil field in a carbonate reservoir. The initial volume of oil in place is estimated at 389 million m^3 (2.45 billion barrels). Recoverable reserves are estimated at 144 million m^3 (904 million barrels). To date 104 million m^3 (652 million barrels) have been produced. In 1985 tertiary recovery (ethane miscible flood) started in part of the field.

INTRODUCTION

The Swan Hills Formation, of the Beaverhill Lake Group (Fig. 1), is composed of stromatoporoid reefal and shoal carbonates of Middle and Upper Devonian age (Givetian-Frasnian). It comprises a basal, laterally extensive platform which is overlain by localized buildups in west-central Alberta and a widespread bank to the south (Fig. 2). Except for some areas where the rocks are pervasively or partially dolomitized (e.g., Rosevear, Kaybob South, Hanlan, Minehead, Blackstone, Ante Creek, Caroline) the Swan Hills Formation consists of limestone with spectacular preservation of original fabric. The Swan Hills Formation is underlain by the limestone, dolomite and anhydrite-bearing Fort Vermilion Formation, and overlain by the shaly limestones and limestones of the carbonaceous-rich Waterways Formation (Fig. 1). A list of references on the past studies of the stratigraphy, paleontology, sedimentology and diagenesis of the Swan Hills Formation is presented in Viau (1987).

Production from the Swan Hills Field and other non-dolomitized Swan Hills fields is mostly from the buildup and upper platform with minor amounts from the lower platform.

The Swan Hills Field is located 115 miles NW of the city of Edmonton in the Plains of west-central Alberta (Fig. 2). The Swan Hills Field comprises two main areas (Fig. 3): the area of the Swan Hills buildup in the south and the area to the north referred to as the House Mountain Field. The Alberta Energy Resources Conservation Board (E.R.C.B.) refers to the first area as the Beaverhill Lake A and B Pools and the House Mountain area as the Beaverhill Lake C Pool. The A and B Pools were originally defined because of production from the upper and lower parts of the formation respectively. They are now considered as one pool. Home Oil Company Ltd. is the operator of the southern area around the Swan Hills buildup which they refer to as Swan Hills Unit No. 1 (Sears, 1978). Shell Canada Ltd. is the operator for the House Mountain area.

The data presented in this paper are all from Township 67-10W5 in the area of the Swan Hills buildup where the Swan Hills Formation is fully

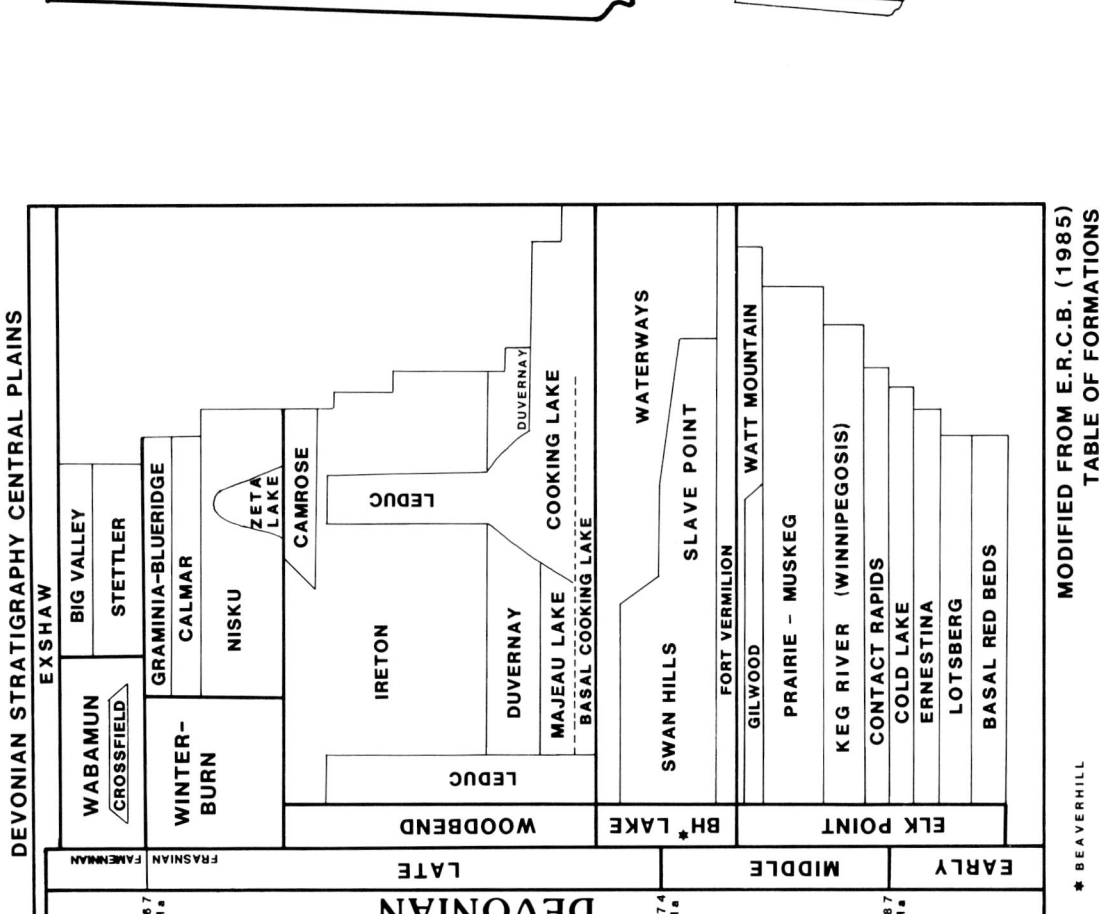

FIGURE 1 STRATIGRAPHIC CHART FOR THE CENTRAL PLAINS IN ALBERTA. AGES FROM HARLAND ET AL. (198

FIGURE 2 PALEOGEOGRAPHY OF THE SWAN HILLS FORMATION IN WEST-CENTRAL ALBERTA

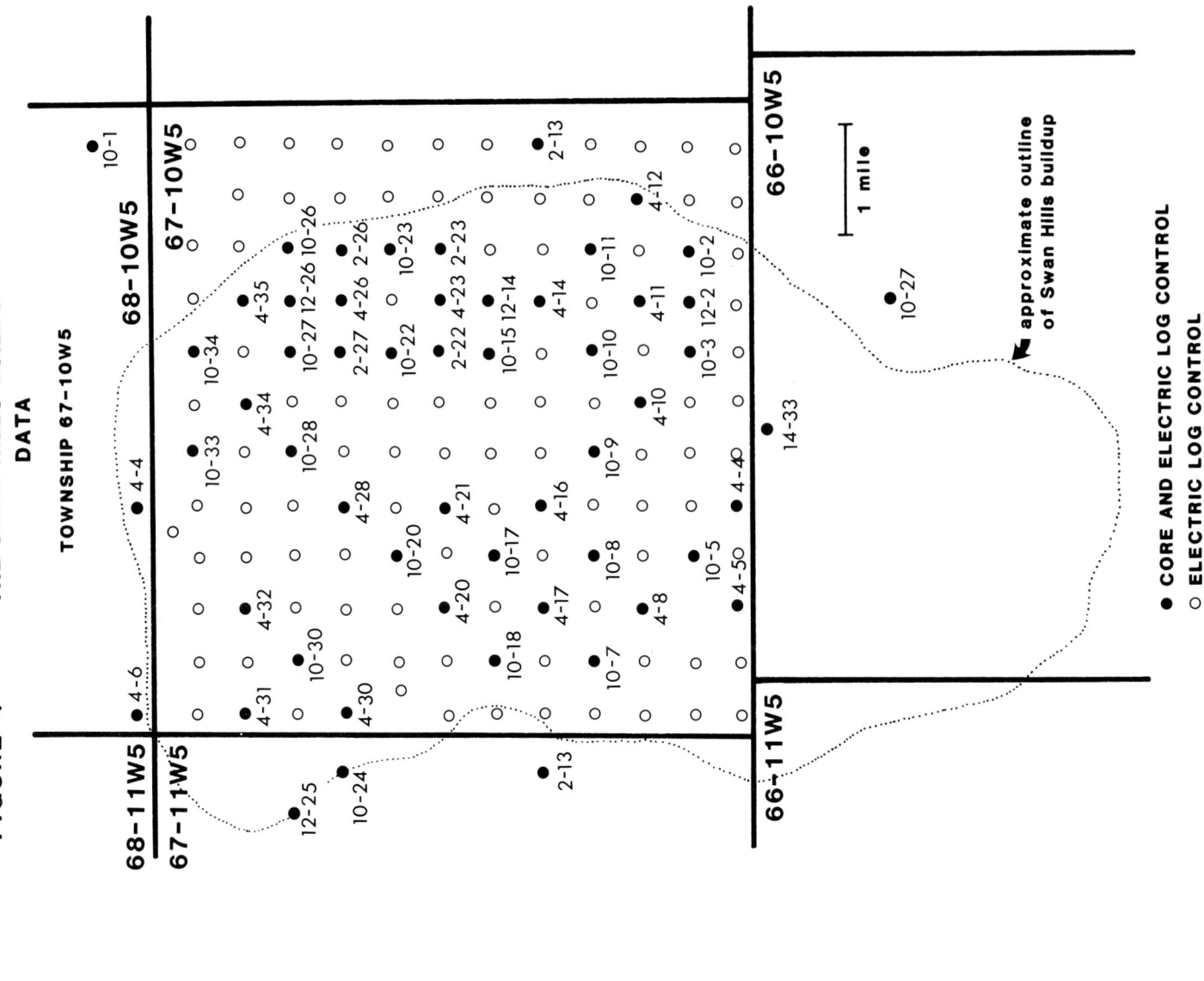

developed (Fig. 4). In the House Mountain area the buildup is not present and production is from the edge of the Swan Hills platform. The Swan Hills platform is commonly referred to as the Slave Point Formation (Figs. 1 and 5) because at least part of the platform is equivalent to that formation north of Township 70 (Craig, 1987).

As shown below (Table 2), production from the House Mountain area accounts for 18% of cumulative production to date (23% of established reserves). By contrast 82% of the cumulative production to date (and 77% of established reserves) is from the area centred at Township 67-10W5 which includes the Swan Hills buildup.

RESERVOIR PARAMETERS AND PRODUCTION HISTORY

Hemphill and others (1970, p.50-51) have summarized the history of Swan Hills reef and field discovery:

> "The Swan Hills Beaverhill Lake reef discovery was made on a large farmout block by the Home Union H.B. Virginia Hills 9-20-65-13W5M well. On January 31, 1957, a drillstem test of a 30-ft interval in the Upper Devonian Beaverhill Lake section flowed 40 °API oil to the surface. Within 30 days another successful Beaverhill Lake well (the Home et al. Regent Swan Hills 8-11-68-10W5M) was drilled 25 mi northeast of the original discovery. This was the first well of the Swan Hills Field. Paradoxically, the confirmation well for the Virginia Hills discovery was a dry hole, even though it was less than 1 mi away; and the 8-11-68-10W5M well at Swan Hills found oil in what since has proved to be one of the poorest producing areas in the entire complex. However, further drilling around Virginia Hills and Swan Hills, and the impressive number of new-field discoveries assured continued activity through the region."

Since then over 1000 wells have been drilled in the 164,115 acre area (66,415 ha) comprising the Swan Hills Field. The field extends from Township 66 to 71 and Ranges 8 to 12 west of the fifth Meridian (Fig. 3). The Swan Hills Formation ranges from a depth of 7400 to 8700 feet and dips towards the SW at a rate of 40 feet per mile.

TABLE 1
LARGEST OIL FIELDS IN CANADA (AS OF DECEMBER 1986)

FIELD / DISCOVERY YEAR		INITIAL ESTABLISHED RESERVES bbl (m^3)	INITIAL VOLUME IN PLACE bbl (m^3)	FORMATION (age)	LITHOLOGY
1- Pembina	1953	1,503,262,000 (239,000,000)	7,421,964,000 (1,180,000,000)	Cardium (Cret)	clastics
2- Swan Hills	1957	904,032,954 (143,730,000)	2,446,266,755 (388,926,000)	Swan Hills (Dev)	carbonates
3- Redwater	1948	805,094,000 (128,000,000)	1,301,988,600 (207,000,000)	Leduc (Dev)	carbonates
4- Amauligak	1984	700,000,000 (111,000,000)	?	(Tertiary)	clastics
5- Hibernia	1979	580,000,000 (92,000,000)	1,400,000,000 (222,583,000)	(Jur-Cret)	clastics
6- Bonnie-Glen	1951	532,746,000 (84,700,000)	786,225,000 (125,000,000)	Leduc (Dev)	carbonates
7- Judy Creek	1959	478,024,800 (76,600,000)	1,078,071,720 (171,400,000)	Swan Hills (Dev)	carbonates
8- Swan Hills South	1959	424,247,010 (67,450,000)	347,865,040 (134,800,000)	Swan Hills (Dev)	carbonates
Ghawar*	1948	83,000,000,000** (13,196,000,000)	?	(Jur)	carbonates
Burgan#	1938	72,000,000,000** (11,447,000,000)	?	(Cret)	clastics

* Largest carbonate oil field in the world (in Saudi Arabia)
\# Largest clastics oil field in the world (in Kuwait)
** From Tiratsoo (1984)

TABLE 2
SWAN HILLS FIELD

	AREA ha	A $10^3 m^3$	B $10^3 m^3$	C $10^3 m^3$	D $10^3 m^3$	E $10^3 m^3$	F $10^3 m^3$
POOL A & B	40666	111100	45200	65900	40100	85300	25800
POOL C	25749	32630	12450	20180	5908	18358	14272
TOTAL	66415	143730	57650	86080	46008	103658	40072

A: INITIAL ESTABLISHED RESERVES
B: PRIMARY PRODUCTION
C: ENHANCED PRODUCTION POSSIBLE
D: SECONDARY PRODUCTION TO DEC 1986
E: CUMULATIVE PRODUCTION TO DEC 1986
F: REMAINING RESERVES

The Swan Hills Field has no gas cap but some solution gas has been produced. The oil-water contact is further downdip to the SW. In the Swan Hills Unit No. 1 the oil has a gravity of 42 °API and less than 0.5% sulphur content. The average porosity and permeability in that area are 8% and 20 md respectively (Sears and Yoeman, 1976, Hemphill and Dunn, 1960). After primary depletion in the early 1960's, the Swan Hills Unit No.1 was put on waterflood by downdip line drive and patterned schemes. Horizontal hydrocarbon (ethane) miscible flooding started on September 1, 1985 at an estimated cost of 2.1 billion CDN dollars (Oilweek, Nov. 29, 1984). It is estimated that 7.9×10^6 m^3 will be recovered by this tertiary project. The House Mountain portion of the field is presently being produced by waterflood on an 80 acre spacing. The feasability and value of tertiary recovery is being evaluated.

RESERVES

The Swan Hills Field is the second largest oil field in Canada in terms of both initial volume in place and initial established reserves (Tables 1 and 2). It is the largest oil field producing from a carbonate reservoir. The Swan Hills Field is the most productive of twenty one oil fields in the Swan Hills Formation (Table 3 and Fig. 5). In fact 37% of the oil reserves in the Beaverhill Lake Group are contained in the Swan Hills Field. The field ranks eighth in terms of gas production from the Swan Hills Formation (Table 3).

The total accumulation of oil in the Beaverhill Lake Group has greater initial volume of oil in place than the Leduc, Keg River or Nisku accumulations (other major producing units in the Devonian of the Alberta Basin): 944.3×10^6 m^3, versus 803.5×10^6 m^3, 487.8×10^6 m^3 and 332.1×10^6 m^3 respectively. However, in terms of initial established reserves, the Beaverhill Lake oil accumulations rank second behind the Leduc accumulations (386.0×10^6 m^3 versus 486.2×10^6 m^3). The higher recovery efficiency for Leduc oil is a reflection of the overall beneficial effects of greater dolomitization.

The total accumulation of gas in the Beaverhill Lake Group ranks second, in terms of initial volume in place, behind the total Leduc gas accumulation:

TABLE 3
DEVONIAN SWAN HILLS AND SLAVE POINT FIELDS
ALBERTA BASIN IN CANADA
(AS OF DECEMBER 1986)

FIELD / DISCOVERY YEAR		OIL INITIAL ESTABLISHED RESERVES 10^3 m^3	OIL RANK	OIL INITIAL VOLUME IN PLACE 10^3 m^3	GAS INITIAL ESTABLISHED RESERVES 10^3 m^3	GAS RANK	GAS INITIAL VOLUME IN PLACE 10^3 m^3
# dolomite present							
#1- Hamburg	1983	------	--	------	1660	14	2305
2- Chinchaga	1973	------	--	------	1000	16	1389
#3- Cranberry	1974	------	--	------	10853	5	15142
4- Sawn Lake	1983	2733	12	12827	------	--	------
#5- Golden	1971	3700	10	3230	------	--	------
#6- Evi	1979	1655	13	5631	------	--	------
#7- Otter	1981	600	19	3000	------	--	------
8- Red Earth	1958	1476	15	16734	------	--	------
9- Loon	1966	1342	16	12782	------	--	------
#10- Slave	1982	3202	11	10843	------	--	------
#11- Seal	1974	602	18	1542	------	--	------
12- Gift	1980	1567	14	11060	------	--	------
13- Snipe Lake	1962	12410	7	31100	294	19	1835
14- Swan Hills	1957	143730	1	388926	8039	8	36601
15- (House Mountain)							
16- (Deer Mountain)							
17- Ethel Lake	1964	13	21	290	221	20	364
18- Swan Hills South	1959	57450	3	134800	6875	10	16272
19- Virginia Hills	1957	25221	5	75536	1433	15	6635
20- (Freeman Lake)							
21- Goose River	1963	8204	8	21167	487	18	2079
#22- Ante Creek	1963	4145	9	7600	974	17	2028
23- Kaybob	1957	20203	6	48970	5232	12	12126
24- (Kaybob East)							
#25- Fox Creek	1976	375	20	1500	------	--	------
#26- Kaybob South	1960	------	--	------	36400	1	104424
27- Judy Creek	1959	76600	2	171400	3803	7	26858
28- (Judy Creek West)							
29- Judy Creek South	1961	673	17	3552	------	--	------
30- Carson Creek (North)	1958	26900	4	57700	6603	11	16009
31- Carson Creek South	1957	------	--	------	3030	9	10941
#32- Rosevear	1971	------	--	------	9600	6	13190
#33- Minehead	1973	------	--	------	2500	13	7142
#34- Hanlan	1976	------	--	------	23940	3	40045
#35- Blackstone	1979	------	--	------	11000	4	18334
#36- Caroline *	1987	------	--	------	??????	2	56000*

* not E.R.C.B. data (Shell preliminary estimate)

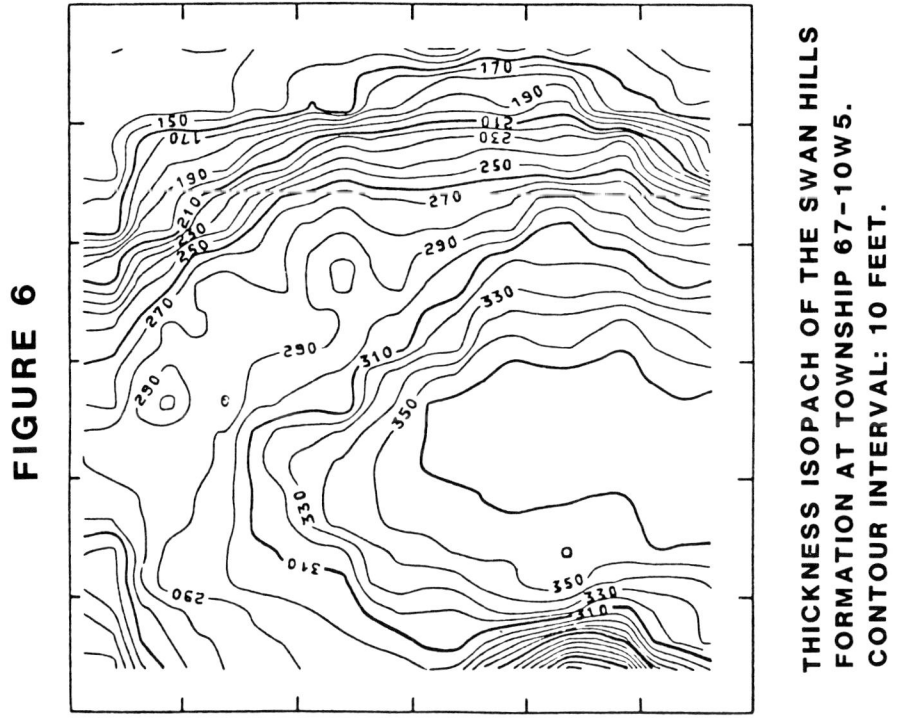

FIGURE 6

THICKNESS ISOPACH OF THE SWAN HILLS FORMATION AT TOWNSHIP 67-10W5. CONTOUR INTERVAL: 10 FEET.

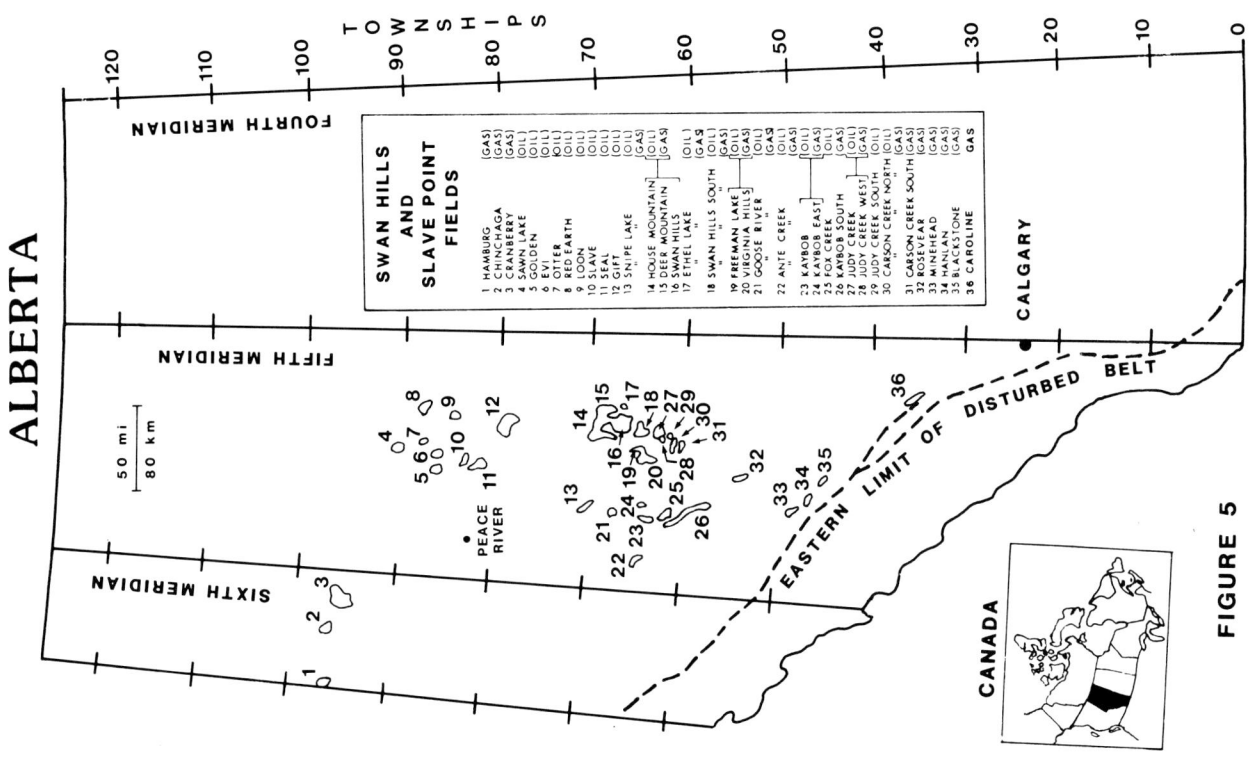

FIGURE 5

340,198 x 10^6 m^3 versus 453,150 x 10^6 m^3. In terms of initial established reserves, the values are 147,777 x 10^6 m^3 versus 239,835 x 10^6 m^3 for the Beaverhill Lake and Leduc respectively.

In the Swan Hills Field (Table 2), production from Pools A & B accounts for 79% of primary production, and 87% of secondary production. The area of the Swan Hills buildup contains 64% of remaining reserves as opposed to 36% at House Mountain.

For the total Swan Hills Field, 15% of the initial volume in place (40% of reserves) has been produced by primary recovery. Secondary recovery, to date, has produced 12% of the initial volume in place (32% of reserves). The remaining 10% of initial volume in place which is accessible (28% of reserves) will be produced mainly via tertiary recovery. In total 37% of the initial volume in place will be produced.

SEDIMENTOLOGY

The Swan Hills Formation in Township 67-10W5 varies in thickness between 36 and 116 m. The Swan Hills Formation is thicker in the southwest and thinner towards the northeast and the north (Fig. 6). Isopach distribution in the northeast shows a "plateau" outlining a ramp morphology. The nature and origin of this ramp are central to the problem of understanding the sedimentological evolution of the Swan Hills Formation.

The Swan Hills Formation contains a wide variety of biogenic components. Stromatoporoids, red algae (Solenopora), green algae (Bevocastria, Ortonella, calcispheres), corals, brachiopods, crinoids, ostracods, gastropods, and conodonts have been identified (see Viau, 1987 for references). The Swan Hills Formation contains a limited number of morphological assemblages of stromatoporoids. Kobluk (1975 and 1978) has described similar assemblages from the Cairn Formation at Miette, Alberta. These assemblages of morphological types and their association with other components have proven useful in terms of distinguishing different biofacies. This is reflected in the lithofacies classification summarized in Tables 4 and 5 for the platform and the buildup respectively. Figures 15 to

24 at the end of this paper show some of these lithofacies. Past authors have arranged these lithofacies into much broader groups to simplify their classifications. Such classifications have not resolved the diversity of rock types in the Swan Hills Formation to a degree where distribution of lithofacies is meaningful for the understanding of lateral and vertical relationships and evolution of these carbonates. The distribution of the lithofacies described in Tables 4 and 5 is discussed in more detail in Viau (1986).

Workers have long been concerned with lateral and vertical internal differentiation of the Swan Hills Formation (Edie, 1961; Thomas and Rhodes, 1961; Murray, 1966; Fischbuch, 1968; Leavitt, 1968; Jenik and Lerbekmo, 1968; Hemphill and others, 1970; Wendte and Stoakes, 1982; Viau, 1986; and others). Extensive core control provides a rich data base to study the nature and evolution of these carbonates. Although it is clear that the Swan Hills Formation comprises a series of superimposed carbonate bodies, the definition of these packages, the environmental interpretation of their different parts and the interpretation of the controls on initiation, development and termination of sedimentation are still subjects of debate.

The Swan Hills Formation at the Swan Hills Field is divided into thirteen major depositional sequences; seven of which are part of the laterally widespread basal platform which is overlain by the six depositional sequences of the more localized buildup. These sedimentary packages are defined on the basis of 1) their internal fossil composition, 2) the types and distribution of lithofacies, 3) their relative thickness and relative position within the overall sequence, 4) the abrupt superposition of deep water lithofacies onto interpreted shallowing-upward sequences, and 5) the nature and continuity or non-continuity of certain contacts between packages. These depositional sequences are interpreted as stages of carbonate growth (chronostratigraphic units). It is important to note that this sedimentological approach is fundamentally different than that used in many past studies of the Swan Hills Formation. It is based on the recognition of sedimentary packages themselves as opposed to the search for surfaces or contacts, such as green shale units, that may separate them and have been interpreted as time equivalent (e.g., Fischbuch, 1968, Murray,

TABLE 4

SUMMARY OF PLATFORM LITHOFACIES

NUMBER SYMBOL	NAME (MAJOR COMPONENTS/ MATRIX)	NC/C	NUMBER OF WELL OCCURRENCE	NUMBER OF OCCURRENCE	BASE PLATFORM OCCURRENCE	TOP PLATFORM OCCURRENCE	DISTRIBUTION	THICKNESS
1P	CORAL, THIN LAMINATED STROMATOPOROID, STACHYODES RUDSTONE AND FLOATSTONE WITH CARBONACEOUS MUDSTONE MATRIX	C	38	APPROX 40	MAJOR	ABSENT	WIDESPREAD	5'-20'
2P	AMPHIPORA RUDSTONE WITH CARBONACEOUS MUDSTONE MATRIX	C	37	> 50	MAJOR	ABSENT	WIDESPREAD	<5'-12'
3P	AMPHIPORA AND STACHYODES FLOATSTONE AND RUDSTONE WITH CARBONACEOUS MUDSTONE MATRIX	C	44	> 50	COMMON	MAJOR	WIDESPREAD	<5'-20'
4P	AMPHIPORA AND OSTRACOD RUDSTONE AND FLOATSTONE WITH CARBONACEOUS MUDSTONE MATRIX	C	17	APPROX 20	MINOR	ABSENT	IRREGULAR	5'-15'
5P	HEMISPHERICAL STROMATOPOROID, AMPHIPORA AND CORAL RUDSTONE WITH CARBONACEOUS MUDSTONE MATRIX	C	35	APPROX 40	MAJOR	RARE	WIDESPREAD	5'-12'
6P	STACHYODES AND AMPHIPORA RUDSTONE WITH CARBONACEOUS MUDSTONE MATRIX	C	15	APPROX 20	MINOR	ABSENT	IRREGULAR	<5'-12'
7P	CARBONACEOUS OSTRACOD WACKESTONE AND PACKSTONE	C	2	2	RARE	ABSENT	PERIPHERAL	5'-20'
8P	AMPHIPORA RUDSTONE AND FLOATSTONE WITH MUDSTONE MATRIX	NC	45	> 50	MINOR	MAJOR	WIDESPREAD	<5'-25'
9P	STACHYODES AND AMPHIPORA RUDSTONE AND FLOATSTONE WITH MUDSTONE MATRIX	NC	15	APPROX 20	RARE	MINOR	IRREGULAR	<5'-10'
10P	AMPHIPORA AND STACHYODES RUDSTONE AND FLOATSTONE WITH PACKSTONE MATRIX	NC	40	> 50	RARE	MAJOR	WIDESPREAD	<5'-12'
11P	STACHYODES AND AMPHIPORA RUDSTONE AND FLOATSTONE WITH PACKSTONE MATRIX	NC	14	APPROX 20	RARE	MINOR	IRREGULAR	<5'-10'
12P	AMPHIPORA RUDSTONE WITH GRAINSTONE-PACKSTONE MATRIX	NC	21	APPROX 30	RARE	MINOR	IRREGULAR	<5'-12'
13P	THIN TABULAR STROMATOPOROID AND CORAL RUDSTONE WITH PACKSTONE MATRIX	NC	32	33	MAJOR	ABSENT	PERIPHERAL WIDESPREAD	8'-15'
14P	ALGAL LAMINATED MUDSTONE	NC	13	APPROX 20	MINOR	COMMON	IRREGULAR	<1'
15P	LAMINATED FENESTRAL POROSITY MUDSTONE	NC	32	APPROX 50	COMMON	COMMON	WIDESPREAD	<1'-5'
16P	MUDSTONE	NC	30	APPROX 50	COMMON	COMMON	WIDESPREAD	<1'-5'
17P	CRACKLED MUDSTONE-INTRACLAST RUDSTONE	NC	21	APPROX 30	RARE	COMMON	IRREGULAR	<1'-10'
18P	PELOIDAL PACKSTONE	NC	6	7	ABSENT	MINOR	IRREGULAR	<1'-8'
19P	STACHYODES AND/OR ONCOLITE RUDSTONE WITH GRAINSTONE MATRIX	NC	2	3	ABSENT	RARE	RARE	5'-8'

NC : PREDOMINANTLY NON-CARBONACEOUS LITHOFACIES
C : PREDOMINANTLY CARBONACEOUS LITHOFACIES

DISTRIBUTION

WIDESPREAD : PRESENT ALL ACROSS THE AREA OF STUDY
IRREGULAR : NO PATTERN IN DISTRIBUTION
PERIPHERAL : ONLY SEEN AT PERIPHERY OF BUILDUP (IN PLATFORM: REFERS TO PERIPHERY OF STUDY AREA)
CENTRAL : ONLY SEEN IN CENTRAL PORTION OF BUILDUP
ALONG RAMP : PRESENT ALONG MORPHOLOGICAL RAMP IN BUILDUP
NOT NE/E : ABSENT NORTH EAST AND EAST OF MORPHOLOGICAL RAMP

OCCURRENCES

MAJOR : > 15
COMMON : 10 - 15
MINOR : 5 - 10
RARE : 1 - 5
ABSENT : 0

TABLE 5

SUMMARY OF BUILDUP LITHOFACIES

NUMBER SYMBOL	NAME (MAJOR COMPONENTS/MATRIX)	NC/C	NUMBER OF WELL OCCURRENCE	NUMBER OF OCCURRENCE	BASE BUILDUP OCCURRENCE	TOP BUILDUP OCCURRENCE	DISTRIBUTION	THICKNESS
1B	AMPHIPORA RUDSTONE AND FLOATSTONE WITH CARBONACEOUS MUDSTONE MATRIX	C	20	>50	RARE	MAJOR	CENTRAL	<1'-5'
2B	STACHYODES RUDSTONE AND FLOATSTONE WITH CARBONACEOUS MUDSTONE MATRIX	C	15	17	MINOR	RARE	CENTRAL	<1'-5'
3B	NODULAR BRACHIOPOD AND CRINOID FLOATSTONE-RUDSTONE WITH CARBONACEOUS MUDSTONE MATRIX	C	3	3	RARE	ABSENT	PERIPHERAL	5'-25'
4B	HEMISPHERICAL STROMATOPOROID RUDSTONE WITH CARBONACEOUS MUDSTONE MATRIX	C	4	4	RARE	RARE	IRREGULAR	<5'
5B	THICK AND THIN TABULAR STROMATOPOROID, BRACHIOPOD AND CORAL RUDSTONE WITH CARBONACEOUS MUDSTONE MATRIX	C	16	18	MAJOR	ABSENT	PERIPHERAL	5'-25'
6B	CRINOID AND BRACHIOPOD CARBONACEOUS PACKSTONE	C	3	3	RARE	ABSENT	PERIPHERAL	5'-20'
7B	NODULAR PELOIDAL CARBONACEOUS WACKESTONE	C	2	3	RARE	ABSENT	PERIPHERAL RARE	5'-15'
8B	AMPHIPORA RUDSTONE AND FLOATSTONE WITH MUDSTONE MATRIX	NC	28	>50 (CO:6)	COMMON	MAJOR	NOT NE/E	<1'-8'
9B	STACHYODES RUDSTONE AND FLOATSTONE WITH MUDSTONE MATRIX	NC	27	>30 (CO:11)	COMMON	MINOR	NOT NE/E	<5'
10B	THIN AND/OR THICK TABULAR STROMATOPOROID, STACHYODES AND BRACHIOPOD RUDSTONE WITH MUDSTONE MATRIX	NC	10	11	COMMON	ABSENT	PERIPHERAL	15'-25'
11B	HEMISPHERICAL AND/OR BULBOUS STROMATOPOROID RUDSTONE AND FLOATSTONE WITH MUDSTONE MATRIX	NC	8	10 (CO:5)	RARE	RARE	PERIPHERAL	5'-10'
12B	AMPHIPORA FLOATSTONE AND RUDSTONE WITH PACKSTONE MATRIX	NC	26	>50 (CO:5)	COMMON	MAJOR	CENTRAL	<5'
13B	STACHYODES AND/OR NOT HEMISPHERICAL STROMATOPOROID AND AMPHIPORA FLOATSTONE AND RUDSTONE WITH PACKSTONE MATRIX	NC	23	>25 (CO:4)	COMMON	MINOR	CENTRAL	<5'-15'
14B	HEMISPHERICAL STROMATOPOROID AND AMPHIPORA RUDSTONE WITH PACKSTONE MATRIX	NC	7	8	RARE	MINOR	IRREGULAR (ALONG RAMP)	<5'
15B	THIN TABULAR STROMATOPOROID, STACHYODES, BRACHIOPOD AND HEMISPHERICAL STROMATOPOROID WITH PACKSTONE MATRIX	NC	6	6	MINOR	ABSENT	PERIPHERAL	5'-45'

NC : PREDOMINANTLY NON-CARBONACEOUS LITHOFACIES
C : PREDOMINANTLY CARBONACEOUS LITHOFACIES

DISTRIBUTION

WIDESPREAD : PRESENT ALL ACROSS THE AREA OF STUDY
IRREGULAR : NO PATTERN IN DISTRIBUTION
PERIPHERAL : ONLY SEEN AT PERIPHERY OF BUILDUP (IN PLATFORM: REFERS TO PERIPHERY OF STUDY AREA)
CENTRAL : ONLY SEEN IN CENTRAL PORTION OF BUILDUP
ALONG RAMP : PRESENT ALONG MORPHOLOGICAL RAMP IN BUILDUP
NOT NE/E : ABSENT NORTH EAST AND EAST OF MORPHOLOGICAL RAMP

OCCURRENCES

MAJOR : > 15
COMMON : 10 - 15
MINOR : 5 - 10
RARE : 1 - 5
ABSENT : 0

CO : LITHOFACIES WITH ALGAL COATING NUMBER AFTER CO INDICATES NUMBER OF OCCURRENCES

TABLE 5 (CONTINUED)

NUMBER SYMBOL	NAME (MAJOR COMPONENTS/MATRIX)	NC/C OCCURRENCE	NUMBER OF WELL OCCURRENCE	NUMBER OF OCCURRENCE	BASE BUILDUP OCCURRENCE	TOP BUILDUP OCCURRENCE	DISTRIBUTION	THICKNESS
16B	AMPHIPORA FLOATSTONE AND RUDSTONE WITH PELOIDAL GRAINSTONE MATRIX	NC	19	26 (C0:7)	MINOR	MINOR	IRREGULAR (AND) (ALONG RAMP)	<5'-15'
17B	STACHYODES AND/OR NOT AMPHIPORA AND/OR NOT HEMISPHERICAL STROMATOPOROID FLOATSTONE-RUDSTONE WITH GRAINSTONE OR NO MATRIX	NC	17	24 (C0:7)	COMMON	MINOR	IRREGULAR (AND) (ALONG RAMP)	5'-20'
18B	STACHYODES, THIN TABULAR STROMATOPOROID AND BRACHIOPOD RUDSTONE WITH PELOIDAL AND SKELETAL GRAINSTONE MATRIX	NC	8	10 (C0:1)	MINOR	RARE	IRREGULAR	10'-25'
19B	HEMISPHERICAL STROMATOPOROID, STACHYODES, CORAL AND BRACHIOPOD RUDSTONE AND FLOATSTONE WITH SKELETAL GRAINSTONE MATRIX	NC	17	26 (C0:4)	COMMON	COMMON	IRREGULAR	10'-25'
20B	THICK TABULAR STROMATOPOROID, HEMISPHERICAL STROMATOPOROID, STACHYODES AND BRACHIOPOD RUDSTONE WITH SKELETAL GRAINSTONE MATRIX	NC	18	24 (C0:1)	MAJOR	MINOR	PERIPHERAL	10'-60'
21B	CORAL, ALGAL-STROMATOPOROID CONSORTIUM, STACHYODES AND IRREGULAR STROMATOPOROID RUDSTONE-FLOATSTONE WITH PELOIDAL AND SKELETAL GRAINSTONE MATRIX	NC	15	18 (C0:6)	RARE	MINOR	IRREGULAR	5'-20'
22B	PELOIDAL PACKSTONE	NC	17	13 (C0:1)	MINOR	RARE	IRREGULAR	<1'-5'
23B	NODULAR PELOIDAL WACKESTONE AND PACKSTONE	NC	4	7	RARE	ABSENT	PERIPHERAL RARE	5'-10'
24B	SKELETAL AND PELOIDAL GRAINSTONE	NC	7	15	MINOR	RARE	PERIPHERAL RARE	2'-10'
25B	CRYPTALGAL LAMINATED MUDSTONE	NC	15	>20	RARE	MINOR	CENTRAL	<1'
26B	LAMINATED FENESTRAL POROSITY MUDSTONE	NC	13	>20	RARE	MINOR	CENTRAL	<1'
27B	MUDSTONE	NC	16	>20	RARE	MINOR	CENTRAL	<1'
28B	CRACKLED MUDSTONE-INTRACLAST RUDSTONE	NC	15	>20 (C0:2)	MINOR	MINOR	CENTRAL	<1'
29B	ALGAL CRACKLED MUDSTONE-INTRACLAST RUDSTONE	NC	13	19	RARE	MINOR	CENTRAL	<1'

1966). The sedimentological evolution of other Swan Hills buildups is similar that of the Swan Hills buildup at Township 67-10W5.

The platform and buildup portions of the Swan Hills Formation are two distinct carbonate bodies at Swan Hills Field. The platform is relatively uniform in thickness, between 45 to 52 m thick, and is relatively extensive, whereas the buildup is more variable in thickness, from 13 to 67 m where present. The buildup is laterally restricted, and, as it becomes younger, it tends to occupy a smaller area. The platform has less diversity in terms of lithofacies (19) as compared to the buildup (29). Sandy lithofacies (calcarenites) are uncommon in the platform whereas in the buildup they tend to be abundant at the periphery of stages and in two cases they are actually present across the entire buildup. In both the platform and the buildup many lithofacies are stratigraphically confined. Lateral extent of lithofacies is significant in the platform. In the buildup, lateral changes of lithofacies are important, and therefore, correlations are short range in nature. In both the platform and the north and east parts of the buildup, lithofacies at the base are relatively deep water, carbonaceous-rich lithofacies with boundstone characteristics. Carbonaceous mud matrix is volumetrically more important in the platform than in the buildup, approximately 50% versus 15%. Crinoids, brachiopods and algal coatings on grains are very abundant in the buildup as compared to the platform. Internal sediment, commonly with geopetal fabric, is abundant and widespread in the buildup whereas it is a minor and stratigraphically confined feature in the platform.

Figure 7 shows an outline of the sedimentological development of the Swan Hills Formation. The Swan Hills platform is composed of relatively uniform and correlatable stages of development. In the lower platform depositional sequences are mainly composed of organic-rich, muddy or silty, deep water coral and stromatoporoid-rich facies and shallow water algal-laminated facies, with shallowing-upward character (stages P1 and P2). The middle portion of the platform consists of two shallowing-upward packages (stages P3 and P4) composed of organic-rich ostracods, Stachyodes and Amphipora mud facies and non-carbonaceous Amphipora rudstones (Dunham modified classification, see Embry and Klovan, 1972) with packstone or grainstone matrices. The upper platform is dominantly composed of Amphipora

FIGURE 7

SCHEMATIC REPRESENTATION OF SEDIMENTOLOGIC AND DIAGENETIC EVOLUTION OF THE SWAN HILLS FORMATION AT TOWNSHIP 67-10W5 IN TERMS OF CHRONOSTRATIGRAPHIC UNITS DEFINED IN CHAPTERS 2 AND 3 IN VIAU (1986) AND THE NATURE OF THE CHANGES AND THE CONTACTS THAT SEPARATE THEM
(BLACK VERTICAL LINES REPRESENT CARBONACEOUS-RICH MUD MATRIX)

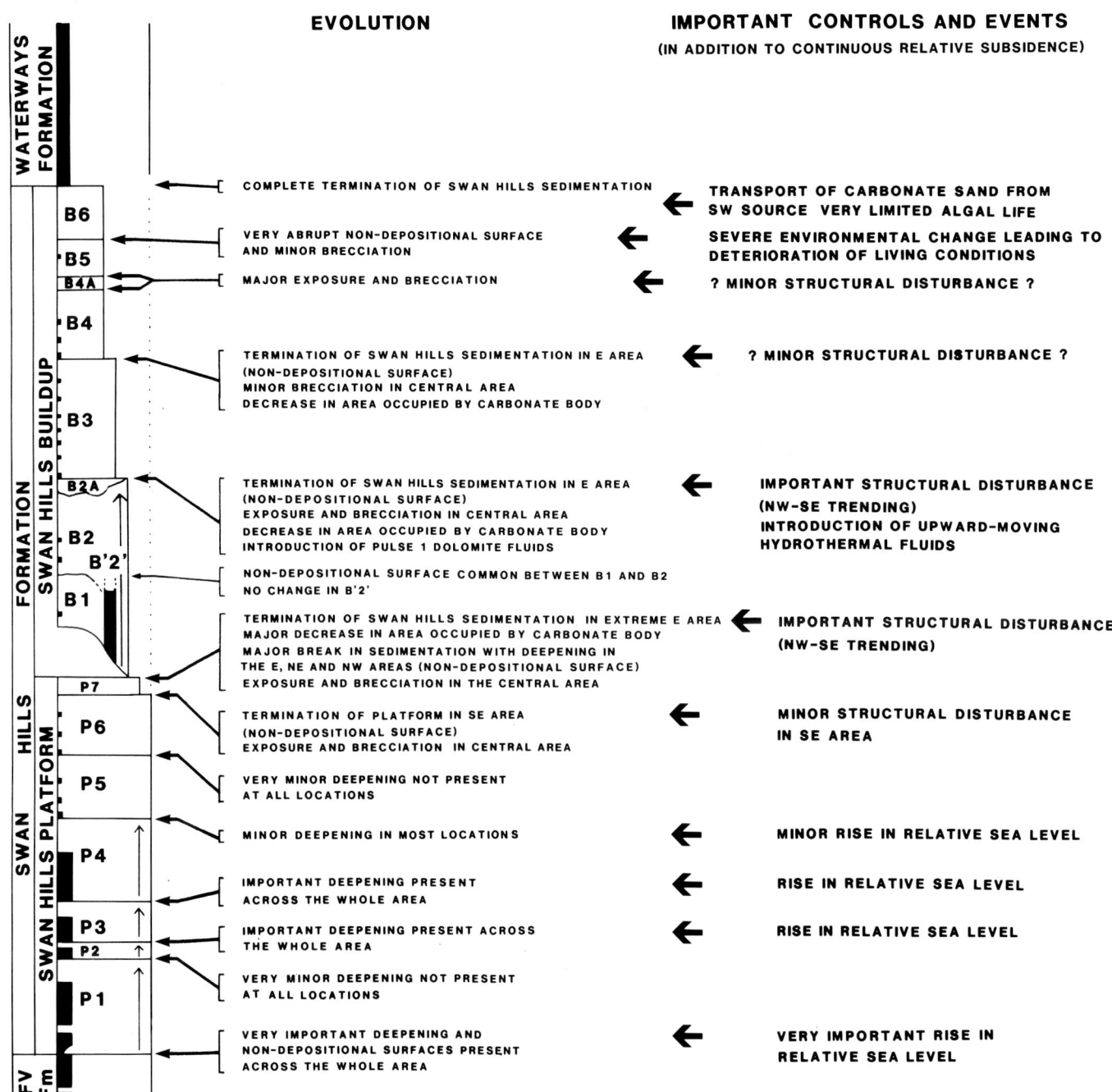

and mud-rich facies (stages P5, P6 and P7) which probably represent a shallow water mud mound environment. The top of the Fort Vermilion Formation, and the tops of Swan Hills stages P2 and P3 are interpreted to reflect important relative rises of sea level due to the presence of deeper water lithofacies over shallow water lithofacies across the entire area. The top of stages P1 and P4 involved only minor deepening. The top of stage P5 is a very minor deepening which does not occur at all locations. The top of stage P6 is the termination of platform sedimentation in the southeastern area of Township 67-10W5 (Viau, 1986). The top of the thin shallow water stage P7 represents termination of Swan Hills sedimentation in the extreme east of Township 67-10W5. It marks a major change in the size of the carbonate body as well as nature of carbonate sedimentation (end of platform sedimentation).

The Swan Hills buildup is composed of a sequence of carbonate sediments whose initial development and distribution was controlled primarily by subtle structural disturbance of the previously relatively undisturbed quiet broad platform. Figure 8 shows the outline of the Swan Hills buildup within Township 67-10W5 and the lithofacies relationships between the top of the shallow Amphipora and mud-rich platform and the base of the buildup. In the central area, the base of the buildup consists of shallow water mud and Amphipora-rich lithofacies. The nature of the contact is commonly abrupt and involves green shale and intraclast material. In the eastern and northern areas, the base of the buildup is composed of relatively deep water crinoid, brachiopod, coral and stromatoporoid-rich carbonaceous lithofacies. The passage is marked by a very sharp non-depositional surface at many locations. Hence, drowning of the platform is only local and cannot be interpreted as resulting from removing the carbonate platform from the photic zone because of a regional rise in sea level. The deepening and non-depositional surfaces in the east and north are chronostratigraphically equivalent to continued shallow water sedimentation or exposure (resulting in brecciation) in the central area.

Computer produced structural cross-sections and maps based on the vertical distribution of 24 electric log markers (ranging from Middle Devonian to Mississippian in age) in 143 wells in Township 67-10W5 show two roughly perpendicular sets of steeply dipping offsets: 1) a NW-SE trending

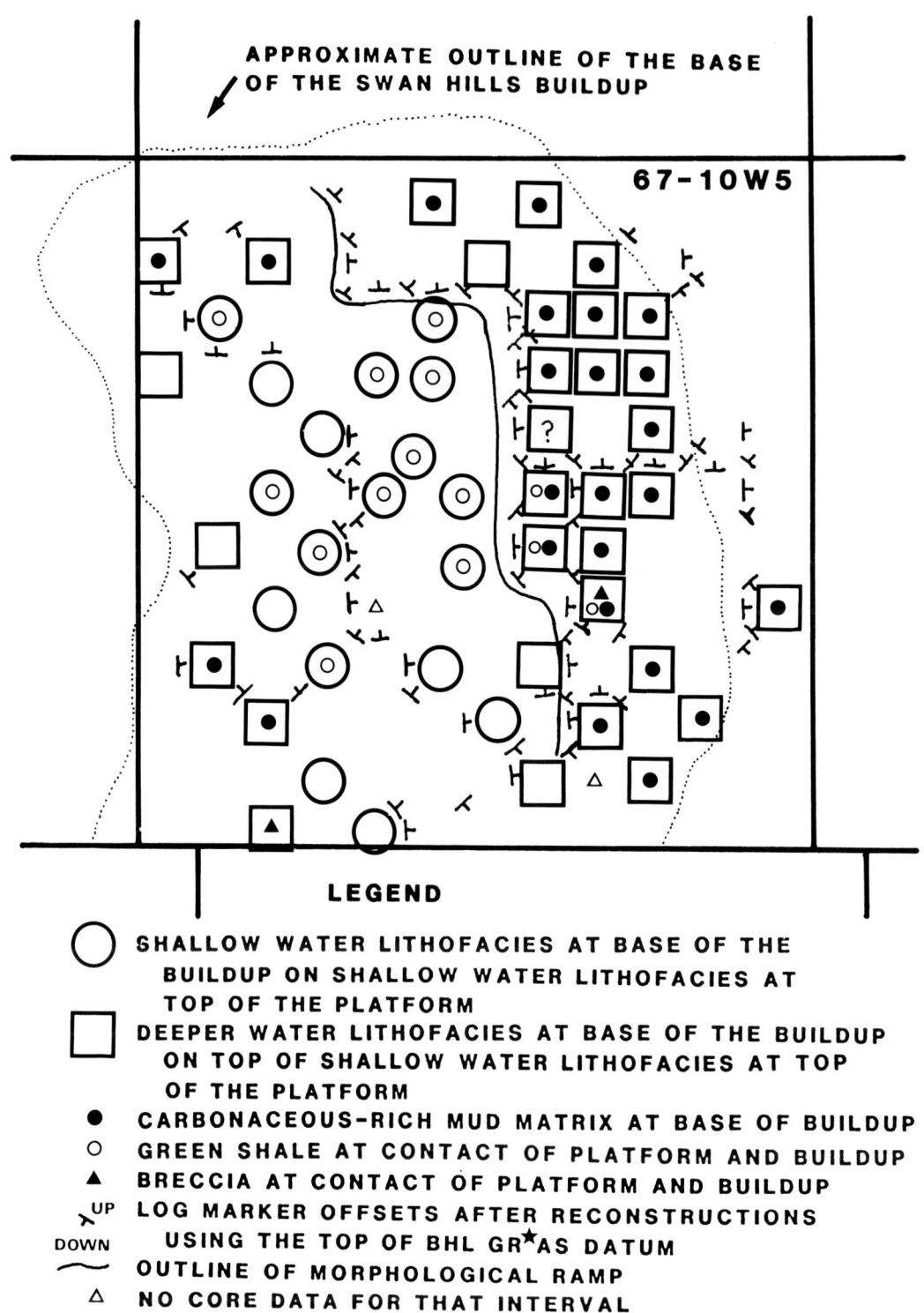

FIGURE 8 LITHOFACIES RELATIONSHIPS BETWEEN THE TOP OF THE PLATFORM AND THE BASE OF THE BUILDUP COMPARED WITH LOG MARKER OFFSETS AFTER RECONSTRUCTIONS

★ BEAVERHILL LAKE GROUP

set which is not pervasive vertically through the section, and 2) a more pronounced SW-NE trending set of offsets seen, in places, all the way up to the Mississippian. Structural reconstructions using a Devonian datum show that the magnitude of the structural offsets is in excess of the effects of compaction, confirming an early timing of the NW-SE trending offsets (Viau, 1986). The offsets are interpreted as deep-seated faults or warping due to such faulting. These reconstructed offsets are shown in Figure 8 to correspond with the sedimentological patterns at the base of the buildup (described above). The isopach of the platform and the sedimentological patterns in the upper platform stages show no relationship to the data shown in Figure 8 (Viau, 1986). Thus relict platform bathymetry cannot explain the sedimentological patterns at the base of the buildup. Hence, at the Swan Hills Field, the differentiation of the Swan Hills Formation from a lower lateraly extensive carbonate body (platform) to an upper succession of localized bodies (collectively the buildup) has resulted from subtle structural disturbance (in the order of 10 to 15 m of vertical displacement). Details of the structural framework at Swan Hills are shown in Viau (1986).

Wendte and Stoakes (1982) concluded that the passage from the platform to the buildup stages at Judy Creek Field (Fig. 2) resulted from a major rise in eustatic sea level and drowning of the platform. Relationships at Swan Hills are incompatible with that conclusion. This is not to say that relative rises in sea level did not occur. In fact it is quite conceivable that pulses in sea level changes could be related to the extensional structural pulses. The structural pulses are most likely responsible for the opening of the NW-SE trending Devonian intracontinental sea-way, extending from NW of the Selwyn Basin to SE Alberta. Consequently sedimentation was affected by sea level changes but the localization of buildups, the nature of sedimentation patterns and the timing of buildup development was primarily controlled by the structural regime.

The presence of faults has been suggested previously by Sikabonyi and Rogers (1959), Martin (1967) and Keith (1970) who have indicated that NW-SE and SW-NE trending deep-seated faults have controlled the localization of channels and buildups in the Swan Hills Formation.

Details of the sedimentological development of the buildup (described below) are critical for oil production at the Swan Hills Field (and at other Swan Hills fields) because the reservoir characteristics are closely related to original sedimentological patterns.

EVOLUTION OF THE SWAN HILLS BUILDUP

The Swan Hills buildup evolved through six major stages of growth. Chronostratigraphic packages in Devonian carbonates of the Alberta Basin are often assumed to be atoll bodies with three major environments, lagoon, margin and fore reef, with the margin rimming the interior. This is misleading because "missing" parts are considered to have been eroded (e.g., Fischbuch, 1968; Viau, 1983). Some Swan Hills buildup stages of growth are sand banks or mud banks with no relationship to the typical slope-rim-interior model. Recognition that this reef model cannot be used to explain all stages of growth (in fact most stages do not fit the model) is very important for reservoir description.

Stages B1 and B2 are two upward shoaling packages that developed only on the upthrown block in the centre of the area and at its edge adjacent to the downthrown block. The chronostratigraphic equivalent of stages B1 and B2, on the downthrown-side is mostly one major shallowing-up package. The base of this package is composed of deep water lithofacies forming a wedge of extra section. This entire shallowing upward package is referred to as stage B"2". This package is interpreted as the equivalent of both stages B1 and B2 because of the presence of two packages of deep water sediment at the area close to the contact between the upthrown and downthrown sides, and the fact that the bottom and top portion of both sequences are well defined. Consequently at the base of the buildup sediments are mostly mud-rich, however, in the central area the muds are shallow water sediments and they are surrounded by deep water muds with no margin type sediment at the interface. The upper portions of stages B2 and B"2" are very similar and consist mostly of shallow water sandy sediments. Hence at that time the buildup was essentially a sand bank or shoal with very little lateral differentiation (Figs. 13 and 14). The contacts at the top of stages B2 and equivalent B"2" have similar relationships to those seen at the top of the platform; brecciation and accumulation of green shale in the central area is chronostratigraphically equivalent to non-depositional surfaces in the eastern area. This represents renewed structural movement along a NW-SE trend. At the limit between the upthrown and downthrown sides at top of stage B2 or B"2" distinct deposits are present. These deposits (lithofacies 17B) consist of a unique mixture of components seen in other facies in both the upthrown and downthrown sides. The deposits have a sharp basal contact onto shallow, quiet water environments, they have mud matrix or no matrix and pendant and fibrous cements are common (Fig. 19). These are interpreted as a combination of tectonic breccias and storm deposits described as stage B2A.

Stage B3 covers a more limited area than stage B2 and its B"2" equivalent (Fig. 14). It is centered around the upthrown block in the center of Township 67-10W5. It is the first part of the buildup which has an "atoll" character in terms of having a ring of sandy marginal lithofacies surrounding, but not rimming, a muddy lagoonal area. The internal mud-rich sequences cannot be correlated with the marginal sands. The cyclic muddy deposits (described in Viau, 1983) represent mounds with a few feet of bathymetry between the highs and lows. The top of many of the muddy sequences shows evidence for exposure or near exposure whereas the marginal sandy units do not show any of these features. This relationship together with the thickness isopach distribution of stage B3 suggest that the interior muds were deposited in shallower water than the exterior sands. Hence stage B3 is best described as a mud bank with a sand apron surrounding it in deeper water (Figs. 13 and 14). The top of stage B3 represents termination of Swan Hills sedimentation in the east and is characterized by many non-depositional surfaces. Minor brecciation is present in the central area. The origin of this contact is not clear but could represent renewed structural disturbance.

Stage B4 developed on the mud interior of stage B3 (relict bathymetric high) and is essentially a mud bank, with minor sandy deposits. Stage B4 represents very shallow, cyclic, quiet water sedimentation. Stage B4A, at the top of stage B4, is a brecciated unit containing pendant calcite cements and green shale. It represents exposure of the buildup to arid or semi-arid climate. The green shales are interpreted as external in origin. They are probably transported either during storms, by normal currents, and/or by eolien transport onto exposed and shallow subtidal areas.

Stage B5 occupies the high left by stage B4 after exposure and relative sea level rise. It shows a gradient in sediment distribution from sandy, in the SW, to silty-muddy towards the NE, to muddy in the extreme NE. The termination of stage B5 is very abrupt everywhere and in places is a non-depositional surface with spectacular truncation. The top of stage B5 marks the position of severe environmental changes of unknown nature leading to deterioration of living conditions.

Stage B6 is a sandy shoal with stromatoporoids of very irregular shapes, crinoids and abundant algae in the form of algal-stromatoporoid assemblages or as coatings on grains. The association is described as an assemblage as opposed to a consortium as described in the past (Klovan, 1964) because there is no evidence that this association was beneficial; in fact the association is interpreted as reflecting a very unhealthy carbonate system where only survivors like algae can thrive. The interpretation is based on the decrease in the area of growth from stages B3 to B4, B5 and B6, the decrease in the abundance of biogenic components, the irregular shapes and small nature of stromatoporoids and the presence of abundant algal life as well as the fact that stage B6 is the last stage of carbonate growth in the Swan Hills Formation. The apron nature of sandy deposits in the SW area of stage B5 suggests that sandy carbonate debris was transported during episodic events from a SW source onto the small area of mud bank of stage B4 and B4A. The depth of water for the bioclastic deposits of stage B6 is unknown. The presence of algal life indicates that the sediments were within the photic zone. However, this could be in a few feet of water or in a deeper location such as that of sea mounds in the modern. It is conceivable that deposits composed of oncolites and brachiopod-crinoid extraclast breccias, which are present in the SW area only on top of the platform (Fig. 23) at the edge of the buildup, be chronostratigraphically equivalent to stage B6 (and not B1). The top of stage B6 does not show harground surfaces such as seen at top of stage B5. This allows, as an alternate working model, for the actual drowning of the Swan Hills complex to have occurred at the top of stage B5 with stage B6 being a bioclastic deposit representing a deep sea mound as opposed to shallow shoal.

POROSITY TYPES AND NETWORK

Most remaining porosity in the Swan Hills Field is primary in origin, predominantly intraparticle and interparticle. Pores filled by dolomite, anhydrite and calcite cements are both primary and secondary: fracture, vug, interparticle, intraparticle, boring, moldic, fenestral and shelter types. The best porosity and permeability development, as seen by core analysis data, is associated with lithofacies that have a sandy matrix as opposed to silt or mud size matrices (Figs. 9, 10, 11, 12). This relationship is due to the fact that most pores are primary in origin and stromatoporoid fossil fragments with good intraparticle pores tend to be associated with lithofacies that have sandy matrix, although there are exceptions. Schultheis (1976), Jardine and others (1977) and Wendte and Stoakes (1982) have shown the same relationships for the Judy Creek and Kaybob reef buildups. This general relationship between porosity distribution and lithofacies types is valid despite cementation and other diagenetic overprint. Consequently an understanding of the distribution and continuity of lithofacies and overall sedimentological evolution of the buildup, as shown in Figs. 13 and 14 is critical to the development of a reservoir model for both primary or enhanced recovery.

The buildup has far better porosity development than the platform because of the abundance of mud-free lithofacies. Lateral continuity is good at the periphery of the buildup in stages B1, B2, and B3. The upper part of stages B2 and B"2" and the entire stage B6 have excellent lateral continuity (Figs. 13 and 14) because they are sand banks (or shoals). The interior of stage B3, the entire stage B4 and the NE part of stage B5 have lower (and in many cases poor) porosity and lateral communication of porosity. Vertical continuity is good at the periphery of the buildup because of stacking of calcarenite facies. In the interior of the buildup, stacking of mud-rich lithofacies (except at the level of stages B2 and B6) makes vertical communication poor. In general, the presence of 1) muddy sediments versus sandy sediments, and 2) intraparticle and vuggy pores versus interparticle pores negatively affects permeability to a greater degree than it affects porosity (Sears and Yoeman, 1976). The detailed description of the vertical and lateral continuity of high porosity and permeability zones, such as those

EXPLANATION AND SYMBOLS FOR CORE DESCRIPTIONS
(FIGURES 9, 10, 11 AND 12)

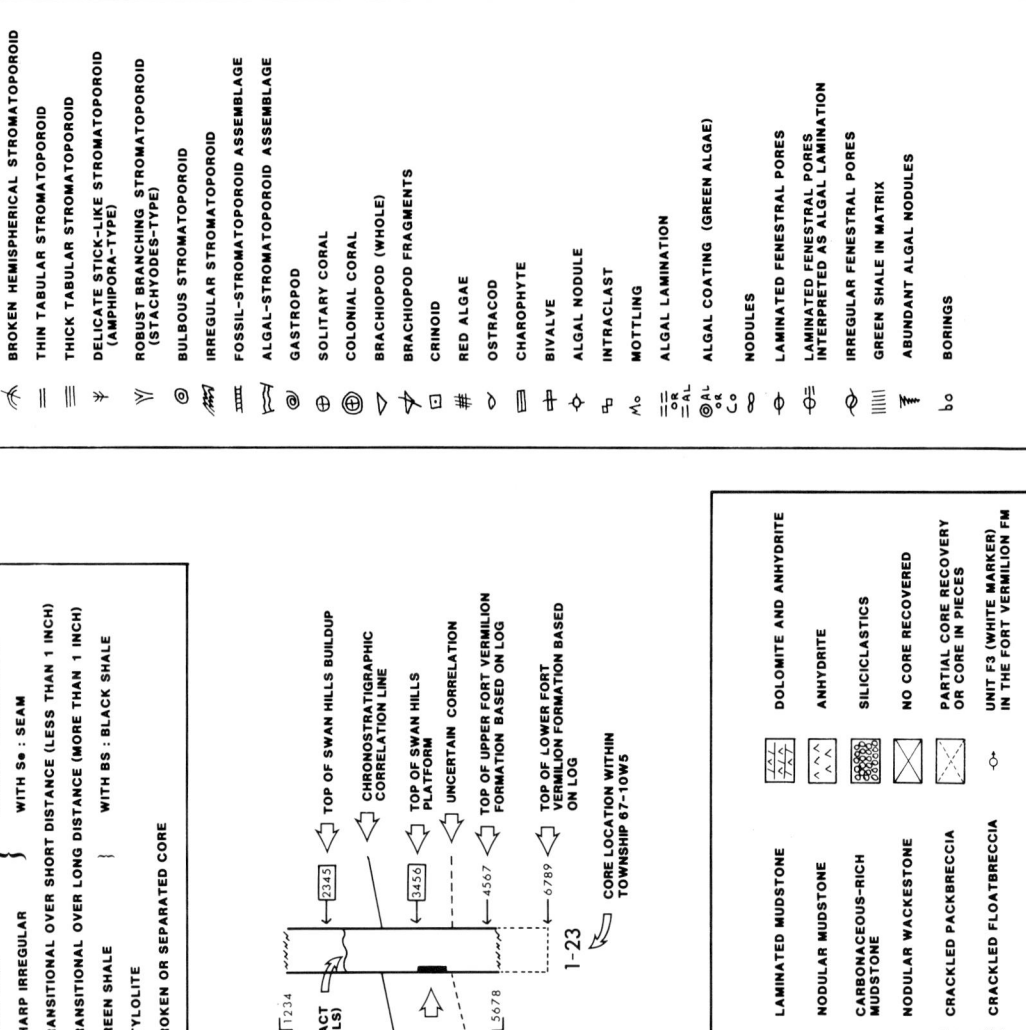

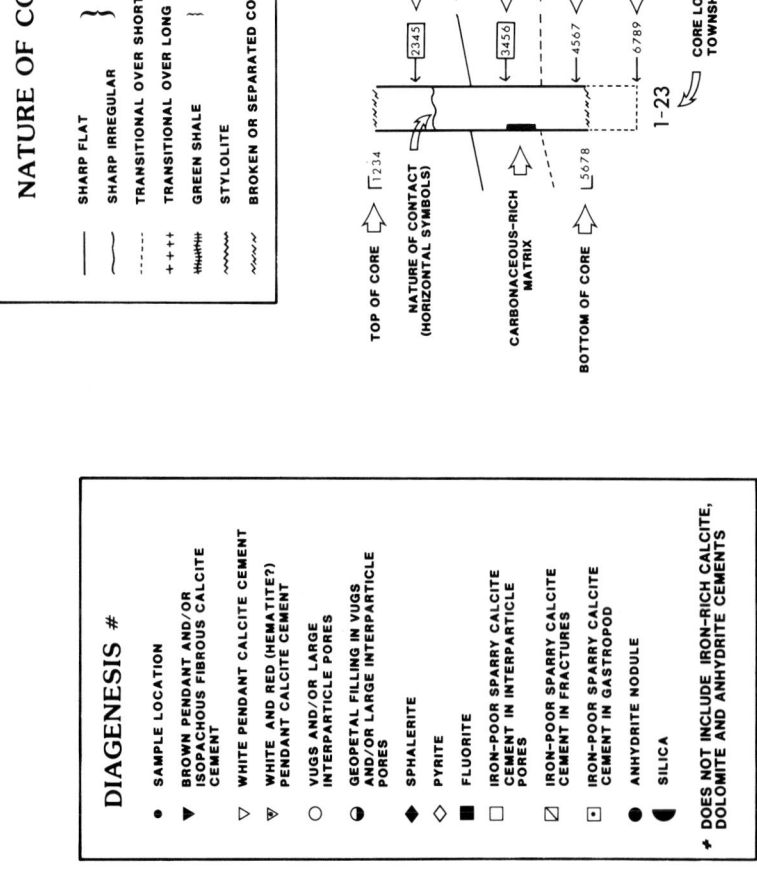

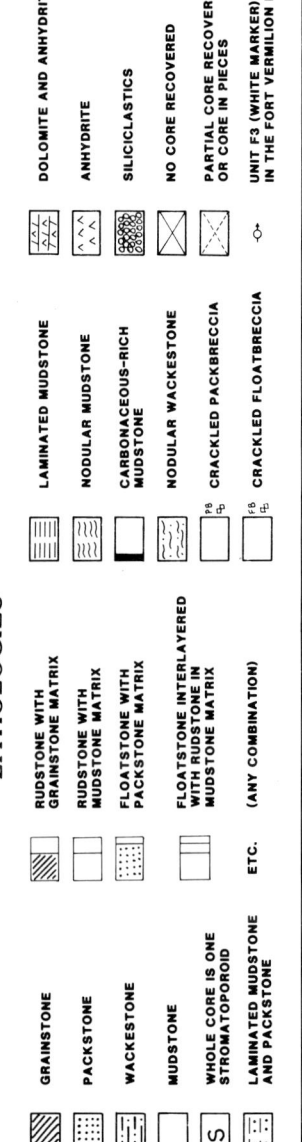

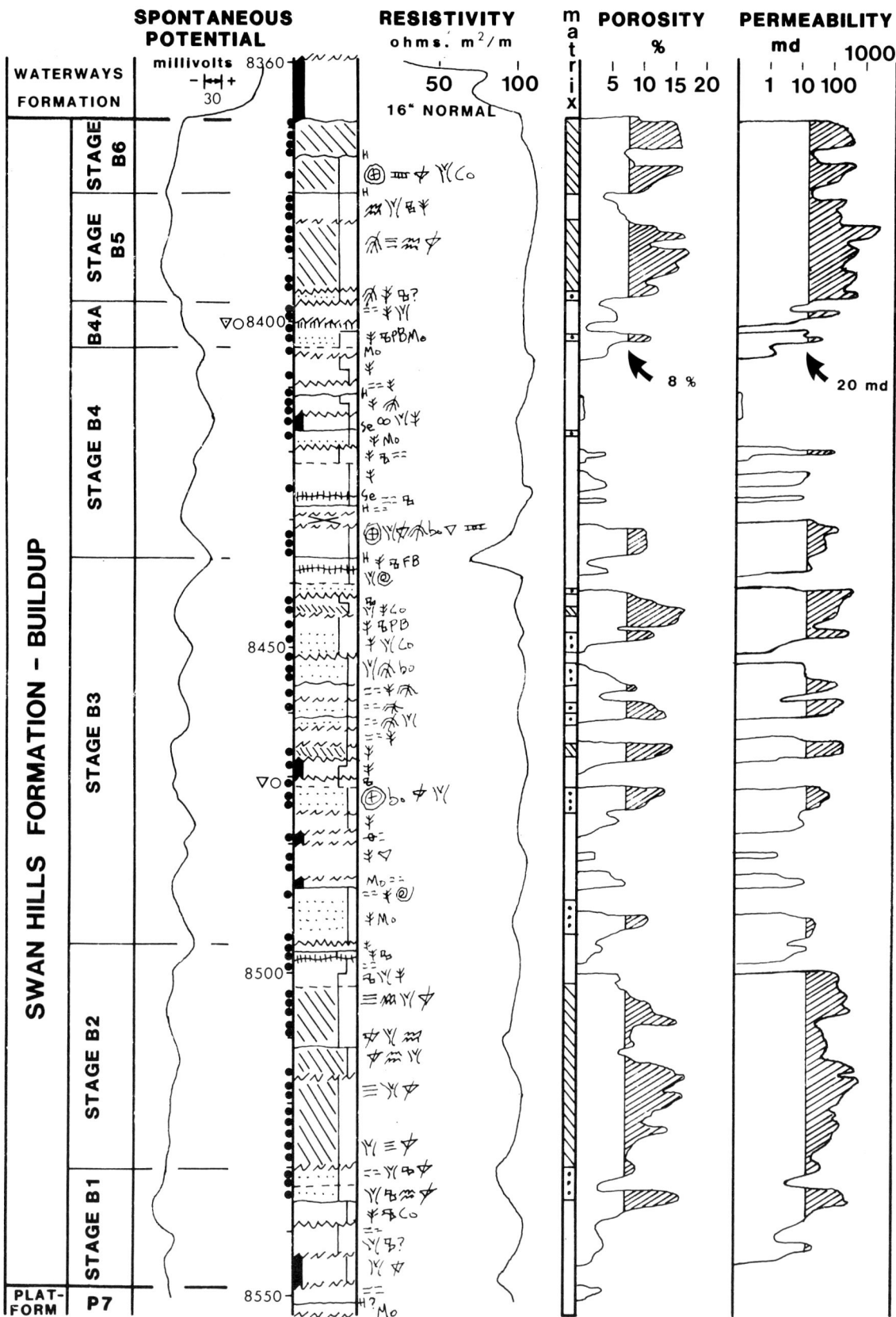

4-8-67-10W5

FIGURE 9

10-5-67-10W5

FIGURE 10

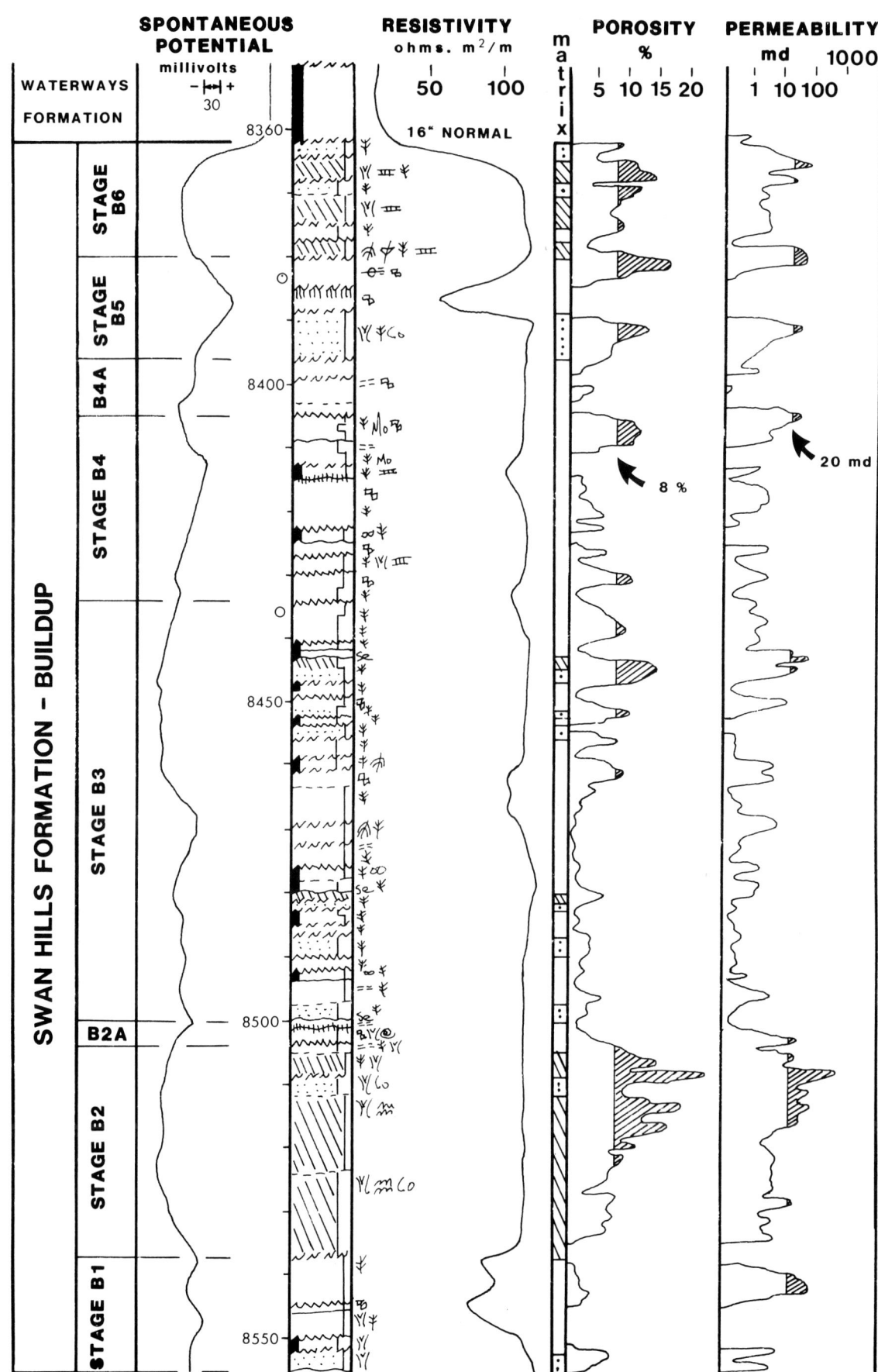

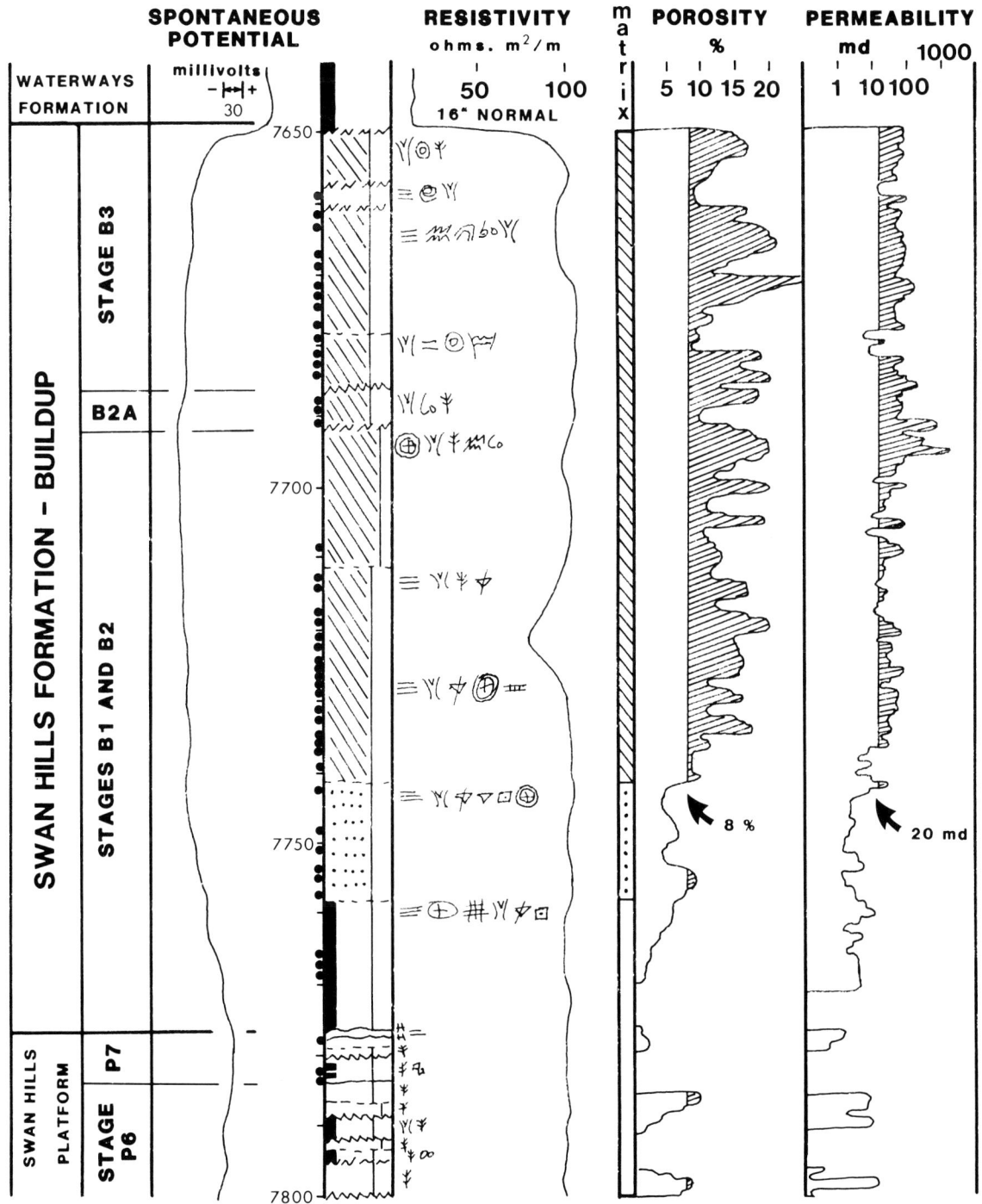

FIGURE 11

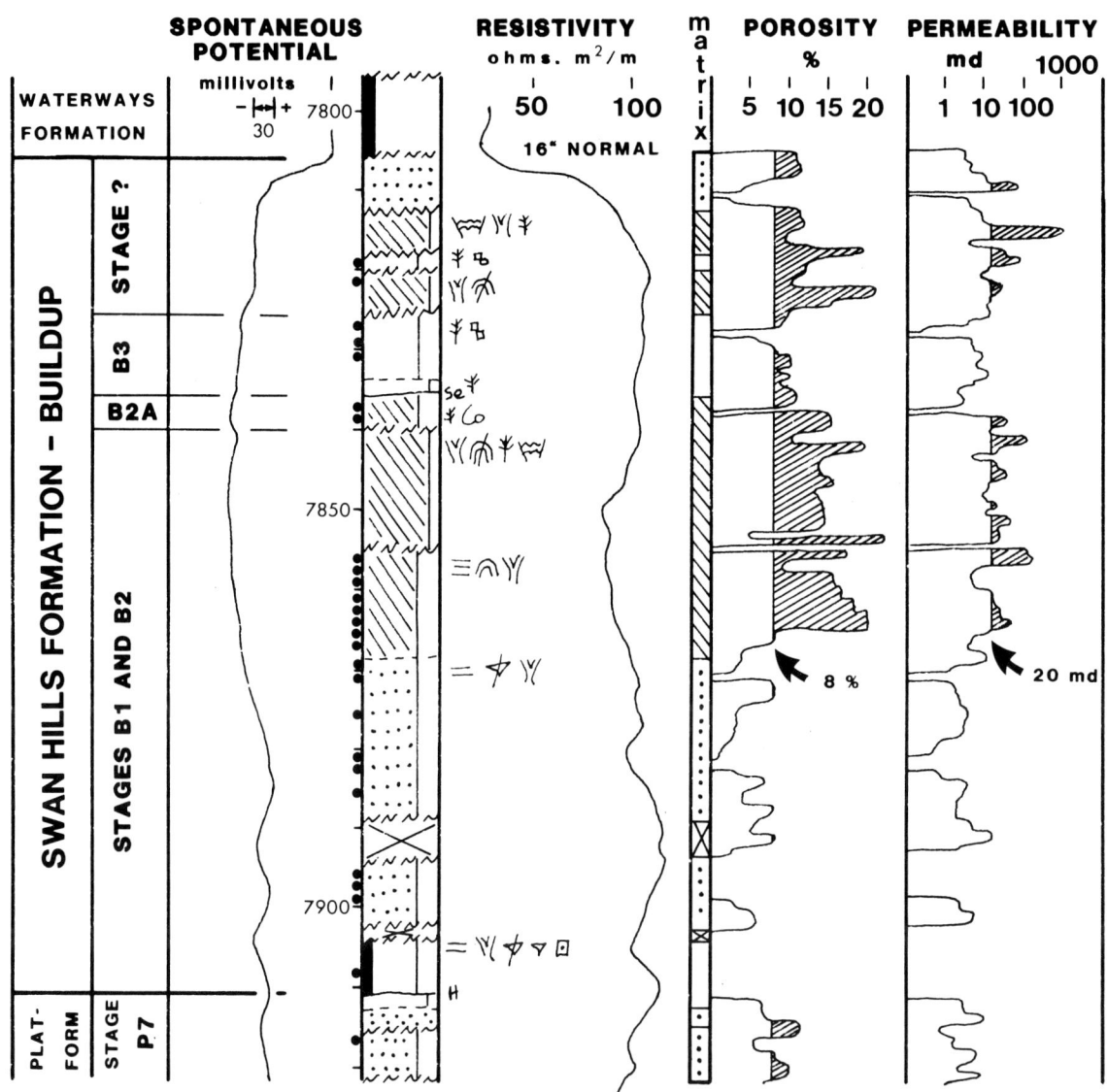

10-33-67-10W5   FIGURE 12

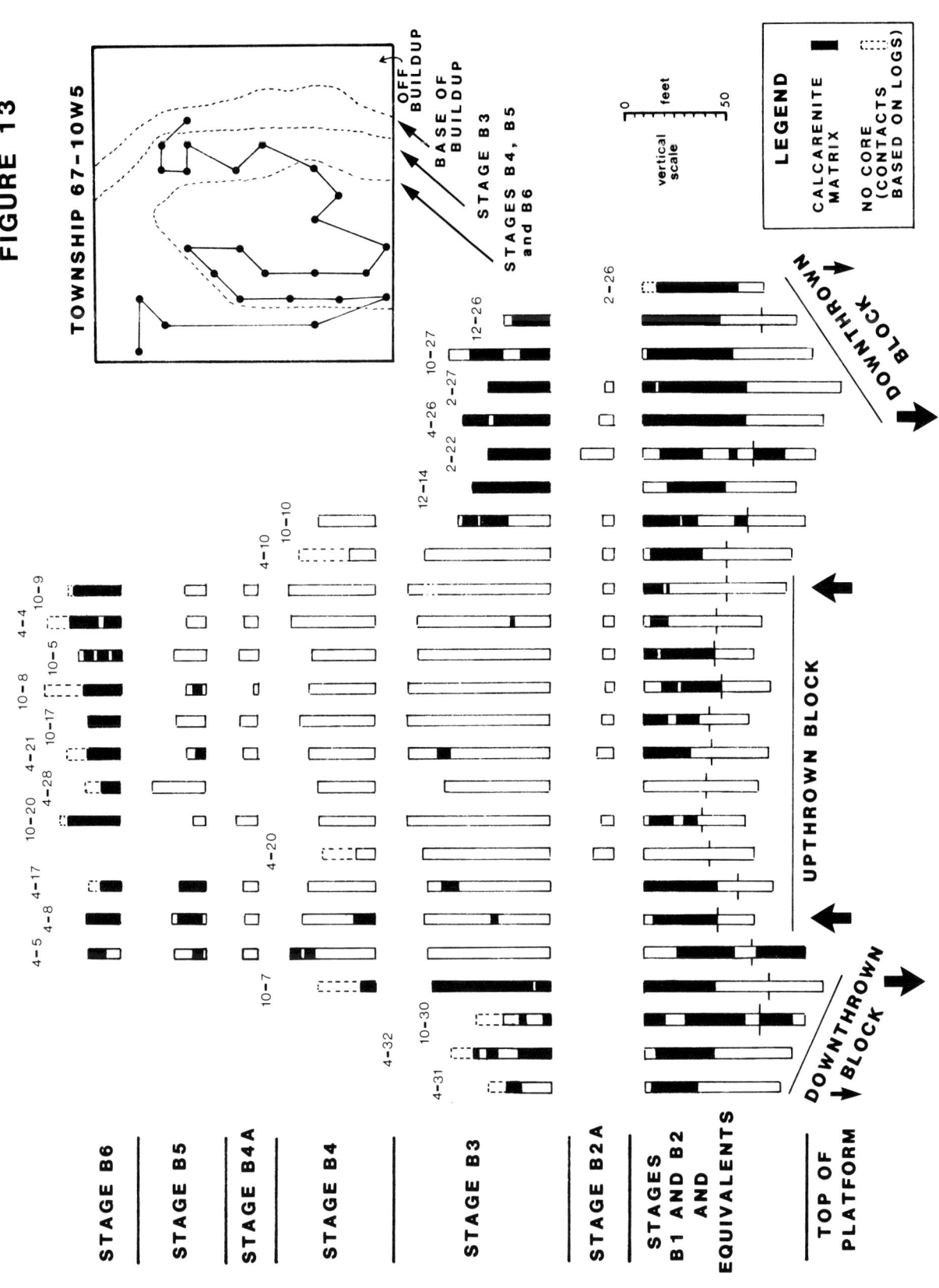

DISTRIBUTION OF DEPOSITS WITH CALCARENITE MATRIX WITHIN EACH CHRONOSTRATIGRAPHIC UNIT OF THE SWAN HILLS BUILDUP AT 25 LOCATIONS IN TOWNSHIP 67-10W5.

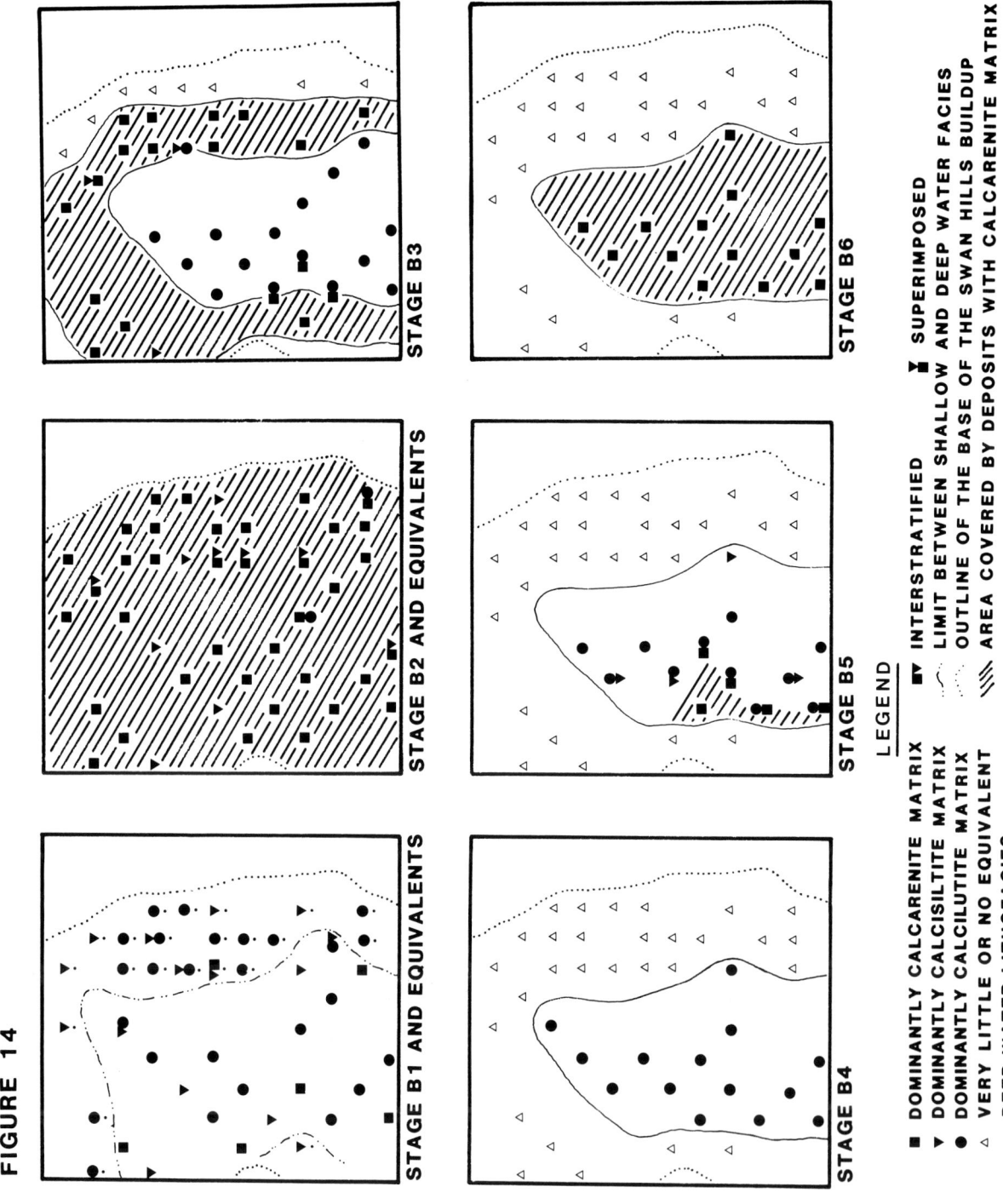

FIGURE 14

MATRIX GRAIN SIZE DISTRIBUTION FOR CHRONOSTRATIGRAPHIC UNITS WITHIN THE SWAN HILLS BUILDUP IN TOWNSHIP 67-10W5

in stages B2, B3, B5 and B6 is essential for enhanced recovery projects. Early breakthrough of injected fluids through such zones leads to poor sweep of lower quality target zones and costly production.

DIAGENESIS

The diagenesis of the Swan Hills Formation has been the subject of very few studies. Havard and Oldershaw (1976), Wong and Oldershaw (1981), Walls and Burrowes (1985) have addressed some aspects of diagenesis in the Swan Hills Formation. In fact, although many models have been proposed, the diagenesis of Devonian carbonates from the Alberta Basin is poorly documented and understood. Very few studies have attempted to integrate sedimentologic data with petrographic and geochemical data for diagenetic elements. Two major assumptions have been part of most diagenetic studies. Firstly, fractures cross-cutting grains, and filled with cements are assumed to be due to compaction, hence they are considered to be a burial phenomenon, and cements filling them are therefore mesogenetic in origin. Secondly, temperature indicators (fluid inclusion and isotopic data) are assumed to reflect depth of burial in terms of normal geothermal gradients and lack of vertical fluid movement through sediment. Data presented in recent years by Aulstead and Spencer (1984), Morrow and others (1986) and Viau (1986) indicate that both assumptions are incorrect in three different Devonian formations (Keg River, Manetoe and Swan Hills respectively) from three different areas.

Cloudy and white sparry, zoned (0.30 to 2.60 mole % FeO), rhombic and saddle shaped dolomite cements are present in the Swan Hills Formation (with anhydrite, calcite, sphalerite, pyrite and minor fluorite) in association with fractures as well as primary intraparticle pores, geopetal fillings and borings. Twelve ubiquitous zones, as revealed by conventional staining and cathodoluminescence, form the framework for the growth stratigraphy of the dolomite cements. Extensive mapping of the zones in 2000 samples from the Swan Hills Field indicates that two distinct pulses of upward-moving fluids occurred during cementation. The absence of pulse one dolomite zones above a distinct chronostratigraphic marker in the buildup (top of stage B2 and equivalent B"2", see Fig. 7) suggests that pulse one was emplaced at the time the buildup was at, or near sea level, before any significant burial and sedimentation above the marker. The steeply dipping NW-SE trending syn-sedimentary structural lineaments described above have controlled the distribution of pulse one dolomite and anhydrite cements.

Carbon isotopic composition of dolomite cements varies from 0 to +4 o/oo PDB. Oxygen isotope values range from -7 to -12 o/oo PDB. $\delta^{18}O$ trend suggests cooling during precipitation of pulse one dolomite. The sulphur isotope composition of sulphides range widely (+7 to -19 o/oo CDT) and are markedly depleted in ^{34}S (50 o/oo) compared to coexisting sulphate cements and early nodules, clearly indicating bacterial sulphate reduction. The sulphur and oxygen isotope compositions of the cements are compatible with Devonian sea water as a source. Limited fluid inclusion data from primary inclusions indicate precipitation in the range of 89 to 105 °C for fluorite and sphalerite and 101 to 110 °C for anhydrite. To date no reliable fluid inclusion data have been obtained from the dolomite and calcite cements. It is concluded that hot, modified Devonian sea water, recirculated upward in the extensional structural framework described above, was responsible for dolomite and anhydrite cementation (Viau, 1986). At least pulse one dolomite cementation occurred penecontemporaneously with the sedimentological development of the buildup. Early mobilization and hydrothermal circulation of sea water through a tectonically active intracontinental sea is compatible with other diagenetic events in the Devonian of Western North America, such as the presence of stratiform lead-zinc-barium deposits in the Selwyn Basin. These deposits are interpreted to be "formed at the sediment-water interface from metalliferous fluids discharged into local basins generated by contemporaneous block faulting" (Goodfellow and Jonasson, 1984, p.583).

SUMMARY

Much of the remaining porosity in the Devonian Swan Hills Formation at Swan Hills Field is primary in origin, predominantly interparticle and intraparticle. Porosity and permeability distribution is primarily controlled by the nature of the matrix between biogenic components. Deposits with calcarenite matrices have better reservoir development than those with calcisiltite. Mud-rich deposits have very poor reservoir development. Fossils with good intraparticle porosity tend to be associated with calcarenite matrices. Reservoir development is better in the buildup of the Swan Hills Formation as compared to the underlying platform because of the abundance of mud-free lithofacies.

The Swan Hills Formation at the Swan Hills Field is divided into six major depositional sequences. The passage from the platform to the buildup, the location of the buildup as well as its subsequent evolution were controlled by subtle structural disturbances related to basement extensional tectonics.

The diagenetic history of the Swan Hills Formation at Swan Hills Field was also affected by the syn-sedimentary structural events. Zoned dolomite and anhydrite cementation was the result of early (pre-burial) recirculation of hot, modified Devonian sea water through the structural framework.

The Swan Hills field is the largest oil field in a carbonate reservoir in Canada with 144 million m^3 of recoverable reserves (904 million barrels). The field has been produced by waterflood for over 25 years. In 1985 tertiary recovery (ethane miscible flood) started in the area of the Swan Hills buildup.

ACKNOWLEDGEMENTS

Shell Canada Ltd. provided financial and technical support for this paper and core/poster display. I am grateful to Susanne Ridley (Esso Resources Canada Ltd.) and Robert Spitzer (Shell Canada Ltd.) for reviewing this paper and making many useful comments for improvement. Jim Craig (S.B. Geological Associates) also helped with parts of the manuscript.

REFERENCES

AULSTEAD, K. L. and SPENCER, R. J., 1984, Diagenesis of the Keg River Formation, northwestern Alberta: Fluid inclusion evidence: Bull. Can. Petrol. Geol., v.33, pp.167-183.

CRAIG, J.H., 1987, Depositional environment of the Slave Point Formation, Beaverhill Lake Group, Peace River Arch: in KRAUSE, F. F. and BURROWES, O. G. (eds.), Devonian lithofacies and reservoir styles in Alberta, 13th C.S.P.G. Core Conference, Calgary, August 19, 20 and 21, 1987, pp.181-200.

EDIE, R. W., 1961, Devonian reef reservoir Swan Hills oilfield, Alberta: Can. Min. Metallur. Bull., v.54, pp.447-454.

EMBRY, A. F. and KLOVAN, J. E., 1972, Absolute water depth limits of Late Devonian paleoecologic zones: Geologische Rundschau, v.61, pp.672-686.

FISCHBUCH, N. R., 1968, Stratigraphy, Devonian Swan Hills Reef complexes of central Alberta: Bull. Can. Petrol. Geol., v.16, pp.444-556.

GOODFELLOW, W. D. JONASSON, I. R., 1984, Ocean stagnation and ventilation by $\delta^{34}S$ secular trends in pyrite and barite, Selwyn Basin, Yukon: Geology, v.12, pp.583-586.

HARLAND, W. B., COX, A. V., LLEWELLYN, P. G., PICKTON, C. A. G., SMITH, A. G. and WALTERS, R., 1982, A geologic time scale: Cambridge Earth Science Series, Cambridge University Press, 131p.

HAVARD, C. J. and OLDERSHAW, A. E., 1976, Early diagenesis in back-reef sedimentary cycles, Snipe Lake reef complex, Alberta: Bull. Can. Petrol. Geol., v.24, pp.27-69.

HEMPHILL, C. R. and DUNN, P. J., 1960, Swan Hills Field: in Oil Fields of Alberta, Alberta Soc. Petrol. Geol., pp.200-201.

HEMPHILL, C. R., SMITH, R. I. and SZABO, F., 1970, Geology of the Beaverhill Lake reefs, Swan Hills area, Alberta: in HALBOUTY, M. T. (ed.), Geology of Giant petroleum Fields, Am. Assoc. Petrol. Geol. Memoir 14, pp.50-90.

JARDINE, D., ANDREWS, D. P., WISHART, J. W. and YOUNG, J. W., 1977, Distribution and continuity of carbonate reservoirs: SPE 6139, Jour. Petrol. Tech., July 1977, pp.873-885.

JENIK, A. J. and LERBEKMO, J. F., 1968, Facies and geometry of Swan Hills reef member of Beaverhill Lake Formation (Upper Devonian), Goose River Field, Alberta, Canada: Am. Assoc. Petrol. Geol. Bull., v.52, pp.1-100.

KEITH, J. W., 1970, Tectonic control of Devonian reef sedimentation, Alberta: Am. Assoc. Petrol. Geol. Bull., v.54, p.854.

KLOVAN, J. E., 1964, Facies analysis of the Redwater reef complex, Alberta, Canada: Bull. Can. Petrol. Geol., v.12, pp.1-100.

KOBLUK, D., 1975, Stromatoporoid paleoecology of the southeast margin of the Miette carbonate complex, Jasper Park, Alberta: Bull. Can. Petrol. Geol., v.23, pp.224-277.

KOBLUK, D., 1978, Reef stromatoporoid morphologies as dynamic populations: Application of field data to a model and the reconstruction of an Upper Devonian reef: Bull. Can. Petrol. Geol., v.26, pp.218-236.

LEAVITT, E. M., 1968, Petrology, paleontology, Carson Creek North reef complex, Alberta: Bull. Can. Petrol. Geol., v.16, pp.288-413.

MARTIN, R., 1967, Morphology of some Devonian reefs in Alberta: A palaeogeomorphological study: in OSWALD, D. H. (ed.), International Symposium on the Devonian System, Alberta Soc. Petrol. Geol., v.II, pp.365-385.

MORROW, D. W., CUMMING, G. L. and KOEPNICK, R. B., 1986, The Manetoe Facies - A gas-bearing, megacrystalline, Devonian dolomite, Yukon and Northwest Territories, Canada: Am. Assoc. Petrol. Geol. Bull., v.70, pp.702-720.

MURRAY, J. W., 1966, An oil-producing reef-fringed carbonate bank in the Upper Swan Hills Member, Judy Creek, Alberta: Bull. Can. Petrol. Geol., v.14, pp.1-103.

OILWEEK, 1984, Swan Hills EOR, project detailed: OILWEEK, Nov. 29, v.35 no.39, pp.12-14.

SCHULTHEIS, N. H., 1976, Kaybob Oil Field, Alberta, Canada: in Braunstein, J. (ed.) North American oil and gas fields, Am. Assoc. Petrol. Geol. Memoir 24, pp.79-90.

SEARS, J. R. and YOEMAN, G., 1976, Swan Hills Field - Beaverhill Lake A and B pools: in CLARK, B. and HUFF, G. (eds.), Joint Convention on enhanced recovery Core Conference, Petroleum Society of Can. Inst. Min. Metallur. and Can. Soc. Petrol. Geol., pp.F1-F17.

SEARS, J. R., 1978, Improved reservoir description: Swan Hills Unit No. 1: Jour. Can. Petrol. Tech., v.17, pp.43-50.

SIKABONYI, L. A. and RODGERS, W. J., 1959, Paleozoic tectonics and sedimentation in the northern half of the Western Canadian Basin: Jour. Alberta Soc. Petrol. Geol., v.7, pp.193-216.

TIRATSOO, E. N., 1984, Oilfields of the world: Houston, Gulf Publishing Company, 392p.

THOMAS, G. E. and RHODES, H. S., 1961, Devonian limestone bank-atoll reservoirs in the Swan Hills area Alberta: Jour. Alberta Soc. Petrol. Geol., v.9, pp.29-38.

VIAU, C. A., 1983, Depositional sequences, facies, and evolution of the Upper Devonian Swan Hills reef buildup, central Alberta, Canada: in HARRIS, P. M. (ed.), Carbonate Buildups - A Core Workshop, Soc. Econ. Paleontologists Mineralogists, Core Workshop No.4, Dallas, April 16-17, 1983, pp.112-143.

VIAU, C. A., 1986, Diagenesis, sedimentology and structure of the Swan Hills Formation, Swan Hills Field, central Alberta, Canada: Ph. D. Thesis, University of Calgary, Calgary, Alberta, Canada, 574p.

VIAU, C. A., 1987, The Swan Hills Formation and the Beaverhill Lake Group at Swan Hills Field and adjacent areas, central Alberta, Canada: in KRAUSE F.F. and BURROWES, O. G. (eds.), Devonian lithofacies and reservoir styles in Alberta, 13 th C.S.P.G. Core Conference, Calgary, August 19, 20 and 21, 1987, pp.201-242.

WALLS, R. A. and BURROWES, O. G., 1985, The role of cementation in the diagenetic history of Devonian reefs, Western Canada: in SCHNEIDERMANN, N. and HARRIS, P. M. (eds.), Soc. Econ. Paleontologists and Mineralogists Spec. Public. No.36, pp.185-220.

WENDTE, J. C. and STOAKES, F. A., 1982, Evolution and corresponding porosity of the Judy Creek complex, Upper Devonian, central Alberta: in CUTLER, W. G. (ed.), Canada's Giant hydrocarbon reservoirs, C.S.P.G. 1982 Core Conference, pp.63-81.

WONG, P. K. and OLDERSHAW, A. E., 1981, Burial cementation in the Devonian Kaybob reef complex, Alberta, Canada: Jour. Sed. Pet., v.51, pp.507-520.

Figure 15 Platform lithofacies.

A. Thin tabular stromatoporoid floatstone-rudstone with packstone matrix. Lithofacies 13P. Stage P1. Scale in cm. (4-4-68-10W5, 8114')

B. *Amphipora* and *Stachyodes* rudstone with carbonaceous mudstone matrix. Lithofacies 3P. Stage P4. Scale in cm. (10-3-67-10W5, 8798')

C. Hemispherical and *Amphipora* stromatoporoid rudstone with carbonaceous mudstone matrix. Lithofacies 5P. Stage P1. Scale in cm. (12-25-67-11W5, 8815')

D. *Stachyodes* and hemispherical stromatoporoid rudstone with carbonaceous mudstone matrix. Lithofacies 6P. Stage P4. Scale in cm. (10-3-67-10W5, 8795')

Figure 16 Buildup lithofacies and equivalents.

A. Nodular, carbonaceous brachiopod wackestone. Lithofacies 3B. Stage ?. Scale in cm. (12-25-67-11W5, 8756')

B. Crinoid and brachiopod packstone-rudstone. Lithofacies 6B. Stage ?. Scale in cm. (12-25-67-11W5, 8718')

C and D. Thin tabular stromatoporoid, brachiopod, crinoid rudstone in carbonaceous mudstone matrix. Red algae (Solenoporacea) present in D. Lithofacies 5B. Stage B1. Scale in cm. (4-26-67-10W5, 7765' and 7770')

NOTE: A and B are lithostratigraphically part of the Waterways Formation but chronostratigraphically interpreted to be slope equivalent of Swan Hills carbonates (Viau, 1986).

FIGURE 15

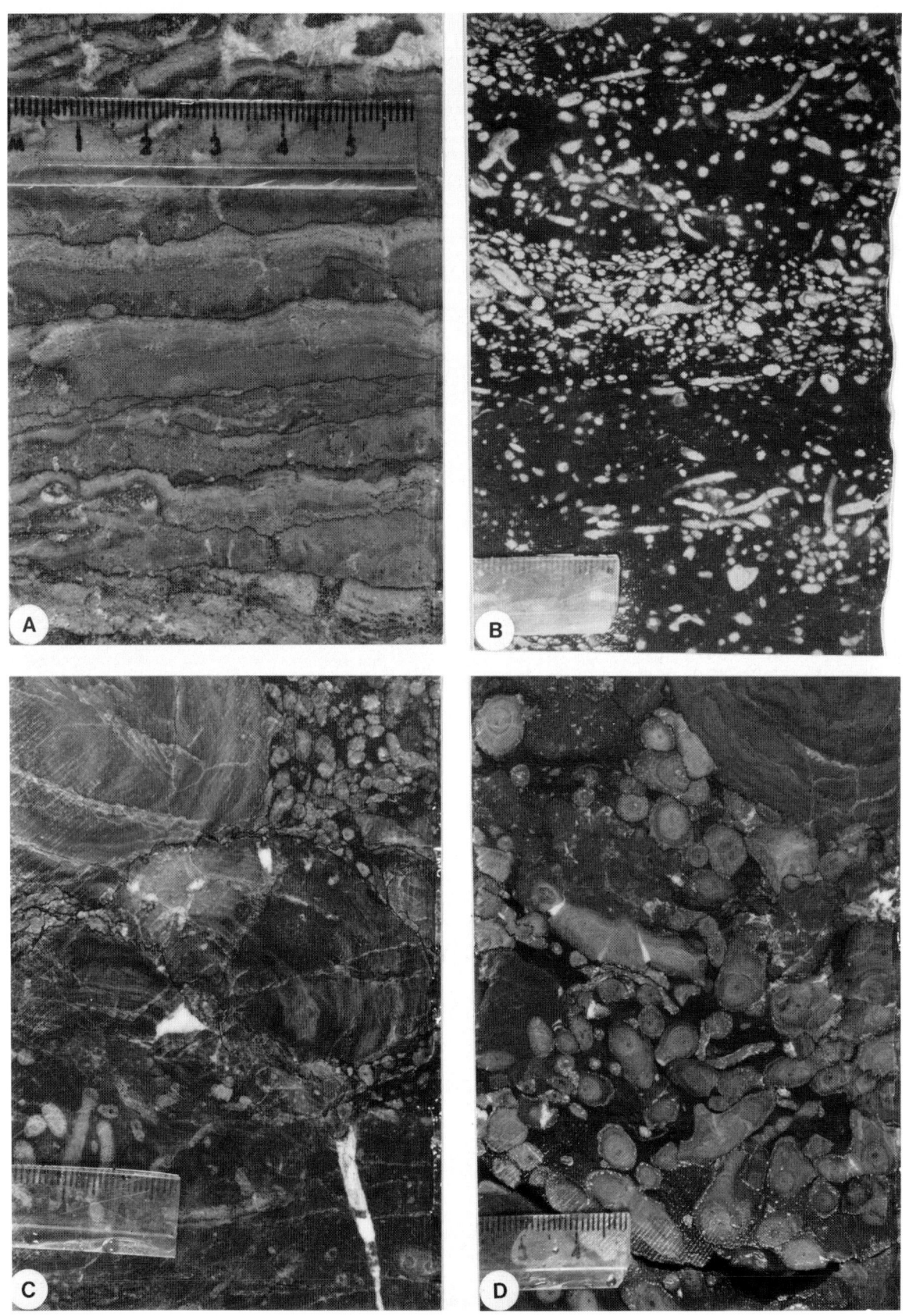

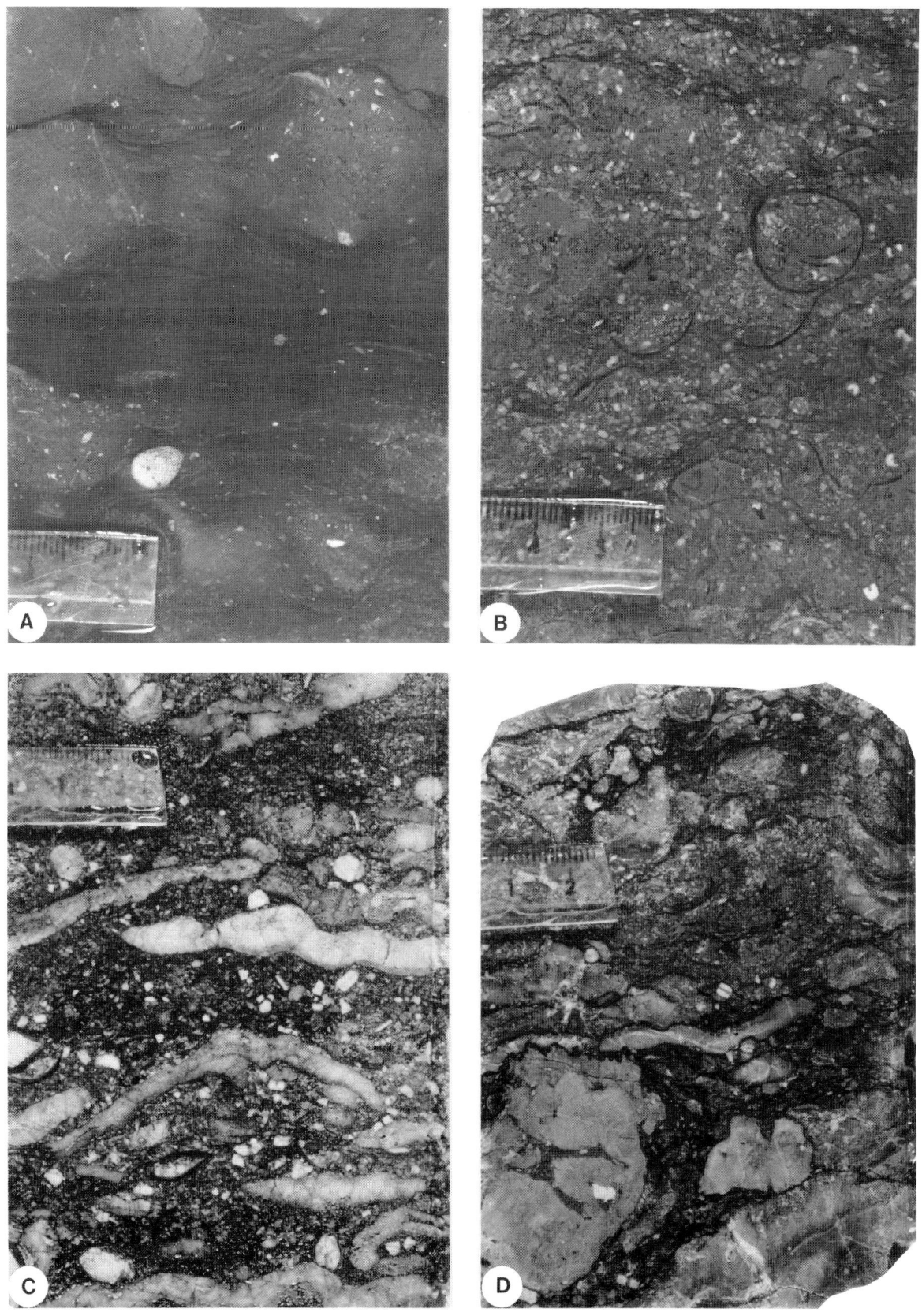

Figure 17   Buildup lithofacies.

A. <u>Stachyodes</u> and irregular shaped stromatoporoid floatstone-rudstone in mudstone matrix. Grains are coated by green algae <u>Bevocastria</u> and <u>Ortonella</u>. Lithofacies 9B. Stage B3. Scale in cm. (4-31-67-10W5, 8324')

B and D. Thin and thick tabular and <u>Stachyodes</u> stromatoporoid, brachiopod rudstone in mudstone matrix. Lithofacies 10B. Stage B"2". Scale in cm. (4-32-67-10W5, 8213' and 8208')

C. <u>Stachyodes</u> rudstone in packstone matrix. Lithofacies 13B. Stage B2. Scale in cm. (10-7-67-10W5, 8455')

Figure 18   Buildup lithofacies.

A. Thick tabular and <u>Stachyodes</u> stromatoporoid and brachiopod rudstone with "boundstone character" in skeletal grainstone matrix. Lithofacies 20B. Stage B5. Scale in cm. (4-8-67-10W5, 8390')

B. Thick tabular and <u>Stachyodes</u> stromatoporoid and brachiopod rudstone in skeletal grainstone matrix. Lithofacies 20B. Stage B"2". Scale in cm. (4-32-67-10W5, 8197')

C. Skeletal and peloidal grainstone. Lithofacies 24B. Stage ?. Scale in cm. (10-24-67-11W5, 8617')

D. <u>Amphipora</u> rudstone in peloidal and skeletal grainstone matrix. Lithofacies 16B. Stage ?. Scale in cm. (10-24-67-11W5, 8656')

FIGURE 17

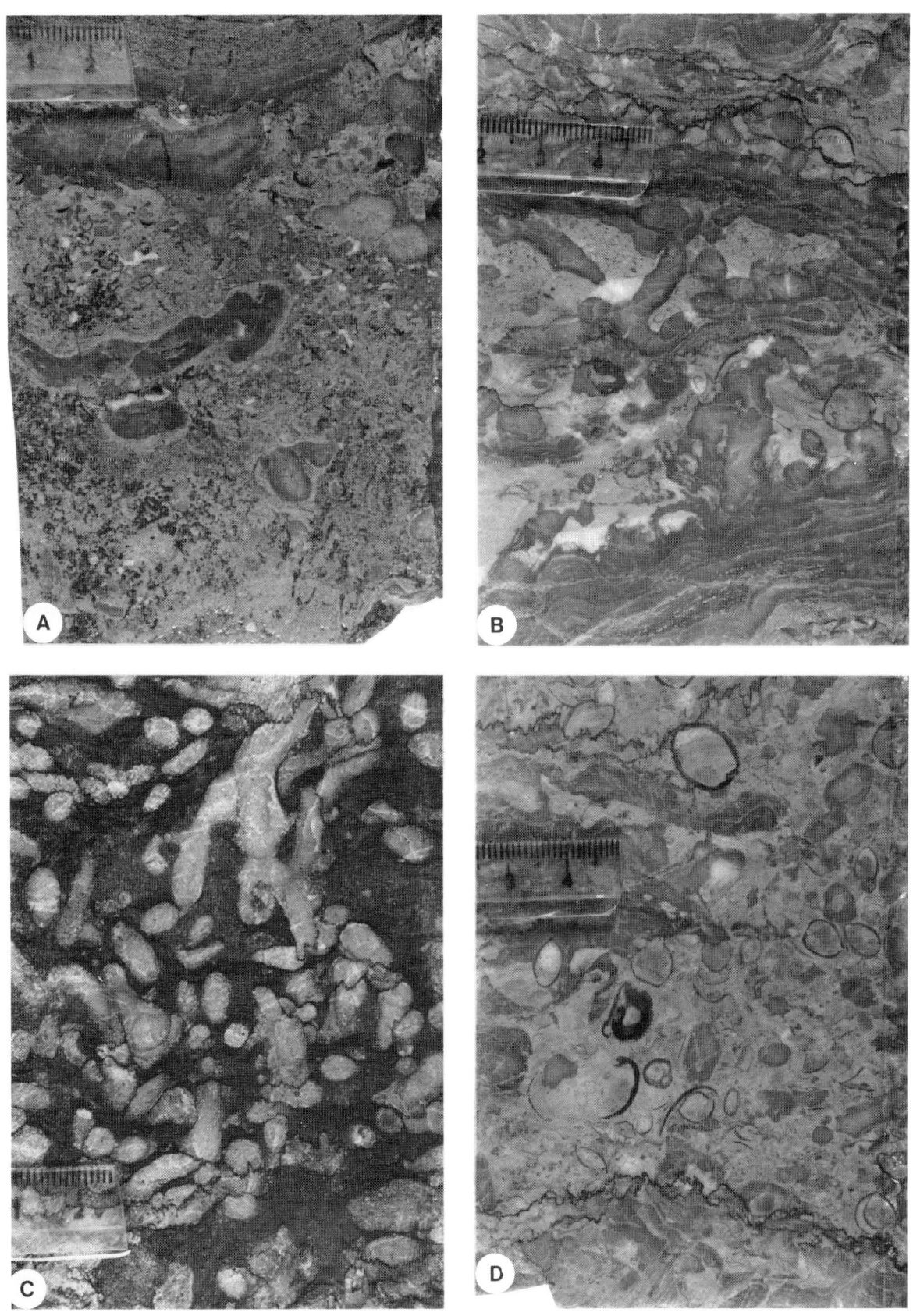

FIGURE 18

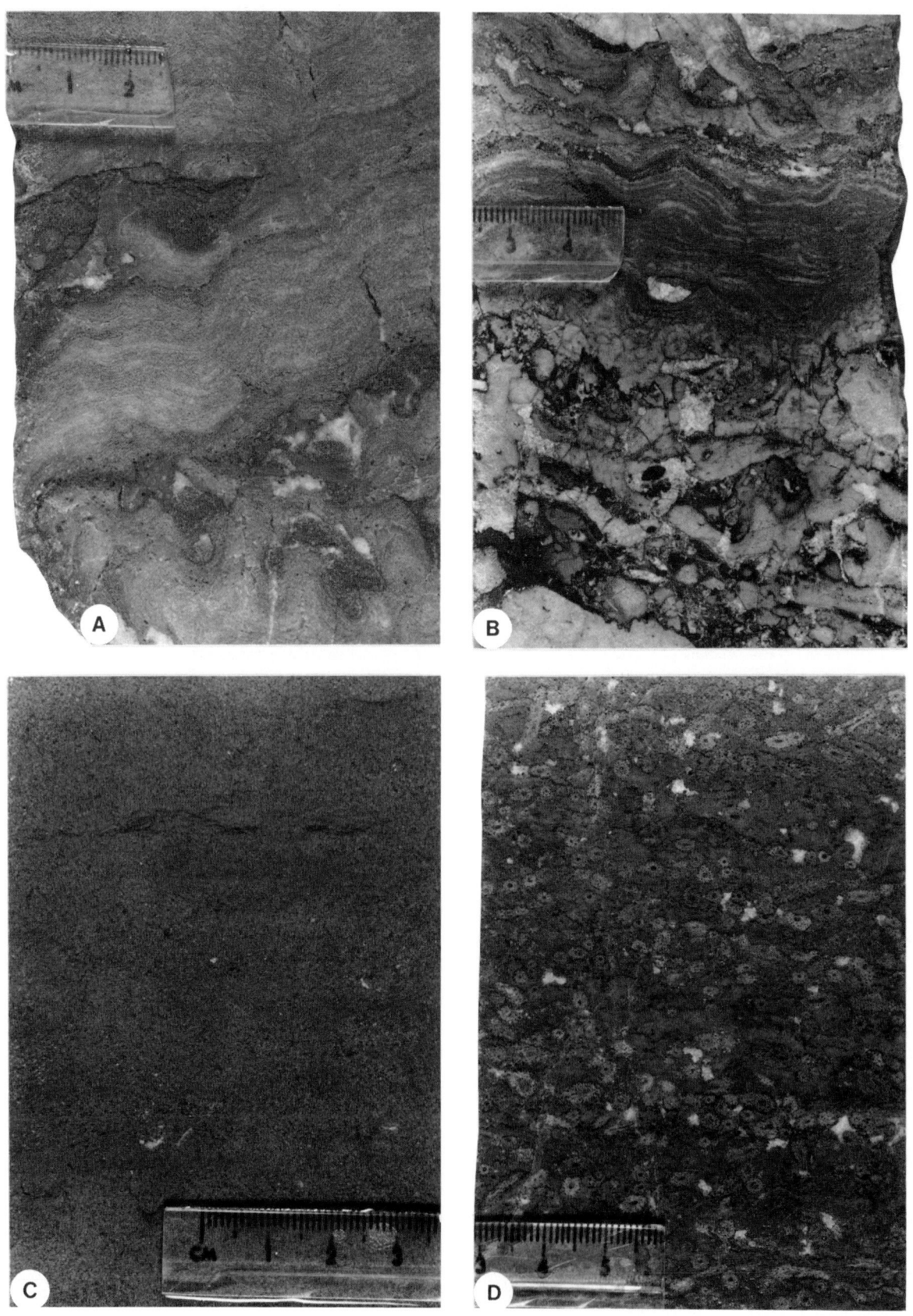

Figure 19 Buildup lithofacies.

A. <u>Stachyodes</u> and <u>Amphipora</u> stromatoporoid, intraclast rudstone-grainstone (no matrix at base). The upper part of this sample is the cemented (non-depositional surface) top of stage B3. RA: red algae. Brown fibrous calcite cement is present in interparticle pores. Arrow shows the base of this deposit on top of peloidal packstone-grainstone. A on top of A . Lithofacies 17B. Scale in cm. (10-22-67-10W5, 7645')

B. <u>Stachyodes</u> stromatoporoid and intraclast rudstone-grainstone (no matrix) sharply overlying fenestral laminated mudstone. Lithofacies 17B. Stage B3. Scale in cm. (4-34-67-10W5, 7868')

Figure 20 Buildup lithofacies.

A. <u>Stachyodes</u> and thin tabular stromatoporoids and brachiopod rudstone in packstone matrix. Lithofacies 15B. Stage B"2". Scale in cm. (4-31-67-10W5, 8337')

B. <u>Amphipora</u> and irregular stromatoporoid rudstone in peloidal and skeletal grainstone matrix. Lithofacies 16B. Stage B3. Scale in cm. (10-7-67-10W5, 8405')

C. Coral and algal-stromatoporoid assemblage rudstone in peloidal and skeletal grainstone matrix. Lithofacies 21B. Stage B6. Scale in cm. (4-8-67-10W5, 8376')

D. Hemispherical and <u>Stachyodes</u> stromatoporoid rudstone with skeletal grainstone matrix. The sample contains some <u>Amphipora</u> and many grains are coated by green algae. Lithofacies 19B. Stage B3. Scale in cm. (4-32-67-10W5, 8136')

E. Algal-stromatoporoid assemblage and <u>Amphipora</u> floatstone in fine peloidal grainstone matrix. Lithofacies 21B. Stage B6. Scale in cm. (10-8-67-10W5, 8172')

FIGURE 19

846

FIGURE 20

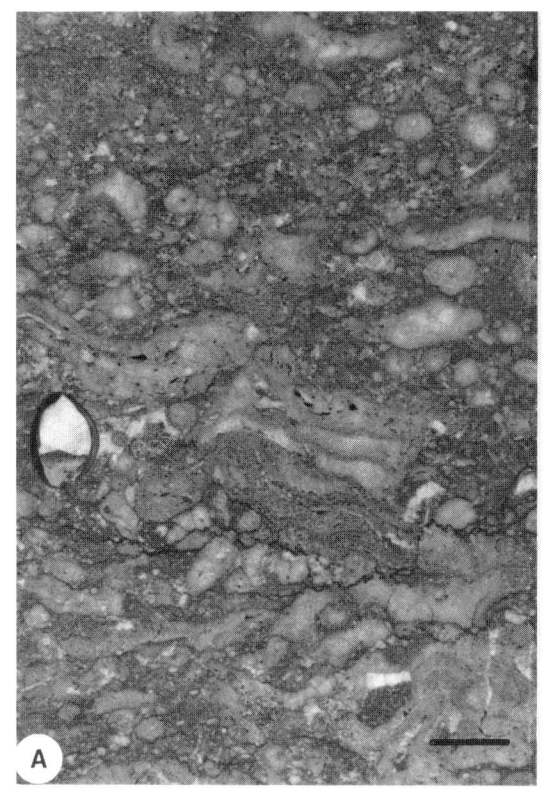

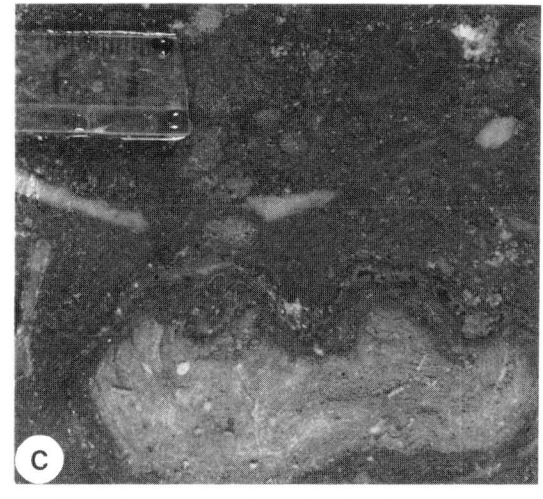
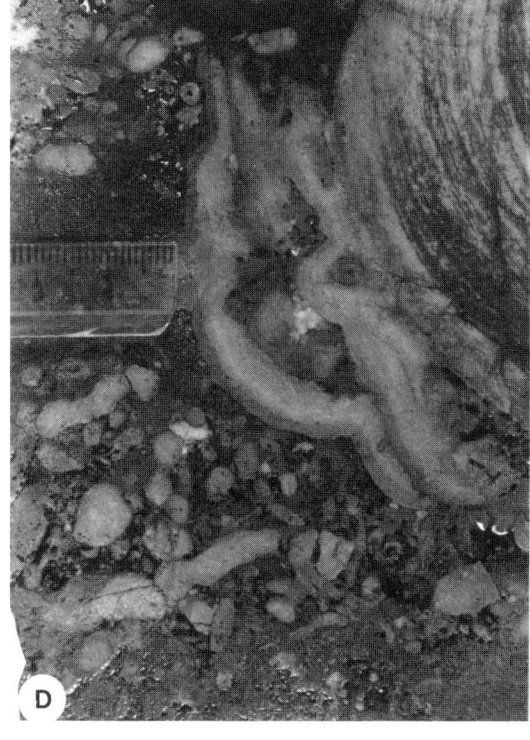
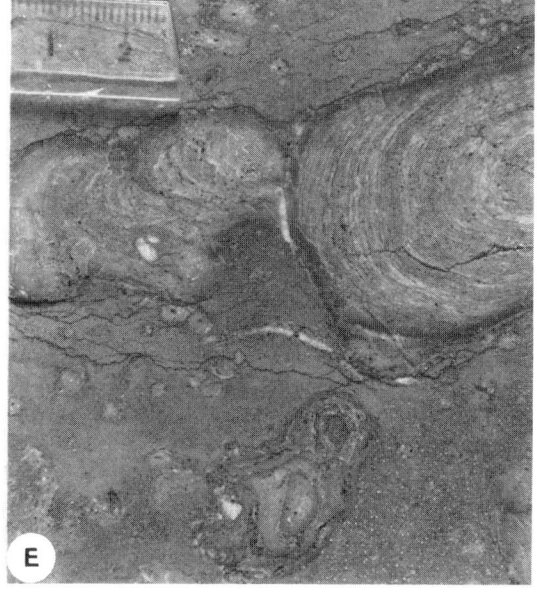

Figure 21 Buildup lithofacies.

A. <u>Amphipora</u> rudstone in peloidal packstone matrix. Lithofacies 12B. Stage B3. Scale: 1 cm. (10-9-67-10W5, 8167')

B. <u>Amphipora</u> rudstone in mudstone matrix. Lithofacies 8B. Stage B4. Scale: 1 cm. (10-9-67-10W5, 8094')

C. Sharp stylolitic (seam) contact between <u>Amphipora</u> floatstone in mudstone (Lithofacies 8B) at base and <u>Amphipora</u> rudstone with carbonaceous mudstone matrix (Lithofacies 1B) at top. Stage B3. Scale in cm. (10-9-67-10W5, 8135')

D. <u>Amphipora</u> (and <u>Stachyodes</u> ?) floatstone at base and rudstone at top with carbonaceous mudstone matrix. Lithofacies 3P. Stage ?. Scale in cm. (4-4-68-10W5, 8025')

Figure 22 Buildup lithofacies.

A. Cryptalgal laminated mudstone. Lithofacies 25B. Stage B4A. Scale: 1 cm. (4-16-67-10W5, 8102')

B. Laminated, fenestral and carbonaceous mudstone. Lithofacies 26B. Stage B3. Scale: 1 cm. (4-8-67-10W5, 8479')

C. <u>Amphipora</u> rudstone in peloidal packstone matrix. Lithofacies 12B. Stage B4. Scale: 1 cm. (4-8-67-10W5, 8403')

D. Packbreccia composed of mudstone intraclasts. Lithofacies 28B. Stage B4A. Scale: 1 cm. (4-16-67-10W5, 8104')

E. Mottled peloidal packstone. Lithofacies 22B. Stage B5. Scale in cm. (10-8-67-10W5, 8192')

FIGURE 21

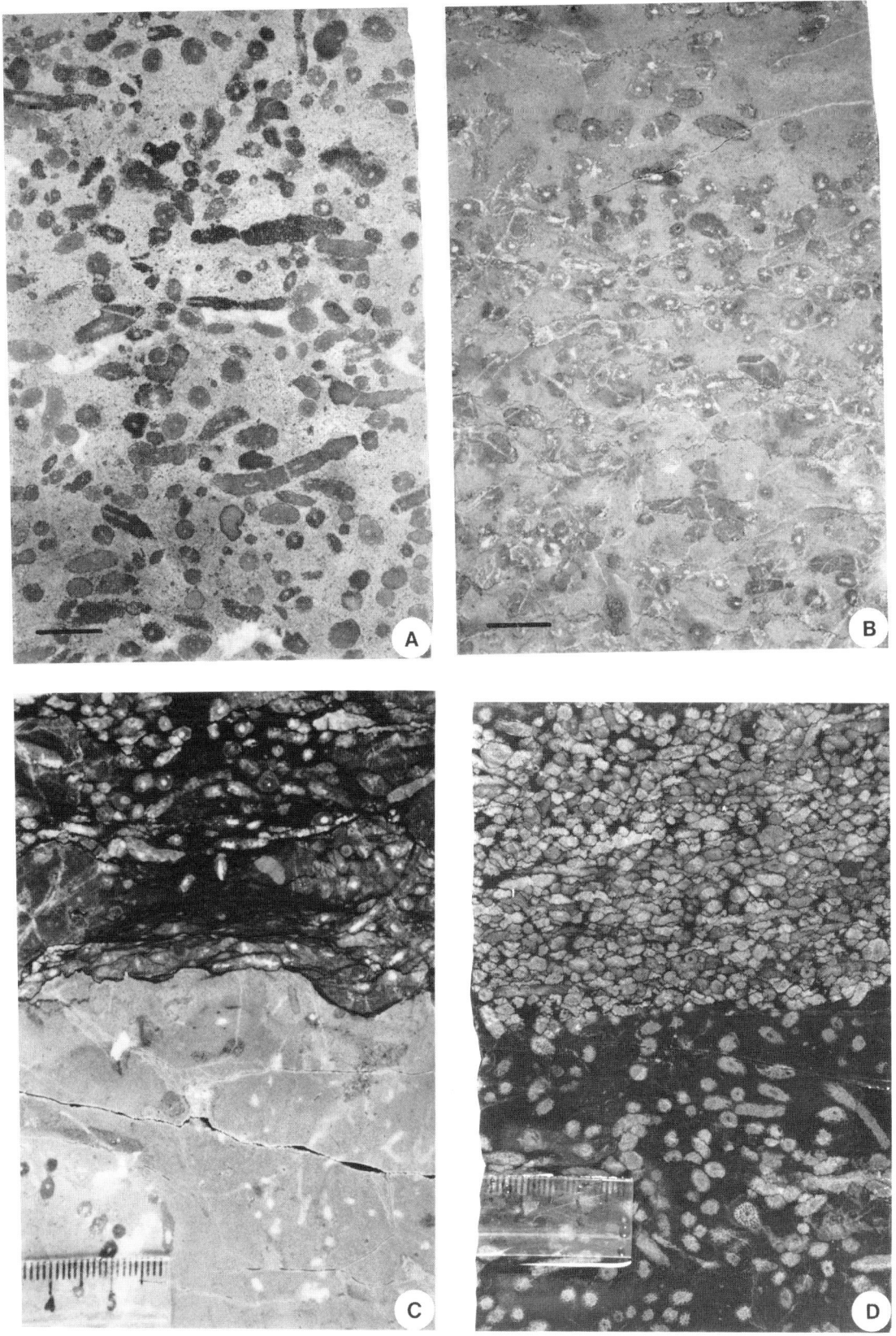

FIGURE 22

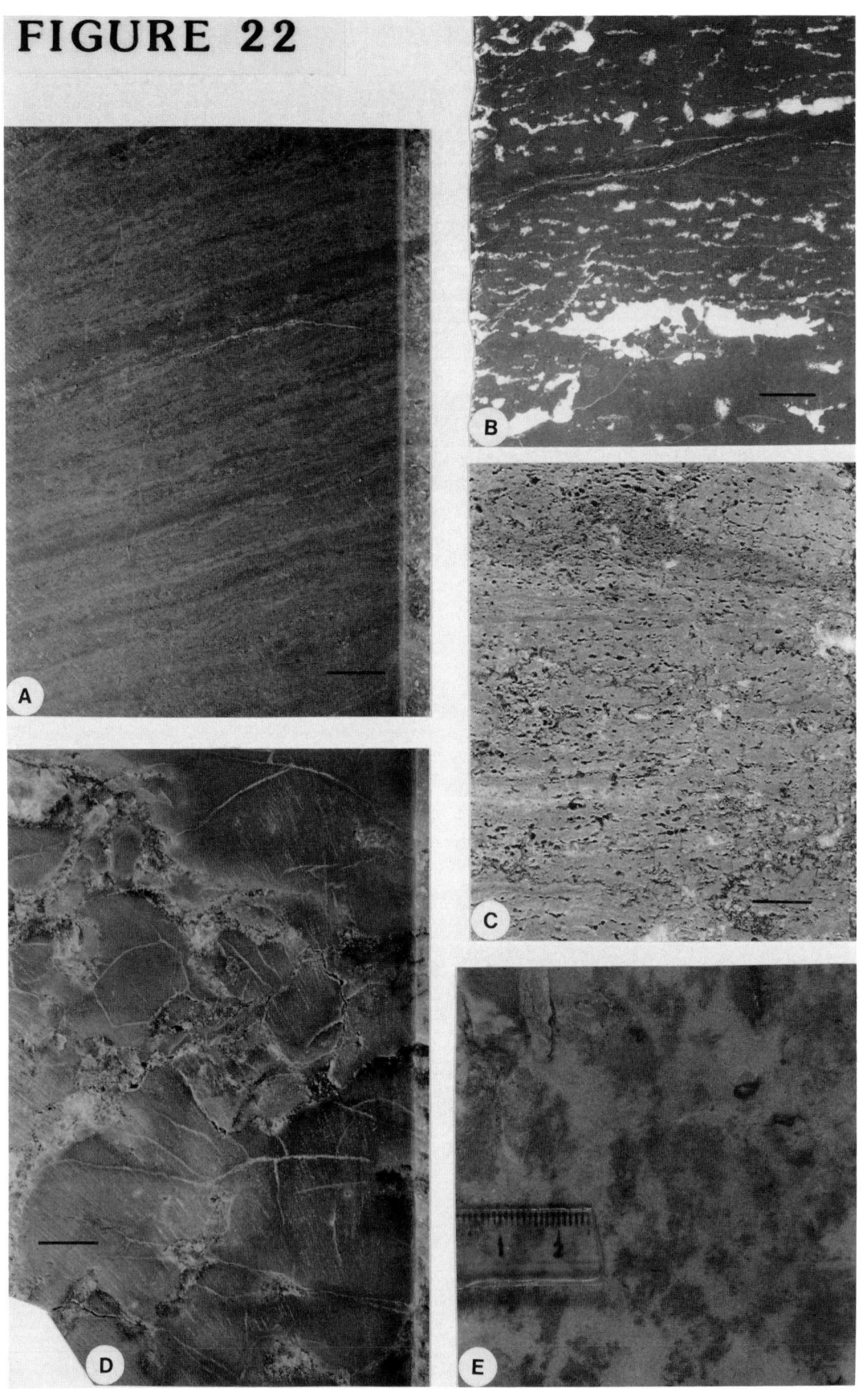

Figure 23   Unclassified lithofacies.

A. Algal coated brachiopod (commonly whole brachiopod) and Stachyodes floatstone-rudstone with mudstone-wackestone nodules. Some carbonaceous-rich muds present. Scale in cm. (10-24-67-11W5, 8613')

B. Brachiopod and crinoid grainstone with large peloidal packstone intraclast with burrows? partially filled with geopetal sediment of identical appearance to sediment in adjacent grainstone. This is the only occurrence seen. Base of stage B1 at periphery of buildup. Scale in cm. (10-7-67-10W5, 8500')

C. Algal coated brachiopods in carbonaceous-rich and poor crinoidal and peloidal packstone matrix. Scale in cm. (10-24-67-11W5, 8611')

Lithofacies in A and C are present in the western portion of the area above the top of the Swan Hills platform in what is lithostratigraphically the Waterways Formation but chronostratigraphically the thin (?) equivalent of the Swan Hills buildup (stage B1 or B6?, Viau, 1986).

D. Oncolitic rudstone with skeletal grainstone matrix at top of the platform. Scale in cm. (4-8-67-10W5, 8570')

Figure 24   Green shale units.

A. Green shale layer in intraclast rudstone at top of stage B2. Scale: 1 cm. (4-10-67-10W5, 8534')

B. Green shale present as geopetal sediment in interparticle porosity as well as within iron-rich dolomite cement filling the upper portions of voids. Scale in cm. (10-3-67-10W5, 8652')

C. Green shale within intraclast floatstone at base of photo and as thin layers in cryptalgal laminated mudstone at top of photo. Scale: 1 cm. (10-8-67-10W5, 8194')

D. Thin green shale occurrences at top of the platform. Scale: 1 cm. (10-30-67-10W5, 8363')

E. Green shale in intraclast and stromatoporoid packbreccia. Scale: 1 cm. (4-5-67-10W5, 8638')

F. Green shale layer in Lithofacies 8B. Scale: 1 cm. (10-3-67-10W5, 8574')

FIGURE 23

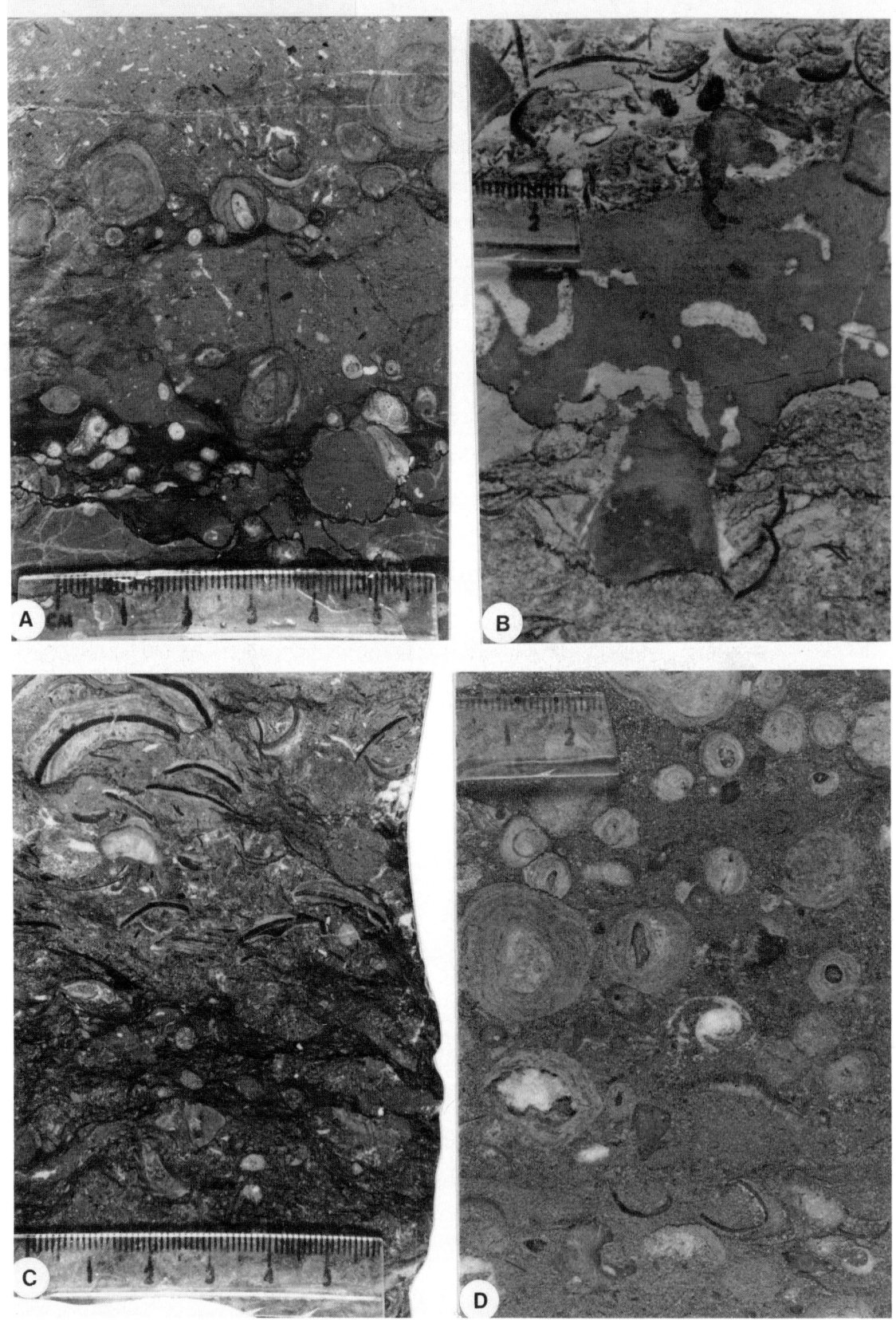

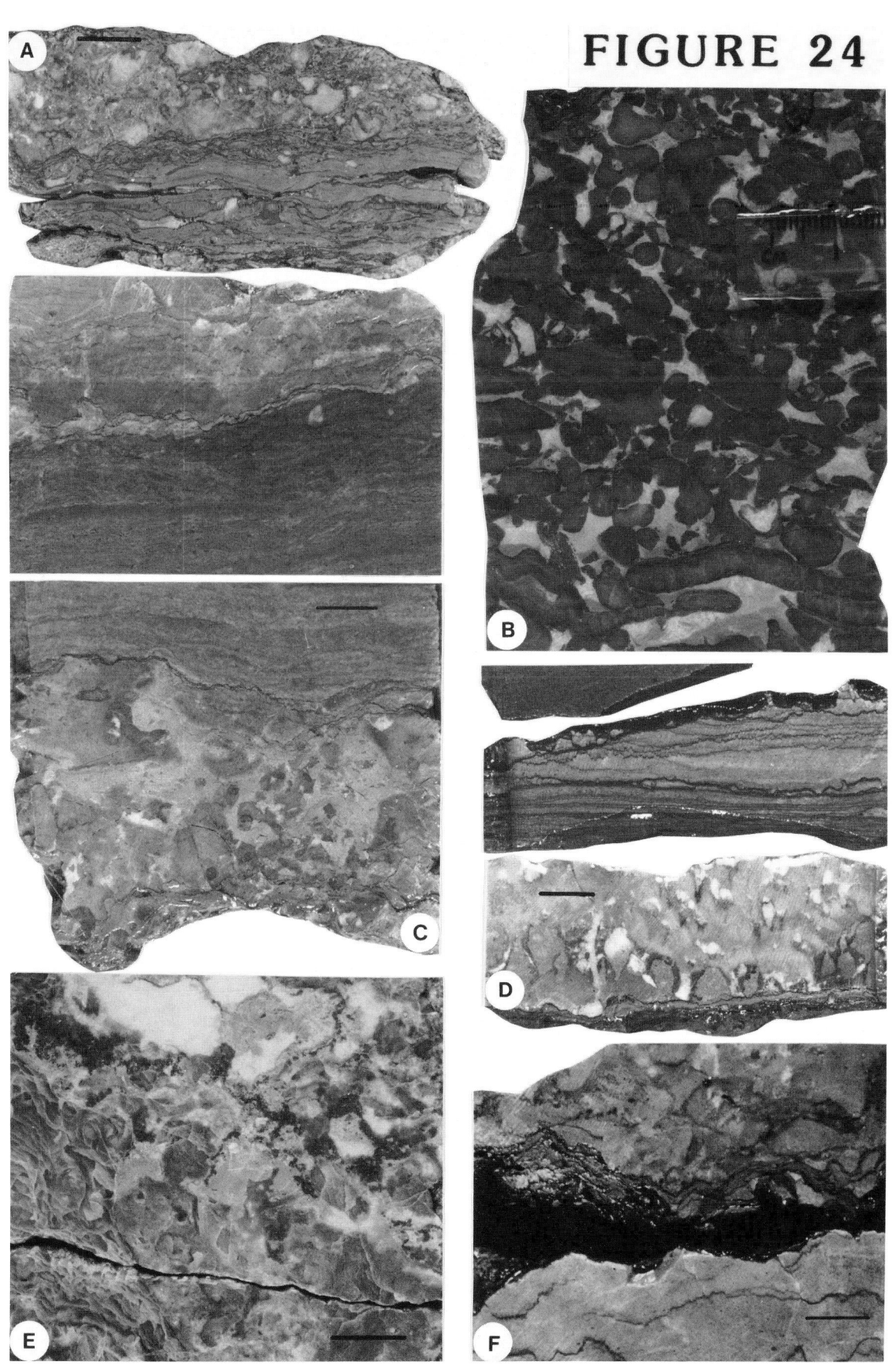